ALLE ZEIT WACH 1842

Hydraulic Design in Water Resources Engineering: Land Drainage

Proceedings of the 2nd International Conference, Southampton University, U.K. April 1986

Editors:
K.V.H. Smith
D.W. Rycroft

Springer-Verlag Berlin Heidelberg GmbH

K.V.H. SMITH

Civil Engineering Department
The University of Southampton
Southampton, U.K.

D.W. RYCROFT

Institute of Irrigation Studies
The University of Southampton
Southampton, U.K.

British Library Cataloguing in Publication Data

International Conference on Hydraulic Design in Water Resources Engineering (2nd; 1986: University of Southampton) Proceedings of the 2nd International Conference on Hydraulic Design in Water Resources Engineering: Land Drainage: University of Southampton, April 1986.
1. Drainage
I. Title II. Smith, K.V.H. III. Rycroft, David W.
631.6'2 TS621

ISBN 978-3-662-22016-0 ISBN 978-3-662-22014-6 (eBook)
DOI 10.1007/978-3-662-22014-6

CONTENTS

SECTION 3. HYDRAULIC ASPECTS OF DRAINAGE DESIGN

SECTION 4. INFLUENCE OF DRAINAGE ON SOILS

SECTION 6. DRAINAGE ECONOMICS

PREFACE

The first International Conference on Hydraulic Design in Water Resources Engineering held at Southampton University in 1984 brought together engineers interested in channels and channel control structures. It was well attended, very successful and generated papers relating to control and diversion structures, sediment control facilities for headworks and intakes, canals under quasi-steady flow conditions, computer simulation of irrigation and drainage canal systems under unsteady flow conditions, and sediment problems in rivers and the effects of engineering works on the regime of rivers.
The success of the first meeting was a major factor in deciding to reconvene the Conference in April 1986, also at Southampton University. The second conference is concerned with the design, constructions and operation of land drainage systems and the wealth of papers received for presentation is an indication of how much this subject has developed in the last few decades. The Conference is intended to bring together as much information as possible in the field of Land Drainage together with forecasts of future developments in this important subject. The Proceedings will provide a unique reference and state-of-the-art presentation to all interested in Land Drainage.
The Proceedings incorporate the text of a keynote lecture given by W.H. van der Molen, an eminent researcher. His participation added to the prestige of the Conference and the Editors would like to thank him most sincerely for his contribution.

K.V.H. Smith
and
D.W. Rycroft
SOUTHAMPTON UNIVERSITY
April 1986

Land Drainage: Art, Skill, Science, or Technology?

W.H. van der Molen
Agricultural University, Wageningen, The Netherlands

1. Art

Could land drainage ever evoke artistic feelings? Could it be called "land art" or "environmental art" in the sense we use these expressions today, as works of art intentionally devised to embellish or modify the landscape? Hardly, I must say; although drainage works have a visual aspect, they are intended and devised to serve other purposes. Still, aerial photographs of drainage operations, especially those involving open drainage, often yield pictures which resemble abstract paintings. The artist Ger Dekkers was so inspired by our "planned landscapes" that he captured their essentials by using series of colour slides as an artistic medium. Such views we all know from our work as drainage engineers, we probably even like them vaguely, but we need the eye of an artist to show us their hidden beauty.

2. Skill

Let us leave art and proceed to skill, a word more familiar to technicians and engineers. The word denotes a practical ability, and drainage started as such. It began with farmers taking measures to protect their fields and their crops from wetness during rainy periods. Often groups of farmers worked together to make ditches and watercourses to improve the outflow. If this was not sufficient, individual farmers shaped their plots in such a way that excess water from the fields was easily led into these ditches.

The farmers' ability to protect their lands from waterlogging is amazing. Ridge and furrow systems exist up to the present day. Such furrows are usually between 5 and 20 metres apart, and the land in between has been shaped by ploughing. The concave strips obtained in this way promote the discharge of water excesses.

More spectacular than these ridges and furrows are the dome-shaped fields, with their crests sometimes more than 1 meter above their rims. We still find them in otherwise flat areas on soils which are very susceptible to slaking and crust-information. Such crusts are rather impermeable and during heavy rains only a small part of the water is able to penetrate into the soil. Because the field is sloping to all sides, the excess is led into the surrounding ditches as diverging overland flow. Such dome-shaped fields are usually about 1 hectare in size and they can still be found in some regions of the Low Countries, and probably also in similar areas elsewhere.

Both ridge and furrow systems and dome-shaped fields are monuments from the past, because modern agriculture needs large and flat surfaces. Underground drainage, followed by land smoothing, merging of several small fields into a single large one, land consolidation schemes and other improvements will soon end their existence in many countries. An exception will probably be the lands in the humid tropics with their high intensity rainfall and a large part of the area being used for cultivating paddy rice. Here, surface drainage systems will remain necessary also under conditions of intensive farming.

Underground transport of water through tubes of fired clay was known and applied by the ancient Greeks and Romans. The main use of such piping systems was the transport of rain water, collected from the roofs of buildings, towards cisterns, underground storage reservoirs in which it was stored for later use. Whether such pipes were also employed for land drainage is doubtful, however. The Romans drained marshy areas in Italy not by the lowering of water, but by raising the land surface. For this purpose silt-laden river waters were brought to such low-lying places, a technique known as "colmatage".

The use of underground conduits for land drainage became common in England in the 18th century. Our colleague Miers has collected a series of plans and drawings of these early works. Subsoil drainage, both by simple and compound systems, was achieved by using conduits made of flagstones, faggots and other materials locally available. In the early 19th century fired clay pipes were manufactured again in England, and produced on an industrial scale. These clay tiles soon replaced the other materials used in land drainage. They remained the dominating material until about 1960, when they began to lose competitiveness with the synthetics, thus marking the end of the stoneware age and the beginning of the plastic era. As both materials are very resistant to decay, this transition will be an ideal marker for future archeologists. In the course of about 150 years of development, the fired clay tiles underwent an evolution which was rather similar to the evolution of living creatures. It started with an amazing variety of shapes, most of which disappeared as rapidly as they came. Soon a few types survived to dominate the scene; in our case the one-foot long tile with or without collar. Near the end, when the genus was threatened by extinction, again a large variety of forms appeared, in a vain attempt to prolong its existence. Clay pipes with spiral-shaped cross-section or tiles provided with longitudinal rills are examples from this late period.

The manual skill of digging trenches with a prescribed depth and slope, and the ways of putting the tiles in, likewise became history. The profession has turned to machinery, in design as well as in the execution of the works. The skills of draughtsmen and workmen have been replaced by those of the computer programmer and the machine operator.

3. Science

As in other branches of engineering, science and technology overtook experience and long practice. Originally the design of a drainage system was based on certain empirical rules obtained from earlier cases. If in a region a certain depth and spacing had given satisfactory results, this practice was continued. If not, it was modified. Such empirical rules, therefore, are children of trial and error. They can be quite useful, although they mostly lead to overdesign. Moreover, they give no insight into the governing processes and cannot be transferred to other regions, nor be adapted to rapidly changing circumstances. For such cases, and for obtaining insight, a more theoretical approach is needed, and this is the point where science takes over from skill.

There is a serious bottleneck, however. The results of theoretical studies are usually not expressed in language understood by the practicians. This easily opens a gap between what is known and what is applied. Already the earliest examples of a theoretical approach to a land drainage problem suffered from this difference in interests. In 1904, the French mathematician Boussinesq found a solution for a certain problem in non-steady groundwater flow towards drains. However, for half a century after its publication, this solution was better known by those working on non-linear partial differential equations than by drainage engineers. Equations derived in the early 19th century by Fourier for other problems in non-steady flow of heat could have easily been applied to analogous problems in land drainage. Their application was delayed for more than a century. Both the Boussinesq and Fourier equations were unearthed by Glover in the USA and written in the easily applicable form of a simple formula yielding the necessary spacing.

The time-lag between formulation and application was much less in steady-state drainage theory. Here Zunker in Germany made the first steps, followed later by Hooghoudt and Ernst in Holland, Childs and Youngs in the UK, Luthin and Maasland in USA, Labye and Guyon in France, and many others. Hooghoudt, for instance, not only derived solutions - which were complicated enough - but he also provided tables to simplify their use. Still more important, a field method was devised to measure permeability in-situ. For this and other solutions graphs and charts soon became indispensable tools for those not interested in complicated mathematical techniques. As a commentator once stated about such a technique: "hodographs are not particularly popular amongst engineers".

Like hand-installed drainage tubes, formulas, tables and graphs also belong to the past. Programmable calculators and hand-held computers have replaced paper and pencil. The art of making nomographs has evolved into computer programming, and user-friendly software has become a must if application is desired. Without such support, the methods will remain good for the classroom, but be ignored in practice.

The classical mathematical methods were elegant, but difficult to handle, and only suited for special cases. Like aristocrats, they were few in number and have lost their once dominating position. Numerical methods have taken over and are able to solve almost any problem conceivable, whereas the computer has taken over the enormous amount of slave labour needed. To the practical man, who never has the necessary data at hand, this improvement, however, is less than might be expected. Given a good and easily manageable method, he will hardly care about its pedigree. Whether it is stemming from analytical results, from finite differences, finite elements or boundary integrals is immaterial as long as its accuracy is better than the reliability of the input data - and this is nearly always true.

A parameter like permeability, for instance, is easily 20% in error and often things are far worse. To measure a permeability tensor in three directions with random orientation would require the work for a PH.D.-thesis, but be out of place for a potato field to be drained. Therefore, in any method, input should be restricted to those quantities which are really indispensable.

The development of reliable and accurate methods for observation in the field has not kept pace with the proliferation of calculation methods. Here, variability and scale effects restrict the obtainable accuracy. Small samples - say rings of 100 cubic centimeters - are inadequate and even auger hole measurements of permeability, where a volume of some 50 litres is involved, still show large variations. Results from duplicate measurements easily differ by a factor of two, and in clay soils even more than that. Newly developed techniques in geostatistics, like kriging, might be of some help, but they do not solve the problem of scale that is obviously behind these phenomena. Methods involving volumes of several cubic metres would be the best approach.

This brings us back to skill. In some areas drainage design criteria are directly derived from observations in profile pits. A pit of 1 m^3 - like those used in the Zuiderzee polders of the Netherlands for this purpose - already reveals much of the variability at first sight. Large cracks are separated about 1m apart, with smaller fissures in between, and a trained observer can make a fairly accurate estimate of the design criteria from their sizes. This decision process may be refined and made less subjective by systematic description of the phenomena observed. Attempts to use these descriptions as a base for a mathematical procedure, however, failed, because the permeabilities obtained were about an order of magnitude too high. Obviously bottlenecks occur in the crack system, which considerably reduce the water transport, but which are being overlooked in the observations. So these profile methods remained a skill and did not become a science. Apart from the Zuiderzee polders, the method is known to me as being applied in England, where it was introduced by Miers.

Other clues for drainage design might be derived from existing maps. The use of soil maps for this purpose has been described, and if the permeability is restricted to the upper layers, such maps could be of value, especially if they show textures rather than soil colour or genetic processes. If, however, deeper layers are involved in water transport - and this is not exceptional - geological maps are most useful. Also here, texture and structure rather than geological age is of most interest to us.

Up to now, we have been dealing with quantities, but, living in the age of quality, we cannot avoid this subject. At least in the UK a vivid debate arose about the influence of land drainage on water quality: favourable, disastrous or absent? This debate will lead us into new fields, like studies about the times of residence in drainage systems of pollutants originating from diffuse sources. Nitrates from agriculture are an example. Both experimental results and theoretical considerations are becoming available, but both are still scarce. It is clear that in this field the scientific approach is the only way to tell fact from fiction.

4. Technology

Drainage machinery has taken the place of handwork since around 1950 in most countries, and the tools have greatly improved since then. The first trenching machines were slow and unwieldy and their adjustment in depth was still done by hand steering. The substitution of the heavy clay tiles by flexible plastic tubing, pre-wrapped with a filter envelope, was already a great step forward. The advent of automatic depth regulation guided by a ray of laser light allowed a much greater speed of operation without sacrificing an accurate depth control. The trenchless machines, direct descendants of the old mole plough, but now able to put the complete pre-wrapped tubing in place, are gradually replacing the slower and most costly trench-diggers. With the advent of this machinery, both trenching and trenchless, the drainage of agricultural fields has become a routine operation, both for large projects and for individual parcels of land. An interesting development is the widening of the application of underground drainage. Quantitatively, the principal new field is drainage for salinity control in arid and semi-arid areas.

The cause of salinization of irrigated lands is the accumulation of groundwater, being recharged by the inevitable water losses. If the natural drainage is insufficient to remove these excesses, the water table will continue to rise until capillary ascent, followed by evaporation sets up a new equilibrium. But even if the groundwater itself is fresh or nearly so, this process leads to a severe accumulation of salts near the soil surface, which is detrimental to plant growth. In cases where the salinization is superficial only, vertical drainage by deepwells may be able to reduce the phreatic level to a safe depth and at the same time provide a source of useful water for extending or intensifying the irrigated area. The only requirement is the presence of a suitable aquifer having sufficient contact with the phreatic water.

If, however, the groundwater to be pumped is salty, or if the aquifer is absent or unsuitable, such vertical drainage is seldom economical and the classical horizontal drainage is preferred. In recent years this type of drainage has increased so much, that the main area of application has shifted from temperate, humid zones to arid and semi-arid regions. The notion that no large irrigation scheme is complete without considering the drainage aspects involved in now accepted generally.

Other applications have been found. Drainage of sports facilities is one of these, although it is far from new. In fact, Hooghoudt's drainage theory was first applied to the construction of golf links near Amsterdam. Drainage of road verges, airfields, cities even, followed. Town extensions are often planned in low places, which are still open because our ancestors carefully avoided building there. Existing ditches were considered useless obstacles and were carelessly filled, thus aggravating the problems. No wonder the new inhabitants complained about mould and dampness.

Also here, drainage proved a good remedy: a circular conduit and in severe cases a double ring around the buildings affected collects the water, which is subsequently drained or pumped into the sewer system. Even salinity control may be aimed at. The ruins of the ancient city of Mohenjo Daro in the Indus Valley in Pakistan suffer badly from salinity, which destroys the bricks. Drainage has been applied here to preserve these valuable remnants of an old civilization.

As a final example of a special application I would like to mention the use of drainage techniques for the installation of large conduits for the transport of natural gas. These pipes, with a diameter often exceeding 1 metre are subject to internal gas pressures of 100 Bar and over. To prevent damage, they are buried in the soil at least 1m deep.

In sandy soils with shallow groundwater they have to be placed in a dry trench, so dewatering the soil is needed prior to their installation.

The problem is similar to the construction of sewer mains in such soils, but the difference is in the scale of the operations. Whereas a main sewer is usually a few kilometers in length, the gas conduits are hundreds of kilometers long. The traditional well-point drainage used for sewers is too costly, too elaborate and too slow in this case. A good solution was the installation of a temporary horizontal drain at a depth up to 4 metres, and located at the place of the future trench. A special trench-digging machine, with a prolonged digging chain to reach the required depth was used for the purpose. It installed a pre-wrapped plastic drainpipe, from which the water was pumped into the existing watercourses. After a few days, the phreatic level had sufficiently dropped to excavate the large trench in which the gas conduit can be installed, after which the pumping of the temporary deep drain is moved to a new section. Many other examples of such special applications could be given, but these will be sufficient to demonstrate the versatility of land drainage, a versatility that certainly will increase in the future.

5. Profession

The question whether drainage is an art, a skill, a science or a technology has been raised, but not answered yet. In fact there is no answer, because land drainage is a profession involving each of these aspects, and still a few more. It has to deal also with agriculture, with civil engineering, with project economy and organization, with machinery and with logistics. It is this strange mixture that makes it attractive enough to devote a large part of your career to it. Many interesting aspects will be presented at this Symposium. Here, the profession will review the "state of the art".

SECTION 1 DESIGN OF FIELD DRAINAGE SYSTEMS

The Modelling of a Drainage System in the River Valley with Ground Water Under Pressure

Waldemar Mioduszewski, Zbigniew Kowalewski
Institute for Land Reclamation and Grassland Farming, Poland

INTRODUCTION

In regulation of water conditions in some river valleys considerable difficulties occur in establishing an optimum ground water level for the agricultural production. The valleys are usually narrow. In the upper part of the valley semi-permeable soils occur, being underlain by permeable formations constituting a main aquifer spreading over the upland area. At such an arrangement free ground water occurs in the upland and ground water under pressure in the valley. The drainage network in the form of e.g. systematic drainage and drainage ditches established in semi-permeable formations cannot fulfil its task.

In this connection it is necessary to analyze other drainage methods. Results of the model investigations on work efficiency of several selected types of drainage systems are presented in the paper. These investigations concerned a valley, in which many years ago the network of drainage ditches was done. None the less, the valley continued to be overmoistened.

DESCRIPTION OF THE RIVER VALLEY AND HYDROGEOLOGICAL CONDITIONS

The area under study /Fig.1/ constituted a fragment of the river valley and the left-bank fragment of the upland adjoining it. The valley is 300-400 m in width, its surface being flat with only one flood terrace. Adjoining uplands are rather high reaching several ten meters a.s.l. Despite the existing network of drainage ditches, the whole valley is wet and swampy, with numerous springs and soakings.

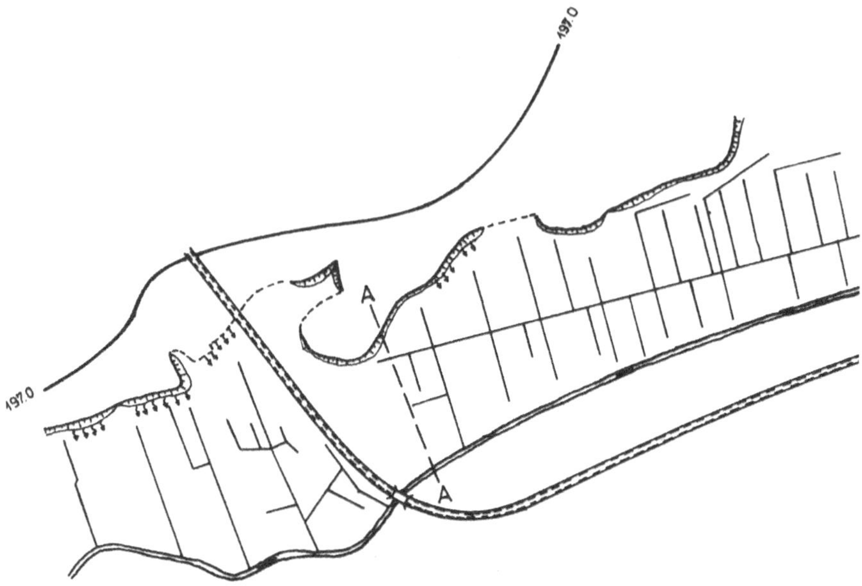

Fig.1. Situation plan of the area investigated.

In the geological structure of this area cretaceous and quaternary formations participate. The cretaceous parent material occurs at the depth of over 20 m and is strongly fissured. Artesian waters occur here and the ground water is stabilized at the level of several to a dozen or so meters above the valley area surface. Cretaceous formazions are covered with quaternary cohesive soils. They are clays and muds in the valley and residual loam on the upland. Over this semi-permeable or impermeable complex of soils the layer of hydrated alluvial sands is to be found, constituting a separate aquifer. It is of quite different thickness varying from 2.0 to 8.0 m. This layer has a hydraulic contact with sands covering the detritus formations in the upland. A separate aquifer occurs here, the ground water level in piezometers stabilizing several meters above the area surface. On these sandy soils organic formations /warps, gyttjas, peats/ of relatively great thickness /5-8 m/ are to be found.

In the valley margins they rise up covering partly also the upland slope. In such situation the

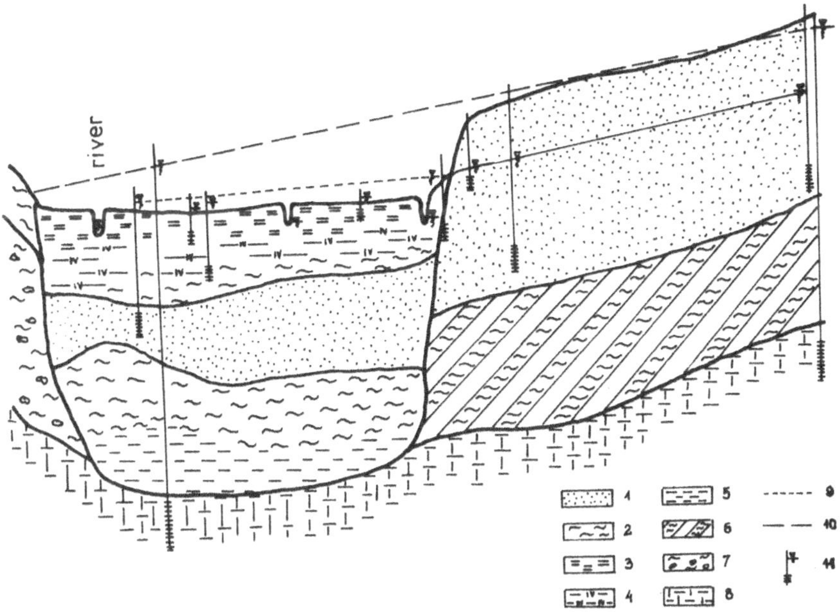

Fig.2. Typical geologic section through the river valley /A-A/: 1 - sands, 2 - silts, loesses, 3 - peats, 4 - muds, 5 - clays, 6 - silty loams, 7 - residual silty-loam, 8 - cretaceous rock, 9 - line of piezometric pressures of the quaternary, 10 - piezometric line of the cretaceous aquifer, 11 - piezometers.

organic formations occur in the form of a fairly dense "coat" for waters filling up alluvial sands laying below. They decrease the outflow of ground waters from the upland to the network of reclamation ditches and to the river.

The arrangement scheme of the geologic layers for the selected section is presented in Fig.2. Also measured piezometric pressure has been marked here.

The performed piezometric measurements have proved that on the valley area under consideration the water pressure in the cretaceous layer is 2 - 4 times higher than in the quaternary aquifer.

The analysis of hydrogeological conditions has proved that the upper organic layer is fed mainly from the first aquifer /quaternary sands/. To check this assumption, model investigations reflecting the actual situation were carried out. These investigations which are not discussed in the present paper, have proved that the layer of cohesive soils separating quaternary sands from cretaceous formations can be regarded as impermeable. Thus one can assume that the value of pressures in cretaceous parent material would not affect directly the feeding of the valley. Therefore, further investigations were restrained to modelling only two layers, viz: organic layer and layer of quaternary sands underlaying the former, connected with sands on the upland.

MODEL INVESTIGATIONS

Model investigations of seepage from the valley area under study were carried out in the section perpendicular to the river /A-A/, at assumption that feeding of the first aquifer, i.e. quaternary sands, would occur only on the side of the upland. Boundary conditions were assumed in the form of suitable piezometric pressures and water levals in open watercourses /river and ditches/. As an external boundary condition of the Ist kind for the calculation model the ground water level discribed by the hydroisochypse of 197 m has been assumed. It is the water level higher than the stabilized piezometric pressure in the middle part of the valley. On the side of the valley the model is limited by a line running along the river, while assuming as the boundary condition of the 1st kind the measured water level in the river. In the calculation section averaged coefficients of permeability resulting from the field measurements have been assumed.

Model investigations were performed on an electric analog-RR network /AP-600/. The continuous area of seepage was represented by a network of resistors. A fragment of the resistor network is presented in fig.3.

The conductivity of resistors has been adopted depending on hydraulic conductivity of the soil medium. A rectangular network of the scale of 5 or 10 m in the horizontal and of 0,5-2,5 m in the vertical direction has been applied. For better modelling of the seepage flow in the region of river and ditches an additional network density increase was applied while connecting the rectangular network with a triangular one. In modelling vertical wells radiality of the water in flow to the well and additional resistances resulting from the arrangement of wells in rows at their assumed diameter have been taken into account.

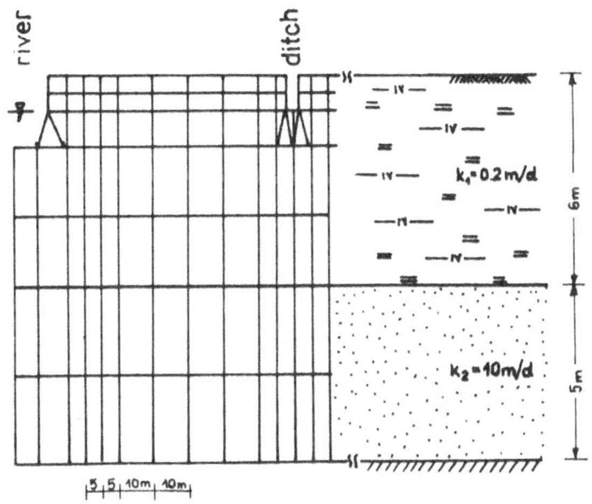

Fig. 3. A fragment of resistance network in the model investigations.

Two series of the experiments were carried out:

Series B Concerned estimation of the efficiency of horizontal drainage systems, in particular:

B-1: two ditches of 1.0 m in depth. It is the model representative for the actual state of the object.

B-2: two ditches by 1.0 m in depth and the 2.5 m deep drainage established along the upland slope /in sandy soils/.

B-3: execution of the ditches and drainage as for B-2. but the drainage depth is 4.0 m.

B-4: execution of a single ditch of 1.0 m in depth and of an "incision" into the upland slope. The incision consists in cutting upper weakly permeable formations to enable the water outflow from sandy soils covering the upland.

B-5: execution of two ditches by 1.0 m in depth and of the "incision" as for B-4.

Series C The aim of investigations within this series was to determine possibilities of drainage of the river valley sector investigated by means of vertical wells. It has been assumed that the diameter of wells would be 0.2 m and that they would reach the roof of impermeable formations. Moreover, it has been assumed that the water outflow from wells would occur spontaneously, without application of pumps. The following inwestigations were carried out:

C-1: one row of wells at the spacing of 150 m situated in the valley middle. The water level in the self-outflow well at the depth of 0.6 m below the area surface.

C-2: one row of wells as for C-1, at the spacing of 50 m between wells.

C-3: two rows of wells along the river valley. The water level in wells at the depth of 0.6 m below the area surface. The spacing between wells in a row 150 m.

C-4: two rows of wells as for C-3 the spacing between wells 50 m.

C-5: two rows of wells at the spacing of 150 m /as for C-3/, water level in the self-outflow well being 1.1 m below the area surface.

C-6: two rows of wells as for C-5 the spacing between wells 50 m.

ANALYSIS OF THE MODEL INVESTIGATIONS

Results of the model investigations are presented in Fig.4 in the form of piezometric lines determined in the by-surface layer of organic soils. It has been assumed that for a correct drainage of the area the piezometric line should be at an appropriate depth below the area surface. The analysis of the presented model investigation results has proved that the drainage in the form of ditches supplemented even by the border drainage /Fig.4a/ does not fulfil its task, as the required drop of the ground water level has not been reached. The piezometric line in the valley is situated in everycase above the area surface /from 0.3 to 1.2 m/ The spacing between pipe drainage ensuring the drop of ground waters below the area surface should amount in this case according to analytical calculations to 2.5 - 3.5 m. At application of the "incision" better effects can be obtained, i.e. the

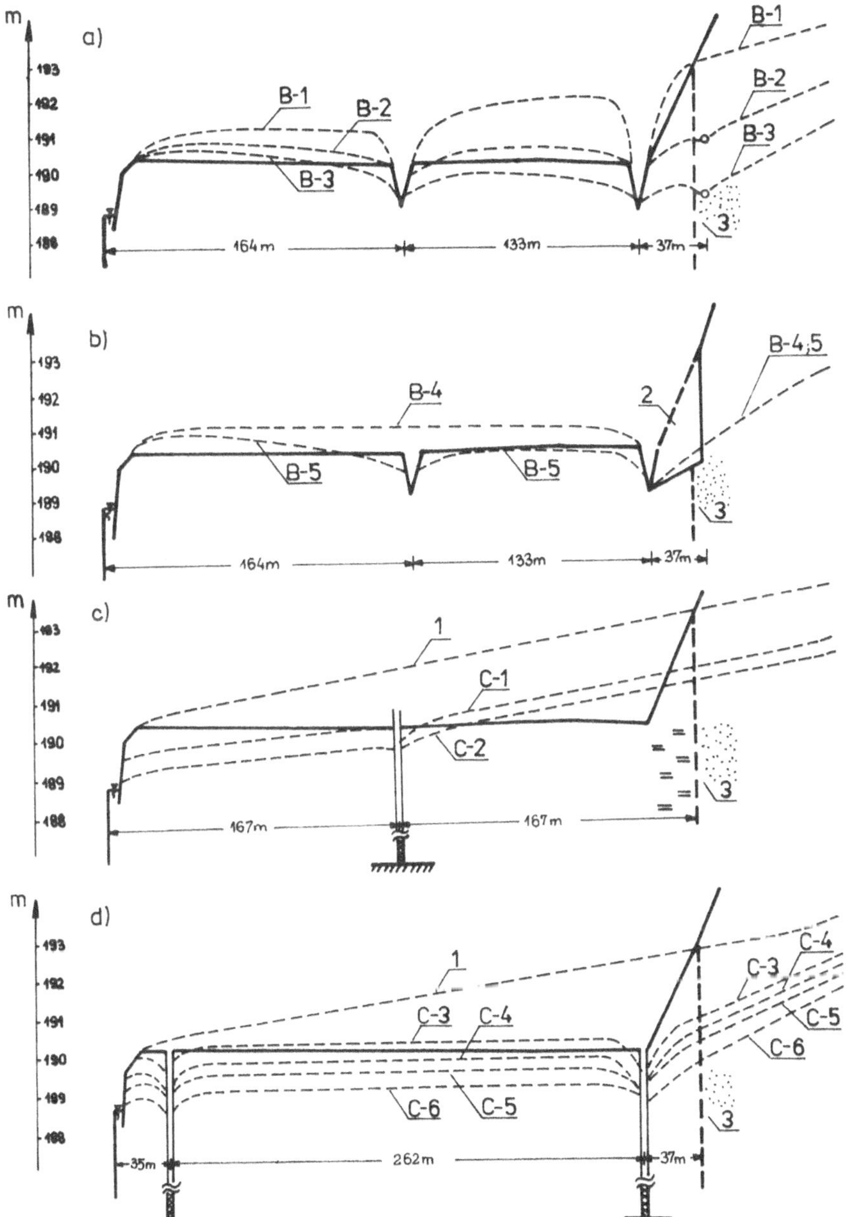

Fig.4. Situation of piezometric lines obtained in the model investigations: 1 - primary piezometric line, 2 - "incision", 3 - line separating semi-permeable soils from permeable ones.

piezometric line will be lowered to the area surface /Fig.4b/. This lowering would be possible on condition of reaching the permeable layer by the "incision". Better effects in functioning of the drainage system can be reached at application of vertical wells. One row of wells, even at narrow spacing between then /Fig.4c/ would be still insufficient. Only two rows of wells could allow to control precisely the ground water level. The curve of depression in the valley is situated at the depth of 0.3 m at the water level in the self-outflow well of 0.6 m or at the depth of 0.7 m at the water level in the self-outflow well of 1.1 m below the area surface.

A significant question to be taken into account in drainage of such valley types is the estimation of the effect of reclamation on the ground water level on the upland. Any efficient drainage methods /incisions, vertical drainage/ lead to a considerable lowering of ground waters to a wide distance from the river valley. Basing on the model investigations it can be assumed that at the distance of 300 - 400 m from the installed structures a lowering of the ground water level on the upland would be practically equal to the pressure lowering depth at the installation point of these structures. This can result in some cases in an unfavourable effect on the natural landscape of the valley and lead to water disappearance e.g. in farm wells.

FIELD OBSERVATIONS

Various attempts of supplementation of the primary drainage network were undertaken in the object, for which the model investigations were carried out. Neither deepening of the river bed even by 0.8 m nor the supplementation drainage at spacing between drain pipelines of 15-20 m gave any positive results, which was confirmed by the model investigations.

In the object concerned so-called "incisions" and two draining wells were executed experimentally. The "incisions" resulted in an intensive outflow of ground waters and in a gradual drying of the valley. Also an intensive outflow from drainage wells and lowering of the ground water lovel in close vicinity of wells was observed.

It was not possible to carry out a full experiment representing the assumptions of the model

investigations. None the less, the observations on functioning of the "incisions" and of drainage wells have proved the possibility of application of such solutions for drainage of the valley.

CONCLUSIONS

Model investigations and field observations have proved that in the object under study, in which the semi-permeable layer of a great thickness is underlain by the aquifer with ground water under pressure, the traditional drainage by ditches or drain pipes cannot ensure correct water conditions.

On the other hand, it is possible to apply a deep drainage with the structures reaching the aquifer layer. Good results can be obtained by application of vertical drains or the so-called "incisions". However, these drainage methods exert an unfavourable influence on adjoining areas leading to considerable lowering of the ground water level on the upland.

The Effect of Drainage Works within the Turu-Lung Somes-Barcau Zone

Iulian Mihnea
Institute of Land Reclamation Studies and Design, Bucharest, Romania

ABSTRACT

It was as recently as the 19th century that drainage works were performed on the Western lowlands of our country where clayey soils, under different degrees of glaciation, are prevailing.

These works allowed for taking back into culture of a surface of 200,000 ha.

The paper proposes to present both the effect of these works on the environment, as well as the different ways of increasing the productivity of clayey soils.

1. LIMITS AND CHARACTERISTICS OF THE ZONE

The territory between the Tur and Barcau rivers in the North Western part of Romania includes an agricultural area of 0.5 mil ha. In this perimeter, limited as such, the Somes and Crasna rivers as well as the Ier Valley divide this territory in hydromeliorative areas or systems, as for example Tur left river bank, Somes river bank, Somes-Crasna etc.

Within the agricultural lands about 75% is arable, about 3% vineyards-orchard plantations and about 22% is covered by meadows.

The pedological studies reveal the fact that within this territory a large soil diversity exists, from most fertile ones as are the chernozems that cover 20% of the arable, to the podzolised and podzolic brown soils representing 38% of the arable, sandy and salty soils occupying 30% and the excessively eroded soil occupying 12% of the arable area. The large annual rain quantities of more than 700mm as well as the soil weak permeability frequently cause water stagnation on large areas.

Over 68% of the agricultural area is affected by moisture excess and about 63% are acid and very acid soils. This concise enumeration of soil conditions also emphasizes the important reclamation problems to be approached and solved.

Moisture excess control was initiated even since 1751-1774 when a series of rectifications and reprofilations were undertaken of the Somes river bed as well as the training of the interior part of the Crasna river. Important works were achieved between 1894-1920 when the Tur and Somes rivers on both banks were partially endiked and the waters in the high Codru zone were collected and evacuated by the achievement of the Homorodu Nou channel and the Ecet pond of about 13,000 hectares was surface drained (1).

Between 1920-1956 more maintenance works of the existing systems were achieved.

Step by step the drainage works were intensified after 1970 floods.

By the works achieved until 1983 there were protected against floods the neighbouring lands of the Crasna and Tur river, the largest part of the Somes river as well as some interior valleys. By the same date about 90% of the moisture excess affected area was surface drained but without succeeding in completely eliminating the moisture excess.

Most difficult problem which arises is due to water stagnation within the plots limited by the channel network as they comprise clayey lands of very low permeability.

2. ANALYSED LAND RECLAMATION ALTERNATIVES

Even since 1970-1980 were taken into study the problems of moisture excess removal from the heavy lands by fitting out three pilot perimeters of 150-300 ha each. Here were experimented alternatives and solutions for land drainage namely:

- tile drainage combined systematically with profound scarification;
- tile drainage combined systematically with mole drainage;
- land ridged strip modelling combined with tile drainage;
- land ridged strip modelling combined with evacuation ditches;
- land levelling with runoff slopes towards the channel network.

The observations carried on in the pilot perimeter during a relatively short period of time have shown to be rather efficient, the solutions comprising the tile drainage combined systematically with cross achieved mole drainage and that of land ridged modelling combined with tile drainage. The first solution requires the achievement of the tile drainage subsurface network at distances of 15-20 between the rows, and the tube covering on the entire trench width is to be done by a filter 25-30 cm height. This solution requires both a large material consumption and a large fuel quantity.

The second solution allows for the locating of the drainage network at higher distances according to the microrelief negative form, with a lesser volume both of material and fuel.

By a comparative analysis also of the investments it results that the second solution requires a lower investment about 18-20%.

Taking into account the present economical conditions it was proposed to adopt as a basic solution that of the land ridged strip modelling combined with tile drainage, a subsurface drainage that operates in the meantime as a collector of the water that runs off through the ditches between the ridges.

On small areas on closed depressions where the moisture excess is also caused by the ground water high level the solution of the tile drainage combined systematically with the mole drainage (about 15-20% of the moisture excess affected area) is to be applied.

3. DRAINAGE PROJECT PROPOSALS

A first measure that was taken into consideration was the analysis of the moisture excess capacity of being evacuated from the agricultural land by the existing surface drainage system.

Accordingly a first category of works to be taken was proposed to be a surface drainage channel network reprofilation existing in the sector where the clogging process was more intense as well as in the zone where the tile drainage had to be introduced. In those zones where the existing network has not the capacity to ensure the moisture excess evacuation it was proposed to increase the network by achieving some new channels. Where the collecting channel length does not allow for achieving sufficiently good water runoff velocities their fragmentation was stipulated as well as achieving some supplementary pumping station or the increasing of the existing station capacities.

Second work category is the land modelling and in the first place the earth spreading of the embankment resulting from the existing channel network reprofilation, and in the achievement of the new network so as to ensure a general slope from the plot towards the surface drainage channels. There where depression zones are met on the plots it was foreseen - according to their depth - either to fill them (when the depth is less than 0.5m) or their drainage by tile drained networks located according to the depression configuration.

The land drainage project type is established in accordance with pedo-hydrogeological conditions. Thus, on low permeability lands there will be achieved a land levelling in order to provide for a general slope towards the surface drainage channels. On lands presenting a very low permeability it was proposed to use land modelling with ridged strips whose ditches discharge into collectors that are closed by means of filtering prisms (Fig. 2).

The modelling and levelling are accompanied by soil scarifying and amendment in order to increase land permeability.

A third category is represented by the agrotechnical works and those of surface and subsurface drainage work maintenance.

At present special concern is given to the channel network maintenance especially by chemical treatment in order to avoid water and soil in the drainage systems, so that the environment should be preserved.

CONCLUSIONS

The Tur-Barcau drainage projects are a model of obtaining a maximum efficiency of low permeability soils affected by moisture excess.

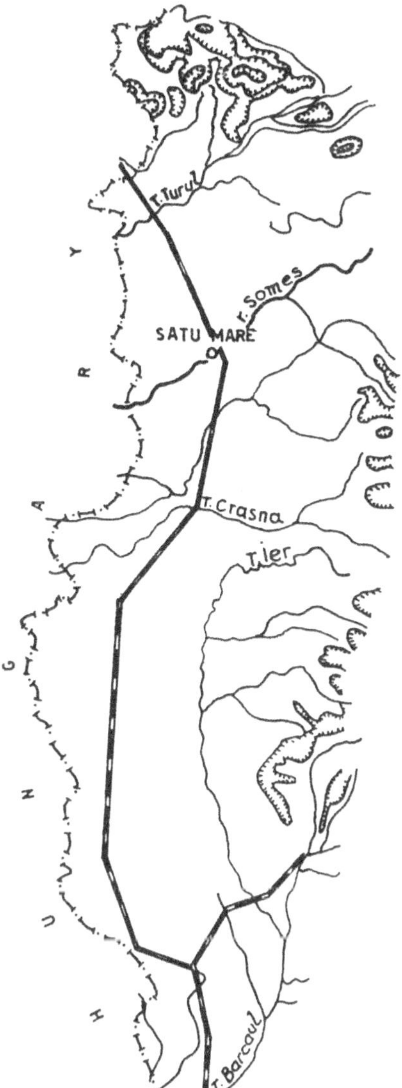

Fig. 1 - Characteristics of the zone under study

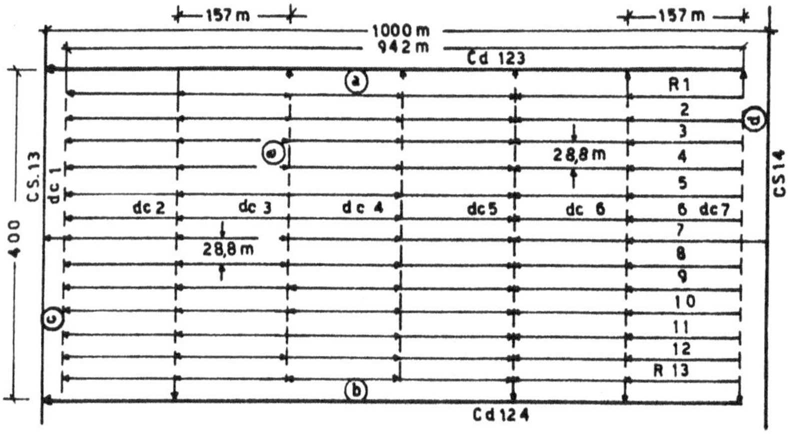

Fig.2 - Ridged strip drainage projected plot

a - Deposit storage planning zone, 30m width
b - Deposit storage planning zone, 30m width
c - End strip
d - End strip
e - Concrete collecting drain Ø 200mm

Surface and Subsurface Drainage Projects Achieved in Stages

Iulian Mihnea
Institute of Land Reclamation Studies and Design Bucharest, Romania

ABSTRACT

Under the current conditions of the energy crisis, the paper proposes to present the conceptions of achieving surface and subsurface drainage works in stages.
These conceptions provide for avoiding higher consumptions of materials, fuel a.s.o., proving themselves economically advantageous.

During years with a normal precipitation regimen, moisture excess phenomenon affects a large agricultural area of about 5.53 mil ha . About 2.5 mil ha out of this area are located within the flooded plains, while the remaining surface of 3.0 mil ha on strongly clayey heavy lands prevailing in Western Plain, on weakly drained depression plains and areas as well as on podzolic soil mountain slopes within the wet zone of the country.
Some sandy or loess soil areas may also be affected by moisture excess under irrigation conditions.

By the end of 1984 the area reclaimed by surface-drainage works included about 2.576 thousand ha.

During the 1986-1990 five years plan the surface-drainage works will be extended all over the zones of the country affected by moisture excess so that the total reclaimed area will reach about 5 mil ha by the end of 1990.

Due to the weak development inherited from the past in the land reclamation field, the necessity of achieving surface and subsurface drainage projects on large territory and in rather different soil relief and climate conditions as well as to the limited technical and financial means, all these reasons have led to adopting a conception for achieving the work complex and the surface and subsurface technical measures in stages.

During the first stage it was provided for executing with priority the main channel network and its adjoining hydrotechnical structures, the pumping stations as well as the strictly required works of emissary development. On the surface-drained plots between the channels it was provided that - by the care of the agricultural units beneficiary of those plots - to be achieved at a lesser scale, leveling, furrow ridge shaping, ditches, mole drainage and deep loosening works, in order to ensure the water excess run-off towards the channel network as well as for soil permeability improvement.

The other works and technical measures for integral reclamation of agricultural land such as: the tile drainage, the rough leveling, salt soil reclamation a.s.o., have been provided to be achieved step by step at other stages according to the possibility of ensuring the necessary material means, as such works require high expenses that are leading to a twofold increase of the average specific investment on the whole system.

The achievement step by step of the surface-drainage work complex and technical measures was necessary in our country as well as in other countries as an objective necessity because the moisture excess soil radical improvement is a long standing process. As a result of the first stage works, important modifications of the territory previous hydrological and hydrogeological conditions are taking place, which must be taken into consideration when establishing the work dimensions for the next stage, in order that the technical material investment be done in strict concordance with the actual requirement and with the highest efficiency. In this manner it was possible to comprise within the surface-drainage systems a higher number of agricultural units to which a minimum of conditions were provided for the increase of their production level.

For the surface and subsurface drainage projects that have to be achieved according to the "National Program for ensuring high, sure and stable agricultural yields" (this program has been approved by the Romanian National Great Assembly (Marea Adunare Naţională) on June 30 1983), it is necessary to adopt some alternative concerning both hydrotechnical schemes and the proper development works.

The hydrotechnical schemes to be proposed within the moisture excess removal projected peri - meters must be thus drawn up as to allow for some of their objects to have multiple functions. An example is given by the utilization of some surface drainage collecting canals as well as for irrigation water supply canals or the use of some pumping stations with another function such as to put under pressure irrigation networks. Another possibility is provided by the using of water volumes proceeding from the surface drainage system or of those that are flowing in natural regimen in order to decrease water pumping expenses in the irrigation project exploitation.

By colligating the land reclamation projects with those of water management ones a multiple use of the natural valleys can be achieved in order to improve the hydrographical basin drainage, the precipitation water evacuation, the maximum flood discharge eventual control, some water for irrigation, fish breeding etc. work storage.

General drainage criteria

Following criteria are taken into consideration in the hydrotechnical scheme conception and achievement:

- natural valleys: either reprofiling or training so that to ensure primarily the excess water collecting and evacuation and then the maximum flow discharge transition and control, water volume storage for irrigation, fish breeding, entertainment and, in case of an economic justification to provide for naviagation possibilities etc;
- the use of surface-drainage collecting canals for water delivery to the irrigation systems or to the piscicultural development;
- pumping station merging by the concentration of the evacuation, repumping, water collecting and irrigation network putting under pressure point;
- the reduction to the essential of the hydrotechnical structure on the main surface-drainage, irrigation piscicultural networks by colligating cum-regulator, valve, fall bridges, etc.;
- the storage of some water volumes proceeded from surface flows or by water lifting on dominant levels to be gravitationally distributed into the irrigation projects;
- the use of the ground layer infiltration discharge water (by recirculation or water supply, irrigation, fish breeding, entertainment a.s.o.);
- the hydrotechnical schemes are also drawn up ac-

cording to the using way of the perimeter to be reclaimed, namely: hydrotechnical scheme specific to a single (agricultural) use or a hydrotechnical scheme specific for two or more (agricultural or piscicultural) uses.

The hydrotechnical scheme alternative analysis must in the end lead to an economical investment and to a minimum fuel and energy consumption both during their execution and especially at the project exploitation.

A case apart is represented by the surface land drainage in the Danube Delta or in the submerged perimeters to be included into the agricultural circuit.

Turning to account of such lands requires special technologies and measures concerning both the land reclamation work nature and complexity and the duration aspect of their achievement. The main factors that require a step by step execution are represented by the lands 60 % submerged swampy regimen and by a reduced state of soil physical ripening.

Our own experience obtained in the Danube Delta within last years has shown that on such territories the complex land reclamation projects require a step by step achievement specific to such lands namely:

- stagnant water evacuation and damming works for land protection against flooding;
- shapings, levelings, irrigation and drainage works on ripening soils;
- shaping, irrigation drainage works on immature soils.

The above mentioned stages are compulsory and are conditioning each other in achieving the investments. In this way the necessary investments required by those project execution must be promoted on basis of studies and design drawn up according to the achievement stages.

First stage comprises water embankment and evacuation works. The execution of such works allows for a partial putting into use of the lands taken off the Danube flood influence.

Second stage comprises mature soil shaping, leveling, irrigation and drainage works.

From by now obtained experience it results that such works execution may be achieved in satisfactory conditions after about 3 years since water damming and evacuation, the required time for ground water stabilization.

Third stage comprises the irrigation and drainage work completion in the immature soil zones.

Such zones cover up to 60 % from the surface drained area and their soils reach their ripening after a period of 5 to 10 years since the surface drainage works were achieved.

From the above mentioned step by step achievement there are excepted those units where there have already been executed both damming and water evacuation works and where shaping, irrigation and drainage works can be implemented.

As regards the proper project, it is necessary that the water evacuation works to be achieved within the channel space (on plots) as well as the leveling, shaping works, ditches, mole drainage, deep loosening and tile drainage, and those meant to convey water from the plot to the emissary (collecting-evacuation canals and their adjacent structures, including pumping stations) should be conceived as a whole. Hence, sure conditions for excess water run off on the entire tract, from the plant root to the emissary, should be provided.

When achieving only the canal network without any excess water removal works from the plot, excess water will be stagnant in the field, even in the channel neighbourhood, while they either are empty or they accomplish their function partially.

A special attention must be given to tile drainage projects as well to heavy soil shaping (furrow ridges). In this respect the agricultural units must organize properly their agrotechnical technology.

The agricultural surfaces to be tile-drainage projected are selected functions of:

a/ The relief conditions (microrelief):
- depression perimeters (microdepressions) placed on river meadows, at morphological contact points, on alluvial low plains, as well as on mountain slopes affected by instability, or by sliding peril.

b/ Hydrogeological conditions:
- ground water level is usually at a medium yearly depth smaller than 1.0 m on low lands or at the mountain slope base, or it may appear under spring or run off forms.

c/ Soil conditions:
- lands affected by moisture excess (irrespective of the supply source) of strong or very strong intensity;
- very strong or temperate salt affected soils;
- lands with swampy potential and/or salts affected soils.

Tile drainage works should be achieved step by step as follows:

In the first stage:
- on lands with ground water at a depth smaller than 0.5 m, or on mountain slopes where springs appear;
- on lands affected by salts and/or strong or very strong alkalic ones, when irrigation-leaching is assured;
- on waste water irrigated lands when ground water depth is smaller than 1 m.

In the second stage (3 to 5 years after surface-evacuation works from the first stage have been achieved):
- on lands where ground water equals a medium yearly depth smaller than 1 m and/or during the characteristic periods at a depth smaller than 0.5 m;
- on lands strongly or very strongly affected by moisture excess (pseudogleyic) on which moisture excess doesn't give in even after surface works have been executed, including also shaping-leveling ones;
- on moisture excess affected lands, placed on microdepression perimeters - saucers - within the irrigated ones, due to ground water secondary level rising, which maintains a depth smaller than 1.0 m even after the measures and works for moisture excess control have been applied;
- on lands affected by moisture excess under water accumulation impact (rice field included) where ground water level is maintained higher (1.00 m) even after frontal or contour surface works have been achieved.

Before finalizing the selection of surface drainage schemes, the most efficient stage must be determined. We have to begin by establishing the characteristics of the perimeter to be surface drained (i.e. excess moisture agricultural lands, swampy or ponding unproductive lands), frequency intensity and duration of the excess moisture phenomenon.

The most complete technical scheme within the maximum work density and sizes will provide for the best result concerning yield increased possibility. In the most cases, nevertheless, such a scheme will not be the most profitable one, due to the high reclamation expenses implied by such a scheme both in execution and operation.

In order to obtain an economic efficient solution we must select and define the economical results (maintenance and operational expenses, repayments) obtained with the complex scheme as compared with more simple ones.

This study implies data concerning the way plants react to different hydric regimen, soils behavior to different climatic conditions. It also requires complex analyses concerning crop development in the benefit influence of the surface and subsurface drainage upon yields within the different possible solutions.

After stating all these data a complex study may be drawn up according to this scheme:

- perimeter area (S) comprises an elementary number of areas covered by different crops (a, b, c...n), the relation being:

$$S = \sum_{1}^{n}$$

S_1, S_n represent different possible crop rotation hypotheses;

- the surface drained perimeter area might be reclaimed by a surface-drainage network (1, 2, 3,..s); r_i showing the net benefit increase between the considered hypotheses (i crop and p type surface-drainage method)

The relation that gives the annual total benefit increase for surface-drained perimeter will be:

$$R_p = \sum_{1}^{n} \overset{M}{} (S_1 \times r_p)$$

By carrying out the study of many crop rotation hypotheses and types of scheme we may obtain the optimal technical and economical hypotheses.

New Drainage Formulas Considering Delayed Gravity Response and Evaporation from Shallow Water Table

Weizhen Zhang
Wuhan Institute of Hydraulic and Electrical Engineering, Wuhan, Hubei, China

INTRODUCTION

Drainage formulas usually can be classified into two categories: steady flow formulas for the cases of long lasting rainfall or closely spaced drains and nonsteady flow ones for the cases of short period rainfall and irrigation water infiltration. In the latter cases the specific yield and evaporation from water table are the important parameters involved in the drainage formulas. As shown by W. Zhang and Y. Zhang (1982) the drainable porosity for a certain soil is a function of rate of drawdown and initial depth to ground water table, as well as intensity of evaporation. The formulas for water table drawdown with delayed gravity response have been developed by Neuman (1982) on the basis of 2-D flow equations. Drainage equations considering evaporation from water table expressed by the formula

$$\varepsilon = \varepsilon_o\left(1 - \frac{\Delta}{\Delta_o}\right)^n$$

and constant drainable porosity have been proposed by H. Hammad (1962), W. Zhang (1963), D. Guo (1965), J. Sha (1972) and X. Qu and Y. Zhang (1981). In this paper formulas considering both delayed yield and evaporation from water table are developed.

DERIVATION OF FORMULAS

In most field conditions the subsurface drainage problem can be simplified as shown in Fig. 1. The water table after the rainfall infiltration is h_i above the impervious barrier and Δ_i below the ground surface. h_n is the elevation of ground surface. The spacing of drains is $B = 2L$, and the thickness and the hydraulic conductivity of soil are M and k respectively. When the amplitude of fluctuation of water table is small compared to the soil thickness, the transmissivity of the aquifer can be approximated by $T = kM$ or $T = k\bar{\bar{h}}$ ($\bar{\bar{h}}$ is the weighted mean thickness both in space and in time).

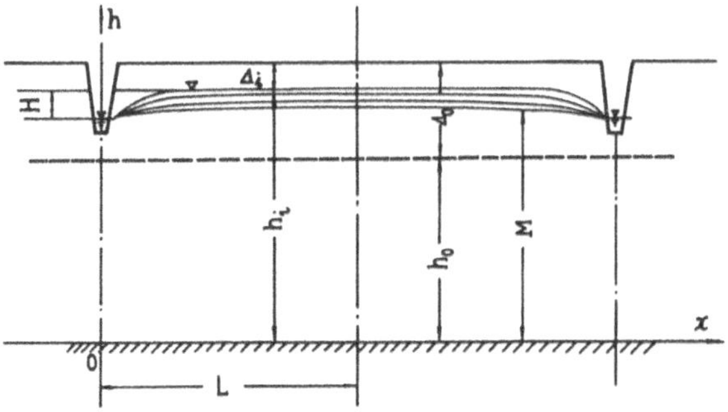

Figure 1

When the evaporation from groundwater varies linearly with water table depth $\varepsilon = \varepsilon_o(1 - \Delta/\Delta_o)$, following the Boulton's empirical expression for the delayed gravity response the linearized flow equation can be written as follows:

$$\alpha\mu_d \int_0^t \frac{\partial h}{\partial \tau}\, e^{-\alpha(t-\tau)}\, d\tau + \mu_e \frac{\partial h}{\partial t} = T\,\frac{\partial^2 h}{\partial x^2} - \varepsilon_o\left(1 - \frac{\Delta}{\Delta_o}\right) \qquad (1)$$

where $h(x,t)$ - the height of water table above the impervious barrier,
- t - time,
- x - the distance from the left drain,
- μ_d - the complete drainable porosity,
- α - the delayed yield index,
- μ_e - the elastic storativity,
- T - the transmissivity of aquifer,
- ε - the evaporation from water table,
- ε_o - the evaporation from water surface,
- Δ - water table depth
- Δ_o - water table depth at which evaporation approaches zero.

as

$$\varepsilon_o\left(1 - \frac{\Delta}{\Delta_o}\right) = \varepsilon_o\left(\frac{\Delta_o - h_n + h}{\Delta_o}\right) = \varepsilon_o\,\frac{h}{\Delta_o} - \varepsilon_o\,\frac{h_o}{\Delta_o}$$

equation (1) can be rewritten as:

$$\alpha \int_0^t \frac{\partial h}{\partial \tau} e^{-\alpha(t-\tau)} d\tau + \frac{\mu_e}{\mu_d}\frac{\partial h}{\partial t} = \frac{T}{\mu_d}\frac{\partial^2 h}{\partial x^2} - \frac{\varepsilon_o}{\mu_d}\frac{h}{\Delta_o} + \frac{\varepsilon_o}{\mu_d}\frac{h_o}{\Delta_o} \tag{2}$$

let

$$\frac{T}{\mu_d} = a_d , \quad \frac{\varepsilon_o}{\mu_d \Delta_o} = b, \quad -\frac{\varepsilon_o}{\mu_d}\frac{h_o}{\Delta_o} = -bh_o$$

equation (2) becomes:

$$\alpha \int_0^t \frac{\partial h}{\partial \tau} e^{-\alpha(t-\tau)} d\tau + \frac{\mu_e}{\mu_d}\frac{\partial h}{\partial t} = a_d \frac{\partial^2 h}{\partial x^2} - bh + bh_o \tag{3}$$

For the drainage problem shown in Fig. 1, the initial and boundary conditions are:

$$h(x,0) = h_i \tag{4}$$

$$h(0,t) = h_i - H \tag{5}$$

$$\left.\frac{\partial h(x,t)}{\partial x}\right|_{x=L} = 0 \tag{6}$$

After applying the Laplace transform to (3), (5), (6) with respect to time t, we obtain:

$$\frac{d^2\bar{h}}{dx^2} - \left(\frac{p+\beta}{p+\alpha} + \frac{(\eta-1)b}{p}\right)\frac{p\bar{h}}{a_e} = -\frac{(\eta-1)bh_o}{a_e\ p} - \frac{p+\beta}{p+\alpha}\frac{h_i}{a_e} \tag{7}$$

$$\bar{h}(0,p) = \frac{h_i-H}{p} \tag{8}$$

$$\frac{\partial\bar{h}}{\partial x}(L,p) = 0 \tag{9}$$

where

$$\bar{h}(x,p) = \int_0^\infty h(x,t)e^{-pt}\,dt$$

$$\eta = \frac{\mu_d+\mu_e}{\mu_e} , \qquad a_e = \frac{T}{\mu_e}$$

Solving differential equation (7) for h, and making use of (8) and)9) we have:

$$\bar{h}(x,p) = \frac{h_i}{p} - \frac{h_i-h_o}{p^2}\,\frac{(\eta-1)b}{\frac{p+\beta}{p+\alpha}+\frac{(\eta-1)b}{p}}\left[1-\sum_{n=o}^{\infty}(-1)^n\left(e^{-(2nL+2L-x)\sqrt{\frac{p}{a_e}\left(\frac{p+\beta}{p+\alpha}+\frac{(\eta-1)b}{p}\right)}} + e^{-(2nL+x)\sqrt{\frac{p}{a_e}\left(\frac{p+\beta}{p+\alpha}+\frac{(\eta-1)b}{p}\right)}}\right)\right] - \frac{H}{p}\sum_{n=o}^{\infty}(-1)^n\left(e^{-(2nL+2L-x)\sqrt{\frac{p}{a_e}\left(\frac{p+\beta}{p+\alpha}+\frac{\eta-1}{p}b\right)}} + e^{-(2nL+x)\sqrt{\frac{p}{a_e}\left(\frac{p+\beta}{p+\alpha}+\frac{\eta-1}{p}b\right)}}\right) \qquad (10)$$

or

$$\bar{h}(x,p) = \frac{h_i}{p} - \frac{h_i-h_o}{p^2}\,\frac{(\eta-1)b}{\frac{p+\beta}{p+\alpha}+\frac{\eta-1}{p}b}\left[1-\frac{\mathrm{ch}\left[(L-x)\sqrt{\frac{p}{a_e}\left(\frac{p+\beta}{p+\alpha}+\frac{\eta-1}{p}b\right)}\right]}{\left[L\sqrt{\frac{p}{a_e}\left(\frac{p+\beta}{p+\alpha}+\frac{\eta-1}{p}b\right)}\right]}\right] - \frac{H}{p}\,\frac{\mathrm{ch}\left[(L-x)\sqrt{\frac{p}{a_e}\left(\frac{p+\beta}{p+\alpha}+\frac{\eta-1}{p}b\right)}\right]}{\mathrm{ch}\left[L\sqrt{\frac{p}{a_e}\left(\frac{p+\beta}{p+\alpha}+\frac{\eta-1}{p}b\right)}\right]} \qquad (11)$$

From equations (10) and (11) we can obtain the following short time and long time approximate solutions for $h(x,t)$.

Solutions for $t < 1/20\beta$

According to the theory of Laplace transformation $t = 1/20\beta = 1/20\alpha\eta$ corresponds approximately to $p > 20\beta$. When $p > 20\beta$

$$p + \alpha \approx p \;,$$

$$p + \beta \approx p$$

equation (11) reduces to

$$\bar{h} = \frac{h_i}{p} - \frac{h_i - h_o}{p}\,\frac{b_1}{p+b_1}\left[1 - \frac{ch\left[(1-\bar{x})\sqrt{\frac{L^2}{a_e}}\sqrt{p+b_1}\right]}{ch\left[\sqrt{\frac{L^2}{a_e}}\sqrt{p+b_1}\right]}\right]$$

$$- \frac{H}{p}\,\frac{ch\left[(1-\bar{x})\sqrt{\frac{L^2}{a_e}}\sqrt{p+b_1}\right]}{ch\left[\sqrt{\frac{L^2}{a_e}}\sqrt{p+b_1}\right]} \qquad (12)$$

where $b_1 = (\eta-1)b$, $\bar{x} = \frac{x}{L}$

After the inverse transformation we have:

$$h(x,t) = h_i - (h_i - h_o) - (h_i - h_o) \sum_{n=1}^{\infty} (-1)^n$$

$$\frac{\cos\left[(1-\bar{x})\frac{2n-1}{2}\pi\right] e^{-\left[\left(\frac{2n-1}{2}\pi\right)^2 + \bar{b}_L\right]\bar{t}}}{\frac{2n-1}{2}\pi}$$

$$+ (h_i - h_o - H)\left\{\frac{ch(1-\bar{x})\sqrt{\bar{b}_1}}{ch\sqrt{\bar{b}_1}} + \sum_{n=1}^{\infty}(-1)^n \frac{\cos\left[(1-\bar{x})\frac{2n-1}{2}\pi\right]\frac{2n-1}{2}\pi}{\left(\frac{2n-1}{2}\pi\right)^2 + \bar{b}_1}\right.$$

$$\left. e^{-\left[\left(\frac{2n-1}{2}\pi\right)^2 + \bar{b}_1\right]\bar{t}}\right\} \qquad (13)$$

where $\bar{b}_1 = b_1 \frac{L^2}{a_e}$, $\bar{t} = \frac{t}{\frac{L^2}{a_e}}$

<u>Solutions for $t < 1/20\alpha$</u>

$t = 1/20\alpha$ approximately corresponds to $p > 20\alpha$, in this case

$$p+\alpha \approx p$$

equation (11) reduces to:

$$\bar{h} = \frac{h_i}{p} - \frac{h_i - h_o}{p}\,\frac{b_1}{p+b_2}\left[1 - \frac{ch(1-\bar{x})\sqrt{\frac{L^2}{a_e}}\sqrt{p+b_2}}{ch\sqrt{\frac{L^2}{a_e}}\sqrt{p+b_2}}\right]$$

$$- \frac{H}{p} \frac{\mathrm{ch}(1-\bar{x})\sqrt{\frac{L^2}{a_e}}\sqrt{p+b_2}}{\mathrm{ch}\sqrt{\frac{L^2}{a_e}}\sqrt{p+b_2}} \tag{14}$$

where $b_2 = b_1 + \beta = (\eta-1)b+\beta$

After inverse transformation we have

$$h(x,t) = h_i - (h_i-h_o)\frac{\bar{b}_1}{\bar{b}_2} + \left(\frac{h_i-h_o}{\bar{b}_2}\bar{b}_1 - H\right)\left\{\frac{\mathrm{ch}(1-\bar{x})\sqrt{\bar{b}_2}}{\mathrm{ch}\sqrt{\bar{b}_2}}\right.$$

$$+ \sum_{n=1}^{\infty}(-1)^n \frac{\cos[(1-\bar{x})\frac{2n-1}{2}\pi]\cdot\frac{2n-1}{2}\pi}{(\frac{2n-1}{2}\pi)^2+\bar{b}_2} e^{-[(\frac{2n-1}{2}\pi)^2+\bar{b}_2]\bar{t}}$$

$$- \frac{h_i-h_o}{\bar{b}_2}\sum_{n=1}^{\infty}(-1)^n \frac{\cos[(1-\bar{x})\frac{2n-1}{2}\pi]}{\frac{2n-1}{2}\pi} e^{-[(\frac{2n-1}{2}\pi)^2+\bar{b}_2]\bar{t}} \tag{15}$$

<u>Solutions for $t > 20/\beta = 20/\alpha\eta$</u>

$t = 20/\beta$ approximately corresponds to $p < \beta/20$, in this case $p + \beta \approx \beta$, equation (11) is simplified to

$$\bar{h} = \frac{h_i}{p} - \frac{h_i-h_o}{p}\frac{(\eta-1)\,b(p+\alpha)}{[\beta+(\eta-1)b]p+(\eta-1)\alpha b}\left\{1 - \frac{\mathrm{ch}\left[(1-\bar{x})\sqrt{\frac{L^2}{a}}\sqrt{\frac{(\alpha+\frac{\eta-1}{\eta}b)p+\frac{\eta-1}{\eta}\alpha b}{p+\alpha}}\right]}{\mathrm{ch}\left[\sqrt{\frac{(\alpha+\frac{\eta-1}{\eta}b)p+\frac{\eta-1}{\eta}\alpha b}{p+\alpha}}\right]}\right\}$$

$$- \frac{H}{p}\frac{\mathrm{ch}\left[(1-\bar{x})\sqrt{\frac{L^2}{a}}\sqrt{\frac{(\alpha+\frac{\eta-1}{\eta}b)p+\frac{\eta-1}{\eta}\alpha b}{p+\alpha}}\right]}{\mathrm{ch}\left[\sqrt{\frac{L^2}{a}\frac{(\alpha+\frac{\eta-1}{\eta}b)p+\frac{\eta-1}{\eta}\alpha b}{p+\alpha}}\right]} \tag{16}$$

where $a = \frac{T}{\mu_e+\mu_d}$

After inverse transformation, we obtain:

$$h(x,t) = h_o + (h_i - h_o - H)\left\{ \frac{ch(1-\bar{x})\sqrt{\bar{b}}}{ch\sqrt{\bar{b}}} \right.$$

$$- \sum_{n=1}^{\infty} \frac{2\bar{\alpha}\cos[(1-\bar{x})\frac{2n-1}{2}\pi]\frac{2n-1}{2}\pi}{\sin\frac{2n-1}{2}\pi\left[(\frac{2n-1}{2}\pi)^2 + \frac{\eta-1}{\eta}\bar{b}\right]\left[(\frac{2n-1}{2}\pi)^2 + \bar{\alpha} + \frac{\eta-1}{\eta} b\right]}$$

$$\left. e^{-\frac{\bar{\alpha}[(\frac{2n-1}{2}\pi)^2 + \frac{\eta-1}{\eta}\bar{b}}{(\frac{2n-1}{2}\pi)^2 + \bar{\alpha} + \frac{\eta-1}{\eta}\bar{b}}\bar{t}} \right\} + (h_i - h_o)\sum_{n=1}^{\infty}$$

$$\frac{2\bar{\alpha}\cos[(1-\bar{x})\frac{2n-1}{2}\pi]e^{-\bar{\alpha}\frac{[(\frac{2n-1}{2})^2 + \frac{\eta-1}{\eta}\bar{b}]}{(\frac{2n-1}{2}\pi)^2 + \bar{\alpha} + \frac{\eta-1}{\eta}\bar{b}}\bar{t}}}{\sin(\frac{2n-1}{2}\pi).\frac{2n-1}{2}\pi\left[(\frac{2n-1}{2}\pi)^2 + \bar{\alpha} + \frac{\eta-1}{\eta}\bar{b}\right]} \qquad (17)$$

where $\bar{b} = b\frac{L^2}{a}$, $\bar{\alpha} = \alpha\frac{L^2}{a}$, $\bar{t} = \frac{at}{L^2}$

Solutions for $t > 20/\alpha$

$t > 20/\alpha$ approximately corresponds to $p < \alpha/20$. In this case $p+\beta \approx \beta$, $p+\alpha \approx \alpha$, equation (11) is simplified to the following form:

$$\bar{h} = \frac{h_i}{p} - \frac{h_i - h_o}{p}\,\frac{\frac{\eta-1}{\eta}b}{p+\frac{\eta-1}{\eta}b}\left[1 - \frac{ch[(1-\bar{x})\sqrt{\frac{L^2}{a}}\sqrt{p+\frac{\eta-1}{\eta}b}]}{ch\left[\sqrt{\frac{L^2}{a}}\sqrt{p+\frac{\eta-1}{\eta}b}\right]}\right]$$

$$- \frac{H}{p}\,\frac{ch(1-\bar{x})\sqrt{\frac{L^2}{a}}\sqrt{p+\frac{\eta-1}{\eta}b}}{ch\left[\sqrt{\frac{L^2}{a}}\sqrt{p+\frac{\eta-1}{\eta}b}\right]} \qquad (18)$$

similarly the solution for large time is:

$$h(x,t) = h_o - (h_i - h_o) \sum_{n=1}^{\infty} (-1)^n \frac{\cos[(1-\bar{x}) \frac{2n-1}{2} \pi]}{\frac{2n-1}{2}\pi} e^{-[(\frac{2n-1}{2}\pi)^2 + \bar{b}_4]\bar{t}}$$

$$+ (h_i - h_o - H) \left\{ \frac{\mathrm{ch}(1-\bar{x})\sqrt{\bar{b}_4}}{\mathrm{ch}\sqrt{\bar{b}_4}} + \sum_{n=1}^{\infty} (-1)^n \frac{\cos[(1-\bar{x})\frac{2n-1}{2}\pi]}{(\frac{2n-1}{2}\pi)^2 + \bar{b}_4} e^{-[(\frac{2n-1}{2}\pi)^2 + \bar{b}_4]\bar{t}} \right\} \qquad (19)$$

where

$$\bar{b}_4 = \frac{\eta - 1}{\eta} b \frac{L^2}{a} \quad , \quad \bar{t} = \frac{at}{L^2}$$

The above equations are derived for fully penetrating drains, in case of partial penetration of drains, the soil thickness should be multiplied by a correction factor proposed by Hooghoudt or Qu.

ANALYSIS OF SOLUTIONS

For most field conditions the drainable porosity μ_d of soil varies from 0.02 to 0.2, while the elastic storativity μ_e usually is of the order of 1×10^{-4}, which is negligible in comparison to μ_d. In this case η can be taken as infinity, and equation (11) becomes:

$$\bar{h}(x,p) = \frac{h_i}{p} - \frac{h_i - h_o}{p} \frac{(p+\alpha)b}{(\alpha+b)p+\alpha b} \left\{ 1 - \frac{\mathrm{ch}[(1-\bar{x})\sqrt{\frac{L^2}{a}}\sqrt{\frac{p(\alpha+b)+\alpha b}{p+\alpha}}]}{\mathrm{ch}[\sqrt{\frac{L^2}{a}}\sqrt{\frac{p(\alpha+b)+\alpha b}{p+\alpha}}]} \right\}$$

$$- \frac{H}{p} \frac{\mathrm{ch}[(1-\bar{x})\sqrt{\frac{L^2}{a}}\sqrt{\frac{p(\alpha+b)+\alpha b}{p+\alpha}}]}{\mathrm{ch}[\sqrt{\frac{L^2}{a}}\sqrt{\frac{p(\alpha+b)+\alpha b}{p+\alpha}}]} \qquad (16')$$

where $a = T/\mu_d$.

Analogous to (17), the solution for h is:

$$h(x,t) = h_o + (h_i-h_o-H)\left\{\frac{ch(1-\bar{x})\sqrt{\bar{b}}}{ch\sqrt{\bar{b}}} + \sum_{n=1}^{\infty}(-1)^n \frac{2\bar{\alpha}\cos[(1-\bar{x})\frac{2n-1}{2}\pi]\frac{2n-1}{2}\pi}{[(\frac{2n-1}{2}\pi)^2+\bar{b}][(\frac{2n-1}{2}\pi)^2+\bar{\alpha}+\bar{b}]} e^{-\frac{\bar{\alpha}[(\frac{2n-1}{2}\pi)^2+\bar{b}]}{(\frac{2n-1}{2}\pi)^2+\bar{\alpha}+\bar{b}}\bar{t}}\right\}$$

$$-(h_i-h_o)\sum_{n=1}^{\infty}(-1)^n \frac{2\bar{\alpha}\cos[(1-\bar{x})\frac{2n-1}{2}\pi]}{\frac{2n-1}{2}\pi[(\frac{2n-1}{2}\pi)^2+\bar{\alpha}+\bar{b}]} e^{-\bar{\alpha}\frac{(\frac{2n-1}{2}\pi)^2+\bar{b}}{(\frac{2n-1}{2}\pi)^2+\bar{\alpha}+\bar{b}}\bar{t}} \qquad (17')$$

Equation (17) is the same as (17'), except in (17') $\eta-1/\eta$ is replaced by 1, and $a = T/(\mu_e+\mu_d)$ in $\bar{b},\bar{\alpha}\bar{t}$ is replaced by $a = T/\mu_d$. Hence (17) is useful for both cases considering and neglecting the elastic storativity. Next we will take equation (17) for the analysis of solutions.

Seepage discharge to drains

As seepage discharge to drains from one side is $-k\,\bar{h}\left.\frac{\partial h}{\partial x}\right|_{x=o}$,

we can easily obtain the equation for seepage rate per unit length of drains:

$$q_o = 2k\bar{h}\,\frac{h_i-h_o-H}{L}\left\{\frac{\sqrt{\bar{b}}\,sh\sqrt{\bar{b}}}{ch\sqrt{\bar{b}}} + \left[\sum_{n=1}^{\infty}\frac{2\bar{\alpha}(\frac{2n-1}{2}\pi)^2}{[(\frac{2n-1}{2}\pi)^2+\frac{\eta-1}{\eta}\bar{b}][(\frac{2n-1}{2}\pi)^2+\bar{\alpha}+\frac{\eta-1}{\eta}\bar{b}]} e^{-\bar{\alpha}\frac{(\frac{2n-1}{2}\pi)^2+\frac{\eta-1}{\eta}\bar{b}}{(\frac{2n-1}{2}\pi)^2+\bar{\alpha}+\bar{b}\frac{\eta-1}{\eta}}\bar{t}}\right]\right. - \frac{k\bar{h}(h_i-h_o)}{L}\sum_{n=1}^{\infty}\frac{2\bar{\alpha}}{(\frac{2n-1}{2}\pi)^2+\bar{\alpha}+\bar{b}\frac{\eta-1}{\eta}} e^{-\bar{\alpha}\frac{(\frac{2n-1}{2}\pi)^2+\frac{\eta-1}{\eta}\bar{b}}{(\frac{2n-1}{2}\pi)^2+\bar{\alpha}+\frac{\eta-1}{\eta}\bar{b}}\bar{t}} \qquad (20)$$

Solutions for the case of negligible evaporation

When evaporation vanishes, b = 0, equation (17) for water table and seepage to drains is:

$$h(x,t) = h_i - H\left\{1 + \sum_{n=1}^{\infty} (-1)^n \frac{2\bar{\alpha}\cos[(1-\bar{x})\frac{2n-1}{2}\pi]}{\frac{2n-1}{2}\pi\,[(\frac{2n-1}{2}\pi)^2 + \bar{\alpha}]}\, e^{-\bar{\alpha}\frac{(\frac{2n-1}{2}\pi)^2}{(\frac{2n-1}{2}\pi)^2+\bar{\alpha}}\bar{t}}\right\} \qquad (21)$$

$$q_o(t) = -2k\bar{\bar{h}}\frac{H}{L}\sum_{n=1}^{\infty}\frac{2\bar{\alpha}}{(\frac{2n-1}{2}\pi)^2+\bar{\alpha}}\, e^{-\bar{\alpha}\frac{(\frac{2n-1}{2}\pi)^2\bar{t}}{(\frac{2n-1}{2}\pi)^2+\bar{\alpha}}} \qquad (22)$$

When gravity response accomplishes instantaneously, the index of delayed yield α tends to infinity, in this case equation (21) becomes:

$$h(x,t) = h_i - H\left\{1 + \sum_{n=1}^{\infty}(-1)^n \frac{2\cos[(1-\bar{x})\frac{2n-1}{2}\pi]}{\frac{2n-1}{2}\pi}\, e^{-(\frac{2n-1}{2}\pi)^2\bar{t}}\right\} \qquad (21')$$

When the instantaneous water level drop H is equal to h-M, let S = h-M, (21') will be:

$$S = -H\sum_{n=1}^{\infty}(-1)^n\frac{2\cos[(1-\bar{x})\frac{2n-1}{2}\pi]}{\frac{2n-1}{2}\pi}\, e^{-(\frac{2n-1}{2}\pi)^2\bar{t}}$$

When $\bar{t} > 0.3$, the second term of the infinite series is negligible. The height of water table at midpoint above the water level in the drains will be:

$$S = H\frac{4}{\pi}\, e^{-\frac{\pi^2}{4}t\frac{k\bar{h}}{\mu L^2}}$$

By taking logarithm we obtain:

$$L = \frac{\pi}{2}\sqrt{\frac{k\,\bar{\bar{h}}\,t}{\mu\,\ln\frac{4H}{\pi S}}} \qquad (23)$$

Equation (23) for this special case is the well-known Glover's equation.

<u>Solution for the case with evaporation and constant drainable porosity</u>

When α tends to infinity, μ_d = constant, equation (17) becomes:

$$h(x,t) = h_i (h_i - h_o) \{1-(\eta') - e^{-\bar{b}\bar{t}} [1-(S')]\} - H(\eta') \qquad (24)$$

where

$$(\eta') = \frac{ch(1-\bar{x})\sqrt{\bar{b}}}{ch\sqrt{\bar{b}}} + \sum_{n=1}^{\infty} (-1)^n \frac{2\cos[(1-\bar{x})\frac{2n-1}{2}\pi]\frac{2n-1}{2}\pi}{(\frac{2n-1}{2}\pi)^2 + \bar{b}} e^{-[(\frac{2n-1}{2}\pi)^2+\bar{b}]\bar{t}}$$

$$(S') = 1 + \sum_{n=1}^{\infty} (-1)^n \frac{2\cos[(1-\bar{x})\frac{2n-1}{2}\pi]}{\frac{2n-1}{2}\pi} e^{-(\frac{2n-1}{2}\pi)^2\bar{t}}$$

Equation (24) is the formula developed earlier by W. Zhang (1963).

COMPARISON OF THEORY TO EXPERIMENT

Experiments on subsurface drainage under the conditions of evaporation and variable drainable porosity have been carried out by Z. Ye (1965). The experiments were conducted in a soil tank 400 cm. long, which simulated fully penetrating ditches spaced at 740 cm. apart (one half of drain spacing L = 370 cm.). The tank was filled with sandy loam, the hydraulic conductivity of which is k = 0.00092 cm/sec. = 0.0552 cm/min. The soil tank was covered by a wind tunnel, where the conditions for artificial evaporation were provided by the ventilators at both ends and electrical bulbs of 200 watts equally spaced in the tunnel. The simulated evaporation from water surface was $\varepsilon_o = 1.595 \times 10^{-3}$ cm/min. The height of soil surface h_n above the bottom of the tank is 92 cm, the initial height of water table h_i = 80 cm. The drainable porosity of the soil μ_d is 0.022, the index of delayed yield is $\alpha = 0.0636\ \mathrm{min}^{-1}$. The water table depth at which the evaporation approaches zero Δ_o = 50 cm (h_o = 42 cm), the relationship between evaporation and water table depth can be expressed linearly by the equation

$$\varepsilon = \varepsilon_o (1 - \frac{\Delta}{\Delta_o})$$

At the beginning of the experiments the water level in the ditches was instantaneously lowered by 23.5 cm. The water table heights and seepage discharge were measured and recorded regularly, and are shown in Table 1 and Fig. 2 respectively.

Table 1 Observed and calculated groundwater levels

Location of observation wells	ground water level (cm) \ time (min)	10	20	30	60	90	120	180	240	300
$\bar{x} = 0.2$	Calculated	71.5	69.5	67.9	64.6	62.6	61.2	59.4	58.2	57.3
	Observed	74.0	69.4	67.0	63.7	61.8	60.5	59.0	57.7	56.5
$\bar{x} = 0.4$	Calculated	76.5	74.9	73.6	70.1	67.4	65.3	62.2	60.0	58.4
	Observed	78.3	75.8	73.5	69.0	66.3	64.1	61.3	59.1	57.4
$\bar{x} = 0.6$	Calculated	78.0	77.0	76.0	73.2	70.6	68.3	64.5	61.6	59.4
	Observed	78.8	78.2	76.7	73.1	70.0	67.3	63.8	60.8	58.4
$\bar{x} = 0.8$	Calculated	78.4	77.7	76.9	74.6	72.2	70.0	66.0	62.7	60.1
	Observed	78.8	78.2	77.6	75.1	72.2	69.3	65.6	62.2	59.5
$\bar{x} = 1.0$	Calculated	78.5	77.8	77.2	75.0	72.8	70.5	66.5	63.1	60.4
	Observed	78.8	78.3	77.6	75.2	72.3	69.4	65.7	62.3	59.5

The parameters corresponding to the experimental conditions used in calculating groundwater level and seepage discharge are

$$\bar{\alpha} = \alpha \frac{L^2}{a} = 52.51$$

$$\bar{b} = \frac{\varepsilon_o}{\mu \Delta_o} \frac{L^2}{a} = 1.197$$

The theoretical values for water table heights and seepage rates calculated from equations (17') and (20) are also given in Table 1 and Figure 2. The agreement between theory and experiment is satisfactory.

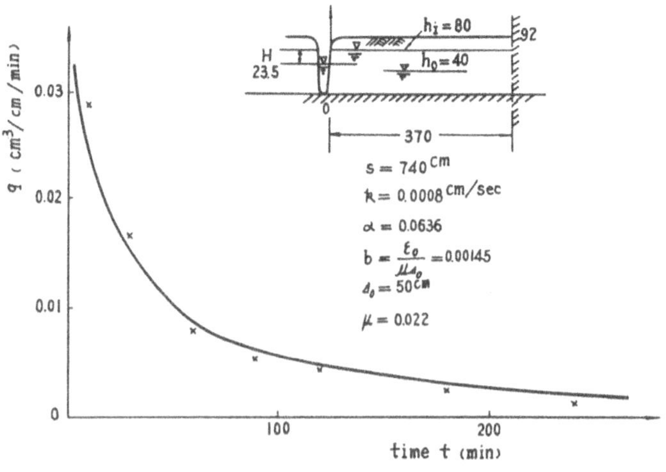

CONCLUSION

Due to delayed gravity response the drainable porosity is a function of depth and rate of water table drawdown. In most of the existing drainage formulas this fact is not considered and it is difficult to select a proper average drainable porosity. In field conditions evaporation from water table plays an important part in lowering groundwater table in drainage areas, in subsurface drainage design this factor should also be taken into account. The formulas developed and verified by experiments in this paper provide a new method for subsurface drainage design which can easily take into consideration the above two factors. The formulas considering only one factor (evaporation or variable specific yield) can be obtained by putting α or $b = 0$ respectively in the solutions. When both of these two factors are neglected the solution is identical to the well-known Glover's equation for falling water table.

REFERENCES

BOULTON, N. (1954) The drawdown of water table under non-steady conditions near a pumped well in an unconfined formation, Proc. Institute of Civil Engineers (London), V.3, Part 3, 564-579.

HAMMAD, H. (1962) Depth and spacing of tile drain systems, ASCE, Proc. 88 (IR1).

NEUMAN, S. (1981) Delayed drainage in a stream aquifer system, Journal of Irr. and Dr. Division, ASCE, V.107, 407-410.

QU, X. and Y. ZHANG (1981) Calculation of field drain spacing in consideration of the influence of evaporation and desalinization requirement, Journal of Hydr. Eng. No.5, 1981.

SHA, J. (1979) Unsteady groundwater flow in a plane area, Journal of Mechanics, No.4, 1979.

YE, Z. (1965) The study of subsurface drainage under the conditions of evaporation, Dissertation.

ZHANG, W. and Y. ZHANG (1983) The study of drainable porosity of soils, Hydrogeology and Engineering Geology, No.5, 1983.

ZHANG, W. (1963) Calculation of the groundwater levels under the conditions of drainage and evaporation, Journal of Wuhan Institute of Hydraulic and Electrical Engineering, No.1, 1963.

Unsteady State Condition Drainage of Sloping Land Theoretical and Experimental Results Concerning the Raising of Water Table Depletion Curve

I. Cojocaru and St. Popescu
Polytechnic Institute of Iași-Romania

INTRODUCTION

Research carried out in an experimental drainage field located in one of the endemic sliding zones of Romania, Dîmbovitza River watershed, Valea Largă sub-basin (Cojocaru, I.1985) has indicated that drainage, due to climatic and hydrogeological characteristics, functions in that zone under unsteady state condition. The dimensioning of drainage for this operational regime consists of finding out those technical characteristics for which the dynamic evolution of the depletion curve, under the influence of water intake and outflow, satisfied the drainage duty permanently. The evolution of the depletion curve may display 3 situations:

- depletion curve raises with time T: $dh/dT > o$
- depletion curve lowers with time T: $dh/dT < o$
- depletion curve remains constant with respect to time T: $dh/dT \simeq o$

In the paper herein it is analysed the raising stage of the depletion curve in the conditions of a systematic drainage located on an impervious layer.

MATHEMATICAL MODEL

Wooding and Chapman have obtained (1966) on the basis of an extended form of the Dupuit-Forchheimer assumption, for a sloping phreatic flow the following equation:

$$\frac{\partial}{\partial x}\left(h \frac{\partial h}{\partial x}\right) - tg\alpha \frac{\partial h}{\partial x} + R\left(1 + tg\alpha \frac{\partial h}{\partial x}\right) = \frac{p}{k \cos\alpha} \frac{\partial h}{\partial t} \quad (1)$$

in which h-height of flow at x distance as against the (x,y) coordinate system origin; α -the angle made by the impervious layer with the horizontal; $R = q_c/k$; q_c-unitary percolated flow; k-hydraulic conductivity;

p-effective porosity; t-time.
Now, if the variable is defined:

$$X = \frac{x}{L} \tag{2}$$

in which L-flow length in x direction and if an unknown function is considered:

$$F = \frac{2h}{(1-R)L} \cos\alpha \tag{3}$$

and an independent variable

$$\tau = \frac{k(1-R)}{2.p.L} t \tag{4}$$

it gives:

$$\frac{\partial F}{\partial \tau} = \frac{4}{(1-R)^2} \cos\alpha \, \frac{p}{k} \frac{\partial h}{\partial t} \tag{5}$$

$$2\sin\alpha \frac{\partial F}{\partial X} = \frac{4}{1-R} \sin\alpha \cos\alpha \frac{\partial h}{\partial x} \tag{6}$$

$$\frac{\partial}{\partial X}\left(F \frac{\partial F}{\partial X}\right) = \frac{4}{(1-R)^2} \cos^2\alpha \, \frac{\partial}{\partial x}\left(h \frac{\partial h}{\partial x}\right) \tag{7}$$

After successive transformations the following general equation is obtained for unsteady state functioning of drainage located on an impervious sloping layer:

$$\frac{\partial F}{\partial \tau} + 2\sin\alpha \frac{\partial F}{\partial X} = \frac{\partial}{\partial X}\left(F \frac{\partial F}{\partial X}\right) + \frac{4R}{(1-R)^2} \cos^2\alpha \tag{8}$$

which, for $\alpha = 0$ becomes:

$$\frac{\partial F}{\partial \tau} = \frac{\partial}{\partial X}\left(F \frac{\partial F}{\partial X}\right) + \frac{4R}{(1-R)^2} \tag{9}$$

Dividing eq.8 by F and developing it we obtain:

$$\frac{\partial^2 F}{\partial X^2} = \frac{1}{F} \frac{\partial F}{\partial \tau} + \frac{2\sin\alpha}{F} \frac{\partial F}{\partial X} - \frac{1}{F}\left(\frac{\partial F}{\partial X}\right)^2 - \frac{\lambda}{F} \tag{10}$$

in which

$$\lambda = \frac{4R}{(1-R)^2} \cos^2\alpha \tag{11}$$

In order to solve eq.(10) the Douglas-Jones method was used (Remson, I, 1971). The Douglas-Jones method makes use of two equations: the former is called predictor and the latter corrector. Each equation forwards the solution by a half time step.
In finite differences for the predictor it can be written:

$$\frac{\partial^2 F}{\partial x^2} = (\Delta x)^2 \, W_{i,j+1/2} \tag{12}$$

in which $W_{i,j+1/2}$ represents the F values in $i, j+\frac{1}{2}$

node (fig.1).

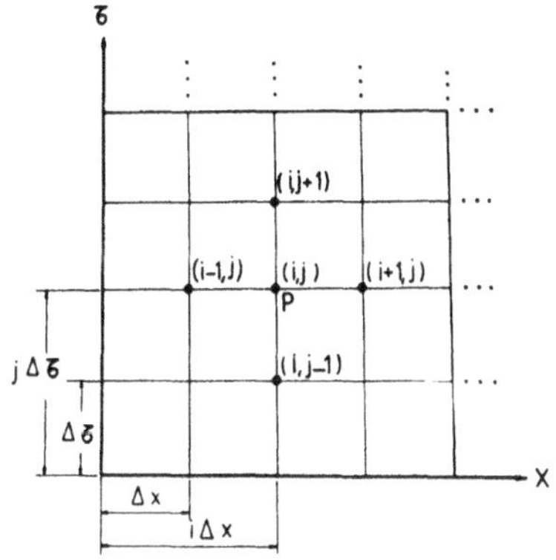

Fig.1. Scheme of node F, in (x_i, τ) coordinates.

$$\frac{\partial F}{\partial x} = \frac{W_{i+1,j} - W_{i-1,j}}{2\Delta x} \tag{13}$$

$$\frac{\partial F}{\partial \tau} = \frac{W_{i,j+1/2} - W_{i,j}}{\Delta\tau/2} \tag{14}$$

If eqs.(12),(13) and (14) are substituted in eq.(10) and if all the substitutions are made according to Douglas-Jones method, the following equation is obtained for the predictor:

$$W_{i+1,j+1/2} - W_{i,j+1/2}\left[2 + \frac{2(\Delta x)^2}{W_{i,j}}\right] + W_{i-1,j+1/2} =$$

$$= \frac{(\Delta x)^2}{W_{i,j}}\left[-\frac{2W_{i,j}}{\Delta\tau} + \sin\alpha\left(\frac{W_{i+1,j} - W_{i-1,j}}{\Delta x}\right) - \right.$$

$$\left. -\left(\frac{W_{i+1,j} - W_{i-1,j}}{2\Delta x}\right)^2 - \lambda\right] \tag{15}$$

For the corrector, in the same way, the following equation will be obtained:

$$W_{i+1,j+1} - W_{i,j+1}\left[2 + \frac{2(\Delta x)^2}{\Delta\tau\, W_{i,j+1/2}}\right] + W_{i-1,j+1} =$$

$$= \frac{2(\Delta x)^2}{W_{i,j+1/2}}\left[-\frac{W_{i,j}}{\Delta\tau} + \sin\alpha\left(\frac{W_{i+1,j+1/2} - W_{i-1,j+1/2}}{\Delta x}\right) - \right.$$

$$\left. -\left(\frac{W_{i+1,j+1/2} - W_{i-1,j+1/2}}{2\Delta x}\right)^2 - \lambda\right] - W_{i+1,j} + 2W_{i,j} - W_{i-1,j}$$

Equations of the type (15) and (16) written in the nodes of the (x,τ) coordinate system, lead to linear system with tridiagonal matricial coefficients. With the aim to evaluate the characteristics of shape and evolution of the depletion curve under the influence of infiltrations,the RENEPA LOADING program was elaborated,which,in principle,solved the linear equation system predictor,corrector respectively,for different initial and boundary conditions.

MODEL TESTING WITH FIELD DATA

Having in view the testing of the mathematical model and the computing program,we have used the results of the measurements made in an experimental drainage field located on a sloping land,already referred to in this paper. The test consisted in finding out the correlation (using the above mentioned mathematical model) between the inflow and the hight of the depletion curve for two drains and in comparing it with the correlation found from experimental measure - ments.

The basic data introduced in the mathematical model where:

-for the first drain (I)

k_1= o,o5 m/day, p_1= 1,35 % -for the first layer

k_2= o,28 m/day, p_2= 6,16 % -for the second layer

α = 5,5o^o and L = 2o m

-for the second drain (II)

k_1= o,61 m/day, p_1= 1o,55 % -for the first layer

k_2= o,o6 m/day, p_2= 1,5o % -for the second layer

k_3= o,3o m/day, p_3= 6,45 % -for the third layer

α = 1o,81° and L = 2o m.

Fig.2 a and 2 b shows the correlation obtained experimentally and those obtained using the mathematical model. Generally,it can be seen a good overlapping of the theoretical results on the experimental ones.

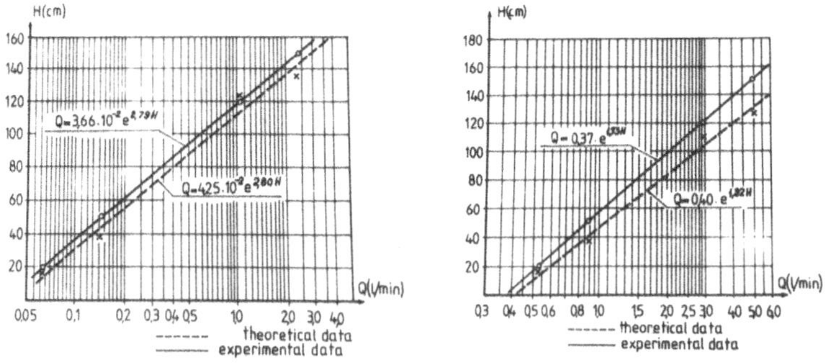

a b

Fig.2. Experimental and theoretic correlation between height of depletion curve and discharge in a) drain I b) drain II.

The existent differences, however, are caused by the fact that the mathematical model applied has considered a homogeneous land with an equivalent hydraulic conductivity found on the basis of real hydraulic conductivities.

SHAPE AND CHARACTERISTICS OF DEPLETION CURVES AT RAISING, FOR SOME NATURAL CONDITIONS

The mathematical model enabled the estimation of:

- -depletion curve at different moments "t" of the raising process;
- -distance in relation to the upstream drain, where the maximum depletion curve is recorded;
- -maximum value of depletion curve height evolution in time.

The computations were made for the following basic data:

$R_1=(\frac{q_c}{k})_1 = o,1$; $R_2= (\frac{q_c}{k})_2= o,o1$

$\alpha = 5^o; 1o^o; 15^o; 2o^o; 3o^o$

L = 2o m -distance between drains.

Fig.3 and 4 show the depletion curve shape at different moments "t" (measured from the beginning of the infiltration process) for three angles of tilt in the impervious layer and two values of the q_c/k ratio. In both cases is evidenced an asymmetry of the depletion curve directly dependent on the slope value and, for the same slope, on the value of q_c/k ratio. The stabilized shape of the depletion curves (stady state condition) are rendered in the uppermost part of each graph. At slope increase a displacement of the depletion curve maximum position towards downstream can be noticed.

This displacement in the moment of steady state condition can be followed in the table 1.

Table 1. Distance at which maximum stabilized value of the depletion curve (in % to L) is obtained, measured with respect to upstream drain

q_c/k \ α	5°	$1o^o$	15°	$2o^o$	$3o^o$
o,o1	6o,o %	7o,o %	8o,o %	82,5%	82,5 %
o,1o	52,5 %	52,5 %	6o,o %	6o,o%	7o,o %

As far as the evolution of the depletion curve maximum with respect to the impervious layer surface is concerned, fig.5 shows the results obtained for the two adimensional factors q_c/k previously stated.

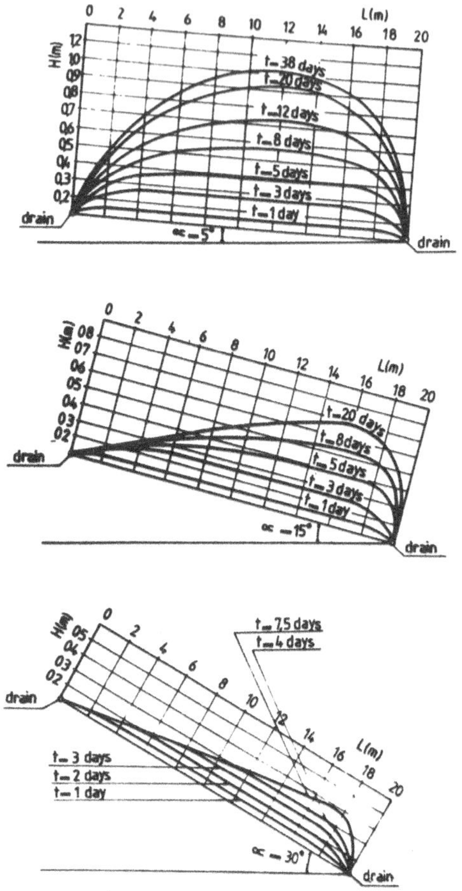

Fig.3 Depletion curves at different moment "t" for a steady and constant loading q_c/k=o,ol and three angles of tilt in the impervious layer (α).

It can be seen that:

-the time lapse when the maximum of depletion curve is stabilized is in an inversely proportional relation to the slope-observation valid especially for the ratio q_c/k=o,ol and

-the decrease of maximum height value of the depletion curve is important at the beginning of the interval,after which the depletion curve approaches asymptotically a height value which,for the same distance between drains,depends on slope and the adimensional ratio q_c/k.

The results concerning the value of the stabilized maximum (steady state condition) of the depletion curve (fig.5) could be compared with the results ob-

tained by Schmit (1966) and Wooding (1966).
Table 2 shows the results given by literature as compared to those obtained by the authors (the H values in the table are given in Wooding coordinates).
As it can be seen the results confirm the validity of the mathematical model,of the algorithm and the computing programme elaborated.
It can be noted that the authors'results correspond, for the low slope areas, with those obtained by Luthin and Schmit and for the high slope areas with the results by Wooding and Chapman.

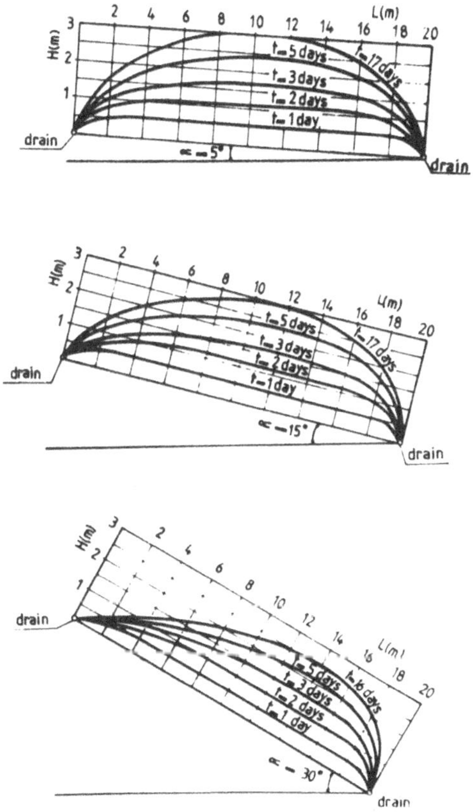

Fig.4.Depletion curves at different moments "t" for a constant and continous loading (q_c/k=o,1) and three angles of tilt in the impervious layer (α).

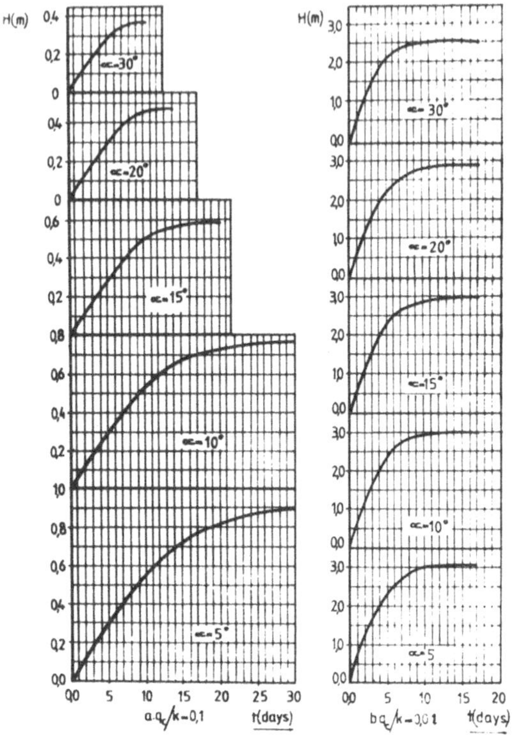

Fig.5.Variation of maximum height of depletion curve with respect the time.

Table 2.Comparative analyses of results obtained using the mathematical model with the steady state condition computed by two methods.

q_c/k		o,1				o,o1			
L(m)		2o				2o			
α (°)		5	1o	2o	3o	5	1o	2o	3o
Schmid and Luthin	L/H	6,5	6,65	7	8,5	22	27	46	65
	H (m)	3,o8	3,o1	2,86	2,35	o,91	o,74	o,43	o,31
Wooding and Chapman	L/H	6,4o	6,4o	6,87	7,78	2o,71	25,12	43,65	65,24
	H (m)	3,13	3,13	2,91	2,57	o,97	o,8o	o,46	o,31
Autors	H	3,o9	3,o6	2,9o	2,57	o,93	o,78	o,48	o,34

BIBLIOGRAPHY

1.Cojocaru,I.(1985)-Contribution to Estimation of Sliding Slope Drainage Functional Parameters with

References to Experimental Platform in Dîmbovitza Watershed Area,Ph.D.Thesis.Polytechnic Institute of Iaşi-Romania.
2.Remson,I.,Hornberger,M.G.,Molt,J.F.(1971)-Numerical methods in subsurface hydrology,Wiley-Interscience
3.Schmid,P.,Luthin,J.(1964)-The drainage of sloping lands,Journal of Geophysical Research,Vol.69,No.8,April 15.
4.Wooding,R.,A.,Chapman,T.,G.(1966)-Groundwater flow over a sloping impermeable layer.1.Application of the Dupuit-Forchheimer Assumption,Journal of Geophysical Research,Vol.71,No.12,June 15.

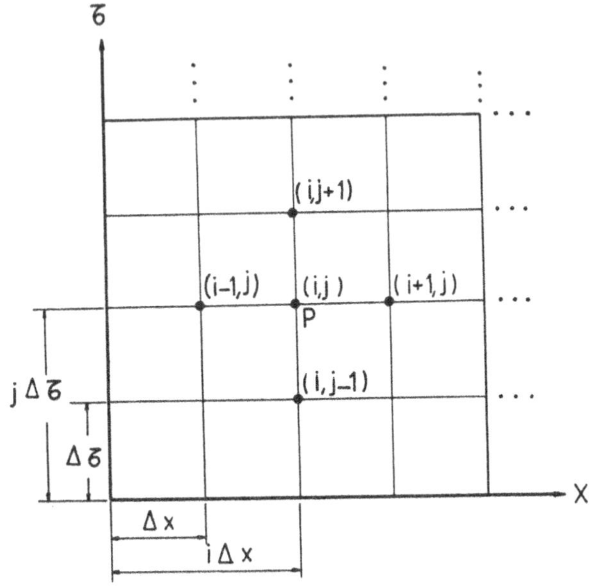

τ
(i,j+1)
(i–1,j)
(i,j)
(i+1,j)
P
(i,j–1)
j Δ τ
Δ τ
Δ x
i Δ x
X

FIG. 1

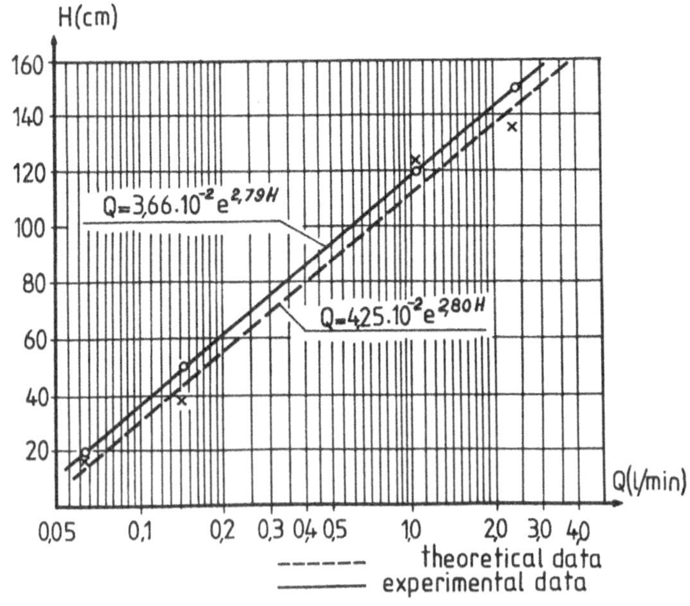

H(cm)
160
140
120
100
80
60
40
20
Q=3,66.10⁻²e^2,79H
Q=4,25.10⁻²e^2,80H
Q(l/min)
0,05
0,1
0,2
0,3
0,4
0,5
1,0
2,0
3,0
4,0
theoretical data
experimental data

FIG. 2a.

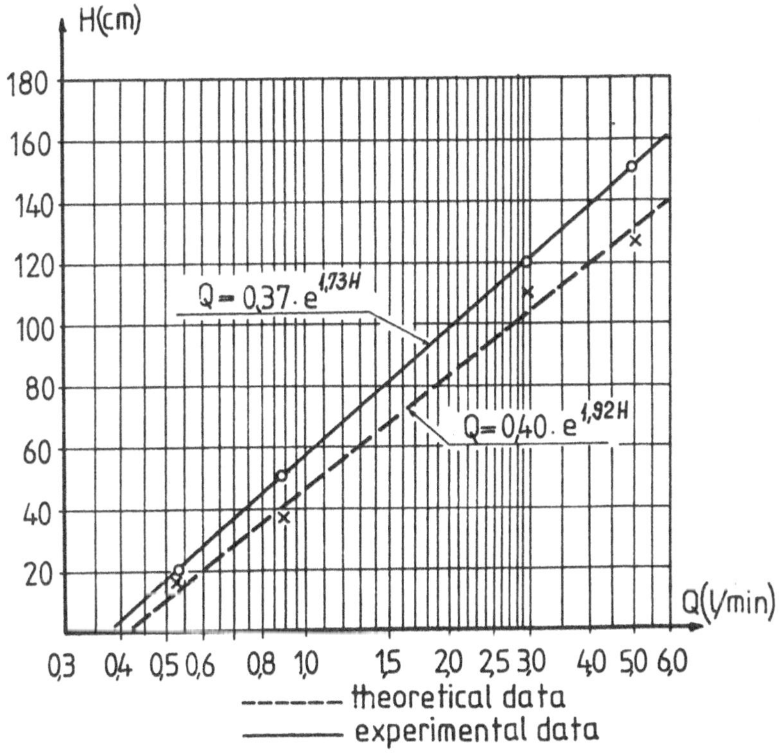
H(cm)
180
160
140
120
100
80
60
40
20
Q = 0,37 · e^1,73H
Q = 0,40 · e^1,92H
Q(l/min)
0,3
0,4
0,5
0,6
0,8
1,0
1,5
2,0
2,5
3,0
4,0
5,0
6,0
------ theoretical data
——— experimental data

FIG. 2 b.

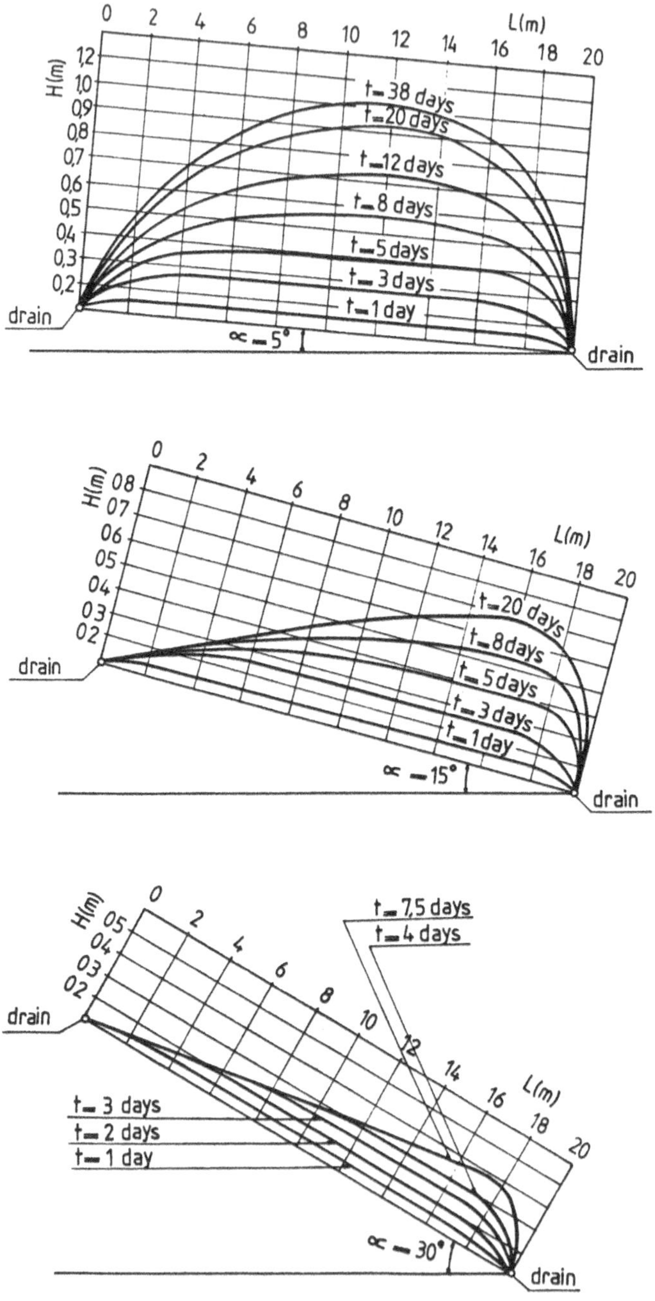

H(m)
L(m)
t = 38 days
t = 20 days
t = 12 days
t = 8 days
t = 5 days
t = 3 days
t = 1 day
drain
∝ = 5°
drain
H(m)
L(m)
t = 20 days
t = 8days
t = 5days
t = 3 days
t = 1day
drain
∝ = 15°
drain
H(m)
L(m)
t = 7,5 days
t = 4 days
t = 3 days
t = 2 days
t = 1 day
drain
∝ = 30°
drain

FIG.3

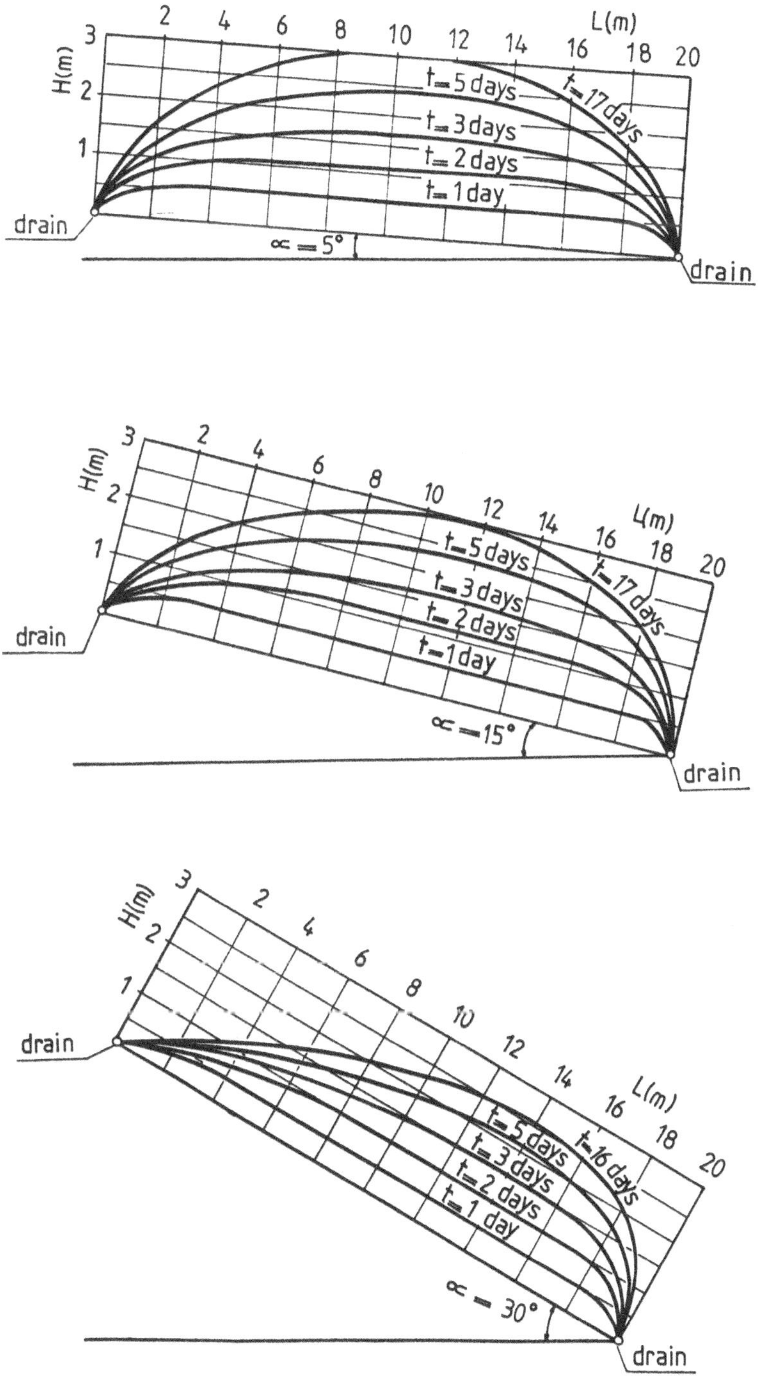
H(m)
L(m)
t=5 days
t=3 days
t=2 days
t=1 day
t=17 days
drain
drain
∝ = 5°
t=5 days
t=3 days
t=2 days
t=1 day
t=17 days
drain
drain
∝ = 15°
t=5 days
t=3 days
t=2 days
t=1 day
t=16 days
drain
drain
∝ = 30°

FIG. 4

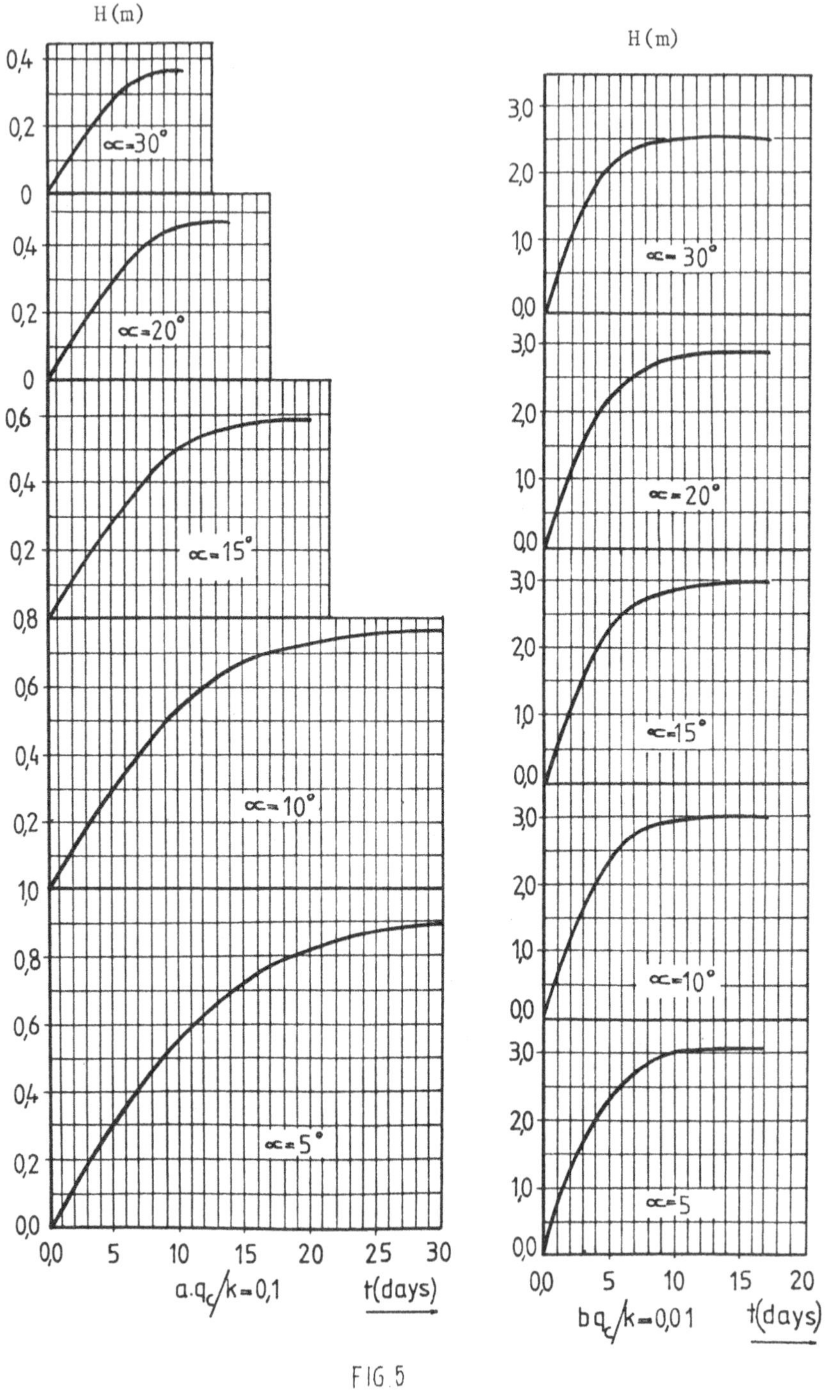
H(m)
α=30°
α=20°
α=15°
α=10°
α=5°
a. qc/k=0,1
t(days)
H(m)
α=30°
α=20°
α=15°
α=10°
α=5
b qc/k=0,01
t(days)

FIG.5

Methods for Estimating the Evaporation from Shallow Water Table in Drainage Design Formulas

Yufang Zhang
Wuhan Institute of Hydraulic and Electrical Engineering, Wuhan, Hubei, China

Wuhan, Hubei, China.

INTRODUCTION

The evaporation plays an important part in lowering the shallow groundwater table. In many areas this factor should not be ignored in subsurface drainage design. Empirical formulas for estimating steady evaporation from shallow water table derived on the basis of lysimeter data have long been used by Averianov (1956), Hammad (1962), Zhang (1963) and others. In this paper an approximate method for determining evaporation from groundwater based on the empirical relationship and groundwater regime observation data is described. Theoretical formulas for computing steady evaporation from water table have been developed by Gardner (1958) for homogeneous soil. Willis (1962) used these formulas to determine the evaporation in two-layered soil by curve-matching. We extended the use of steady solutions to multilayered soils (Y. Zhang 1966). Experiments and theoretical study shows that the evaporation depends upon not only the climatic conditions (usually characterized by potential evaporation), but also the capacity of soil to transport water from groundwater to soil surface. The evaporation can never exceed the smaller value between the above two limitations. In this paper graphs based on experimental data showing the above relationship are given. Under the condition of subsurface drainage the water table is always descending, hence the evaporation is also changing with time. As the steady evaporation is the maximum value for a given soil at a definite depth to groundwater and under definite climatic conditions, when this value is used in design the effect of drainage will be overestimated. With the aim of clarifying the process of evaporation during drainage, in this paper through numerical simulation the nonsteady evaporation under typical pattern of drawdown of water table for some soils is investigated.

EMPIRICAL METHOD FOR ESTIMATING EVAPORATION FROM GROUNDWATER

Based on the data obtained from the lysimeters, where the groundwater table is kept constant, the following relationship between the steady evaporation ε from shallow water table, potential evaporation ε_o and groundwater depth Δ_o has been developed by Averianov (1956) and others:

$$\varepsilon = \varepsilon_o(1 - \frac{\Delta}{\Delta_o})^n \qquad (1)$$

where Δ_o - the groundwater depth at which the evaporation approaches zero,
n - empirical soil index, n varies from 1 - 4.

Some authors prefer an exponential formula:

$$\varepsilon = \varepsilon_o\ e^{-c\Delta} \qquad (2)$$

Experimental and theoretical study shows that for a definite soil the evaporation from water table depends upon two factors: climatic factor which is usually represented by potential evaporation and the water transportation capacity of the soil from the water table to the ground surface, which is controlled by the capillary conductivity and groundwater depth. The potential evaporation and the water transportation capacity of the soil are the two limiting factors of evaporation. These circumstances can well be illustrated by the graphs in Fig. 1 showing relationship between evaporation and water table depth based on lysimeter data of the Liuhu experimental station.

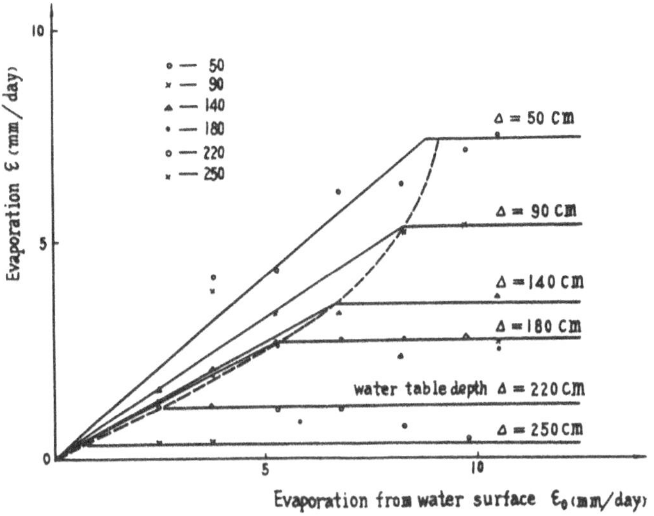

Figure 1

On the right side of the dotted line the evaporation is controlled by water transportation capacity of the soil, for a definite groundwater depth, despite the increase of ε_o , the evaporation remains unchanged. On the left side of the dotted line the evaporation is controlled both by potential evaporation and transportation of water. Strictly speaking, the empirical equation (1) can be applied only to the conditions corresponding to that part of graphs on the left side of the dotted line. In areas where vertical recharge and discharge predominate, the empirical constants in equation (1) can be determined from the groundwater level observation data. From the empirical equation (1)

$$\frac{\varepsilon}{\mu} = \frac{d\Delta}{dt} = \frac{\varepsilon_o}{\mu}\left(\frac{\Delta_o-\Delta}{\Delta_o}\right)^n$$

$$-\frac{d(\Delta_o-\Delta)}{dt} = \frac{\varepsilon_o}{\mu}\left(\frac{\Delta_o-\Delta}{\Delta_o}\right)^n$$

integrating equation (1) with respect to t, we obtain:

$$(\Delta_o-\Delta_i)^{-(n-1)} - (\Delta_o-\Delta)^{-(n-1)} = \frac{(n-1)\varepsilon_o t}{\mu\,\Delta_o^n} \qquad (3)$$

If a period lacking in rainfall and irrigation could be found, by fitting the observed groundwater level data in this period, the parameters Δ_o, n and μ can be determined. In areas where the index n can be approximated by 1, the procedure for identifying the parameters will be greatly simplified. Since

$$\frac{\varepsilon}{\varepsilon_o} = 1 - \frac{\Delta}{\Delta_o}$$

$$\frac{\varepsilon}{\mu\,\varepsilon_o} = \frac{1}{\mu} - \frac{1}{\mu}\frac{\Delta}{\Delta_o} \qquad (4)$$

By plotting $\frac{\varepsilon}{\mu\varepsilon_o}$ = $d\Delta/\mu dt$ against corresponding Δ as shown in Fig. 2, the intercept on the ordinate, when $\Delta = 0$, is equal to $1/\mu$, and the intercept on the abscissa will be Δ_o. Since the parameters are determined on the basis of groundwater regime observation data and prescribed empirical expression, the other factors not entering in the formula, such as nonhomogeneity, delayed yield of gravity drainage, etc. are all included in the former parameters. The Δ_o, n and μ thus obtained are the apparent ones. As the parameters are ascertained from the descending groundwater level data due to evaporation, it should not bring about remarkable error when the same empirical expression for evaporation with these parameters is used in

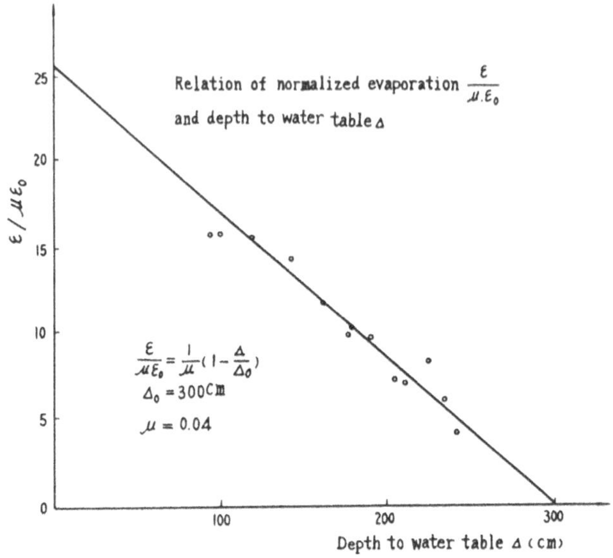

Figure 2

unsteady drainage design formulas.

THEORETICAL METHOD FOR ESTIMATING EVAPORATION

The one-dimensional differential equation for saturated-unsaturated flow in soil is:

$$c(h)\,\frac{\partial h}{\partial t} = \frac{\partial\left[k(h)\,\frac{\partial h}{\partial z}\right]}{\partial z} + \frac{\partial k(h)}{\partial z} \tag{5}$$

In homogeneous soil the equation for steady evaporation from shallow water table can be obtained directly from the Darcy's law:

$$\varepsilon = -\,k(h)\,\frac{d(h+z)}{dz} \tag{6}$$

where,

- $c(h)$ - water capacity of soil,
- h - pressure head of soil,
- z - ordinate, taken positive upward,
- t - time,
- k - hydraulic conductivity of soil,
- ε - flux of soil moisture (evaporation).

Replacing pressure head h in equation (6) by suction head ψ, we have

$$\varepsilon = - k(\psi) \frac{d(-\psi+z)}{dz} = k(\psi) \left(\frac{d\psi}{dz} - 1\right)$$

$$z = \int \frac{d\psi}{1 + \frac{\varepsilon}{k}} \tag{7}$$

Usually the hydraulic conductivity k in terms of ψ can be expressed in the following typical form:

$$k = \frac{a}{\psi^n + b} \tag{8}$$

$$k = a\, e^{-c\psi} \tag{9}$$

where a, b, c and n are arbitrary constants. The value of n varies from 2 (for clayey soil) to 4 (for sandy soil). The equations for homogeneous soil when n = 2, 3, 4 in (8) and for (9) have been obtained by W.R. Gardner (1958). W. Willis (1960) used these solutions to determine the evaporation from two-layered soil by curve-matching. With the aid of dimensionless parameters by constructing graphs depicting the relationship between dimensionless pressure head $\bar{h}$, and the thickness of the soil layer $\bar{d}$ for n = 2, 3, 4, we extended the use of the Gardners solutions to multi-layered soils (Y. Zhang 1966).

In areas with drainage facilities, after rainfall and irrigation under the influence of drains the groundwater level will be declining, the evaporation from the falling water table will also be changing with time. To find the evaporation, the saturated-unsaturated flow equation (5) should be solved with the corresponding initial and boundary conditions. The initial condition depends upon the depth to water table and the initial water content θ or pressure head h:

$$\theta(z,0) = \theta_i(z)$$

or (10)

$$h(z,0) = h_i(z)$$

Upper boundary conditions

Numerous experiments carried out in laboratory and in the field show that both evaporation from soil surface and evapotranspiration can be approximately expressed as a function of soil moisture content θ (or pressure head h) schematically shown in Fig. 3.

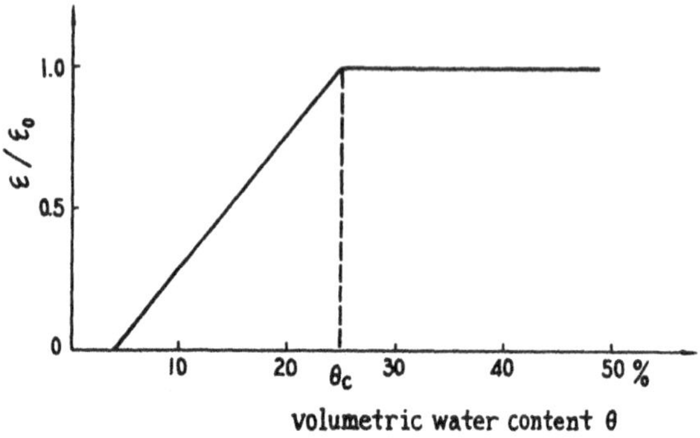

Figure 3

In this case the upper boundary conditions will be (Y. Zhang 1966):

$$\theta(0,t) \geq \theta_c \,, \qquad \varepsilon = \varepsilon_o$$

$$\theta(0,t) < \theta_c \,, \qquad \varepsilon/\varepsilon_o = a\theta + b \tag{11}$$

or

$$h(0,t) \geq h_c \,, \qquad \varepsilon = \varepsilon_o$$

$$h(0,t) < h_c \,, \qquad \varepsilon/\varepsilon_o = af(h) + b \tag{12}$$

$$h(0,t) \geq h_c \,, \quad k(h)\left(\frac{\partial h}{\partial z} - 1\right) = \varepsilon_o$$

$$h(0,t) < h_c \,, \quad k(h)\left(\frac{\partial h}{\partial z} - 1\right) = \varepsilon_o[af(h) + b] \tag{13}$$

Lower boundary conditions
Depending upon the pattern and the rate of drawdown of ground water table frequently encountered in drainage area, we have the following boundary conditions:

Instantaneous drawdown of water table
This happens in the soil profile in the nearby vicinity of a drainage ditch after a peak storm flow. The boundary condition is:

$$z = L \,, \qquad h(d,t) = 0 \tag{14}$$

where d - depth to water table kept after the instantaneous drawdown.

Time-dependent drawdown of water table The boundary condition for time-dependent water table drawdown is:

$$z = d(t), \quad h(d,t) = 0 \tag{15}$$

In case of drawdown of water table at constant rate v

$$d(t) = d_o + vt \tag{16}$$

where d_o --- the initial water table depth

In case of drawdown induced by drains or tube wells, d(t) can be determined approximately from the corresponding formulas for groundwater drawdown.

Drawdown of groundwater table due to evaporation When the soil profile is located beyond the scope of the influence of drains and there is no cross flow to lower aquifer, the drawdown of groundwater table is mainly due to evaporation; in this case the lower boundary can be taken as an impervious barrier:

$$z = D \ , \qquad \frac{\partial h}{\partial z} = 1 \tag{17}$$

Drawdown of groundwater table due to leakage flow to lower aquifer When the soil profile is underlain by a semi-pervious layer with hydraulic conductivity k_2 and thickness δ, and the piezometric pressure in the underlying aquifer is h_3 , the boundary condition at the bottom of the soil profile, z = L, is:

$$q = -k(h)\left(\frac{\partial h}{\partial z} - 1\right) = \frac{k_2}{\delta}\,[h(L,t) + \delta - h_q] \tag{18}$$

Owing to the nonlinearity of equation (5) and its complicated initial and boundary conditions, the solution for evaporation can only be obtained by numerical methods.

The finite difference equation

When the implicit finite difference scheme is used, discretized form of equation (5) can be written as follows:

$$C_i^{j+\frac{1}{2}}\frac{h_i^{j+1}-h_i^j}{\Delta t} = k_{i+\frac{1}{2}}^{j+\frac{1}{2}}\frac{H_{i+1}^{j+1}-h_i^{j+1}}{(\Delta z)^2} - k_{i-\frac{1}{2}}^{j+\frac{1}{2}}\frac{h_i^{j+1}-h_{i-1}^{j+1}}{(\Delta z)^2} - \frac{k_{i+\frac{1}{2}}^j-k_{i-\frac{1}{2}}^j}{\Delta z} \tag{19}$$

or

$$E_i\, h_{i-1}^{j+1} + F_i h_i^{j+1} + G_i h_{i+1}^{j+1} = H_i \tag{19'}$$

where i, j are the sequential numbers of nodes on the space and time axes,

$$E_i = -k_{i+\frac{1}{2}}\ , \qquad F_i = k_{i+\frac{1}{2}} + k_{i-\frac{1}{2}} + rc_i^{j+\frac{1}{2}},$$

$$G_i = -k_{i-\frac{1}{2}}\ , \qquad H_i = rc_i^{j+\frac{1}{2}}h_i^j - \Delta z(k_{i+\frac{1}{2}}^j - k_{i-\frac{1}{2}}^j)$$

Verification of mathematical model

Experiments conducted on soil column packed with Xuchang sandy loam are used for the verification of the mathematical model. The hydraulic conductivity and soil water characteristic curve of the soil can be expressed by the following equations respectively:

$$h \geq 0, \quad k(h) = 7 \text{ cm/d} , \quad 0 > h \geq -50 \text{ cm}, \quad k(h) = 7e^{0.0316k} ,$$

$$h < -50 \text{ cm} , \quad k(h) = 2.79e^{0.0132k} \quad ;$$

$$h \geq 0, \quad \theta = 0.448, \quad 0 > h \geq -200 \text{ cm}, \theta (h) = (194785-h^2)/434782,$$

$$h < -200 \text{ cm}, \theta (h) = 0.4615e^{0.0044176h} + 0.182844$$

At the beginning of the experiments the water table in the soil column was instantaneously lowered from the soil surface to a depth of 100 cm. and the conditions for potential evaporation of 1.1 cm/d were provided by an electrical bulb. The intensity of gravity drainage and evaporation were observed periodically and are shown by crosses in Fig. 4.

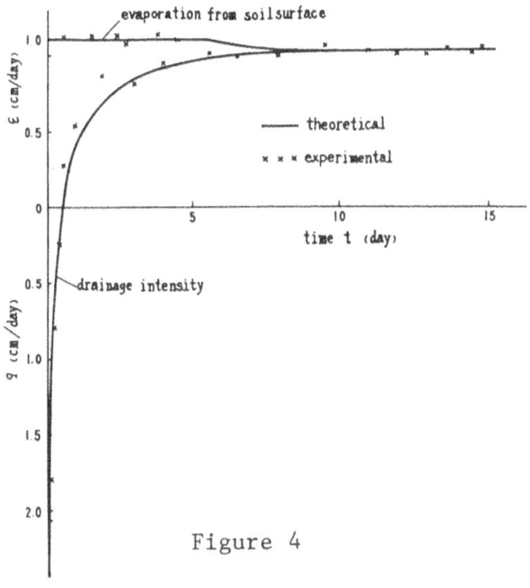

Figure 4

The intensity of drainage and evaporation rate were calculated by numerical simulation under the boundary conditions:

$$z = 0 , \qquad -k(h)\left(\frac{\partial h}{\partial z} - 1\right) = -\varepsilon$$

$$h \geq -300 \text{ cm}, \quad \varepsilon = 1.1 \text{ cm/d} ,$$

$$h < -300 \text{ cm}, \quad \varepsilon = 0.00333h + 2.0599 \text{ cm/d} ,$$

$$z = 100 \text{ cm}, \qquad h(z,t) = 0$$

The results of calculations are shown in Fig. 4 by solid lines. It can be seen that the agreement between the theoretical curves and the observed data is satisfactory.

The application of numerical method

For a soil profile with initial groundwater table near the ground surface, the trend of variable of evaporation under the climatic condition characterized by potential evaporation $\varepsilon_0 = 6.5$ mm/d and the conditions of instantaneous drawdown of groundwater level has been investigated. The hydraulic conductivity k(h), retention curve θ(h) and the water capacity c(h) of the loam soil studied are expressed by the following equations respectively:

$$h \geq 0, \quad k(h) = 7/1440 \text{ cm/min},$$

$$h < 0, \quad k(h) = 7/1440\ e^{0.0255h},$$

$$h \geq 0, \quad \theta = \theta_s = 0.452, \quad c(h) = 0,$$

$$h < -50, \quad \theta = 0.7333 - 0.090074 \ln|h|, \quad c(h) = -0.090074/h,$$

$$0 \geq h \geq -50 \text{ cm}, \quad \theta = 0.452\ e^{0.00342}, \quad c(h) = 0.00155e^{0.00342h}$$

For the problem studied the upper boundary condition is

$$h \geq -213.917 \text{ cm}, \quad \varepsilon = 0.65/1440 \text{ cm/min}$$

$$h < -213.917 \text{ cm}, \quad \varepsilon = [3.25(0.7333-0.090074 \ln|h|)-0.1625]/1440 \text{ cm/min}.$$

the lower boundary conditions
for instantaneous drawdown:

$$z = 150 \text{ cm}, \quad h(150,t) = 0;$$

for constant rates of drawdown:

$$z = d(t) = vt, \quad h(d,t) = 0,$$

$$v = 0.04, \quad 0.02, \quad 0.01, \quad 0.005 \text{ cm/min}.$$

for drawdown induced by subsurface drainage:

$$z = d(t) = 254.65e^{-1.364 \times 10^{-5} t}, \quad h(d,t) = 0$$

By finite difference method considering the above initial and boundary conditions we obtained the graphs of variation of evaporation with time and water-table depth for different pattern of drawdown shown in Fig. 5.

The graphs show that under the same climatic conditions at the beginning of the water-table drawdown, there exists a period of constant evaporation. Then it begins to decrease, the greater the rate of drawdown, the shorter the period of constant evaporation and the quicker the evaporation diminishes. In case of drawdown induced by drains, the rate of water table drop decreases with time, hence the graph of evaporation intensity intersects with the graphs of different rates of drawdown.

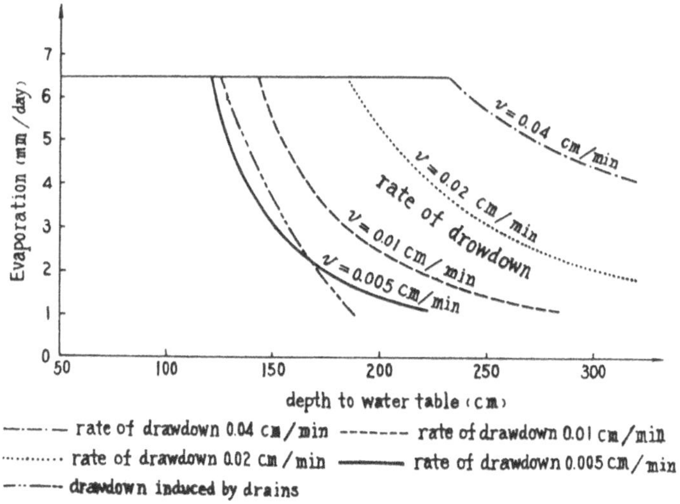

Figure 5

CONCLUSION

In areas where vertical recharge and discharge are the predominant factors of groundwater budget, and the regime observation data are available, the empirical formula can be used to determine the evaporation. When groundwater level data are not available and the soil property data, such as the unsaturated hydraulic conductivity and the soil water characteristic curve, etc. can be found, the numerical method for theoretical estimation of evaporation under different prescribed design conditions may also be used. However, in this case, the drainable porosity should be determined accordingly.

REFERENCES

AVERIANOV, S. (1956) Seepage from irrigation canals and its influence on regime of groundwater, "Influence of irrigation system on regime of groundwater", Academic Press USSR 140-141.

GARDNER, W. (1958) Some steady-state solutions of the unsaturated moisture flow equation with application to evaporation from a water table, Soil Science 85: 228-232.

HAMMAD, H. (1962) Depth and spacing of the drain systems, ASCE Proc. 88 (IR1).

WILLIS, W. (1960) Evaporation from layered soil in the presence of a water table, SSSA Proc. 24: 239-242.

ZHANG, W. (1963) Calculation of the groundwater regime under the conditions of drainage and evaporation, Journal of Wuhan Institute of Hydraulic and Electrical Engineering.

ZHANG, Y. (1966) The study of evaporation in salt-affected areas, Dissertation.

Tubewell-Spacing Formulas for Subsurface Drainage

R.J. Oosterbaan
International Institute for Land Reclamation and Improvement, Wageningen, The Netherlands

INTRODUCTION

Well-spacing formulas are not commonly used in practice. For example McCready (1978), reviewing various drainage techniques by wells, does not mention well-spacing formulas.

The well-spacing formulas presented in this paper are based on the assumption that the aquifer is recharged only by the vertically downward percolation of rain or irrigation water. Further, the wells are thought to be arranged in a square-grid pattern, but the formulas can be adapted to other patterns (e.g. hexagonal) by a conversion factor whose value is usually close to unity.

The amount of water pumped from the wells during a certain period is thought to be equal to the recharge during the same period, so that the change in storage of water in the soil is zero. Thus, on average, the level of the watertable is constant, although fluctuations may occur within the period considered.

Further, the wells are assumed to be placed in flat land. In sloping land, the hydraulic head and the zone of influence in upslope direction are greater than those in flat land, whereas in downslope direction they are smaller than in flat land. However, since the land surface is also higher upslope and lower downslope, the depth of the watertable below the soil surface between the wells will not be appreciably different from the depth in flat land. Hence, the formulas derived for flat land are also valid for land slopes of up to several per cent, especially if the depth of the impermeable layer is reasonably constant.

All formulas can be expressed explicitly in the spacing (L), so the spacings can be calculated with any pocket calculator, provided that it contains logarithmic function keys.

FULLY-PENETRATING WELLS IN HOMOGENEOUS AQUIFERS

The spacing formulas of fully-penetrating wells in homogeneous aquifers can be derived from the Darcy equation of motion, a water balance, and the Dupuit assumptions, which imply a negligible resistance to vertical flow compared with the resistance to horizontal flow. Therefore there is a maximum depth, of wells and aquifer alike, to which the formulas are applicable. The differential equation is (see figure 1):

$$\pi . q(R^2 - x^2)dx = 2\pi . x . H . K . dH, \tag{1}$$

where:

- π is the ratio between circumference and diameter of a circle;
- q is the average well discharge (m^3/day per m^2 surface area served by the well), equalling the average deep percolation (m/day);
- R is the radius (m) of an equivalent circular area of influence of the well : $\pi . R^2 = L^2$;
- L is the spacing between the wells arranged in a square-grid pattern (m);
- x is a horizontal distance from the well (m);
- dx is a small increment of x;
- H is the height of the watertable above a horizontal impermeable layer (m);
- dH is the increment of H over dx;
- K is the hydraulic conductivity of the aquifer (m/day).

Equation (1) can be integrated from r to R, where $H_r = F-g$ and $H_R = F-i$ respectively, and:

- r is the radius of the well, including a permeable filter (m);
- F is the depth of the well below the soil surface, equalling the depth of the aquifer (m);
- g is the average depth of the waterlevel in the well below the soil surface (m);
- F-g is the depth over which water enters the well if no impermeable casing is present (m);
- i is the average depth of the watertable below the soil surface, midway between the wells (m).

Using the following relations:

$$(F-i)^2 - (F-g)^2 = 2(F-g)(g-i) + (g-i)^2,$$
$$\ln 1.7 = 1/2,$$
$$1.7\sqrt{\pi} \approx 3,$$
$$L^2/\pi - r^2 \approx L^2/\pi,$$

the integration gives:

$$L^2 . \ln L/3r = Y, \tag{2}$$

where: $$Y = 2\pi . K(F-g)(g-i)/q + \pi . K(g-i)^2/q \tag{3}$$

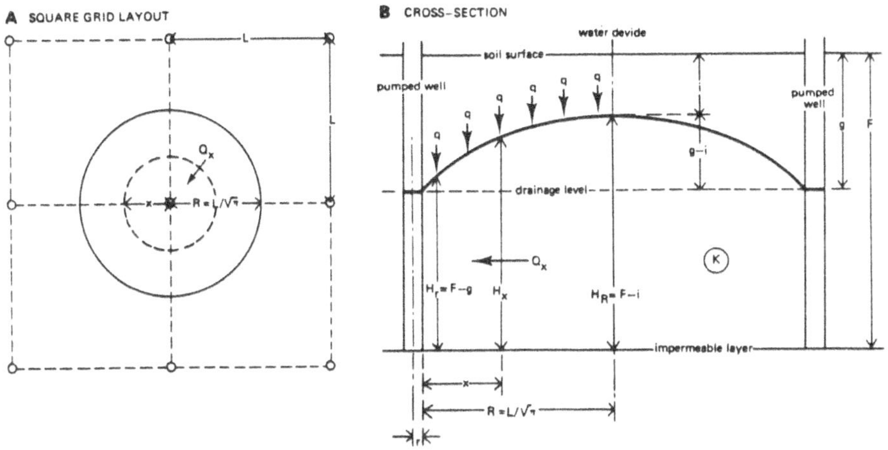

Figure 1. Pumped well system for subsurface drainage, fully penetrating

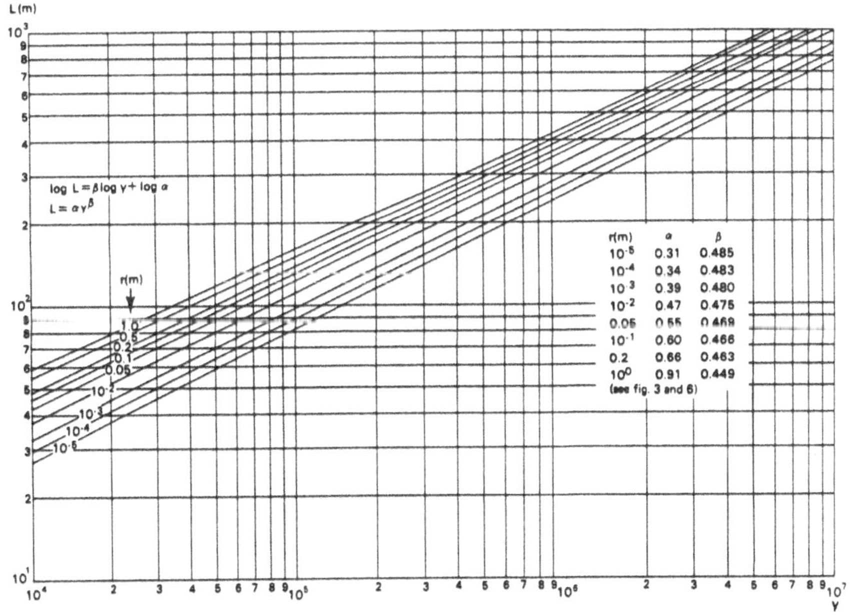

Figure 2. Plot of L versus Y from Equation 2.

A graph of L versus Y (for different values of r) on two-sided logarithmic paper (figure 2) shows that their log values are linearly related. Hence: log L = b.log Y + log a, or:

$$L = a.Y^b \qquad (4)$$

The values of a and b depend upon r as shown in figure 3. For values of r between 0.5 m and 0.05 m, an approximation of the relation between a, b and r is:

$$a = 0.18 \log r + 0.78, \qquad (5)$$

$$b = 0.465 \qquad (6)$$

Hence, using equations (3), (4), (5) and (6), we can calculate the required well spacing with a pocket calculator.

WELLS IN VERY DEEP AQUIFERS (Z > 0.2 L)

In very deep aquifers, the flow around the wells is almost spherical, and almost all the resistance to the flow is in the region of the spherical flow. According to Hooghoudt's principles (Wesseling, 1973), we can assume that spherical flow occurs if the depth (Z) of an impermeable layer below the bottom of the well is greater then 20% of the well spacing (Z > 0.2 L).

Figure 4 depicts the flow from a spherical equipotential with radius s and potential P_s to a concentric spherical hole with radius c and potential P_c. The discharge (Q, m^3/day) through any concentric sphere with radius x, potential Px, and surface area $4\pi.x^2$ can be expressed as: $Q = 4\pi.x^2.K.dP_x/dx$. Integration from x = c to x = s yields:

$$P_s - P_c = Q\ (1/c - 1/s)\ /\ 4\pi.K \qquad (7)$$

Setting $P_s - P_c$ = g-i (as defined before), considering the flow in a half sphere so that $Q = 2q.L^2$, and assuming that the well can be replaced by a half sphere with equivalent radius G/3 = (E-g)/3, we can change equation (7) into :

$$g-i = q.L^2\ \left(3/(E-g) - 1/s\right)\ /\ 2\pi.K$$

The equivalent radius G/3 represents the mean value of the three dimensions of the well: G/3 = (G+r+r)/3 in which r has been taken negligibly small compared with G.
Neglecting the relatively small term 1/s, we get:

$$g-i = 3q.L^2\ /\ 2\pi.K(E-g), \qquad (8)$$

so that L can be calculated from:

$$L = \sqrt{2\pi.K\ (g-i)(E-g)\ /\ 3q} \qquad (9)$$

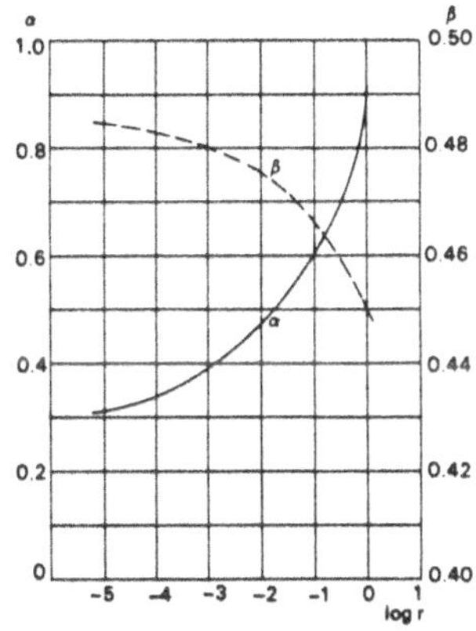

Figure 3. Plot of data from Figure 2.

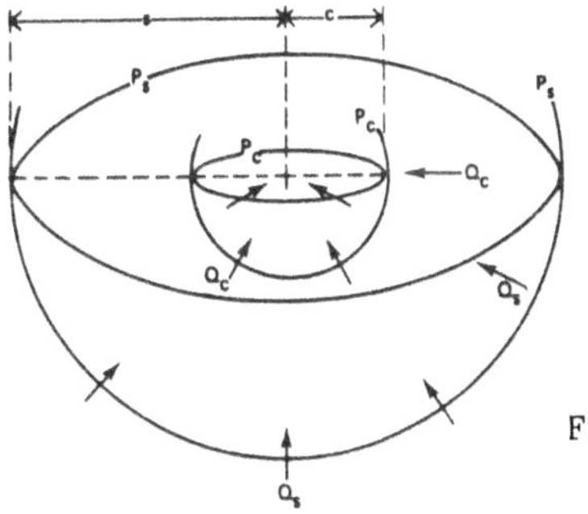

Figure 4. Spherical flow

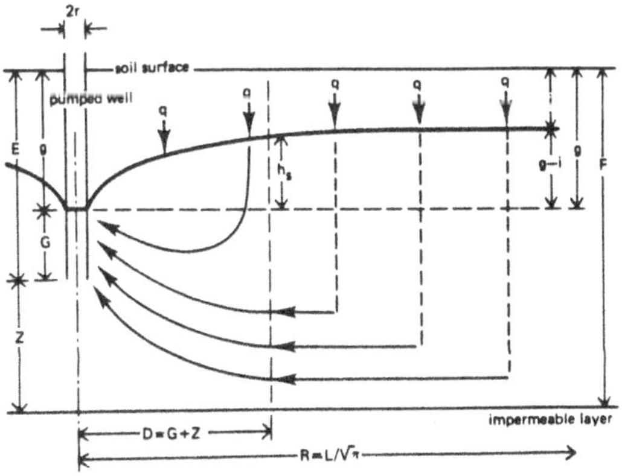

Figure 5. A partially penetrating pumped well.

The procedure followed in the derivation of Equation (9) is similar to that followed by Hooghoudt in deriving a drain-spacing equation with non-horizontal cylindrical flow to the drain.

PARTIALLY-PENETRATING WELLS ($G < Z < 0.2\ L$)

Figure 5 shows a partially-penetrating well in an aquifer of medium depth, so that neither the assumption of fully spherical flow nor the assumption of Dupuit (only horizontal flow) is valid. In conformity with the principle of Hooghoudt, the flow can be considered spherical in a region between the well and a distance $D = Z + G = Z + E-g$ from the well, and horizontal in a region between a distance D from the well and midway between the wells. The total hydraulic head (g-i) can be subdivided into a hydraulic head (h_s) required for spherical flow and a hydraulic head (h_o) for horizontal flow:

$$g-i = h_s + h_o, \tag{10}$$

$$h_s = 3q.L^2 / 2\pi.K(E-g), \tag{11}$$

$$h_o = q.L^2 (\ln L/3D) / 2\pi.T, \tag{12}$$

where:

$T = K(F-g) + K(g-i)/2$ (the transmissivity, m^2/day).

Equation (11) is in conformity with equation (8) and equation (12) is in conformity with equation (2), with r replaced by D. Equations (10), (11), and (12) can be combined into:

$$g-i = \frac{q.L^2}{2\pi} \left\{ \frac{\ln L/3D}{T} + \frac{3}{(E-g)K} \right\} \tag{13}$$

Equation (13) can also be written as:

$$L^2.\ln L/3r_e = Y, \tag{14}$$

where: $Y = 2\pi.T\ (g-i)/q,$ (14a)

and: $r_e = D.\exp\{-3T/(E-g)K\}$ (14b)

To solve L from equation (14), we can use the procedures for fully-penetrating wells described in section 2. For values of re < 0.1 (which is generally the case), the relation between a, b, and re can be linearized as demonstrated in figure 6. Hence, equation (4) is also applicable, with equations (14a,b), and:

$$a = 0.60 - 0.41 \log(-\log r_e), \tag{15}$$

$$b = 0.495 - 0.058 / (1 - \log r_e) \tag{16}$$

If $G/3 < D < G$, we can regard the well as fully penetrating without making a large relative error. For the situation that $0.1\ L < D < 0.3\ L$, equations (9) and (14) will usually produce relatively small differences.

THE EQUIVALENT MAXIMUM DEPTH OF FLOW

In the formulas for fully and partially penetrating wells, we have ignored the resistance to vertical flow, but if the aquifer is quite deep, we cannot ignore it altogether.

If we assume that the soil has a vertical hydraulic conductivity (K_v) and a horizontal conductivity (K_h), the hydraulic head (h_v) needed to overcome the resistance to vertical flow is approximately $h_v = q(F-g)/K_v$, where the relatively small value of $h = g-i$ has been neglected.

Using equation (3), but neglecting the relatively small second term on its right-hand side, we find that the hydraulic head (h_o) to overcome the resistance to horizontal flow to a fully penetrating well is approximately:

$$h_o = q.L^2 (\ln L/3r) / 2\pi.K_h(F-g)$$

Setting the total hydraulic head $h = h_o + h_v$, we get:

$$h/q = L^2 (\ln L/3r) / 2\pi.K_h(F-g) + (F-g)/K_v$$

Minimizing h/q as a function of F-g yields:

$$d(h/q)/d(F-g) = -L^2(\ln L/3r)/2\pi.K_h (F-g)^2 + 1/K_v = 0$$

so that the maximum value of F-g, represented by M, is:

$$M = L \sqrt{(\ln L/3r) K_v / 2\pi.K_h}, \qquad (17)$$

M being an equivalent maximum depth of flow under the Dupuit assumptions. If F-g > M, then the factor F-g in the well-spacing formulas should be replaced by M.

WELLS IN A SEMI-CONFINED AQUIFER

Figure 7 shows a pumped well penetrating into the lower, more permeable layer of a semi-confined aquifer. The upper, slowly permeable layer extends to a depth F_s below the soil surface. In that layer, there may be a considerable resistance to vertically downward flow. The hydraulic head required to overcome this resistance equals $h_s = q (F_s-i)/K_s$, where K_s represents the vertical hydraulic conductivity of the upper layer. We can now apply the foregoing well-spacing formulas, provided that:

- g-i is replaced by $h' = g-i-h_s$;
- F-g is replaced by $D' = F-F_s$;
- E-g is replaced by $G' = E-F_s$;
- K is replaced by K', the hydraulic conductivity of the deeper layer.

WELLS IN ANISOTROPIC SOILS

If the soil is uniformly anisotropic, with K_v and K_h as defined before, the well-spacing formulas can be adapted as follows (Boumans 1979):

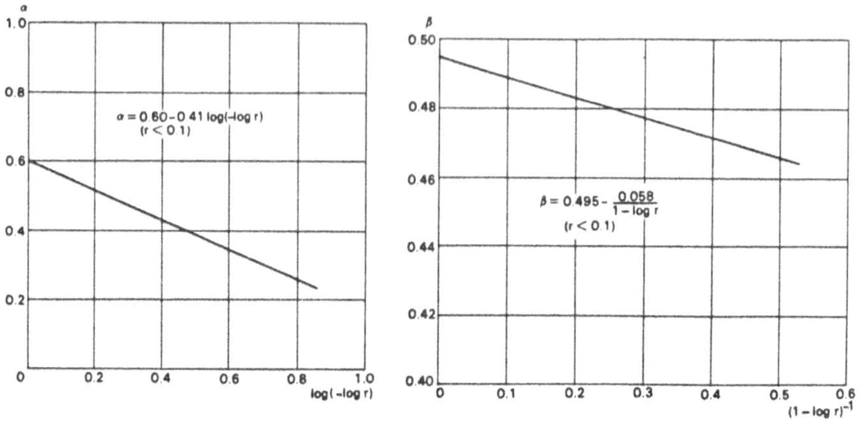

Figure 6. Linearization of data from Figure 3.

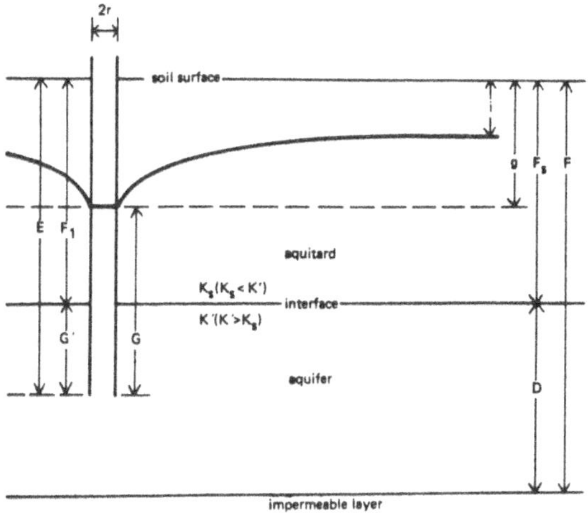

Figure 7. Partially penetrating well in semi-confined aquifer

- In all parts representing horizontal flow, replace K by K_h;
- In all parts representing spherical flow, replace K by $K_g = \sqrt{K_h . K_v}$;
- In all parts representing vertical flow, replace K by K_v;
- If $K_v < K_h$ and the well is partially-penetrating, assume different values Z' for Z (Z' < Z) and calculate the corresponding L values with the formulas of Section 3 or 4, whichever are applicable, using:
  $h' = h - q(G + Z' + h)/K_v$ instead of g-i, and
  $F' = F - Z + Z'$ instead of F;
  choose the maximum value of L as the final result. (One will usually find that this occurs when Z' = 0 because, with small K_v values, the flow will not go deeper than the bottom of the well.)

EXAMPLES

In all the following examples, the average discharge of the pumped wells is taken as q = 0.002 m/d and the desired average depth of the watertable as i = 3 m below the soil surface. Four aquifer profiles will be considered:

1. A homogeneous profile with a depth of F = 10 m, K = 2 m/d
2. A homogeneous profile with a depth of F = 100 m, K = 2 m/d
3. A uniform anisotropic profile with a depth of F = 100 m and K_h = 5 m/d, K_v= 0.2 m/d
4. A semi-confined profile with an upper layer of thickness Fs = 10 m, Ks = 0.2 m/d, and a deeper layer of thickness $D' = F - F_s$ = 90 m, K' = 5 m/d.

The average depth of the waterlevel in the well is assumed to be g = 5 m below the soil surface in profile 1 and g = 10 m in the other profiles. The depth of the wells is E = 10 m below the soil surface in profile 1 and E = 40 m in the other profiles. The radius of the wells is taken as r = 0.1 m, including the filter.

The calculation of the well spacings for the different aquifer profiles is as follows.

——— Profile 1 (fully-penetrating wells) ———

According to equation (3), Y = 62800 + 12600 = 75400. Hence equation (4) with a = 0.60 and b = 0.466 gives: L = 112 m.

Conclusion: The area served by one well (1.25 ha) is too small for practical purposes. The transmissivity of the aquifer is not sufficiently large.

——— Profile 2 (partially-penetrating wells) ———

G = E-g = 30 m, Z = F-E = 60 m, D = G+Z = 90 m, check: G < Z. Equation (14) will be used with: T = K(F-g) + K(g-i)/2 = 180 + 7 = 187 m^2/d (the term K(g-i)/2 appears to be comparatively small), Y = 4100000 (eq. 14a), r_e = 0.0078 (eq. 14b). Equations (15) and (16) or figure 6 yield: a=0.467, b=0.476. Equation (4) gives: L = 656 m. (Check: Z < 0.2 L.)

The application of equation (9) would have given almost the same result (L = 663 m) because 0.1 L < Z < 0.3 L.

Conclusion: The area served by one well (43 ha) is almost 40 times greater than that for profile 1 because the transmissivity of the aquifer is much higher.

——— Profile 3 (anisotropic) ———

Assuming different values of Z' < Z, and calculating the corresponding L values, we find that L is maximum (663 m) when Z' = 0 because, for other values of Z' > G, the effective radius r_e becomes very small. The resistance to spherical flow is so great compared with the resistance to horizontal flow, that the water will not go deeper than the bottom of the well. We can then calculate the well spacing as if the well were fully penetrating, taking into account the resistance to vertical flow, as explained before.

Conclusion: The area served by one well (44 ha) is almost the same as for profile 2. The positive effect of a high K_h value is neutralized by the negative effect of a low K_v.

——— Profile 4 (semi-confined) ———

F = 100 m, F_s = 10 m, D' = F-F_s = 90 m, G' = E- F_s = 30 m, Z= 60 m, T = K'.D' = 450 m^2/d and r_e = D'.exp{-3T/G'.K'} = 0.011, Y = 2π.T(g-i)/q = 9900000. (See equation 14a.) According to figure 6 or equations (15) and (16), a = 0.48, b = 0.475. Hence, L = 1010 m (equation 4).

Conclusion: The area served by one well (100 ha) is the largest of the four profiles. Drainage by pumped tubewells thus becomes feasible in semi-confined aquifers with an upper layer of considerable depth and slow permeability and a lower layer that is deep and permeable (if not under great artesian pressure or below sea level) because:

- The properties of the upper layer prohibit the economic spacings of open ditches or horizontal pipe drains, and
- The properties of the lower layer permit a wide and economic spacing of wells.

REFERENCES

J.H. Boumans 1979. Drainage calculations in stratified soils using the anisotropic soil model. In: Proceedings of the International Drainage Workshop (Ed. J. Wesseling). Publ.25, ILRI, Wageningen.

J. Wesseling 1973. Subsurface flow into drains. In: Drainage Principles and Applications, Vol.II. Publ.16, ILRI, Wageningen.

W. McCready 1978. Drainage construction for vertical tubewell drainage. I.C.I.D., New Delhi.

SECTION 2 MAIN DRAINAGE AND FLOOD CONTROL

Area Reduction Factors for Land Drainage Design in Flat Basins in Moderate Temperate Climates

L.K. Smedema, J.H.M. Mooren and C.F.J.M. Veraa
Department of Civil Engineering, Delft University of Technology, The Netherlands

INTRODUCTION

High values of specific discharges from a drainage basin (= discharge per unit basin area, e.g. expressed in $m^3.s^{-1}.km^{-2}$ or in $l.s^{-1}.ha^{-1}$) generally tend to decrease somewhat with increasing size of the basin. This phenomenon is explained as being due to:
- dynamic storage in the drainage system resulting in attenuation of discharge hydrographs as basin size increases
- non-uniform coverage of large basins by the heavy rains producing high discharges.

In land drainage design, this phenomenon is taken into account by applying a so-called area reduction factor to the specific design discharge:

$$q_A = ARF.q_o \qquad (1)$$

where q_A = specific design discharge for area of size A ($m^3.s^{-1}.km^{-2}$ or $l.s^{-1}.ha^{-1}$)

q_o = 'point' specific design discharge ($m^3.s^{-1}.km^{-2}$ or $l.s^{-1}.ha^{-1}$)

ARF = area reduction factor ($\leq$ 1.0, dimensionless).

The specific design discharge (q_{design}) is also commonly referred to as the drainage coefficient; q_o is the q_{design} of a 'small area' for which ARF = 1.0 (compares to the concept of 'point' rainfall as used in watershed hydrology).

Values of q_a are used to determine the required capacity of a canal or structure:

$$Q_A = q_a \,.\, A \qquad (2)$$

where Q_A = design discharge for area A ($m^3.s^{-1}$ or $l.s^{-1}$)

A = area size (km^2 or ha)

The area reduction factor as defined by equation 1 is most applicable to flat areas. In sloping areas, the influence of the size of the drainage basin is generally otherwise, implicitly or explicitly taken into account in the formulae and calculation procedures by which design discharges are determined for such areas (rational formula, curve number method, etc.)
Area reduction factors used in design practice are mostly based on experience. This paper presents some results of a model study on area reduction factors in a flat basin.

MODEL STUDY

The model study involved the routing of discharges generated by different typesof (design) storms through a drainage system and analysing the observed discharges in relation to the corresponding area sizes.

The drainage system used in this model study covers a maximum area of 200,000 ha and consists of a hierarchical structured, rectangular pattern of tertiary, secondary and primary drainage canals (see figure 1). The area is assumed to be slightly sloping from the sides towards the centre and to the outlet point, slopes being assumed to be in the order of 0.01-0.02 % (10-20 cm per km). The primary canal, running through the centre, receives drainage discharges from 10 secondary drainage canals, each serving a secondary block of 20,000 ha. Each secondary block comprises 20 tertiary blocks of 1,000 ha. These tertiary blocks are conceived to be made up by a large number of fields, all provided with some form of field drainage system and internal system of drainage ditches which discharges at one central point into a secondary drain.

Rainfall-discharge model for tertiary block
The tertiary block of 1,000 ha has been taken as the 'small area' for which ARF = 1.0. All ARF values determined in this study, therefore, are relative to the q_{design} of this block. The transformation of rainfall into drainage discharge from such a tertiary block is assumed to be integrally described by the linear reservoir model (Smedema and Rycroft 1983). According to this model, the specific discharge at any time t (q_t in $mm.day^{-1}$) is linearly dependent on the not yet discharged excess rain in the considered area (referred to as the current dynamic storage S_t, expressed in mm water depth):

$$q_t = \alpha . S_t \qquad (3)$$

The proportionality coefficient α in this relationship is called the reaction factor. The discharge q may be related to the excess rain R ($mm.day^{-1}$) by noting that:

$$\frac{dq}{dt} = \alpha \frac{dS}{dt} = \alpha (R - q)$$

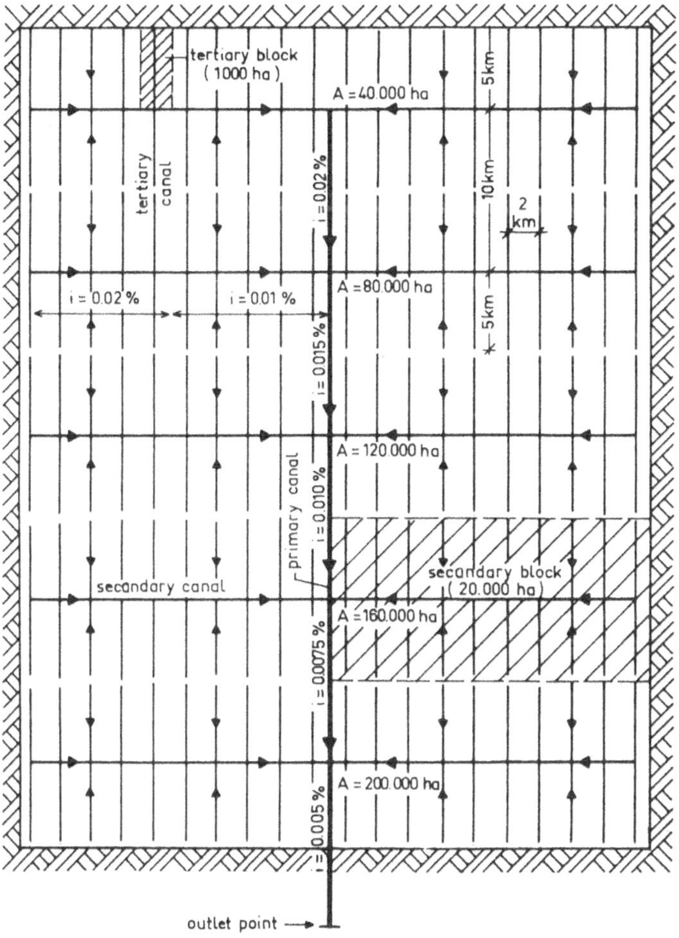

Figure 1 : Model drainage system

and integrating this differential equation over the time interval Δt from t=t-1 to t=t :

$$q_t = q_{t-1} \exp(-\alpha\Delta t) + R_{\Delta t} (1 - \exp(-\alpha\Delta t)) \qquad (4)$$

It has been shown (Smedema 1985) that equation 4 quite well describes the drainage discharges generated by rainfall under a range of field drainage conditions, including overland flow drainage, interflow drainage and groundwater drainage. For different types of field drainage, different α-values apply. Typical examples of each of the above types of field drainage, together with indicative values of α, are given in figure 2. It should be noted that when, as done in this study, these values of α are applied in equation 4 to describe discharges from tertiary blocks of 1,000 ha, it is assumed that the discharge transformation in the flow through the internal ditch system can be neglected or is included in the applied α-value.

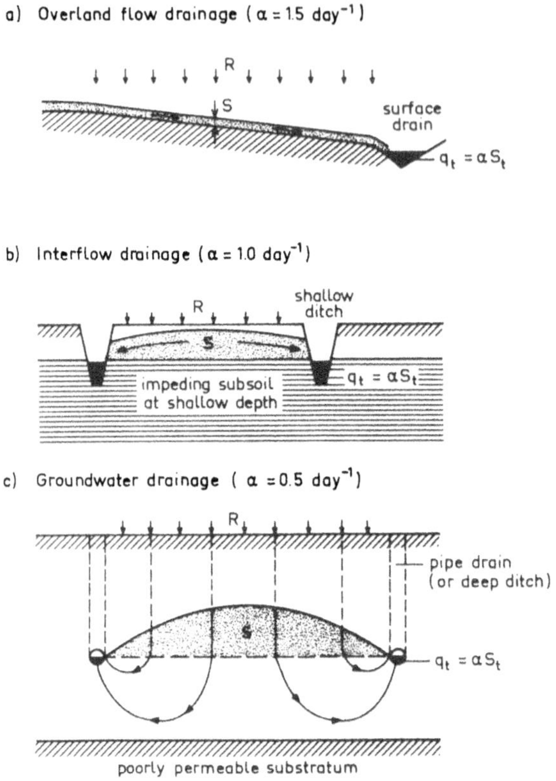

Figure 2 : Types of field drainage

Routing model for secondary and primary system

Discharges entering the secondary canals from the different tertiary blocks are routed through these canals, and subsequently through the primary canal, to the outlet point by means of the so-called ICES-FLOW model developed by the Department of Civil Engineering of Delft University (Booy 1978 and 1982). The model is based on simultaneous numerical solution of the two basic differential equations which jointly describe non-steady, varied (non-uniform) flow through a canal system:

flow equation

$$\frac{\partial h}{\partial x} = -\frac{1}{gA_f}\frac{\partial Q}{\partial t} - \frac{1}{C^2 A_f^2 R} Q|Q| + \frac{2b_f}{gA_f^2} Q \frac{\partial h}{\partial t} \qquad (5)$$

continuity equation

$$\frac{\partial Q}{\partial x} = -b_s \frac{\partial h}{\partial t} \qquad (6)$$

where: h = water depth (m)
Q = discharge ($m^3.s^{-1}$)
x = longitudinal distance (m)
t = time (s)
g = gravitational acceleration ($m.s^{-2}$)

A = crossectional area (m^2)
R = hydraulic radius (m)
b = canal width (m)
C = Chézy roughness coefficient ($m^{-\frac{1}{2}}.s^{-1}$)
subscripts: f = for flow, s = for storage

Application of this routing model requires the canal system to be schematised in sections, bounded by nodal points. Latter points were conveniently chosen to coincide with canal junction points, while additional nodal points were added to arrive at section length < 2 km throughout the system (see figure 3)

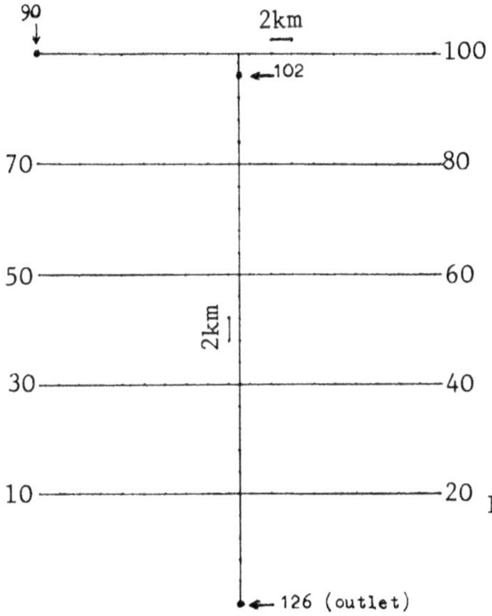

Figure 3 : Schematization of the model drainage system for the ICES FLOW routing model

The crosssectional and longitudinal dimensions of the different canal sections were arrived at by designing the system for steady uniform discharge, using the Manning formula and following criteria:
- specific design discharge 1.5 $l.s^{-1}.ha^{-1}$ (for the entire system, ARF = 1.00)
- free board 1.40 m
- Manning roughness coefficient
  . small canals (water depth ≤ 1.50 m) K_m = 30
  . large canals (water depth > 1.50 m) K_m = 40
- width/depth ratio = 1-2 for small canals (depth < 0.75 m)
  = 2-3 for medium canals (depth 0.75-1.50 m)
  = 3-4 for large canals (depth > 1.50 m)
- max flow velocity 1.10 $m.s^{-1}$
- free outflow at outlet point

MODEL RUNS AND RESULTS

The concept of the area reduction factor applies to the conditions under which the drainage system is designed to function, i.e. when the system carries its design discharge. In this study the specific design discharge was assumed to be 1.5 $l.s^{-1}.ha^{-1}$ and the canal system has been dimensioned accordingly. Care has been taken that the rainfall input to the model was always compatible with this design.

Model runs have been made for three different types of rainfall input:
- rainfall intensity constant in time and uniform in space
- rainfall intensity variable in time but uniform in space
- rainfall intensity constant in time but variable in space

Area reduction factors were determined by relating the discharges as determined with the model at the different nodal points to the corresponding upstream areas. Following nodal points were used for this purpose (see figure 3):
- nodal point 90 : upstream area 1,000 ha ($q = q_o$; ARF = 1.0)
- nodal point 102 : upstream area 40,000 ha
- nodal point 126 : upstream area200,000 ha

For the different nodal points, area reduction factors were determined as (compare with equation 1) :

$$ARF = q_A / q_o \quad (7)$$

where in this case q_A refers to areas of resp. 40,000 ha and 200,000 ha and q_o to the area of 1,000 ha. Furthermore, values of q used in calculating ARF values by means of equation 7, were the average high values over a period equal to half the critical storm period, in conformity with the practice of basing design discharges for flat basins on average high discharges rather than on short lasting peak discharges which can be accomodated by the freeboard (see figure 4).

Critical storm periods vary with the value of α used in the rainfall discharge model for the tertiary block. When this value is low, the discharge is more delayed and attenuated than when the α-value is high. These characteristics make that for the same return period, high intensity, short duration storms are most critical (requiring highest q_{design}) when α has a high value while for a low α-value this applies to lower intensity, longer duration storms. Critical storm durations under the considered conditions were determined with the model as follows:
- α = 0.5 day^{-1} , critical storm duration T = 2.0 days
- α = 1.0 day^{-1} , critical storm duration T = 1.0 day
- α = 1.5 day^{-1} , critical storm duration T = 0.5 day

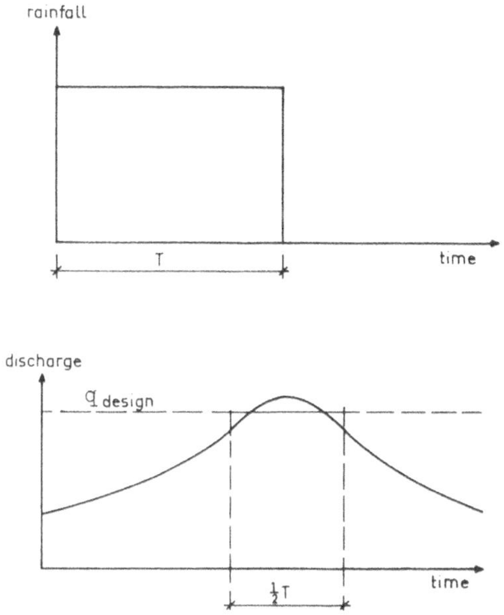

Figure 4 : Design discharge approach for flat basin

Rainfall intensity constant in time and uniform in space
In these runs design storms of constant intensity uniformly covered the entire basin of 200,000 ha. Storm durations varied for different α-values as indicated before, while the rainfall intensity was such that the resulting drainage load was compatible with the drainage coefficient of 1.5 $1.s^{-1}.ha^{-1}$ for which the canal system was designed.

An example of the transformation of the discharge hydrograph taking place in the main system is shown in figure 5. Area reduction effects are small (see high ARF values in table 1). A slight decrease in the ARF values with increasing α-values can however be noticed which is of course quite plausible as the q_o - hydrographs for high α-values will be more peaked than for low α-values, and therefore peak attenuation in the main system will have more impact in the first case.

Table 1 : Area reduction factors (ARF) for different values of α under constant, uniform rainfall intensity

	α = 0.5 day^{-1}	α = 1.0 day^{-1}	α = 1.5 day^{-1}
A = 1,000 ha	1.00	1.00	1.00
A = 40,000 ha	1.00	0.98	0.95
A = 200,000 ha	0.99	0.96	0.90

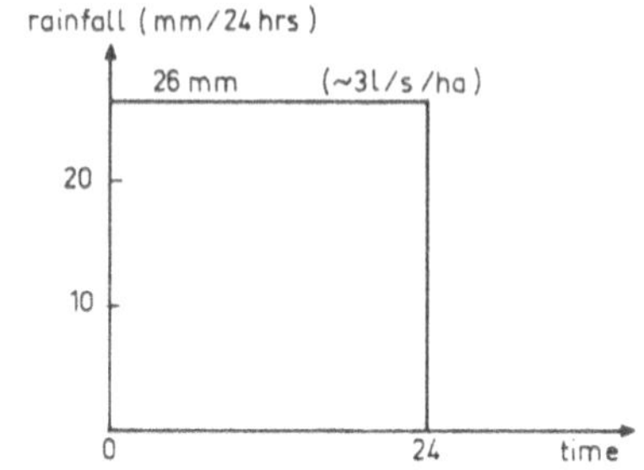

	0.5 day average high discharges		peak discharges	
	q	ARF	q	ARF
A = 1,000 ha	1.72	1.00	1.90	1.00
A = 40,000 ha	1.69	0.98	1.79	0.94
A = 200000 ha	1.65	0.96	1.73	0.91

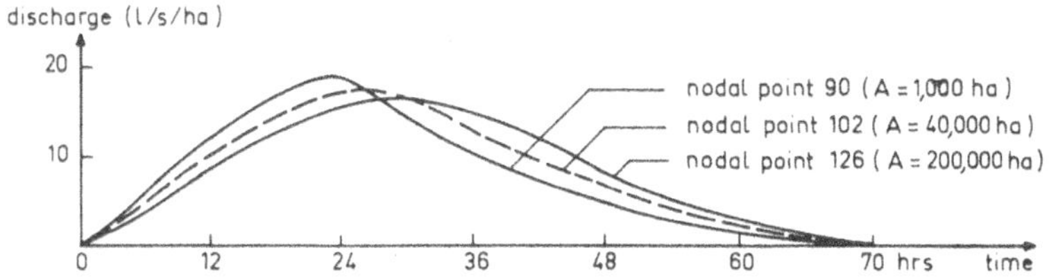

Figure 5 : Example of model run
(constant rainfall, α = 1.0 day^{-1}, T = 1.0 day)

Rainfall intensity variable in time but uniform in space

In these runs two types of storms with variable intensity were used as the rainfall input, but for both types the area coverage was uniform. The characteristics of these two types of storm are indicated in figure 6. The storm duration (T) is always equal to the earlier mentioned critical storm duration for the applies α-value, while the intensity (I) variation within the storms is based on the ratio $I_{\frac{1}{2}T}/I_T$ = 1.5, a common ratio for the used types of design storms.
Table 2 compares the ARF values of these two types of variable intensity storms with the corresponding constant intensity storm for two values of α whereby the tertiary unit discharge generated by the constant intensity storm is taken as the standard (q_o). It shows that in all cases the increasing intensity storms generated the highest discharges, higher than the constant intensity storms and therefore resulting in values of ARF > 1.00).

Table 2 : Area reduction factors (ARF) for two different α-values under time variable rainfall intensity

	α = 0.5 day^{-1}			α = 1.0 day^{-1}		
area (ha)	constant intensity rainfall	increasing intensity rainfall	decreasing intensity rainfall	constant intensity rainfall	increasing intensity rainfall	decreasing intensity rainfall
A= 1,000	1.00	1.07	0.99	1.00	1.09	1.02
A= 40,000	1.00	1.07	0.99	0.98	1.06	0.99
A=200,000	0.99	1.05	0.98	0.96	1.03	0.97

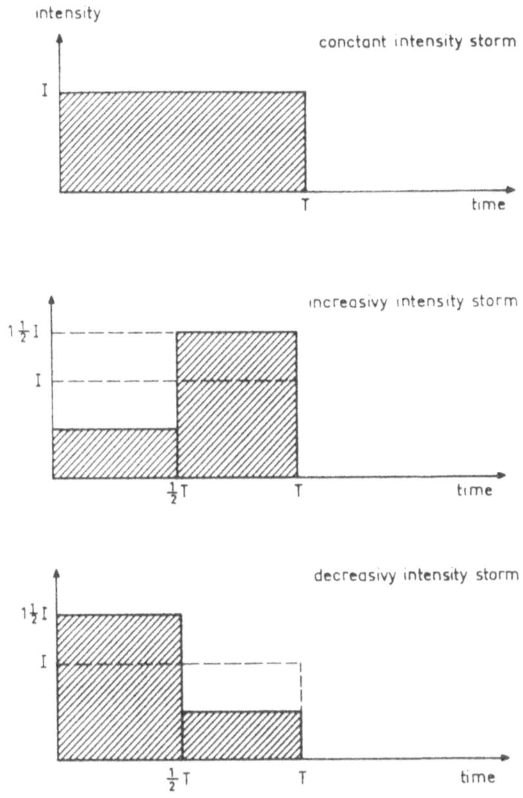

Figure 6 : Types of variable intensity storms

The ARF decreasing effect of higher α-values can still be noted although with variable intensity storms, this effect is largely negated by the peak discharge increasing effect of the short lasting higher intensity rainfall periods of the variable storms.

Rainfall intensity constant in time but variable in space

In these runs, different parts of the basin were covered by storms of different intensity but coverage within each sub-area was uniform and also the intensity was constant throughout the storm duration. Three different spatial distribution patterns were considered (figure 7) whereby the highest rainfall intensities were respectively conceived to be concentrated in the upper, middle and lower part of the basin. Furthermore, rainfall intensities are related to sub-area size conform the relationships given in table 3 such that the average intensity over the entire basin is the same as used in the constant uniform rainfall intensity runs.

The results are presented in table 4 whereby the calculated ARF values are again related to the tertiary unit discharge generated by the constant uniform intensity rainfall. It shows that the applied spatial variability in rainfall intensity has no influence on the ARF.

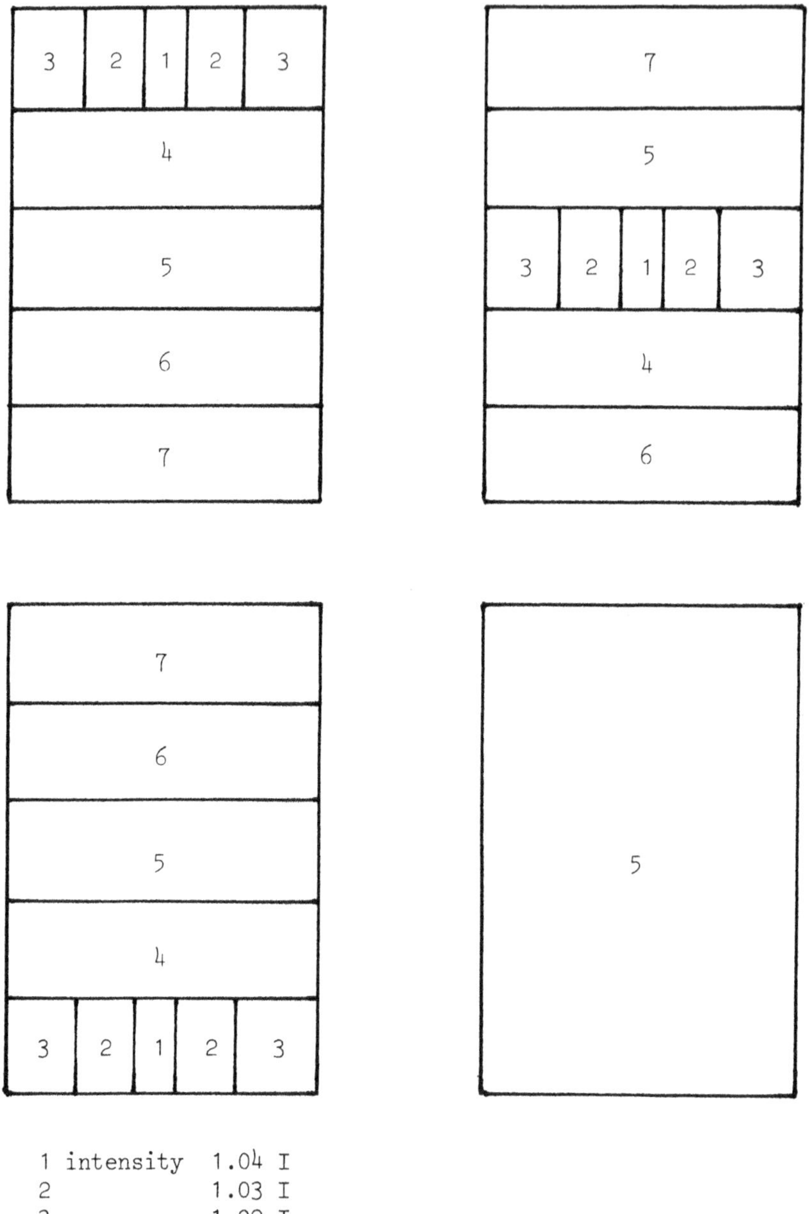

1	intensity	1.04 I
2		1.03 I
3		1.02 I
4		1.01 I
5		1.00 I
6		0.99 I
7		0.98 I

Figure 7 : Spatial rainfall distribution patterns

Table 3 : Area reduction factors for point rainfall (Netherlands 24 hrs rainfall, Witter 1983)

	1 x 1 yr	1 x 5 yrs	1 x 50 yrs
A = 100 ha	1.00	1.00	1.00
A = 5,000 ha	0.98	0.96	0.96
A = 100,000 ha	0.96	0.94	0.92

Table 4 : Area reduction factors (ARF) for $\alpha = 1.0$ day^{-1} under spatially variable rainfall intensity

	A = 200,000 ha
- highest intensity in the upper part of the basin	0.96
- highest intensity in the middle part of the basin	0.96
- highest intensity in the lower part of the basin	0.96
- uniform intensity	0.96

DISCUSSION AND CONCLUSIONS

Conclusions drawn from this study are strictly limited to flat drainage basins in moderate temperate climates, which are the conditions for which the model drainage canal system was designed and the conditions under which the different model runs were operated. The $q_{design} = 1.5$ $1.s^{-1}.ha^{\cdot 1}$ on basis of which the canal system was dimensioned and to which the rainfall input of the model was adapted, is a typical value for these conditions (e.g. prevailing in most of N.W. Europe). This also applies to the used rainfall variability in time and in space: the used distribution models are admittedly rather crude but are conceived to capture the main characteristics although here obviously lie possibilities for further improvements of the model.

Under the above defined conditions, the area reduction effect appears to be small, the q_{design} reductions being < 5-10% (ARF > 0.90-0.95) for A = 200,00 ha. This finding is in fair agreement with design practice in the Netherlands where for polder type basins of up to 50,000 - 100,000 ha hardly any area reduction effect is taken into account. The results suggest that under the suppressed variability of depression type rainfall, the area reduction effect is roughly as much a hydraulic effect (discharge attenuation due to canal storage) as a rainfall variability effect.
The ARF concept, as stressed before, applies to the design discharges and has been determined accordingly in this study by expressing ARF relative to average high discharges rather than relative to peak discharges as the latter discharges are generally not taken as the design discharges in flat basins. Lower

ARF values (more reduction) are obtained when these values were related to the observed peak discharges (see table 5) but these values are less relevant to design.

Table 5 : Comparison of ARF values relative to average high discharges and peak discharges

	$\alpha = 0.5\ day^{-1}$		$\alpha = 1.0\ day^{-1}$		$\alpha = 1.5\ day^{-1}$	
	average	peak	average	peak	average	peak
A = 1,00 ha	1.00	1.00	1.00	1.00	1.00	1.00
A = 40,000 ha	1.00	0.97	0.98	0.94	0.95	0.89
A = 200,000 ha	0.99	0.96	0.96	0.91	0.90	0.84

Area reduction effects in empirical design discharge formulae are often taken into account by reducing the basin area as in:

$$Q_A = q_o \cdot A^{i} \tag{8}$$

where the coefficient $i < 1.0$, meaning that the discharges Q increase less than proportional with the size of the area A. This type of formulae evolved from plotting observed high/peak discharges vs the corresponding basin areas.

Combining the equations 2 and 8, it follows that:

$$q_A \cdot A = q_o \cdot A^{i}$$

$q_A = A^{i-1} \cdot q_o$, so according to eq. 1, $ARF = A^{i-1}$.

Using this relationship, ARF values have been calculated for $i = 5/6$ which is the value of the coefficient i in the well-known Cypress Creek formula. Results are given in table 6.

Table 6 : Area reduction effects taken into account in the Cypress Creek formula : $Q = q_o \cdot A^{5/6}$ (where A in sq. mile and Q in cusec per sq-mile)

A =	1 sq mile	(~ 250 ha)	ARF = 1.00
A =	5	(~1300 ha)	ARF = 0.88
A =	10	(~2600 ha)	ARF = 0.69
A =	20	(~5100 ha)	ARF = 0.62

The Cypress Creek formula originates from the USA and is applied widely in Central, Eastern and Southern States for basinsup to some 5000 ha and for field drainage ranging from groundwater drainage to overland drainage. It shows that this formula allows for considerable area reduction effects. The same applies for the empirical design discharge formulae used for irrigated areas in India (see Singh 1983):

- $Q = q_o\ A^{3/4}$ (orthern and Central India)
- $Q = q_o\ A^{2/3}$ (Southern India)

where Q in cusec, q_o in cusec per sq. mile and A in sq. miles.

For A = 40,000 ha and A = 200,00 ha, the area reduction effects are in the order of 70% and 80% for i = 3/4 and even more for i = 2/3. These high reduction percentages may be partly explained by the fact that they relate to peak discharges but most of the reduction must be explained as being due to rainfall and other variability within the basin (in irrigated areas, e.g. there is considerable variability in antecedent soil moisture conditions). As yet not enough information on this variability is available to make ARF model studies opportune for these types of climates.

References

- Booy, N.G. (1978) Report on the ICES subsystem FLOWS, Communication on Hydraulics, Dept. of Civil Engineering, T.H. Delft
- Booy, N.G. (1982) FLOWS user's guide, Dept. of Civil Engineering, T.H. Delft
- Smedema, L.K. and D.W. Rycroft (1983) Land Drainage, Batsford, London
- Smedema, L.K. (1985) Drainage coefficients for heavy land, J of Irr. and Drainage, ASCE, 111, 2 : 101 - 112
- Witter, J.V. (1983) The area reduction factor in hydrological design due to temporal and spacial differences in rainfall, Nota 62, Dept. of Hydraulics & Hydrology, L.H. Wageningen (in Dutch)
- Singh Bharat (1983) Fundamentals of Irrigation Engineering, Nem Chand & Bros, Roorkee.

An Integrated Surface and Groundwater Flow Model for the Design and Operation of Drainage Systems

E.P. Querner
Institute for Land and Water Management Research (ICW), Wageningen, The Netherlands

INTRODUCTION

In The Netherlands dense channel systems are present for land drainage. In periods with a shortage of water these systems often serve to supply water for agricultural purposes as well. Because of the double function of these channels it is very important that they are maintained properly. Weed control, for instance, is necessary to prevent waterlogging or flooding in the case of high discharges. Maintenance costs are a heavy burden for the local water authorities. The frequency of maintenance activities is usually determined by a rule of thumb. This may result in too much maintenance. On the other hand the frequency of maintenance may be too small giving a greater danger for flood damage or insufficient water supply. The situation is even more complicated because water levels vary during the year. For a good water management the water level in winter is generally kept low to allow a quick drainage, whereas the level is raised in spring to conserve water for dry periods in summer or to make subsurface irrigation possible.

The design of drainage systems is often based on analytical models that predict the discharge from an area without considering typical local conditions. Storage of water in the ground or in open channels may differ, however, from generalized or average conditions. Also, pattern, size and flow resistance of the drainage system can have a great effect on the discharge rate.

In order to overcome the shortcomings of empirical and analytical methods I have developed an integrated surface- and groundwater model: SIMPRO, simulation of hydrological processes. A relatively simple one-dimensional saturated-unsaturated groundwater model controls the inflow and/or the outflow per nodal point of the surface water system, depending on conditions within the area. With criteria such as minimum and maximum water levels during the year and an assumed maintenance program, it is now possible to calculate whether water levels are low enough to prevent flooding, but also high enough to allow for water conservation and/or subsurface irrigation.

DESCRIPTION OF THE FLOW MODEL

Model concept

The discharges and water levels in a channel system depend on vegetation and on land drainage conditions. With the aid of analytical methods the discharges and water levels can be estimated for a certain region. These methods cannot take into account the channel network or conditions where drainage is alternated by water supply. If one considers a summer half year and a winter half year the processes of drainage and subsurface irrigation are depending on hydrological conditions, such as heavy rainfall or long dry periods. A hydraulic model for channel networks connected with a groundwater flow model has then to be used to simulate drainage and supply conditions. The groundwater model must yield the boundary conditions for the open channel flow. It should simulate the rather complex processes involved in such a manner that it is accurate enough without requiring too much input data and computer time.

The groundwater flow model used here consists of a relatively simple one-dimensional model of the saturated and unsaturated zone. The unsaturated zone is an important reservoir that acts as a transition zone between the atmosphere and the subsurface water movement (rainfall, possible sprinkling and evapotranspiration). The interaction between surface and groundwater is a function of the ditch density and geohydrological properties of the soil profile.

Unsaturated zone

The unsaturated zone is modelled as a reservoir. Water is considered to be stored in the root zone until a certain equilibrium is reached. If the equilibrium is exceeded, excess water will percolate to the unsaturated zone between root zone and phreatic surface. If the water content in the root zone is below its equilibrium, capillary flux from the phreatic surface will take place. The height of the phreatic surface is calculated by means of a storage coefficient which is dependent on the groundwater depth. The model is described in detail elsewhere (Querner and Van Bakel, 1984). A schematization of the groundwater model is shown in Figure 1 and the major functions are summarized below.

The change of water content ΔV_w over a timestep Δt due to rainfall P_r, sprinkling P_s and evapotranspiration E_a is:

$$\Delta V_w = (P_r + P_s - E_a) \cdot \Delta t \qquad (1)$$

Evapotranspiration is a function of the crop and of the water content in the root zone. Without considering percolation or capillary rise the water content for the next timestep would be:

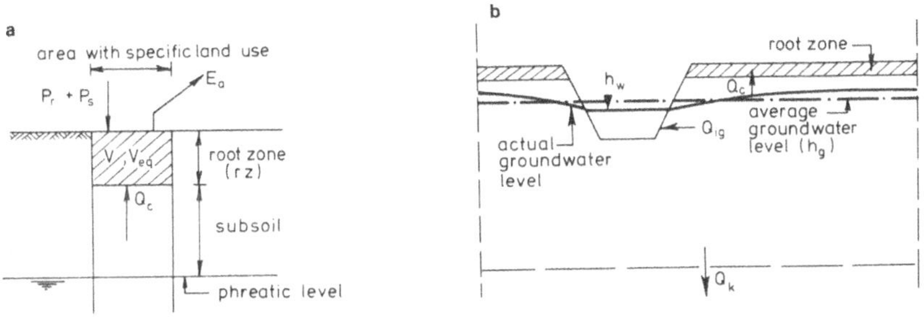

Figure 1. Schematization of groundwater model. Nomenclature used: P_r, rainfall; P_s, sprinkling; E_a, evapotranspiration, V, water content; V_{eq}, equilibrium water content; Q_c, capillary rise; Q_{ig}, drainage; Q_k, leakage

$$V = V_w(t) + \Delta V_w \tag{2}$$

The equilibrium water content in the root zone is a function of the soil physical properties, s, the groundwater depth h* and the thickness of the root zone, rz. If the water content V is above equilibrium percolation occurs, otherwise capillary rise is effective. Therefore the capillary rise is defined as:

$$Q_c = f\{s, h^*, rz\} \qquad V < V_{eq} \tag{3}$$

or

$$Q_c = \frac{V_{eq} - V_w}{\Delta t} \qquad V > V_{eq} \tag{4}$$

The water content for the next timestep can now be calculated as:

$$V_w(t + \Delta t) = V + \Delta t . Q_c \tag{5}$$

This model concept was verified with results from a more accute model SWATRE (Belmans et al., 1983). For a sandy soil evapotranspiration and capillary rise in winter did not differ, but for the summer half year the simple model gave 3-8% lower actual evapotranspiration, so it underestimated capillary rise.

Saturated zone

The saturated zone has interactions with the unsaturated zone and the surface water, while there may be leakage or seepage over the lower boundary (Figure 1b). By means of a central time difference such as the Crank-Nicholson approximation a change in groundwater depth, Δh_g, over a timestep Δt can be written as:

$$\mu \frac{\Delta h_g}{\Delta t} = \theta Q_i^{t+\Delta t} + (1 - \theta) Q_i^t \tag{6}$$

where μ is the storage coefficient and Q_i represents all flow to or from a nodal point. All variations in flow over a timestep are assumed to vary linearly with the change in groundwater depth. The lower boundary flux can be written explicitly as a function of the groundwater depth.

Although the timestep for the groundwater flow model can safely be chosen up to one week, a timestep of one day is generally used, because meteorological data is often available on a daily basis.

Interaction between surface and groundwater

The interaction between the surface water and the groundwater system can be described as:

$$Q_{ig} = \frac{(h_w - h_g)}{\alpha T} \tag{7}$$

where h_w is the water level in the surface water system and T is the so-called drainage resistance; the geometry factor α is necessary to convert the head midway between two parallel ditches to the average hydraulic head used in this equation (Ernst, 1978). The drainage resistance is either kept constant or used as a function of the groundwater depth, dependent on the number of ditches per unit area and the geohydrological properties of the subsoil.

The ditches that actually are involved in the interaction between surface and groundwater are divided into a primary and a secondary system. The primary system consists of the larger channels that always contain water. The secondary system consists of shallow ditches which are intermittently filled with water. All water stemming from the secondary system in a certain area is allocated to a certain point of the primary network.

Surface water model

For the simulation of the water movement in the open channels the Saint Venant equation is used (Chow, 1959). Because the change in flow rate is very small, gravitational forces can be neglected, so that a simple computational scheme suffices without strict limitations on the timestep to be used. The primary open water system is divided into sections with nodes on either side (Figure 2). For each node a water level h_w is calculated and for each section a discharge Q. For a section with nodes i and j the discharge is:

$$Q_{ij} = K_{ij} (h_{wi} - h_{wj}) \tag{8}$$

where K_{ij} represents the roughness and geometry of a channel section. Using the continuity principle and the above relation for all nodal points a set of equations is obtained in the form:

$$\{Q\} = [K] \cdot \{h_w\} \tag{9}$$

where the matrix [K] can be considered as a resistance and storage matrix. It contains all contributions to the flow resistance between point i and its adjacent nodes and the storage capacity at node i. Using matrix inversion, equation (9) can be solved to give the water levels in all nodal points. Because the resistance factor in the [K] matrix is a function of the water level the solution is obtained by successive approximation. In general only a few iterations per timestep are required.

The timestep can be chosen freely in principle, but for numerical stability it is limited by factors like section length, change in flow rate, channel geometry, etc. In practice the maximum timestep is between a quarter of a day and half a day. The consequence is that different timesteps occur in the groundwater and the surface water model. The drainage flow is therefore calculated for every timestep used in the surface water model, keeping the groundwater level constant. The total drainage flow over a number of timesteps is used for the next timestep of the groundwater model.

SPECIAL FEATURES OF THE MODEL

Because of vegetation growth, the channel roughness varies with time. The vegetation growth is a function of time, temperature, rainfall, soil type, water depth, water quality, etc. To account for these roughness changes, the growth rate - as a function of time - and a correction factor on the growth rate - depending on the length of the vegetation - are used as input of the model. A relation between type, density and length of the vegetation and the roughness coefficient can be obtained from methods like those reported by Kouwen, Li and Simons (1981). The drag force due to the water flow results in a bending of the vegetation. The flexural rigidity and density of the vegetation resists the bending. Therefore the roughness coefficient must be a function of flow velocity, water depth, roughness height and vegetation characteristics.

For water management purposes it is necessary that during the year the target level in the system can be changed, to allow for water conservation and subsurface irrigation. At present weir levels over the calculation period can be specified. It is also possible to set the target levels as a result of certain criteria calculated during the simulation period, e.g. groundwater depth and water content in the root zone. If a specific criterium is met, the target level should be raised or lowered in small steps of say 0.05-0.10 m once per week (Van Bakel, 1986). In this way the model can be used for simulating the practical and real water management.

Maintenance of a channel system in the form of weed control is carried out in summer to prevent too high water levels and flooding. The occurrence of flooding depends on a number of factors, such as the channel system, the actual vegetation in it, the maintenance program, the meteorological and the hydrological conditions. The hydrological conditions can be the occurrence of drainage and subsurface irrigation and the storage capacity in the unsaturated zone.

To find acceptable and also practical solutions for the maintenance program and/or channel sizes, it is necessary to simulate the involved processes and to determine if the chosen conditions are acceptable with respect to hydraulic and hydrological criteria, like velocity, water level and groundwater level.

EXAMPLE

To illustrate its applicability, the model SIMPRO was tested on a hypothetical channel network as shown in Figure 2. At node 1 the water level is kept constant in order to allow excess of water to discharge and a shortage to be compensated. At node 10 a weir is situated where water can leave the region in times of extremely high water levels. In the example a specific vegetation growth rate and maintenance program is stipulated. In the program the actual vegetation length is converted to the Manning roughness coefficient in order to relate it to the hydraulic loss in a channel section. The model was run for the year 1981 (Fig. 3). Because the Manning coefficient is a function of the flow velocity, the decrease in roughness at the second and third clearance of weeds is not that pronounced as that for the first one (Fig. 3a). In spring and autumn the water level is high and drainage dominates due to heavy rainfall. The summer period shows a demand for sub-

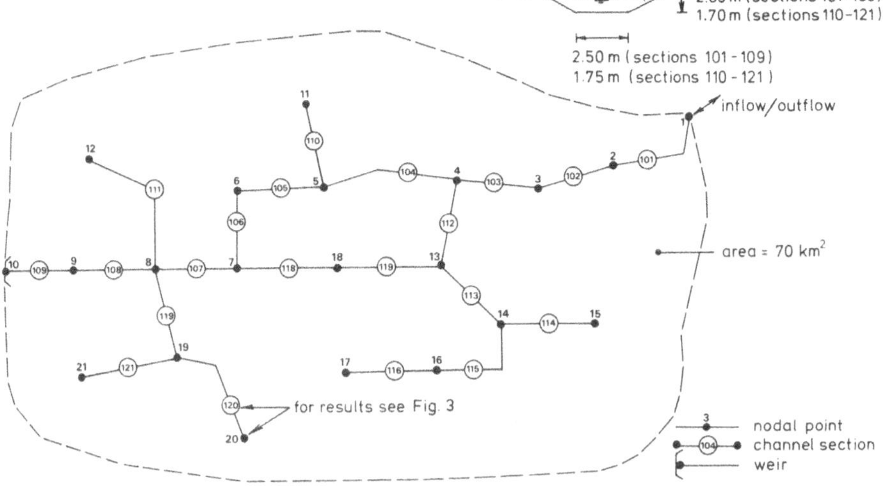

Figure 2. Schematization of a hypothetical open channel system

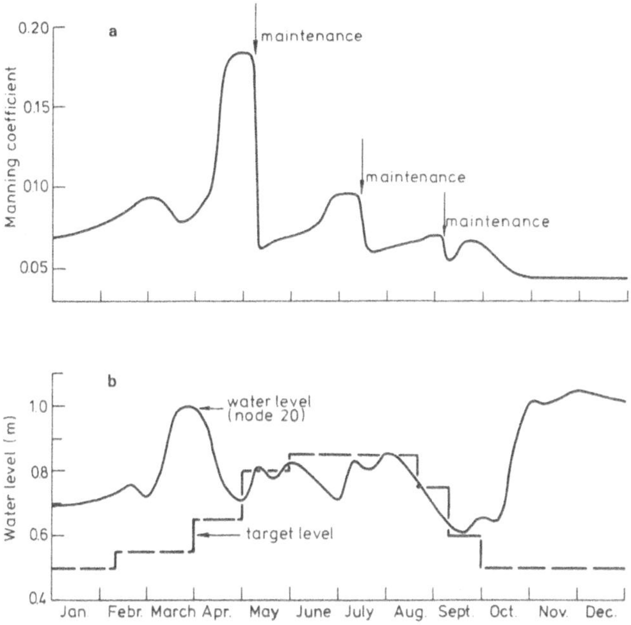

Figure 3. Calculated results for a hypothetical channel network as shown in Figure 2 for the simulation period 1981. In a the variation of the Manning roughness coefficient over the simulation period for section 120 is shown together with the executed maintenance. In b the water level for nodal point 20 is given, together with the prescribed target level

surface irrigation, when the water level is below the target level (Fig. 3b).

CONCLUSION

As shown by the example the integrated surface water and groundwater flow model is suitable to assist in design and operation of channel networks. The example focused on weed control, but the model can also be used to determine required channel size or water level management for water conservation. Although not discussed in this paper, the model can also be used in relation with a crop production model and cost model (Querner and Feddes, 1985). In this way the effect of certain decisions in water management can be related to crop production on a regional level. From the cost model and the crop production model the cost benefit follows.

The model in its present form can be used in regions where manipulations of water levels have no important influence on the regional groundwater flow, because the amount of seepage or leakage is used as a boundary condition. In the near future a regional groundwater flow model will be included to allow for these cases.

REFERENCES

Bakel, P.J.T. van (1986) Planning, Design and Operation of Surface Water Management Systems. A Case Study. Report 13. ICW, Wageningen, The Netherlands. 114 pp.

Belmans, C., J.G. Wesseling and R.A. Feddes (1983) Simulation Model of the Water Balance of a Cropped Soil: SWATRE. J. Hydrol. 63, 3/4: 271-286. Techn. Bull. n.s. 21. ICW, Wageningen, The Netherlands.

Chow, V.T. (1959) Open Channel Hydraulics. Mc Graw-Hill Book Co. Inc., New York, 680 pp.

Ernst, L.F. (1978) Drainage of Undulating Sandy Soils with High Groundwater Tables. I. A Drainage Formula based on a Constant Hydraulic Head Ratio. II. The Variable Hydraulic Head Ratio. J. Hydrol. 39, 3/4: 1-50. Techn. Bull. 106. ICW, Wageningen, The Netherlands.

Kouwen, N., R.M. Li and D.B. Simons (1981) Flow Resistance in Vegetated Waterways. Trans.ASAE 24, 3: 684-690, 698.

Querner, E.P. (1984) Program FEMSAT, Part 1. Calculation Method for Steady and Unsteady Groundwater Flow. Nota 1557. ICW, Wageningen, The Netherlands. 23 pp.

Querner, E.P. and P.J.T. van Bakel (1984) Description of Second Level Water Quantity Model, Including Results. Nota 1586. ICW, Wageningen, The Netherlands. 67 pp.

Querner, E.P. (1985) The Influence of Vegetation on the Discharge in Open Channels. Nota 1600. ICW, Wageningen, The Netherlands. 58 pp. (in Dutch).

Querner, E.P. and R.A. Feddes (1985) Program SIMCROP - Calculation of Agricultural Production and Cost on a Regional Level. Nota 1666. ICW, Wageningen, The Netherlands. (in press).

Application of the Hydrological Model GRODRA for the Design of a Water Management System in the Land Consolidation Project Giethoorn-Wanneperveen

J.W. Wesseling
Delft Hydraulics Laboratory, The Netherlands
J.M.L. Jansen
Government Service for Land and Water Use, The Netherlands

INTRODUCTION

Both in wet regions where excessive precipitation has to be discharged and in arid regions where irrigation is applied, the most common form of water resources management is control of the surface water system. In the Netherlands, where much of the land lies below sea level and a substantial inflow of water occurs via precipitation and a number of major rivers, the traditional water resources management aimed at drainage and discharge of excess water. In the course of time an intricate system of waterways, pumping stations, sluices and weirs, combined with a dense network of ditches and drains developed. Nowadays the system is essential for protection against inundation and for control of the groundwater table in order to provide soil moisture conditions amenable to agricultural production in both wet and dry periods.

For this reason particular attention has been paid to the interaction of groundwater and open channel flow in a physically based hydrologic response model that has been completed recently at the Delft Hydraulics Laboratory. This model, which has been named PREDIS for its description of precipitation - discharge processes has been constructed as a series of self-contained computer modules each concerned with a specific part of the hydrological cycle. The modular interpretation of the classical portrait of the hydrological cycle underlying the structure of PREDIS has been depicted in Figure 1. The model PREDIS (Gilding and Wesseling, 1983) and the coupling of groundwater and open channel flow (Crebas et al., 1984) have been presented earlier. The present paper focuses on the modules for saturated groundwaterflow and drainage to open channels. The combination of these modules, named GRODRA, provides a valuable tool in situations where the groundwater table is controlled by management of the open channel system.

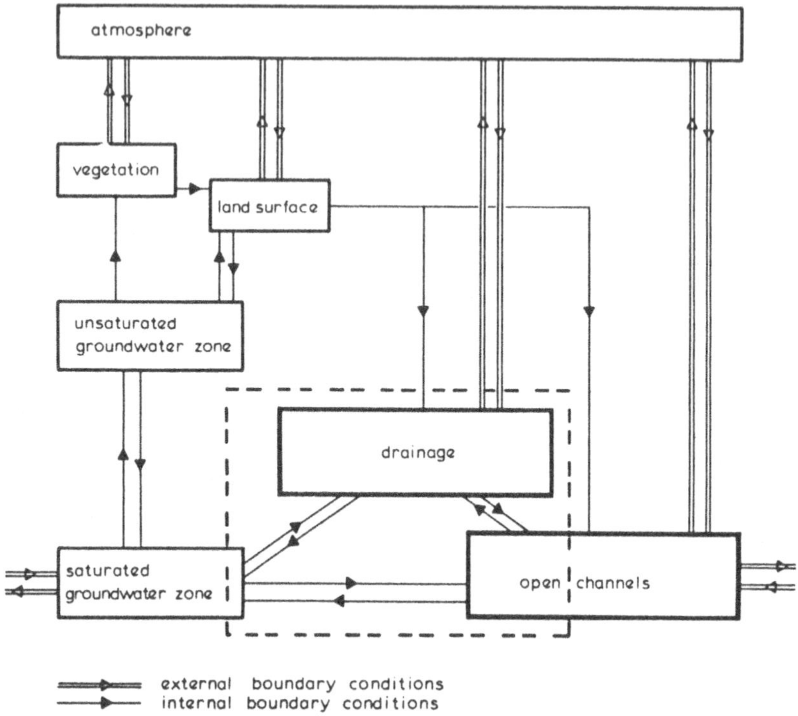

Figure 1 Modular interpretation of the hydrological cycle in the hydrologic response model

CONCEPTUAL BASIS OF GRODRA

In the general structure of the hydrologic response model, the surface water system has been divided into two subsystems.
The first subsystem comprises the large waterways in which the flow and storage effects are significant. The main function of this system, which will be referred to as the open channel system, is to collect the water from the smaller conduits and discharge it to one or more points outside the region concerned. The water level in the open channel system provides a boundary condition to the smaller conduits that discharge to it. PREDIS comprises a special subroutine to compute the water level in the open channel system as a function of flow, storage and the operation of regulating devices. However, very often the water level in the open channel system can be fully controlled by opreation of regulating devices. In those cases, for computational purposes, the water level in the open channel system can be assumed to be a given state variable.
The second subsystem, which will be referred to as the drainage system, is concerned with all the small conduits discharging to the open channel system. These conduits would require too much refinement to be considered individually on a regional basis, but amalgamated they have an important in-

fluence on the hydrologic behaviour of a region.
GRODRA has been designed for those situations where the water level in the open channel system is fully controlled. The saturated groundwater flow and seepage to the drainage system then are the two processes that govern the hydrologic behaviour of the region.

The module for saturated groundwater flow is based upon a schematization of the flow regime into a number of highly-permeable horizons alternating with poorly permeable layers. In the main aquifers it is supposed that the Dupuit-Forcheimer assumption of essentially horizontal flow may be applied. The interaction with the drainage system is supposed to take place in the uppermost phreatic layer.
The exchange between this aquifer and a drainage conduit is formulated as a function of the groundwater head and a drainage resistance.
For the mathematical description of the exchange function an expression, based on the formulae for drainage design, is used in which the most important physical properties of the drainage system are represented.

MATHEMATICAL REALISATION

By analogy to Darcy's law (or Ohm's law) the exchange between the phreatic aquifer and a drainage conduit is formulated as:

$$Q = (\phi - h)/R \qquad (1)$$

in which d denotes the groundwater potential, h the given open conduit water level and R the drainage resistance.
In the module for saturated groundwater flow the spatial component is treated by a Galerkin finite-element technique, using quadrilatic elements.
The method yields an approximate solution of the form:

$$\phi(x,y,t) = \sum_{i=1}^{n} \phi_i(t)\, v_i(x,y) \qquad (2)$$

where $\phi_i(t)$ are the computed groundwater potentials in the nodal points of the grid and v_i is a series of basic functions.

If for each element of the grid the drainage resistance can be computed from the physical characteristics of the drainage system in the element and the open conduit water level is given, the exchange can be computed by combination of equations (1) and (2) with the equation for groundwater motion.
To define the drainage resistance and the open-conduit water level for the various elements of the grid, the following assumptions have been made:
- the effects of storage in the drainage system may be neglected
- the drainage system can be divided into various subsystems,

each discharging to a point of the open channel system with a defined potential
- the water level within each subsystem may be regarded as uniform
- the drainage systems can be characterized by a limited number of prototype drainage conduits which may be regarded as parallel and equidistant
- the overall influence of the drainage system can be incorporated from the prototype conduits by the principle of superposition.

In GRODRA, the drainage resistance for a prototype conduit has been based upon the Hooghoudt-Ernst formula and reads (Ernst 1962, Streltsova 1974, Van Beers 1976):

$$R = (1/\pi K_2) \ln(\eta D_2/p) + \rho/p \qquad (3)$$

where η is a geometrical shape factor, ρ is an entrance resistance accounting for the hydraulic impedance of material lining the bottom of the drain, p is the wetted perimeter of the drain and K_2 and D_2 refer to the hydraulic conductivity and the thickness of the aquifer below the line of drainage.

APPLICABILITY OF GRODRA

GRODRA can be applied for those situations where a strong interaction between groundwater flow and open channel flow occurs provided that the potential in the open channel system can be controlled.

The program contains a formulation to compute a drainage resistance based on the physical characteristics of the system. For practical applications one is however free to define the various drainage parameters in such a way that the behaviour of the system under consideration is simulated in the best possible way.

As an example of the approach the Giethoorn-Wanneperveen investigation will be presented.

APPLICATION IN THE GIETHOORN-WANNEPERVEEN PROJECT

The Giethoorn-Wanneperveen project area is situated in the northern part of the Netherlands. The area comprises some 5000 hectares of marshland with a rich flora and fauna and is of particular interest from a point of view of nature and landscape conservation. High groundwater tables and small and dispersed agricultural plots are hampering a proper cultivation of the land and limit the agricultural development. The land consolidation project aims at improvement of the conditions for agricultural development in one half of the area, while in the other half the requirements of nature and landscape prevail. Figure 2 shows the lay-out of the project area.

Improvement of agricultural conditions can be realised by adaptation of the surface water system, aimed at enlargement of agricultural plots and reduction of the groundwater head.

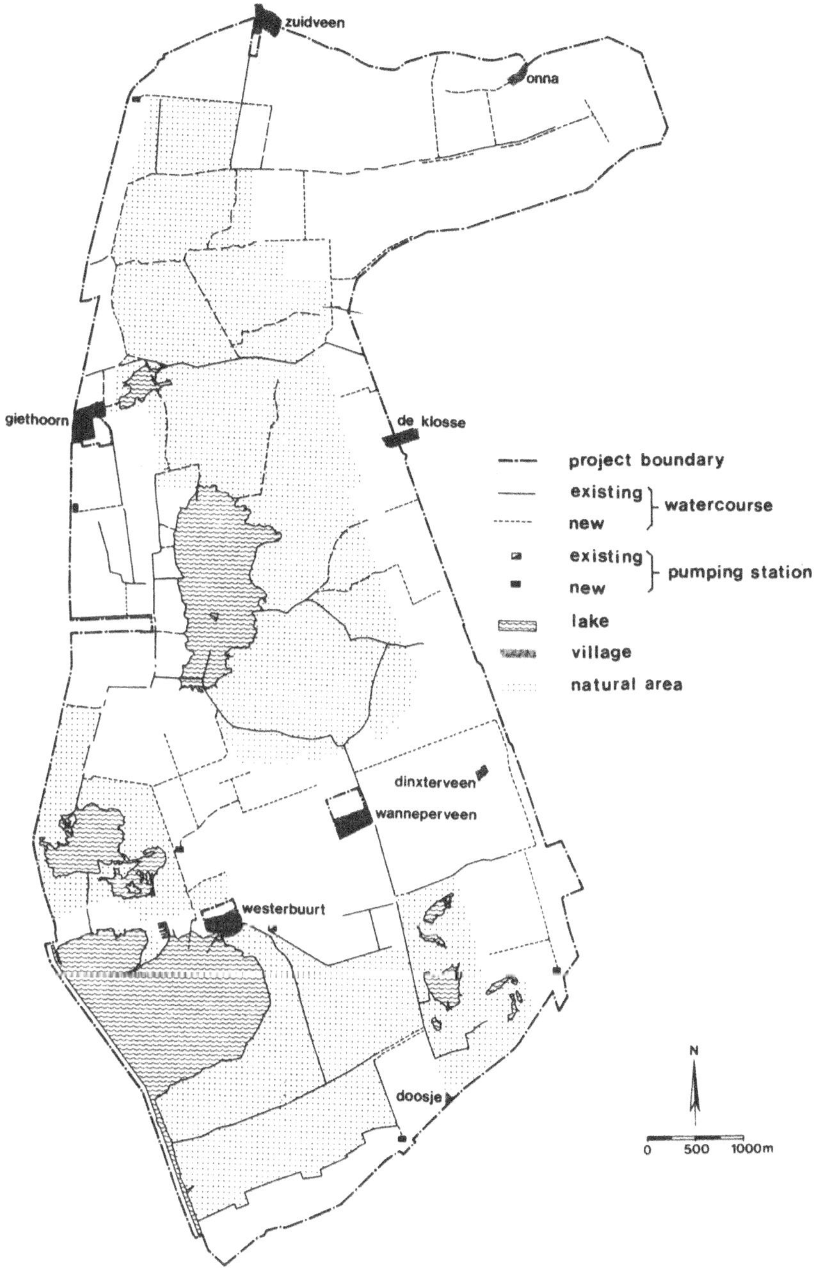

Figure 2 Landconsolidation project „ Giethoorn-Wanneperveen

Conservation of nature and landscape however, require a constant or even increased groundwater head. The conflict between the interests of agriculture and nature conservation is deteriorated by the geohydrology of the area; since the peaty toplayer is underlain by a permeable sandy aquifer (Fig. 3). Possibilities for recharge to the nature area by external supply are limited by the chemical characteristics of the available water which originates from Lake IJssel.
Though the water quality of Lake IJssel is satisfactory for agricultural use, abundant supply to the nature area may cause a considerable change of its vulnerable vegetation.

The conflicting interests of agriculture and nature conservation and the strong hydrological interactions between the various parts of the project area required a careful design of the water management system. During the process many alternatives have been evaluated. The program GRODRA has been used to simulate the hydrological effects of the various project alternatives. Subsequent post processing was done to translate the hydrological effects in terms of agricultural production and natural values.

The design of the model has been based on the geohydrological schematization presented in Figure 3. The model comprises a phreatic aquifer and a semi confined deep aquifer separated by an aquitard.

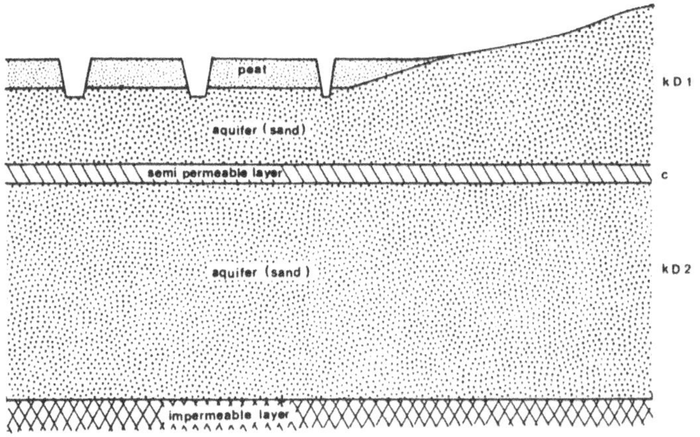

Figure 3 Geohydrological schematization

The influence of the peaty toplayer has been neglected in the groundwater model; but has been accounted for in the interaction with the surface water system, by specification of an increased entrance resistance term in Equation (3).
In the model the surface water system has been represented by four classes of open conduits. The characteristic parameters

for each of the classes are presented in Table 1.

class	bottom depth (m)	bottom width (m)	side-slope
1	1,20	0,50	1:1,25
2	1,40	1,40	1:1,25
3	1,50	2,00	1:1,50
4	1,90	2,50	1:1,50

Table 1 Subdivision of drains in classes

The spatial variation in the surface water system has been accounted for by specification of a drain-density parameter for each of the four classes in the various elements of the finite element grid.

In the first phase of the study a model has been designed covering the entire project area and some 3400 hectares of adjacent land. Thus the boundary conditions of the model would not be influenced by changes of the groundwater and surface waterhead within the project area. The finite element grid of this model comprised 131 elements. The model has been used for calibration of the geohydrological parameters and for calculation of the overall effects of various project alternatives. For model calibration observed groundwater heads and measured discharges of three pumping stations were available. In particular the drainage resistance appeared to be a sensitive model parameter. The calibration runs demonstrated the model's ability to give a satisfactory representation of the hydrologic behaviour of the area.
In subsequent phases of the investigation more detailed models have been developed, covering only part of the project area, to simulate the effects of local measures. The boundary conditions for these simulations have been derived from the results of the over-all model.

The model simulations showed that in the present situation only little of the strong seepage flow that once caused the development of the typical nature in the northern part of the area has remained. Only the drains along the eastern border are fed by seepage from the higher sandy soils adjacent to the project area. From the major part of the surface water system in the northern natural area infiltation to the groundwater occurs. In summer periods the water demand for the entire project area amounts to about 50,000 m^3/day.
Implementation of the project will cause a decrease of the surface water head in the agricultural land of the northern part of the area. As a consequence the phreatic level in the natural part will decrease with about 30 cm in those parts adjacent to agricultural plots, reducing to less then 5 cm at a distance of 1000 m from the boundary between natural and agricultural land. The water demand for the northern part of the project area will increase by some 6,600 m^3/day.

In the southern part of the project area, the surface water head will locally be increased by implementation of the project. Here the project will have positive effects on nature, since infiltration of surface water with a poor quality will be reduced.

A hydrological and ecological survey has been prepared in order to monitor the effect of the projects' implementation. Though the main drainage system has only been completed in 1982/83 and the adaptation of field drains is still in progress, some preliminar conclusions can be drawn. The groundwater flow can be represented by the scheme drawn in Figure 4. The observed chemical composition of the groundwater indicates that underneath the natural area the groundwater draining from the high sandy soils at the eastern side of the project area mixes with the infiltrating surface water.
Detailed observation of the groundwater head showed that the interaction between the groundwater and a deep canal along the western border of the project area is less strong then has been assumed in the model. Apparently the permeability of the canal bottom which was supposed to penetrate the peat layer is still relatively high.
In general however, the observations show that the predicted effects seem to agree with reality.

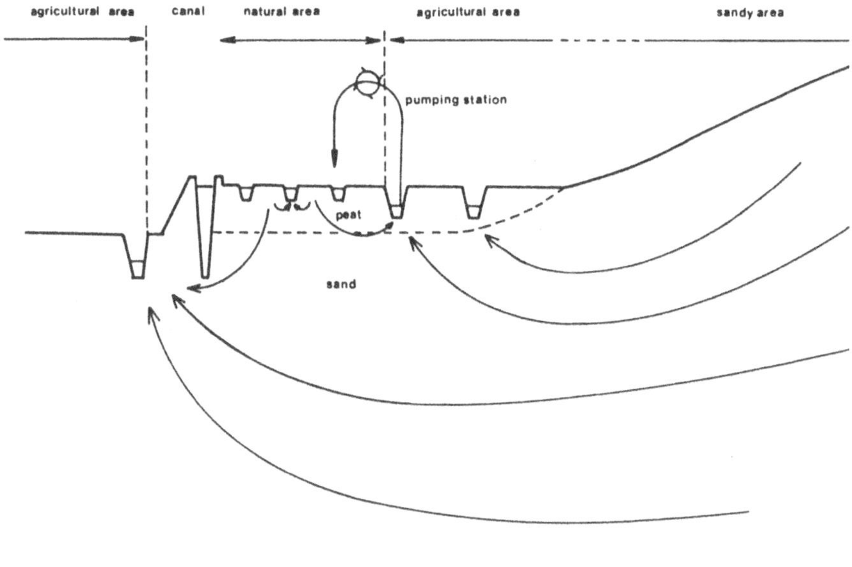

Figure 4 Schematical presentation of the groundwater flow

ACKNOWLEDGEMENT

The authors wish to thank Dr. B.H. Gilding for his participation in the development of GRODRA and the application in the Giethoorn-Wanneperveen project.

REFERENCES

Beers, W.F.J. van, 1976, Computing drain spacings, Int. Inst. Land. Recl. Improvm., Wageningen, Bull. no. 15

Crebas, J.I., Gilding, B.H. and Wesseling, J.W., 1984, Coupling of groundwater and open-channel flow, Journal of Hydrology, 72, pp. 307-330

Ernst, L.F., 1962, Groundwater flow in the saturated zone and its calculation when horizontal parallel open conduits are present, Cent. Agric. Publ. Doc. (PUDOC), Wageningen (in Dutch)

Gilding, B.H. and Wesseling, J.W., 1983, PREDIS - a physically based hydrologic response model, 20th Congr. Int. Assoc. Hydraul. Res., Proc., Subj. D, Vol. V, Int. Assoc. Hydraul. Res., Moscow, pp. 26-32

Streltsova, T.D., 1974, Method of additional seepage resistances - theory and application, J. Hydraul. Div. Proc. Am. Soc. Civ. Eng., 100, pp. 1119-1131

CALDRA: A Software Package for Drainage Design

P. Pauwels, J. Feyen and G. Wyseure
Laboratory for Soil and Water Engineering, Katholieke Universiteit te Leuven, Belgium

INTRODUCTION

Like in other areas of agricultural research, soil and water engineers try by technical means to free food production from variation-inducing factors like climatic and soil conditions. To avoid or minimize the effect of rainfall excess and of prolonged dry periods on crop growth they make use of drainage and irrigation networks.

An important aspect concerning these systems is the ability for the designer to determine the appropriate dimensions because it is those which count for the quality of the installation as well as the economic acceptability of the investment. For a considerable time, a number of theoretical insights and practical relationships have been available in this area. They include the flow models for saturated and unsaturated ground-water flow, the models describing the drainage and infiltration processes, and the models for approximation of the discharge processes in simple and more complex drainage systems. These models make accurate determination of dimensions possible for every situation and every desired change of condition, be it a lowering of the ground water by drainage or a raising of the soil water content by irrigation.

Much of this knowledge did not yet find its way into every day design practice. The most used design method for soil and water technical operations is, and remains, the application of rules derived from experience. The adaptation of the plan to the terrain data can, therefore, only be done approximately, which renders the potential for an economic evaluation of the plan too limited. A possible prognosis of the effect on crop growth and production is completely out of question with present design methods.

These deficiencies are, in part, due to the lack of technical resources in the design offices, as modern design procedures for drainage and irrigation projects require both an extended data structure and a complex calculation method. Recent developments in data processing and computer applications, however, can radically change this situation. The designer can now already make use of rapid and relatively price-friendly units of data handling as well as processing apparatus for the solution of graphic problems. It would seem no more than reasonable to place these new facilities at the service of the soil and water engineer. For some time now, software packages have been designed especially for the design of drainage and irrigation projects. In this form, the general application of complex but valuable calculation techniques can be accelerated and lead to a general improvement of the quality of soil and water engineering designs.

Not only is the faultless adaptation of the design to the measured situation and to the desired criteria thus made possible, but it also opens the way to efficient development of the design activity as such. The total number of hours needed to transform a specific set of data into an acceptable plan can be significantly reduced by the elimination of the time consuming work of calculating, plotting, and correcting. As a result this increases the processing capacity of a design office, and the time saved provides an opportunity to devote more attention to the recording of field data. It is obvious that this shift of the work from the drawing board to the reality in the field will certainly be beneficial to the quality of the plan.

Finally, these improvements are no longer far from the ultimate step, the optimization of the design, namely the determination of all the design dimensions adapted to all relevant field data, such as soil type, predominant weather conditions, and farming practice, as well as to the cost-benefit ratios of the change of condition that is planned. Up to the present, this has remained something for the future for two reasons. First, realization of all the possible design variants down to the level of cost is impossible to manage without technical support. Second, the estimate of the benefits of a particular soil and water technical operation remains more of an empirical operation rather than a practical aid. It is primarily the latter that stands in the way of the introduction of true optimization into design methods. Indeed, in spite of a great deal of research devoted to crop simulation, very few, if any, practically applicable yield relationships have been established in function of soil moisture conditions.

Nevertheless, this last obstacle cannot be a hindrance for the introduction of CAD techniques in soil and water engineering. The computer possibilities, the use of improved formulas for dimensioning and cost calculation, and the increase of efficiency within the planning office itself will lead to more justified design.

In the next section, we will try to illustrate the application possibilities of CAD by means of the software package CALDRA, which was set up as a computer aided drainage design system. The designing of a drainage network requires both a number of calculations and graphic analysis of the input data and is, therefore, almost a typical case in which CAD can lead to improved design.

THE STRUCTURE OF CALDRA

The concept preliminary for the global structure of the system included the idea that the designer was to remain in absolute control during the execution of the design. The reason for this is obvious : a drainage design takes account of many factors. Rarely it is a question of a univocal and optimal solution. The system, therefore, is conceived as a group of independent procedures, each on itself able to execute a facet or a form of calculation or plotting. From this point of view, the designer can choose is the most reasonable solution, according to his opinion.

The practical application of these design procedures is attended to by a number of interactive systems, which, as it were, enclose the central solution procedures. These interactive systems consist of an entire series of instruments with which the designer can execute the various segments of the design. In the first place, a part has been built in with which he can define any field or arbitrary surface. Thereafter he can transmit this defined field with ancillary and, in his opinion, relevant data to the processing unit. The results are then graphically presented by means of plotting subroutines.

In addition, the interactive peripheral procedures have the potential of adding special structures such as collectors, manholes, and linings to the design. And at last some features are included permitting the designer to provide the final plan with a frame and a legend.

By constructing the system in such a way, the designer preserves his important function but is dispensed with the time-consuming plotting and calculating. In addition, he has the capability of applying design procedures, which previously were too complex to use manually, such as detailed cost calculations, the non-steady state design algorithms, and the design of several variant solutions.

THE MOST IMPORTANT PROCESSING PROCEDURES

The automatic design system uses the field as a structural unit. This means that all operations, both the dimensioning calculations and the plotting, are executed for each field separately. Both large-scale and small-scale design tasks are solved by the repetition of this solution system for each field.

The first requirement for the dimensioning of a field is the collection of all the necessary data. The system was conceived in a way that a maximum amount of data can be processed. This means that the following data can be taken into account :

- planographic data of the field :
    - the form of the field
    - the slope : percentage and direction
    - specification of the drainage discharge network : location and depth of the ditches
- soil data :
    - the saturated conductivity
    - the presence of impervious substrata
    - the storage coefficient
- climate data : criterion figures or daily data of the precipitation
- crop data

This does not mean, however, that the system wil stop if one of these data is missing. In such a case, either the program will supply acceptable default values or the system will request the necessary data from the designer interactively. Obviously, the correctness of the generated solutions depends largely on the accuracy and the completeness of the field data supplied.

After the collection of the data, the design is limited to the determination of three quantities : the drain base, the drain spacing, and the drain direction. Since these three together determine the quality of the drainage, they are also determined as a group, i.e., the possible drain directions in the field are determined first. For each drain direction that emerges from this selection process, the maximum drain depth is determined from the planographic data of the drainage discharge routes. This drain depth together with the soil characteristics, the precipitation data, and the crop data determines the combination of drain direction, depth, and the spacing which will be proposed as the final design solution. All these derivations for the field are made completely automatically in the central section of the design procedure. The following scheme gives an overview of what happens.

1 Start

2 Input of the planographic data
Interactive determination of the planographic data per parcel or block

3 Determination of the possible DRAIN LAYOUTS within the geometry of the parcel or block

4 Procedure to be executed for each selected layout

4.1 Determination of the drain depth

4.2 Choice of the calculation method (block A or B)

A · Interactive input of :
- Daily precipitation in winter for several return periods
- The average water table to be obtained in function of the benefit that is expected from the drainage

· Input of soil data :
- Saturated hydraulic conductivity
- Storage coefficient
- Depth of the impervious substratum

· Calculation of the drain spacing by means of the Krayenhoff Van De Leur formula

B . Interactive input of :
- Drainage discharge criterion
- Water table depth "
Both drainage criteria are selected as functions of the benefits effect that is intended by the drainage

. Input of soil data :
- Saturated hydraulic conductivity
- Depth of the impervious substratum

. Calculation of the drain spacing by means of the Hooghoudt or the Ernst formula

4.3 Calculation of the cost of drainage

5 For alternative design return to step 4

6 Selection of the layout on the basis of the cost formula

7 Plotting of the drainage plan

8 End

Figure 1 : Flowchart of the CALDRA design system

DETERMINATION OF THE DRAIN DIRECTION AND DRAIN DEPTH

The problem here is situated primarily in the graphic character of the information on which the drain direction is based. Within the field, the drain direction will be adapted to the location of the field ditches, the slope of the terrain, the form of the field, and the bordering rows of trees, if any. Normally, the designer will judge these data visually and so derive the possible directions. The programming language, however, is limited to type data in the form of integers or real numbers that permit elementary tasks, as input and output, control commands, and simple mathematical operations.

Within the solution system, therefore, the graphic information must be presented numerically. This is done by assigning the field borders a sequential number. The coordinates of the corners of the field will indicate the location of the borders, and, finally, a code number is given to each border line that specifies its nature, be it a ditch, a road, or a row of trees. The graphic data are transformed in this way into numerical data. The slope is also presented in the algorithm as a degree of slope and a direction of slope.

The determination of the drain direction now merely involves the running of a number of tests. These tests deal sequentially with the drainage possibilities, the slope, the form, and the special border structures of the parcel in function of each possible drainage direction.

From this selection process, several possible directions will remain for which the drain depth, the spacing, and the cost are then calculated.

CHOICE BETWEEN STEADY STATE AND NON-STEADY STATE APPROACH

For the calculation of the drain spacing there are a number of implicit and explicit formulas. Many of these formulas are solved iteratively and demand a great deal of calculations, but they are no problem for a computer. The software package offers the possibility of choosing between a steady state and a non-steady state approach.

The difference lies in the integration of precipitation data. In a steady state formula, these are introduced in the form of a drainage discharge criterion (e.g. 10 mm per day). The formula then determines the drain spacing necessary to satisfy this criterion. In the non-steady state formulas, one may choose the data of an actually observed year with a given return period. The formulas provide the height of the water table for each day and for various drain spacings. These results can be combined, for example, to give the average water table during a critical period or a frequency of exceedance for a critical depth. This figure can then serve for making an evaluation on the postulated drain spacing.

DEFINITION OF THE DESIGN CRITERIA

For steady state methods, these criteria consist of a drainage discharge criterion, which is a measure of the maximum discharge the network should be able to drain off, and a depth criterion for the water table. The latter is generally established in function of the crop. On this level, when one wants to go to the limit, one can make use of the curves constructed by Van Wijk and Feddes (1975). These curves give the yield reduction for various water tables for wheat and sugar beets. When these results are combined with a calculation of the annual cost for a particular drainage, then the drain spacing can be determined on the basis of a cost-benefit analysis. This approach, however, can hardly be considered to have a general validity, for the results of the published curves are only valid for the experimental situation. Therefore, this method is not included in the flowchart.

The use of criteria is also possible for the non-steady state approach. The criteria are a depth criterion for the average water table and a return period for the precipitation data used, the latter indicating the percentage of the years in which the average water table may be higher than the criterion. Again, the work of Van Wijk and Feddes (1975) offers an alternative. For grassland, they constructed a curve that presents the reduction of the yield in function of the average water table for the period of October to May. The non-steady state calculation technique, which calculates the average water table for various drain spacings then can be combined with a procedure determining the corresponding yield reduction for grassland. This method permits the depth criterion to be selected in a more justified manner. The graphic output of such an automated procedure is presented in Figure 2. The designer can, in the given system decide by himself if a simple steady state criterion (a constant drainage discharge and average depth) suffices or if a more extended analysis of the criterion determination is required.

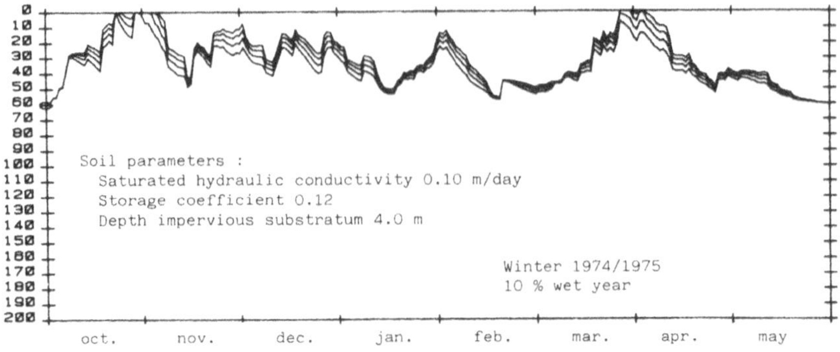

Drainspacing in m	Costs in Bfr/ha	Average water table depth in cm	Yield reduction for N-dose in kg/ha		
			300	150	0
8	50,000	37	13	20	40
10	40,000	34	16	24	46
12	33,333	31	19	27	51
14	28,571	30	20	30	55

Figure 2 : Calculated variation of the water table in peat for various drain densities with the yield reduction levels expected for three nitrogen fertilization doses

THE ECONOMIC EVALUATION CAPABILITY

In each field, the drain dimensions are economically justified by the automatic calculation of the cost of each alternative design. The cheapest solution is automatically sought and plotted. This method has an advantage over the manual approach as there will never be time to execute the cost calculation several times. Under these circumstances, the decision will be rather made on the basis of a qualitative judgment than on the result of a calculation.

GRAPHICAL SYSTEM

This section of the program employs the major principles of computer-aided design. This new design technique derives its efficiency from the use of two forms of expression of the data processed. On one hand, a CAD system will always have the data in digital form; on the other hand, the designer can also have a visual picture of the data on a graphic screen. It is primarily the direct link between the two forms of expression that is so typical for a CAD design system and that makes it such a powerful technique since the designer can immediately select and separate via the screen specific parts of the digital information. The data no longer need to be looked up in all sorts of documents, and specific characteristics of the selected portion do not have to be calculated. Everything is immediately available in digital form. With these numerical data, the designer can have calculations carried out in function of the desired changes in the prevailing condition. The final data set can be supplemented, and the designed changes visualized directly on the screen.

The structure of CAD programs is, therefore, always interactive. Generally, they consist of several independent modules, each of which performs the design of a specific element. Thus, a number of procedures have been developed especially for the design of drainage. The possibilities that are provided are summarized below.

First, the designer must be able to visualize the digital data in a relevant manner. Each kind of data, therefore, is plotted by a separate subroutine. These subroutines all work with a different manner of presentation, for example, a parcel border is presented by an ordinary line, and a ditch by an arrow. The designer can in this way distinguish the various data groups on the screen.

The second and, for drainage, very critical capability is a procedure that permits to define arbitrary surfaces or fields. This consists of the localization of the field and the grouping of the information important for this field. These data are then transmitted in group to the main module of the program where the drain density and the drain direction are calculated.

As a direct consequence the designer gets the capability of testing out various geometric configurations. For drainage work this is critical because the cost and the practicability largely depends on the size and the form of the individual parcels. Thus, in the predesign phase, the designer can divide the area into parcels that are as large as possible. The program will automatically report if this parcel is too large to drain as a whole and also give the reasons why. The designer can adapt the form and the size of the parcel in function of this response.

In drainage work, it is also of great importance to design the peripheral systems like collectors, manholes, and linings correctly. In the automated design procedure presented here, there is no separate design module built in for all these elements. Attention was rather given to constructing the program in a modular manner so that the desired modules could be built in easily and in function of the need. In principle, the design of these systems remains completely in the hands of the designer. The advantage is that he no longer needs to do the plotting or to calculate the cost. This is done completely automatically at his command.

The last module of the CAD procedure takes care of the definitive plan. This can include several things, depending on the specific requirements of the design office. The reproduction of the plan at an appropriate scale with a frame and a legend are essential.

CONCLUSIONS

One of the major advantages of the computerized approach is the capability of executing a cost-benefit analysis for each variant solution and for each parcel. For the division of the area into blocks, various geometric variants can be tested with the aid of the interactive input.

A second gain is the capability of taking into account a maximum amount of data. Together with the preceding, this can lead to a more justified design.

The interactive character allows the designer to execute his ideas immediately and to test them for economic feasibility. The designer can thus work in a much more creative manner and is no longer hampered by the need to perform the calculations or do the plotting. Suggestions and comments from the client can be worked out without any problem.

In addition, the procedure leads to a much more rapid treatment of the area. This is enhanced if the data bank of the area is also used in other aspects of reallocation.

The Controller of Pumping Aggregates in Draining Pumping Stations in Poland

D. Dejas, A. Reinhard
Institute for Land Reclamation and Improvement, Agriculture University, Wroclaw, Poland

INTRODUCTION

From many parts of the farming areas in Poland water must be periodically or permanently drained off by the pumping stations. The tidal flats of the Vistula river delta cover the area of about 170 000 ha and there are 115 pumping stations. The tidal flats are about 8 m over sea level to about 2 m below sea level. The pumping stations have there very important functions as the elements of the water-economic system. The efficient and reliable control of the pumping aggregates in these stations enables its economic exploitation, higher durability and efficiency.

THE PURPOSE OF THE PAPER

The devices used so far for turning on and off the pumping aggregates have many important technical imperfections, such as: deceptiveness of the mechanical sub-assembly, low sensitivity, occurring of electrolysis on the electrode detectors and the use of such voltage and current which are dangerous for the servicing personnel. In 1980 in the Institute of Land Reclamation and Improvement at the Agriculture University in Wroclaw research work was carried out successfully and a device was constructed for control of the pumping aggregates.
vice is based on the transistor keys, which stimulate the timelag relay transmitting the signal to the distribution board and turning off and on the pumping aggregate. This device has been used in Polish drainage pumping stations for four years. Because of the constantly growing demands of the pumping station users the Institute starts the re-

search work on a small-size device of higher sensitivity, reliability and ability of the long-distance signal transmission from the electrode detectors to the relay system.

THE WORK OF THE CONTROLLER

The main function of the device is signalling the particular water levels at which the pumping aggregate should be turned off or on and transmitting the signal to the distribution board. The water-level sensors are electrodes made of acid resistant steel, between which occurs 15 V voltage. The electrodes are in the perforated plastic guard which protects the system against undesired shorting. The measuring-turning system controlling one pumping aggregate is placed in the cover of the following size: 80x100x45 mm to which the signals from electrodes are transmitted. The initial signals from the device are transmitted to the distribution board. The basic element of the device is MCY 74011 system which consists of four NAND gates. Two of these gates are used for constructing trigger RS, one for reversing the signal from electrode 1, and the second for reversing the signal from trigger RS /Figure 1/. The rising water-level in the reservoir starts the current flow while the falling water-level stops the flow of current between electrode 2 of minus potential to electrode 1 and electrode 3. The signals from electrodes 1 and 3 are transmitted to the input $\bar{S}$ and input reset $\bar{R}$ trigger RS. These signals at the proper water-levels in the reservoir start changes of the logic states on outputs Q and $\bar{Q}$ of trigger RS. The signal from output $\bar{Q}$ trigger RS is transmitted by the gate to the dilatory system 4 constructed on resistors R_9, R_{10}, R_{11} and condenser C_3. With the capacitor R_{10} is regulated the required delay of the turning on of the pumping aggregate. The possibility of time regulation of turning on the pumping aggregate is very important in the case of installing some aggregates when the electric current cut-off is possible. The different starting time of the pumping aggregates protects the electroenergetic device against its overcharging in case of the simultaneous start. From the dilatory system 4 the signal is transmitted by system 5 to the electroenergetic devices 6 starting the pumping aggregate 7. The potentiometers R_2 and R_5 are used for sensitivity regulation of trigger RS.

THE RESULTS OF RESEARCH

The research work shows great usefulness of MCY 74011 system for devices controlling the pumping aggregates. This system increased sensitivity so much that the resistance in the input circuit /conductors resistance, electrodes and water/ can be 1 $M\Omega$, while the transistors allowed elimination of transmitters, which increased reliability and efficiency of the device, and also reduced its size. This device, based on the logic elements, has been patented and after successful work in Poland and GDR is now mass-produced. It is used not only in the drainage pumping stations but also in water lines sewage-treatment plants and hydraulic engineering. The work of this device has changed the exploitation of the stations and brought great economic effects.

REFERENCES

Dejas,D. Reinhard,A. Trzeciak,W. /1981/ Nowoczesne rozwiązania w zakresie automatycznego sterowania i pomiaru natężenia przepływu oraz mechanizacji pracy w przepompowniach melioracyjnych. Materiały na konferencję naukowo-techniczną n.t. : Odwadniające przepompownie melioracyjne. 111-136,Elbląg.

Dejas,D. Reinhard,A. Trzeciak,W. Medyk,J. Bondar-Nowakowska,E. /1980/ Urządzenie do automatycznego sterowania pracą agregatów pompowych oraz wyniki badań nad jego wdrożeniem w przepompowni melioracyjnej Osiek. Seminarium naukowo-techniczne "Wdrożenie badań nad możliwością automatyzacji przepompowni melioracyjnych" 6-14, Wrocław.

Dejas,D. Reinhard,A. Trzeciak,W./1982/ The control of pumping aggregates and the measurement of discharge in polder pumping stations. Papers International Symposium "Polders of the World". Vol. 2, 114-119 The Netherlands.

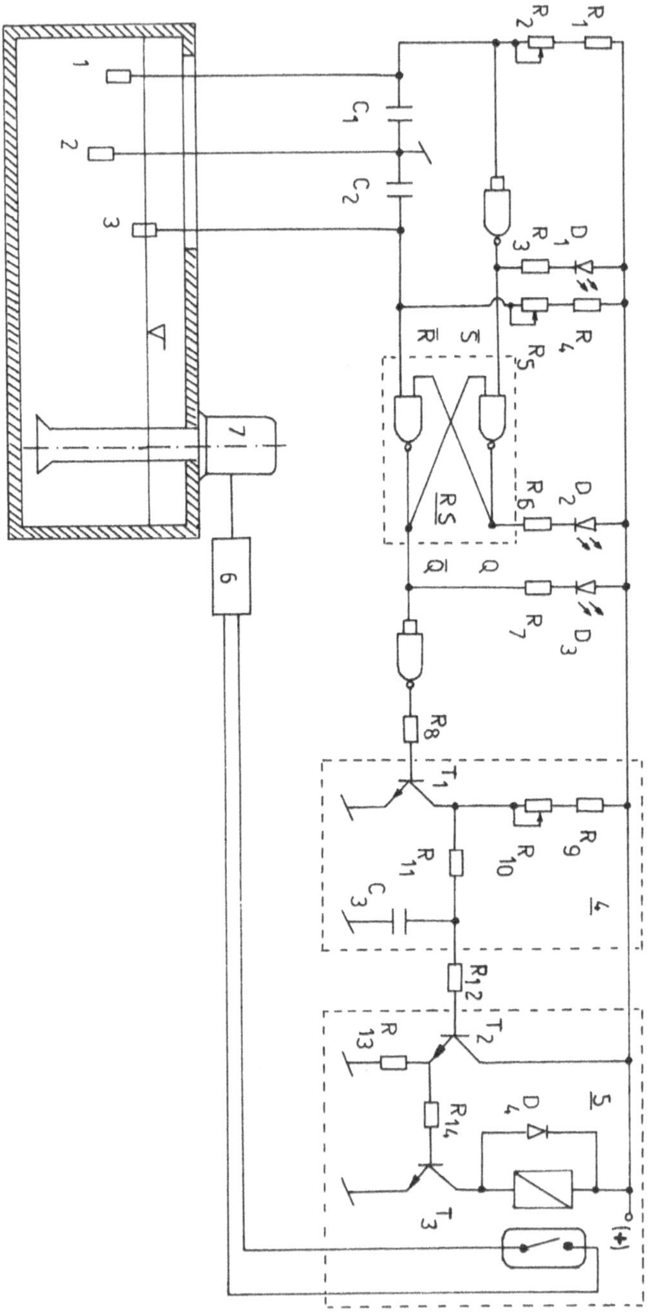

Figure 1. Diagram of the controller of pumping aggregates

A Method of Flood Routing in Ungauged Catchments with Particular Reference to Proposed Changes. A Case Study.

K.J. Riddell and I.S. Heijne
Partner and Senior Engineer, C.H. Dobbie and Partners, Croydon, U.K.

ABSTRACT

A method based on Muskingum techniques using variable coefficients has often been suggested in the past as being potentially as accurate as the best alternative methods for river flood routing, whilst offering significant advantages in speed of computation and simplicity. A method to predict variable coefficients from steady-flow backwater curve analysis has been developed which, when used in a Muskingum flood-routing routine, compares favourably against other techniques. It is suggested that such a method may even be applied using a programmable calculator in simple cases and may have benefits in more complex situations when assessing potential changes or where physical parameters can be defined but there is an absence of hydraulic and hyrological data. Test results are presented and a case study, corresponding to the latter situation mentioned above, is described.

INTRODUCTION

When faced with problem of flood routing along a relatively short reach of river, the standard methods of analysis did not offer the required flexibility or accuracy we required. It was therefore appropriate to assess the viability of the existing methods and conclude our own technique for applying these methods. A factor in our decision was the existence of a detailed and accurate topographic survey of the existing channel and flood plain.

The River Arrow Flood Study included the assessment of flooding on an existing river reach against the changes in flooding resulting from partial development of the flood plain. The paper indicates the methods of flood routing used in the study.

BASIS FOR ANALYSIS

Our method of analysis is based on the Muskingum flood routing technique as developed by G.T. McCarthy (1935) and modified by J.A. Cunge (1969).

The Muskingum Technique

By stating that the difference between inflow and outflow for a reach is equal to the change in reach storage.

$$\frac{dS}{dt} = Q_j - Q_{j+1} \qquad \text{...(1)}$$

where Q_j is Inflow rate to reach
Q_{j+1} is Outflow rate from reach

During the advance of a flood wave, the inflow will exceed outflow thus producing a wedge of storage. Conversely, during the recession of a flood wave the outflow will exceed the inflow resulting in a negative wedge storage. The storage volume of a reach when inflow equals outflow and dS/dt=0 is then described as prism storage.

The Muskingum technique requires two parameters to be determined which identify the storage characteristics of the reach. The first 'K' represents the storage volume available; the second 'x' represents the relative importance of the wedge storage to the prism storage.

$$S = K[xQ_j + (1-x)Q_{j+1}] \qquad \text{...(2)}$$

where K,x are Muskingum Coefficients
S is Storage volume of reach

From equations (1) and (2) it follows that

$$\frac{Kd}{dt}[xQ_j + (1-x)Q_{j+1}] = Q_j - Q_{j+1} \qquad \text{...(3)}$$

This equation can be rewritten in the finite difference form.

$$Q_{j+1}^{n+1} = C_1Q_j^n + C_2Q_j^{n+1} + C_3Q_{j+1}^n \qquad \text{...(4)}$$

where
$$C_1 = \frac{Kx + 0.5\Delta t}{K(1-x) + 0.5\Delta t} \qquad \text{...(5)}$$

$$C_2 = \frac{0.5\Delta t - Kx}{K(1-x) + 0.5\Delta t}$$

$$C_3 = \frac{K(1-x) - 0.5\Delta t}{K(1-x) + 0.5\Delta t}$$

The Muskingum technique then bases its determination of the parameters K and x on the analysis of previous flood events. This facet obviously limits the viability and accuracy of the method.

The Muskingum Cunge Technique

The improvement of the Muskingum method as developed by Cunge (1969) introduced the facility to calculate the parameters K and x. Cunge observed that the finite difference equations 3 and 4 are representations of the equation.

$$\frac{\partial Q}{\partial t} + \frac{\omega \partial Q}{\partial x} = \frac{\mu \partial^2 Q}{\partial x^2} \qquad \ldots(6)$$

By relating the constants and to the Muskingum parameters K and x it is possible to determine their values as follows;

$$K = \frac{\Delta x}{\omega} \qquad \ldots(7)$$

$$x = 0.5 - \frac{\alpha \bar{Q}p}{L\omega \Delta x} \qquad \ldots(8)$$

Where α = Attenuation parameter
L = Length of reach
Δx = Length of each subreach
ω = convection speed

The calculation of the attenuation parameter is discussed fully in the Flood Studies Report (1975).

Modifications to the above methods for our analysis

For the purposes of our study it was necessary to predict flood routing parameters by purely hydraulic means. The use of implicit numerical techniques was not appropriate and for this reason we investigated the relationship between the flood routing parameters K and x and the steady flow backwater curve analysis which was necessary in any case for the accurate prediction of flood levels at specific locations.

For the determination of variable parameters, consider ones viewpoint as moving along the reach at the speed of the flood peak. By ignoring the acceleration component associated with changing discharge, the flow can be considered to be steady subject to inflow/outflow equivalent to a lateral flow situation.

Parameters K and x are determined for a range of flowrates and therefore reflect the changing physical and hydraulic characteristics of the reach according to the nature of flooding during the flood event.

The concept of variable parameter Muskingum has been suggested in the past but to our knowledge has not been tested to date. This method incorporates the use of variable parameters and yields an improvement in accuracy compared with traditional methods.

Calculation of the Muskingum parameters K and x

Muskingum parameters are calculated for a set of discrete flowrates within the range of interest. Standard hydraulic details such as channel and flood valley shape, frictional losses, channel gradient will have already been determined so that a backwater profile may be determined.

For each discrete flowrate it is now necessary to carry out a set of four backwater analyses. These are:

	Inflow	Outflow	
1.	$Q_{I1} = Q + \Delta Q$	$Q_{O1} = Q + \Delta Q$	)...(9)
2.	$Q_{I2} = Q - \Delta Q$	$Q_{O2} = Q - \Delta Q$	)
3.	$Q_{I3} = Q + \Delta Q$	$Q_{O3} = Q$	)
4.	$Q_{I4} = Q - \Delta Q$	$Q_{O4} = Q$	)

Where ΔQ is small compared with Q.

In the backwater analysis where flowrate is changing along the reach, the change in flowrate is accounted for by using a constant inflow/outflow along the reach and using momentum as the criteria for calculation of the backwater profile.

From equation No. 2 the following applies:

$$S_1 = KQ_{O1} \quad \ldots(10)$$
$$S_2 = KQ_{O2} \quad \ldots(11)$$

Therefore $$K = \frac{S_1 - S_2}{2\Delta Q} \quad \ldots(12)$$

Also

$$S_3 = K[Q_{O3} + (Q_{I3} - Q_{O3})x] \quad \ldots(13)$$
$$S_4 = K[Q_{O4} + (Q_{I4} - Q_{O4})x] \quad \ldots(14)$$

Therefore $$S_3 - S_4 = K(Q_{I3} - Q_{I4})x \quad \ldots(15)$$

$$x = \frac{S_3 - S_4}{2K\Delta Q} \qquad \ldots(16)$$

where S_i is the storage volume for backwater profile i
K,x are the Muskingum parameters for flowrate Q.

By carrying out this approach on a series of subreaches and various flowrates a set of Muskingum parameters may be determined.

These values are then used in the finite difference equation (equations 4 and 5)

Determination of Backwater Profiles

For the calculation of the backwater profile a computer program was expressly developed. The program employs a momentum and flow force approach to account for variations in width and depth of the channel and valley. Energy loss due to friction is calculated by Manning's equation and is included as a friction force term, lateral inflow/outflow is accounted for by an increase/decrease in the momentum term.

The use of actual survey data rather than idealised typical cross sections, enables alterations or additions to be made readily to reflect proposed works such as construction or removal of embankments or channel diversions.

The cross sections also take account of differential flow conditions by identifying a set of slices. Although the program assumes one dimensional flow, the identification of a set of slices and the inclusion of a meander factor allows the user to identify the primary channel and make allowance for overland flow along the flood valley with appropriate values of Manning's 'n' in different slices.

The program was further developed to calculate the storage beneath the profile and automatically determine the four backwaters and the calculation of K and x parameters for a selected flowrate.

For the purposes of our investigation, it was necessary to use an advanced method of backwater profile analysis. However, it should be noted that the complexity of the backwater calculations is independent of the method of calculation for the Muskingum parameters. Providing the backwater calculation can take account of lateral inflow/outflow and the total volume of storage can be determined from the backwater then the data is available for calculation of the Muskingum parameters K and x.

Comparison of results with alternative methods

For the purposes of comparing the results of our method against existing published methods, it was decided to route a synthetic flood in a uniform rectangular channel, 100 km long and 50 m wide with a Manning's roughness coefficient 0.035 and a bed gradient of 1×10^{-3}. The synthetic flood hydrograph at the upstream section of the reach is defined by

$$Q(t) = Qbase + Qamp\left[\frac{t}{tp}\exp\left(1 - \frac{t}{tp}\right)\right]^{\beta} \qquad ...(17)$$

where β = 16
Qbase = 100 cumecs
Qamp = 800 cumecs
tp = 24 hours

The Flood Studies Report (1975) makes available the results of routing this synthetic hydrograph for the following methods.

(i) Leap Frog finite difference scheme.
(ii) Muskingum-Cunge.
(iii) Linear diffusion.
(iv) Variable parameter diffusion.

Using our method, the channel was divided into 10 subreaches each of 10 km length and each having the same flood routing parameters. A set of Muskingum parameters K and x were calculated as follows.

OUTFLOW Q (cumecs)	MUSKINGUM K	PARAMETERS x
100	5118	.439
200	3944	.412
300	3437	.385
400	3112	.364
500	2893	.345
600	2728	.329
700	2599	.313
800	2494	.299
900	2416	.285

Using a time step of 30 minutes the results were almost identical to those shown for the Leap Frog and the Variable Parameter Diffusion Method. See Figure 1 and Table 1.

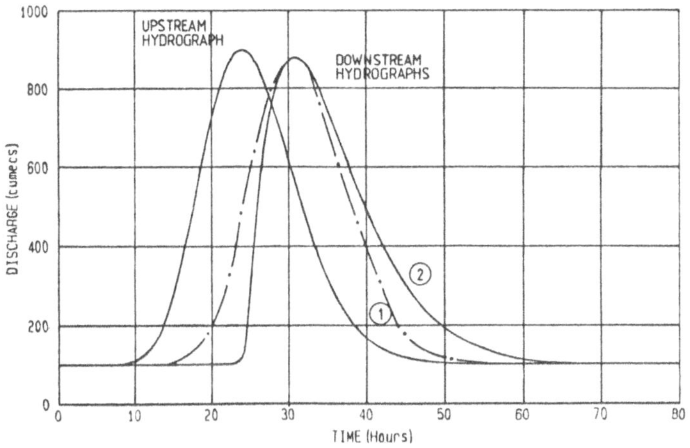

(1) Muskingum - Cunge and Linear Diffusion.

(2) Our variable Muskingum method; the Leap Frog method and the Variable Parameter Diffusion method.

FIGURE 1: Comparison of hydrographs for a regular channel.

Method	Subreach Length (Km)	Time Step (sec)	Error in Peak Discharge (%)	Error in Speed of Flood Peak (%)
Proposed	10	1800	0.09	2.50
MC	10	3600	-0.03	-1.57
LD	5	720	-0.08	-1.07
VPD	5	720	-0.02	0.02

TABLE 1: Comparison of errors in predicted hydrographs for a regular channel.

Based on the quality of the results of this comparitive test and the ease of application to a range of situations, we consider this method to have much potential

THE RIVER ARROW FLOOD STUDY

The methods described have been successfully applied to the flood study of the River Arrow. The area of interest was a four kilometre section where the possible construction of road

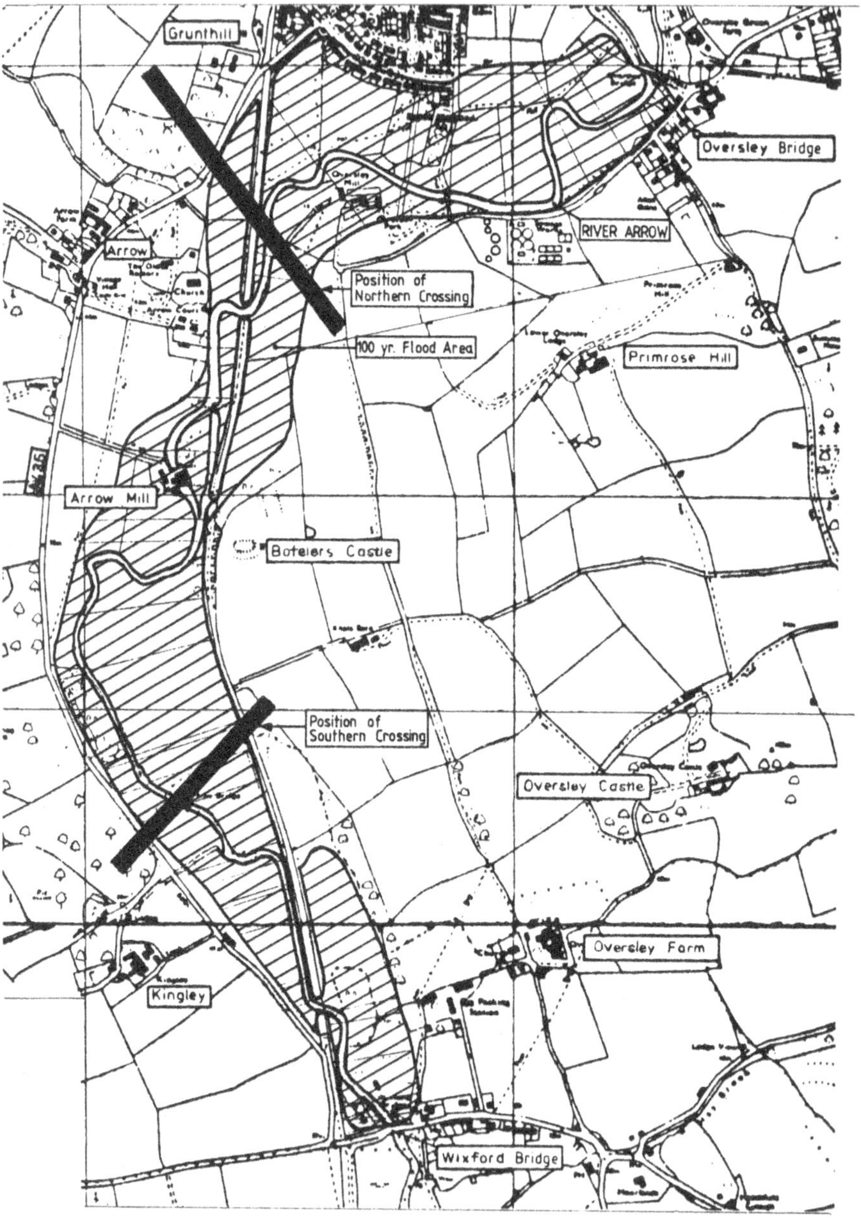

FIGURE 2: Plan of Study Area showing
proposed embankments.

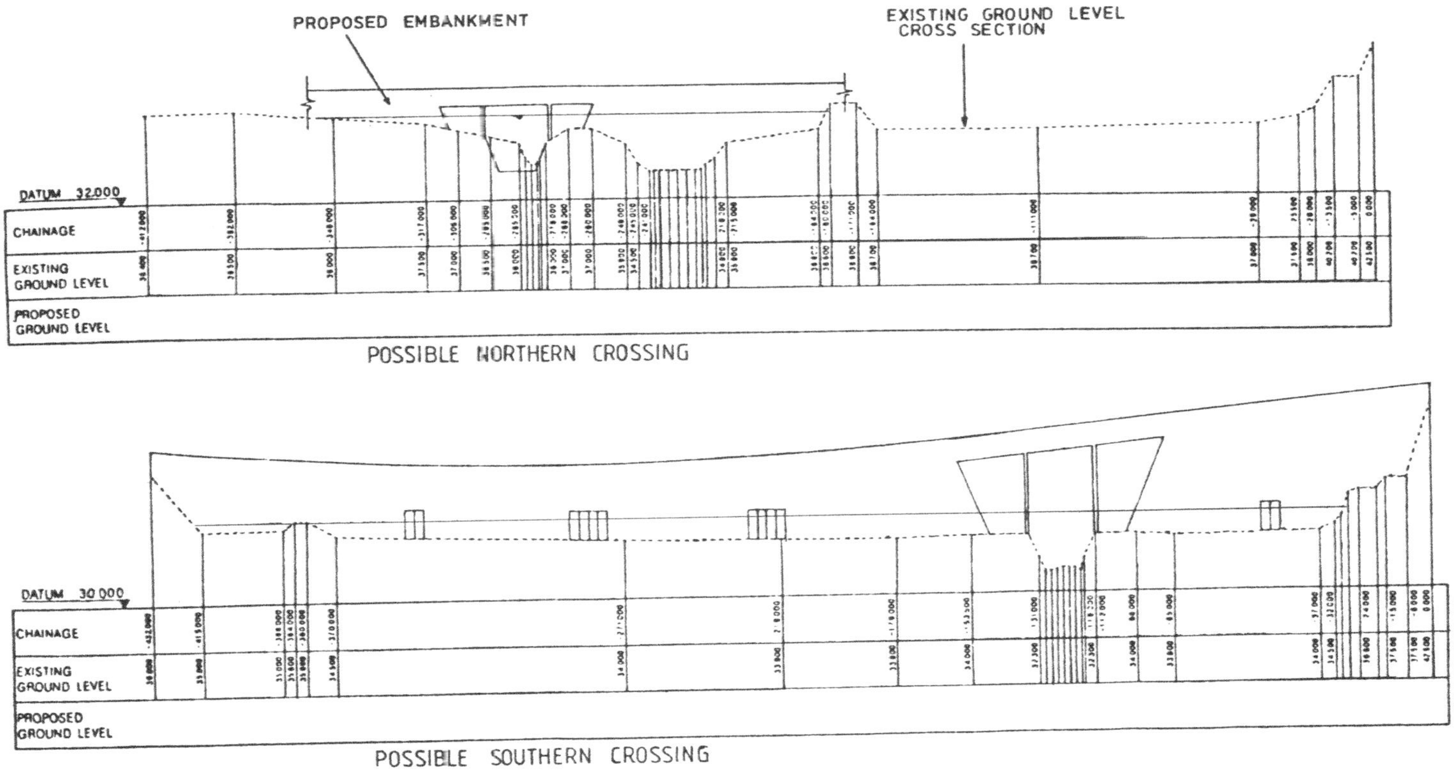

FIGURE 3: Typical channel and flood plain cross sections including proposed embankments.

embankments on the flood plain was being considered. A plan of the river showing the position of the embankments is shown in Figure 2. Typical valley sections and possible embankment solutions are shown in Figure 3.

For the study we were equally interested in the actual flood levels along the river as we were in the attenuation of the peak flow due to the flood plain. The conditions found suggested a relatively high proportion of flow along the flood plains. It was therefore necessary to carry out a sophisticated backwater analysis.

The comparitive assessment of flood attenuation for the present conditions as compared with potential solutions was carried out by assigning eleven subreaches along the reach. For each subreach K and x parameters were calculated for a range of flowrates. Typical results of this procedure are shown in Figure 4. It is of interest to note the distinct change in variables as bankfull capacity is exceeded.

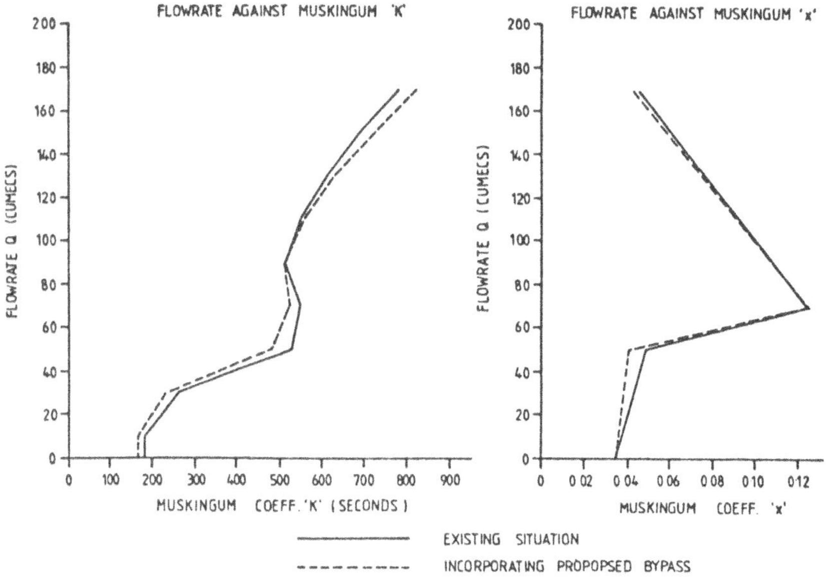

FIGURE 4: Example of variations in 'K' and 'x' with depth .

Flood hydrographs were developed according to Flood Studies Report Methods for a range of flood events. The flood routing procedure was then carried out in sequence through all the subreaches.

Inclusion of proposals was achieved by making the relevant adjustments to the river cross sections and re-calculating the flood routing parameters. The method allowed for easy modification of cross sections and therefore facilitated the analysis of a variety of alternatives. A typical output of the flood routing is shown in Figure 5.

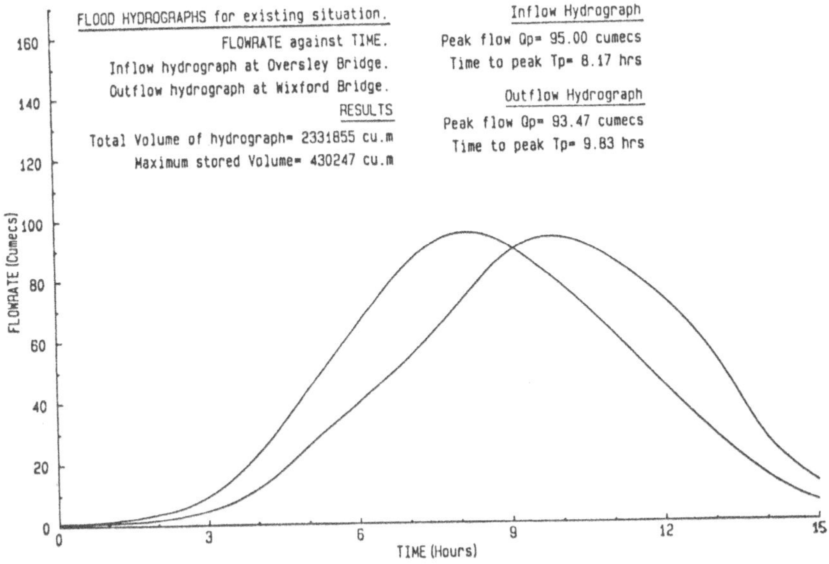

FIGURE 5: Typical output of routed hydrograph.

By development of the proposed design, it was possible to conclude a solution which offered the optimum trade off between loss of storage area and increased afflux at the constrictions.

CONCLUSION

This technique does not supersede the existing traditional and advanced methods in existence, it does, however, make available a suitable method of flood routing analysis or flow attenuation prediction, to engineers who may otherwise be wary of the time and cost involved in the modelling of unsteady flows, especially if this exercise is regarded as peripheral to another major activity.

The method is easily applied to a situation where the engineer has limited hydraulic and hydrological data but nevertheless requires an indication of the flow attenuation achieved.

Because the method can be based on a straightforward backwater profile analysis, the user would give a typical cross section and hydraulic details to establish the flood routing parameters.

There is no method constraint applied and therefore the criteria for assessing the hydraulics can be made appropriate to the problem, e.g. The method would be equally suited to the use of Colebrook-White for the calculation of flow resistance as the use of Manning's equation.

In the same way, the application of the method is equally general. As a tool to land drainage problems, the ability to consider flood attenuation is a great advantage. The effect of proposed major land drainage works on run-off to a receiving water course maybe assessed from purely physical characteristics rather than the statistical techniques used previously (p.131 ICE 1981) However, the method can also be applied to pulse flow problems, e.g. intermittant pumping from pumping stations or the variation flow through a sewage treatment works.

As a tool available to engineers faced with varied flow problems, this method offers a quick and efficient method of predicting peak flow attenuation and travel time. With the use of more advanced backwater curve analysis the method is capable of being used for the comparison of alternative strategies in complex situations.

ACKNOWLEDGEMENTS

The work described was undertaken by C.H. Dobbie and Partners, Consulting Engineers to The Department of Transport, West Midlands Regional Office.

The authors wish to thank all persons who assisted in the preparation of this paper.

This paper is given by permission of the Director (Tp) West Midlands Region, Department of Transport.

REFERENCES

1. Natural Environmental Research Council (1975) Flood Studies Report Parts I - IV. N.E.R.C. London.

2. Chow, V.T. (1964) Handbook of Applied Hydrology.

3. Cunge, J.A. (1969) On the subject of a flood propagation computation method (Muskingum method) Journal of Hydraulic Research, Vol. 7, No. 2.

4. Shepherd, K (1979). Some applications of flood routing methods with particular reference to small streams, Kington Polytechnic, Kingston.

5. Institution of Civil Engineers (1981) Flood Studies Report - Five Years On.

6. C.H. Dobbie and Partners (1984) Report on a Flood Study of the River Arrow in the region of the proposed crossings by the Alcester Bypasses. Unpublished.

7. C.H. Dobbie and Partners (1972) Crawley/Gatwick sub-regional study. Unpublished.

Design Study Main Drainage System Markerwaard

E. Schultz
IJsselmeerpolders Development Authority
G. Severs
Rijkswaterstaat, Zuiderzeewerken Directorate
G.A. Ven
IJsselmeerpolders Development Authority

INTRODUCTION

The Zuiderzee project was founded by Dr. C. Lely. His plan was published as an act in 1918 and comprised (figure 1):
- the construction of a 32.5 km barrier dam from North Holland to Friesland;
- the reclamation of the polders Wieringermeer, Northeastpolder, Eastern and Southern Flevoland and the Markerwaard.

The enclosure and partial reclamation of the Zuiderzee was undertaken for four reasons:
- to increase protection against flooding;
- to improve watermanagement in the adjacent 'old' land;
- to enlarge the area of farmland to increase food production;
- to improve the traffic connection with the northern part of the country.

In accordance to the latter, the Wieringermeer and the North-east-polder are mainly used for agriculture. The growing shortage of space in Netherlands and the fact that Flevoland is situated close to the densely populated western part of the country, have brought about change in the original ideas about land use. Urban development, recreation and nature reserves are making increasing demands on space. The land use in the realized IJsselmeerpolders is given in table 1.

Table 1. Land use in the IJsselmeerpolders as a percentage of the total area.

	Wieringer-meer	Northeast-polder	Eastern Flevoland	Southern Flevoland
Farmland	87	87	75	50
Residential areas	1	1	8	25
Woods/nature reserves	3	5	11	18
Infrastructure	9	7	6	7
Total area (ha)	20,000	48,000	54,000	43,000

At the moment it is intended to reclaim the last polder the Markerwaard, with an area of 41,000 ha. Ahead of the final decision studiés have been carried out in order to design a drainage masterplan for the polder. As the execution of the work starts with the construction of the dikes and main drainage system, the plan is worked out in more detail in relation to:
- the alignment and dimensions of the dikes;
- the water levels in the main drainage system;
- the alignment and dimensions of the primary drains;
- the location and capacity of the pumping station(s) and ship lock(s).

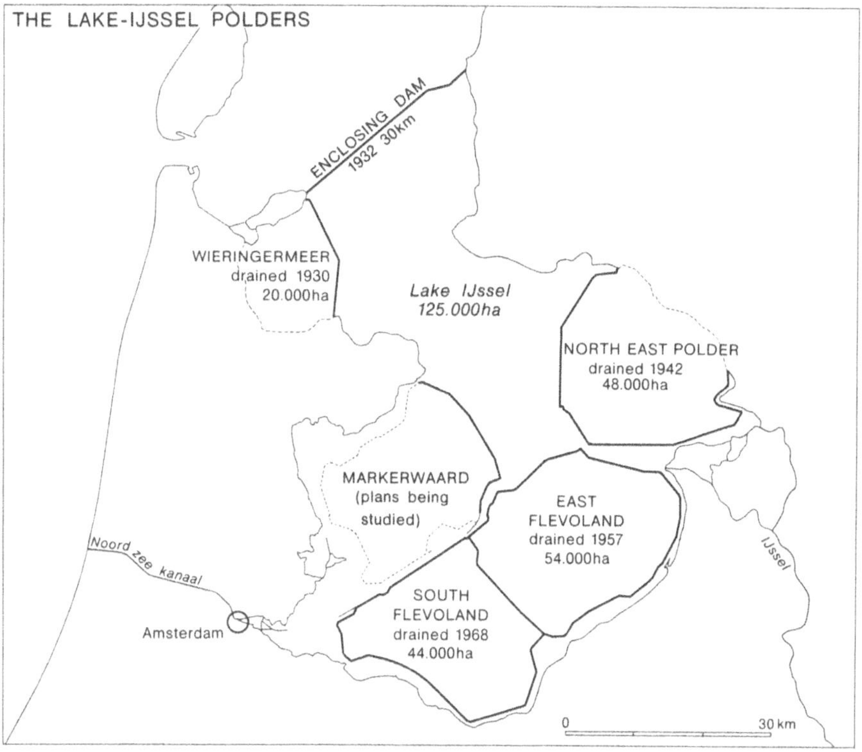

Figure 1. The Zuiderzee project.

In this paper emphasis is given to the design of the main drainage system. The studies are carried out by the two authorities belonging to the Ministry of Transport and Public Works, which are responsible for the project:
- Rijkswaterstaat, Zuiderzeewerken Directorate;
- the IJsselmeerpolders Development Authority.

DRAINAGE SYSTEM IN THE IJSSELMEERPOLDERS

Efficient drainage in the IJsselmeerpolders is ensured by the following components of the drainage system (figure 2):

- open field drains, after some years replaced by subsurface pipe drains;
- ditches along the long sides of the rectangular parcels to collect water from the pipe drains;
- secondary drains (main ditches) to collect water from the ditches;
- primary drains (canals) to transport the water towards the pumping stations;
- pumping stations.

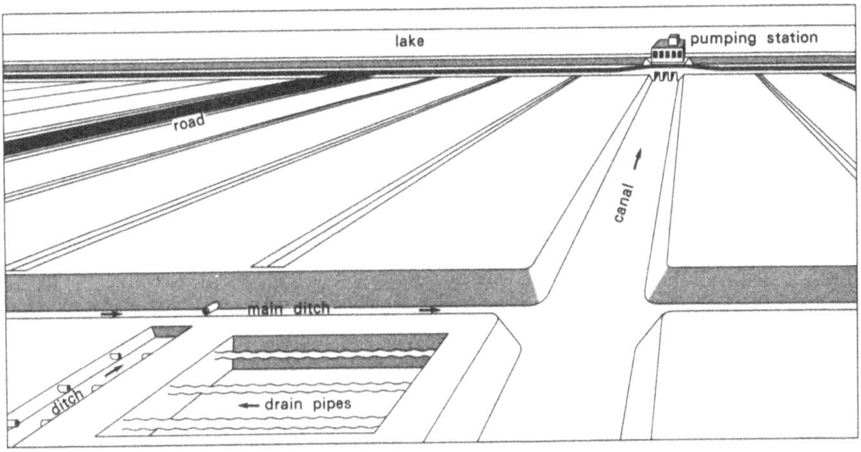

Figure 2. Schematic lay-out of the drainage system in the IJsselmeerpolders.

Pumping stations, which pump the polders dry and subsequently keep them drained, are built at the same time as the dikes. The water levels in the drainage system (polder water levels) are determined by the degree of drainage required for the different uses of the land. Depending on the surface elevation the polders are divided in two or more sections with different polder water levels and pumping stations. These polder water levels vary from 4.50 to 6.20 m below water level in the surrounding lake which is about mean sea level (m.s.l.). The primary drains are navigable and dredged while the polder is still under water.

Main characteristics of the drainage system in the existing polders are listed in table 2.

The main part of the seepage is intercepted by the open drains, adjacent to the dikes. The seepage is almost constant over the year and is estimated from waterbalance data. Depending on the

Table 2. Main characteristics of the drainage system in the existing polders.

	Wieringer meer	Northeast polder	Eastern + Southern Flevoland
Total area (ha)	20,000	48,000	97,000
Average seepage (mm/d)	0.7 - 1.2	0.9	1.0
Number of sections	3	2	2
Number of pumping stations	2	3	4
Open water storage (%)	1.8	1.0	1.2
Pumping capacity (mm/d)	14.7	14.0	11.7

geohydrological conditions the seepage water is locally to salt, thus affecting the quality of the open water in the polder. Also the quality of the water in the IJsselmeer is affected in a negative way, because of the salt load of the pumping stations. The percentage of open water storage is determined at polder water level. By exceedance of the polder water level the open water storage increases with about 0.1 % per 0.1 m exceedance.

Because of the climatic conditions, the type of land use (mainly arable land) and the soil texture (mainly loam to clay) irrigation is far less important than drainage. Water for sprinkler irrigation can be delivered by the open drains, as far as the quality permits. Because of the continuous seepage, intake of water to maintain the polder water level in dry periods is not necessary. An exception is found in the Northeastpolder where light sandy soils are present, which have water deficiency during the growing season. In an area of about 10,000 ha a separate distribution system for water supply has been constructed. The pipe drainage system is also used to maintain ground water levels at the desired depth for the uptake of water by the plant roots (subsurface irrigation).

DRAINAGE SYSTEM MARKERWAARD

Physical conditions

For the design of the main drainage system (primary drains and pumping stations) of the Markerwaard the following relevant information was available:

- the total area to be reclaimed is 41,000 ha.
- the main type of land use will be agriculture (> 50 % of the total area). Depending on the soil suitability arable farming, dairy farming, horticulture and fruit growing are planned. Considerable parts will be used for afforestation and the creation of nature reserves.

- there is a large variation in the texture of the top soil. In the northeastern part medium fine to locally coarse sand is found. In the southwestern direction the texture changes gradually into loam and clay. Locally old marine clay and peat is present.
- based on calculations with computermodels for ground water flow (finite elements method) the average seepage after reclamation is estimated at 0.7 mm/d.
- in the northern part of the Markerwaard the deeper groundwater has a chloride content ranging from 1000 to 3000 mg Cl/l. In the southwestern part 100 to 1000 mg Cl/l is found.
- the elevation of the lake bottom at emergence ranges from about 2.0 to 4.8 m - m.s.l. Because of ripening and dewatering of the loosely packed marine sediments a considerable subsidence will occur after emergence. On base of the available data an accurate prognosis can be given for the subsidence. For the design of the drainage system the surface levels 80 to 100 years after emergence are taken as decisive.

Boundary conditions for the design

Based on information and practice in the existing IJsselmeerpolders the following items were taken into consideration:

- for agriculture the minimum required depth of the subsurface pipe drains at the outlet is 1.10 m below soil surface. For fruit orchards this depth is 1.30 m. To avoid frequent submerging of the subsurface drain outlets the polder water level should be at least 1.40 m below soil surface. A freeboard of 0.30 m is sufficient to account for flow resistance and temporary higher water levels due to wind effects and rainy periods;
- the acceptable exceedance frequency for the polder water level in the main drainage system is:
  0.40 m once in 2 years
  0.65 m once in 10 years;
- the maximum flow velocity in the open drains is 0.25 m/s to prevent erosion;
- the primary drains should be navigable for ships up to 1350 tons. This means at minimum a wet cross section of 145 m2, a water depth of 3.65 m, a bottom width of 30.4 m, side slopes 1:3 and a water line width of 52.3 m;
- the pumping station(s) should be located at the south eastern dike of the polder, because of the salt load. From here the water pumped out can be discharged via Amsterdam to the Northsea. By this negative effects on the quality of the fresh water in the surrounding lakes can be minimized;
- because of the uncertainties in the final land use and allocation of the types of land use the drainage system can be designed for agriculture, with respect to the polder water level and pumping capacity. This gives enough flexibility for the final land use planning.

The polder water levels

It can be concluded that about 25 % of the Markerwaard will have an elevation between 5.0 and 5.2 m - m.s.l. Therefore the main polder section should have a polder water level of 6.6 m - m.s.l. (figure 3). The deeper part of the polder should have a

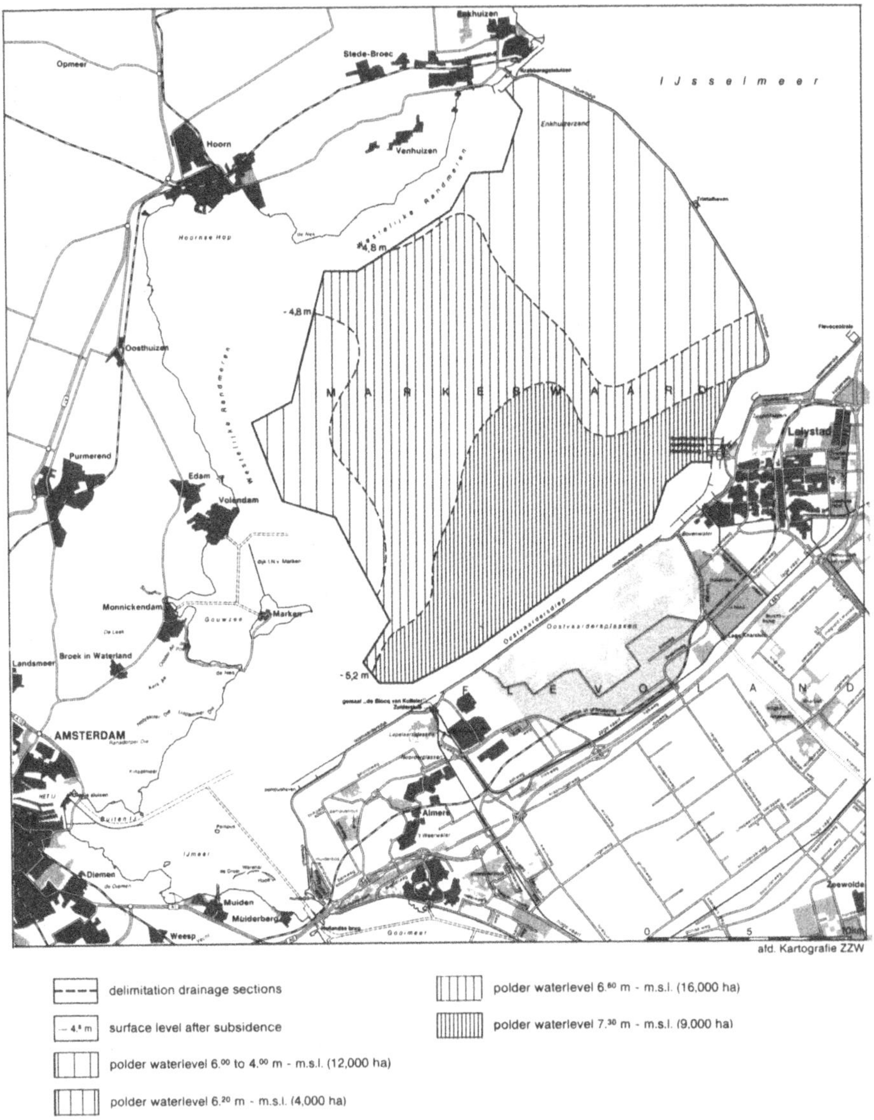

Figure 3. Main poldersections in the Markerwaard.

polder water level of 7.3 m - m.s.l. The higher elevated part in the west may get a level of 6.2 m - m.s.l. to reduce the excavation costs of the open drains. In the higher elevated north-eastern part of the polder with light loamy to sandy soils different polder water levels are required. Depending on the final land use intake of water will be necessary, especially for agriculture. In this area the water levels during the growing season will be about 0.40 m higher than in winter time. This higher elevated areas can discharge the excess water by gravity to the main polder section at 6.6 m - m.s.l.

Pumping capacity

The required pumping capacity to maintain the water level in the main drainage system within the limits mentioned before, depends on:
- climatical conditions;
- soil conditions;
- land use;
- drainage conditions;
- seepage;
- storage capacity in the open drains.

For this study data have been used of the Northeastpolder. The size of this polder and most of the physical conditions are quite similar to those of the Markerwaard. For the Northeastpolder time series of daily amounts of water pumped out by the pumping stations (outflow) are available. To find the daily inflow of excess water into the main drainage system, the daily changes in open water storage are added to outflow data. This is possible by using water level recordings at different points in the primary drains and the known relation between water level and open water storage. The inflow data are corrected for seepage and intake of water from outside the polder. This results in a time series of daily net inflow data over a period of 21 years.

With these data a frequency analysis with Gumbel's extreme value method has been applied. This results in a frequency of exceedance of net inflow as listed in table 3.

Table 3. Frequency of exceedance of net inflow in mm of the Northeastpolder, period 1959 - 1979.

	Frequency		
Days	once in 2 years	once in 10 years	once in 100 years
1	10	19	28
2	17	28	40
3	22	34	50
4	26	40	58
5	29	45	66
6	33	51	75
7	36	56	82
8	38	60	88
9	41	64	94
10	44	68	100

The data from table 3 have been used to find the required pumping capacity for the Markerwaard by using the following water balance equation:

$I + S = Q + B$

I = net inflow of excess water into the open drains (mm)
S = seepage inflow (mm)
Q = discharge of the pumping station (mm)
B = storage capacity in the open drains (mm)

The calculations have been executed for different durations and frequencies once in 2 and 10 years. For S a value of 0.7 mm/d was taken. The storage capacity in the open drains was based on 1 % of open water at polder water level, increasing to 1.4 and 1.65 % at a rise of the water level to 0.40 and 0.65 m. A percentage open water of 1% was considered to be a minimum in order to realize the required discharge capacity and for navigation. A higher percentage was not economic because of land losses. The calculated net pumping capacity was 11.9 mm/d or 57 m3/s. The frequency of once in 10 years was decisive. The calculated capacity is based on 100 % reliability and direct switching on of the pumps in wet periods.

<u>Lay-out main drainage system</u>

For the lay-out four alternatives have been subject of study (figure 4). The alternatives have been compared with respect to:
- water management aspects;
- investment and maintenance costs;
- vulnerability (demolition, sabotage);
- navigation and location of ship locks;
- distribution of brackish seepage in the polder.

Taking all these factors into consideration preference was given to one pumping station (alternative 4 on figure 4). The pumping station will be equipped with common type of pump units with a capacity of 11.5 m3/s each. To realize the net pumping capacity of 57 m3/s five units will suffice. To have a safety margin and to compensate in case of maintenance six units with a total capacity of 69 m3/s or 14 mm/d will be installed. This capacity is in fair agreement with an evaluation of the water management in the existing polders. Two units will be used for the lowest polder section (9000 ha) with a polder water level of 7.3 m -m.s.l. One of these two units can also be used to assist the other four units to maintain the water level of 6.6 m-m.s.l. As energy sources both electricity and diesel will be used. Detailed studies have proved that the use of wind energy for pumping is not compatitive with conventional energy at current prices.

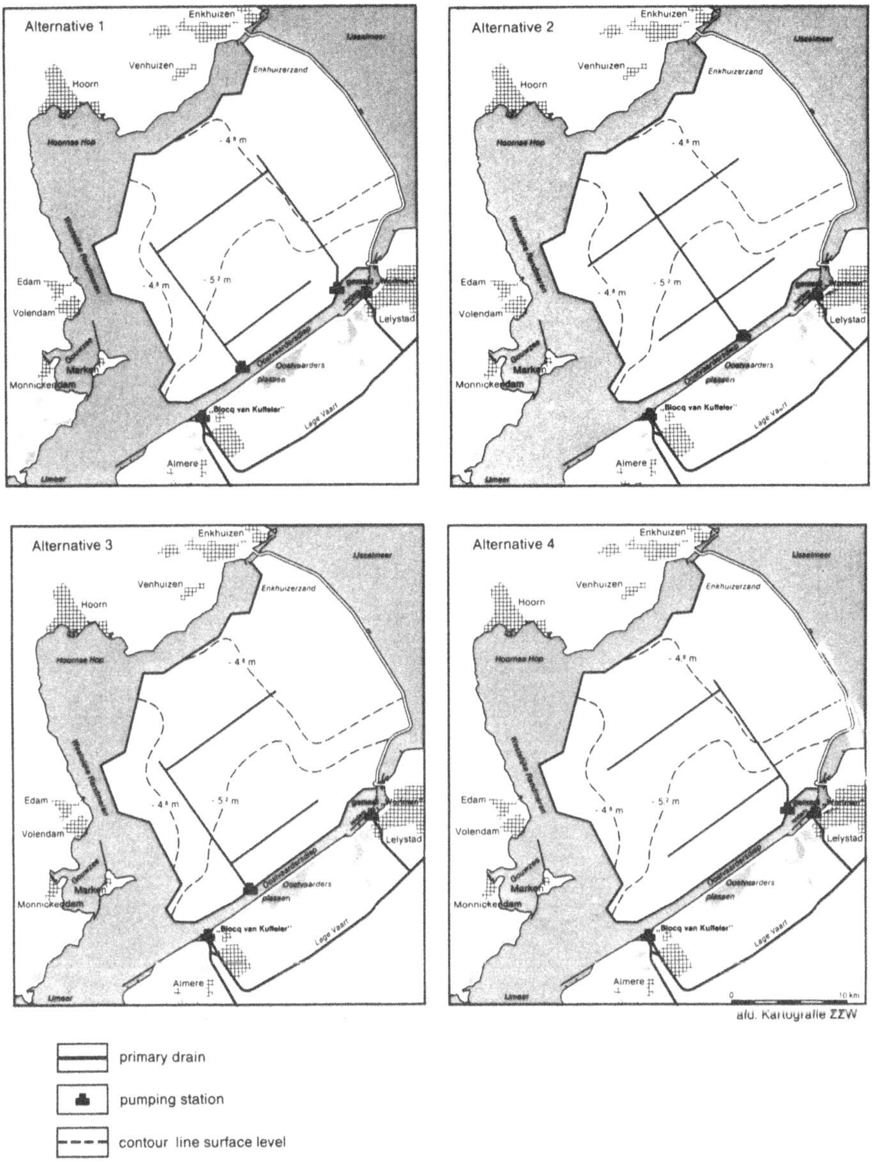

Figure 4. Alternative lay-outs for the main drainage system of the Markerwaard.

In figure 5 the dimensions of primary drains necessary to realize the required discharge capacity are indicated. Dimensions are determined by the maximum flow velocity at the design discharge of 0.25 m/s. Requirements for navigation have to be superimposed on the given dimensions in figure 5.

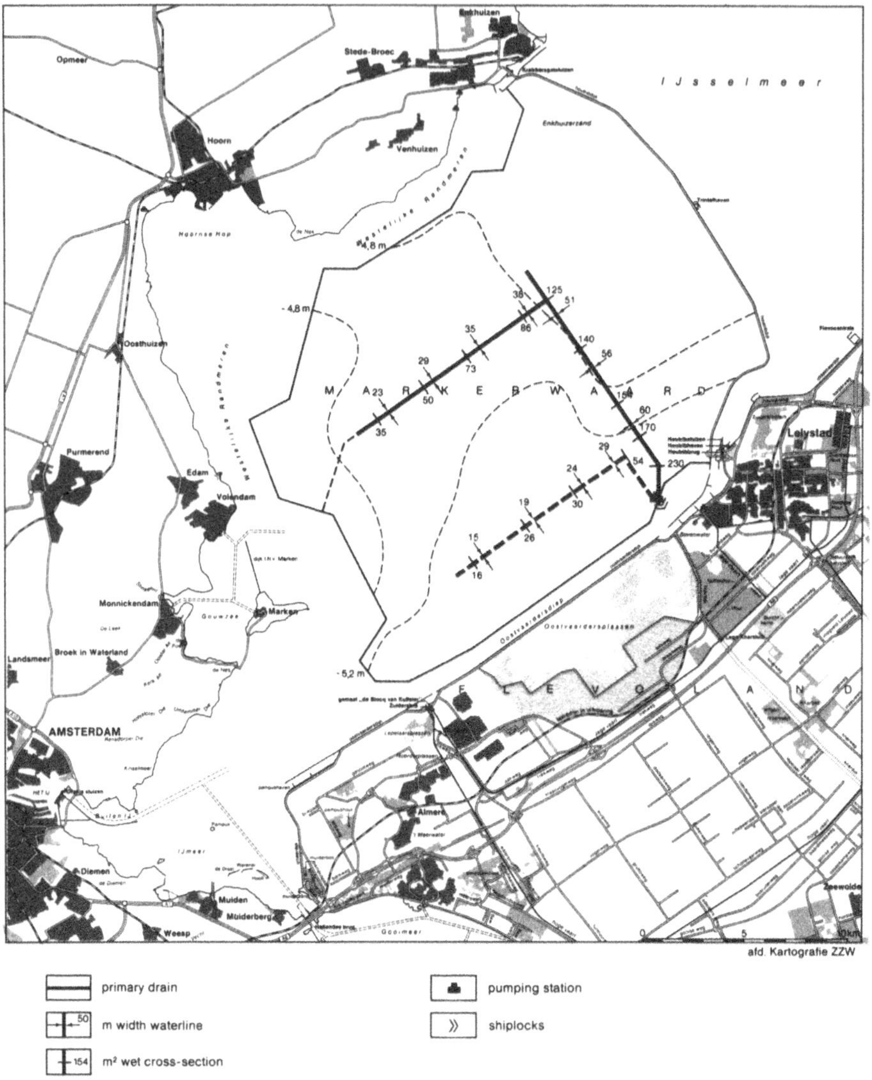

Figure 5. Minimum dimensions of primary drains.

ABSTRACT

The Zuiderzeeproject comprises the reclamation of polders with a total area of over 200,000 ha in the central part of the Netherlands. To complete the project it is intended to reclaim the last polder Markerwaard (41,000 ha) in the near future. Agriculture will be the most important type of land-use. Also afforestation, recreation, nature reserves and human settlements will play an important role. The paper summarizes the characteristics of the main drainage system in the existing polders.

For the Markerwaard the structure, dimensions and capacity of the main drainage system have been subject of study. Time series of daily inflow of excess water into the open drains have been derived from historical data of an existing polder. These daily figures have been converted to frequency exceedance of inflow with Gumbel's extreme value method. After setting criteria for the acceptable rise in water level in the open drains, the required pumping capacity could be established.

Land Drainage and Flood Prevention in the Central Somerset Lowlands with Particular Reference to the South Drain Scheme Stage III.

N.P. Harrison
Watson Hawksley

INTRODUCTION

The West Country is fortunate in the geological formations making up the Somerset lowlands, and mans efforts to drain these areas, now make a visual patchwork of fields, rhynes, trees and hedges presenting a fascinating area of study for conservationists and engineers alike.

The main body of the Somerset levels is flanked on the north by the Mendip Hills, on the south-west by the Quantock Hills and to the south by the Blackdown Hills and rising ground to the south-east. Past erosion of the Triassic and Jurassic rocks have produced an uneven surface of generally reducing level towards the Severn Estuary. This erosion has left prominent ridges of Curry Rivel, Stathe and the Polden Hills and the 'islands' of Brent, Glastonbury, Meare, Burtle, Wedmore and many others. Within the basins, between the ridges, post glacial deposition has produced the basic land as we know it today. The depositions have been laid down in four principal stages, firstly gravels, sands and clays fill the valleys to approximately 0.0m AOD, secondly peat then covers the greater part of the lowland, thirdly the peat is overlayed by aluvial clay adjacent to the natural river courses and other areas of habitual flooding. Lastly there is a belt of marine clay laid down along the coastline of up to 8 km in width.

The peat deposits vary in depth from under 0.5 m in parts of the Axe valley to up to 7m in the Brue valley. The general level of the moors is between 3m and 4m AOD but parts of the South Drain area are only 2m AOD. The coastal clay belt averages 5.5m to 6.5m AOD which presents a barrier through which the rivers must pass.

The principal rivers that pass through these low lying moors are the Parrett, Tone, Isle, Yeo in the south and the Brue, Axe and Congresbury Yeo north of the Polden Hills. The rivers have very flat gradients such as 1:6000 for the River Parrett and 1:7500 for the River Brue in the lower reaches. Due to a tidal range of up to 12m in the Bristol Channel and the High Spring Tides of 6.8m AOD the rivers have been embanked through the levels. Drainage, in the past, has been strictly tidal with periods of tide lock of up to 4½ hours.

These physical features combined with an annual rainfall of 1000mm over the hills have led to extensive flooding problems.

DRAINAGE HISTORY

Due to the adverse conditions in the levels settlement was very slow to occur. Several neolithic lake villages have been discovered on the fringes of the levels but little in the way of organised settlement occurred until the establishment of the early Christian Church at Glastonbury in the fourth century and later at Athelney and Muchelney. The influence of these ecclesiastical estates grew such that by the 1300's most of the Somerset levels were under the control of the church. As sole owners of these vast and wealthy estates the church was in the forefront of all drainage activity undertaken during the subsequent centuries.

The river system has been subject to many improvement schemes over the last 800 years some of which have been beneficial some less so. Early schemes concentrated on the River Brue, the excavation of new cuts, embanking sections, the draining of Meare pool and eventually the erection of tidal doors at Highbridge in about 1485. Apparently little in the way of further improvements was carried out until the 1600's when the disastrous flood of 1607 struck Somerset. An extreme tide inundated the Brue valley to a depth of 4m, the floodwater reached the foot of Glastonbury Tor 22 km inland. Many villages were flooded and countless numbers of livestock were lost together with houses and ricks. A record of the flood quoted the following; " a co'pany of Hogs and Pigs, being feeding upon one on the ricks perceiving it to go away more and more fro' the', they got up to ye top, and there maintained the' in eating. Nay which is more strange, conies in great numbers being driven out of their burroughes by the tyde, were seen to sit for safety on the backs of sheepe, as they sworm up and down and at last were drowned with them ".

This was a tidal flood but one which illustrates some of the problems of the Somerset levels. Following this event extensive improvements were made to the sea defences with the aim of preventing a repeat. Reclamation and drainage continued, over the next 150 years, around the perifery of the moor in a rather piecemeal manner. It was not until the Drainage Acts of the 1770's to 1830's that interest was rekindled and the finance found.

The schemes for the Brue Valley resulted in the draining of Westhay Godney, Mark and Glastonbury Moors and the cutting of a rhyne that was to become the North Drain in later years. The initial results were good, many thousands of acres previously waterlogged and of little use became fine grazing and dairying land. The success was short lived due to to the flood of 1794 when the whole of the valley was flooded. The principal cause was the inadequate capacity of the major watercourse the River Brue and its outfall at Highbridge Clyse. These defects resulted in high water levels in the Brue during periods of heavy rain or tide lock which then overspilled into the surrounding moors.

Following much discussion a new outfall was constructed at Highbridge in 1802 in a new cut with a cill level some 3m lower than the old clyse and several sections of the river straightened. Despite the expenditure of some £60,000 (1802) the Brue scheme was not a success, the outfall doors leaked, the lower channel suffered sever siltation problems and the river channel still proved to be inadequate.

The South Drain was first cut in 1804 to drain the moor between the Brue and the Polden Hills. Due to its connection to the Brue some 6 km inland drainage could only occur when the level in the Brue was low, in fact often the level in the Brue was such that the water flowed back down the South Drain and into the moors.

To sum up by 1830 a great deal of work had been carried out but the results were disappointing. Many kilometers of drain had been cut but the area suffered continually from both flooding and inadequate maintenance. The only positive achievement was the establishment of the pattern of drainage channels that was to form the basis of all subsequent activity. The Commissioners of Sewers realised that the deficiencies of the river system could only be overcome by extensive works to the river Brue, the channel required both widening and deepening particularly through the coastal clay belt.

Both the North and South Drains remained a problem to which there was no agreed solution other than pumping. The Commissioners considered procuring an Act for improving the drainage but this was not pursued because of the advent of the Glastonbury Canal Scheme which offered the opportunity of combining both drainage and navigation.

A new outfall and lock were constructed in the old river channel at Highbridge, the canal then followed first the River Brue, along Cripps river then the South Drain to Ashcott corner. Here it branched off in a new channel through Sharpham Heath and crossed the River Brue by an aqueduct before it reached Glastonbury.The promoters of the canal had assumed that by penning the water at Highbridge sufficient depth would be available for navigation to Glastonbury. The rise in ground level leading to Glastonbury was found to be greater than expected,therefore,locks were constructed at both Shapwick and Maze Wall and the banks of the canal raised. Navigation was successfully completed but not for long due to the unforseen problems in the peat area of Shapwick Heath. Due to the high penning levels in the canal the whole drainage system in the area was rendered practically useless, also the peat, through which the canal had been cut, became waterlogged and swelled. Eventually conditions became so bad that navigation practically ceased and in 1848 the canal was purchased by the Bristol and Exeter Railway Company, a track was laid alongside the redundant canal, the locks were removed and once again drainage had priority over navigation.

No real improvements in the area were realised for nearly a further 100 years until the second World War.

THE SOUTH DRAIN SCHEMES Figures 1 and 2

The South Drain basin is approximately 18 km in length and 6.5 km in width. The catchment area is 11700 hectares of which 64% lies below High Water Ordinary Spring Tides. The main drain leading to Gold Corner Pumping Station is 15 km in length. The water is then lifted into the Huntspill River and then gravitates 8 km through the coastal belt to outfall into the Parrett Estuary.

Gold Corner Pumping Station and the Huntspill river were completed in 1942 in a wartime scheme to combine drainage of the levels with the provision of a freshwater reservoir for the Royal Ordnance Factory at Puriton. The full $17m^3$/sec potential of the pumping station could not be utilised however until the South Drain was sufficiently enlarged to enable the moorland water to drain to the station. This was achieved in a series of schemes, Stage I in 1962, Stage II in 1969 and Stage III in 1978.

Stage III, the main subject of this paper, involved the improvement of 10 km of main drain, the reconstruction of three road bridges, accommodation bridges and other structures.

Gold Corner Pumping Station had the capacity to pump $17m^3$/sec from the area, this relates to $0.14m^3$ sec/km^2. The main drain was designed to convey this quantity of water to the pumping station whilst maintaining the top water level just below ground level at the lowest point in the moor.

The area suffered floods on a probability of 1 in 2 year for winter flooding and 1 in 5 years for summer flooding. In addition the River Brue overtopped its banks in the Glastonbury area which then drained into the lower peat moors causing further flooding.

The Scheme involved the widening of the main drain through the extensive peat deposits of the Brue valley. These deposits are being dug for horticultural use, to annual production being approximately 100,000 tonnes.

The close proximity of these workings to the main drain together with the failure of the Glastonbury Canal bank in 1976 caused concern over the stability of the proposed channel.

The Glastonbury Canal failure comprised of a 34m section of peat bank that was pushed back approximately 4m on a horizontal plane towards the drained and excavated moor. Breaches occured at either end of the failed section allowing water to flood into the peat workings. The area flooded in this incident was 110 ha.

On first sight the bank appeared to have slid back on the peat to clay interface at approximately 0.0m AOD. Tests carried out on site with pumping, the measurement of water levels through the bank and the determination of the bulk density of the peat suggested that the bank was liable to 'float' under certain conditions.

Wessex Water Authority appointed Professor J.N. Hutchinson as Consultant to comment upon the stability of peat banks with particular reference to the South Drain Scheme.

Trial pits were dug, samples taken and tests carried out by Dr. D. Petley of Warwick University. These test identified a very weak 'leaf' layer in the peat approximately 10mm above the clay.

Triaxial tests indicated the effective angle of internal resistance $\phi' = 16.5°$ minimum and $\phi' = 20°$ average. Permeability and shear tests were also carried out on the samples.

From the result of these tests and the canal bank failure Professor Hutchinson developed an expression enabling the factor of safety to be calculated for the banks. The critical condition was calculated from:-

$$F = \frac{(W - U)\tan \phi'}{P}$$

where W = the gross weight of the bank
U = the uplift calculated from the piezometric level in the bank.
P = the horizontal force due to the differential head exerted on the bank.

This simple calculation was then applied to the scheme banks and a 'safe' distance for excavation of peat established. Where extraction of the peat had already taken place within the safety distance, remedial work was carried out under the scheme. The stability of these banks was improved by a combination of the following methods.

a) The formation of a clay filled cut-off trench in the bank to limit the extent of the rise in piezometric head through the bank and consequently reduce the uplift.

b) Fill the adjacent peat excavation with clay to butress the bank.

c) Surcharge the banks with clay or other imported material to increase the effective weight of the bank.

Where the area behind the bank had not been excavated the onus was placed upon the owners of the land not to excavate in such a way that would endanger the bank.

Consideration was given to settlement of the bank, particulary where the banks were to to be surcharged. This has not been a problem so long as the weight of surcharge was not excessive. Also shrinkage of the peat following drainage has caused widespread drainage problems in the Fens. In Somerset, with its moderately high rainfall, carefully controlled summer penning levels and relatively small area of arable farming, shrinkage has not been of great concern.

THE EXECUTION OF THE WORKS

Approval to start the scheme was obtained from the Ministry of Agriculture, Fisheries and Food in September 1978.The scheme was estimated to cost £479,500 (1978).

The scheme was carried out by Wessex Water Authority D.L.O. with contracts being let for bridge piling and reconstruction where necessary.

Structures.

The road bridges were founded on H Section steel piles 15m long driven to set in the lias. Stone faced concrete abutments were constructed to carry the prestressed concrete beam deck.

Two ancient masonry arch bridges near Street area were replaced with similar arched structures based upon an Armco pipe arch with masonry headwalls and a concrete deck.

Two bridges on the disused Highbridge to Glastonbury railway were flumed by sheet pile using walls and a lowered invert.

Channel Works.

Excavation of the channel was carried out in the 'dry' from the downstream end of the scheme. A temporary sheet steel pile dam was installed together with a 24" diesel pump set and two 6" pumps. The larger pump enabled the drain to be emptied rapidly while the smaller pumps were sufficent to cope with the normal seepage flow. Dragline excavators working on mats were used for the bulk excavation of the channel. The sequence of operaton went as follows, topsoil strip and stockpile at rear of working area, excavation of peat and stockpile adjacent to topsoil, excavation of clay and tip adjacent to line of drain. The peat was then spread back over the clay and finally the bank topsoiled. Several of the smaller peat operators removed some of the surplus peat in lieu of compensation payments.

Work continued throughout 1979 and 1980, briefly stopping during the winter months where uneconomic. The line of the disused railway along the lower portion of the drain proved invaluable as an all weather access road.

During 30th May 1979 heavy rain caused extensive flooding of the area, work was brought to a standstill on the structures and channel works. The River Brue overspilled into Sharpham Moor and flooded the peat workings over a large area.

The owner of one peat area hired pumps to remove this floodwater from his own pit. The pump discharged to a Drainage Board Rhyne which ran bank full. After several days continous pumping when the level in the pit was approximately 1.4m lower than the rhyne, the bank failed in the now well established manner. A portion of the bank was moved sideways into the excavated area leaving vertical shear cracks down to the clay. Water poured through the breach and again flooded the workings.

This further evidence of the potential instability of peat banks reinforced the Authorities approach to the land negotiations and very few objections were raised over the necessity to restrict peat extraction adjacent to the main drain.

The scheme was completed in 1982 with the reseeding of all spoil areas and boundary fences completed.

CONSERVATION

The whole of the South Drain and surrounding rhyne system is man made. The channels are generally trapizodal and canal like. One area of Shapwick Moor has been designated an SSSI to preserve the natural regrowth of a manmade landscape. Every effort was made to preserve trees and copses and other attractive features, tree planting was carried out where possible in the corners of fields adjacent to the drain with the landowners permission. The rough stone pitching under the road bridges also provided a ledge for otters as well as its primary function of protection and the sighting of these increasingly rare animals continues. The lack of natural features such as meanders is a minor disadvantage compared to the peace and solitude of the area.

OBSERVATIONS AND CONCLUSIONS

Judged from an agricultural point of view the scheme has been a success. Flood water is rapidly removed from the benefit area and water levels in the South Drain much reduced under high flow conditions. Summer penning levels are maintained at a fairly high level to provide water and fence and also will serve to minimise peat shrinkage.

ACKNOWLEDGEMENTS

In the preparation and submission of this paper I wish to express my thanks particularly to:-

Mr. B.A. Tinkler) of Wessex Water Authority
Mr.D.G. Alsop)

Professor J.N. Hutchinson Imperial College, London

Dr. D. Petley Warwick University

Mr. M. Roberts) MAFF Taunton
Mr. P. Evans)

Mr. M. Williams Who's book 'The Drainage of the Somerset levels' provided most of the historical information.

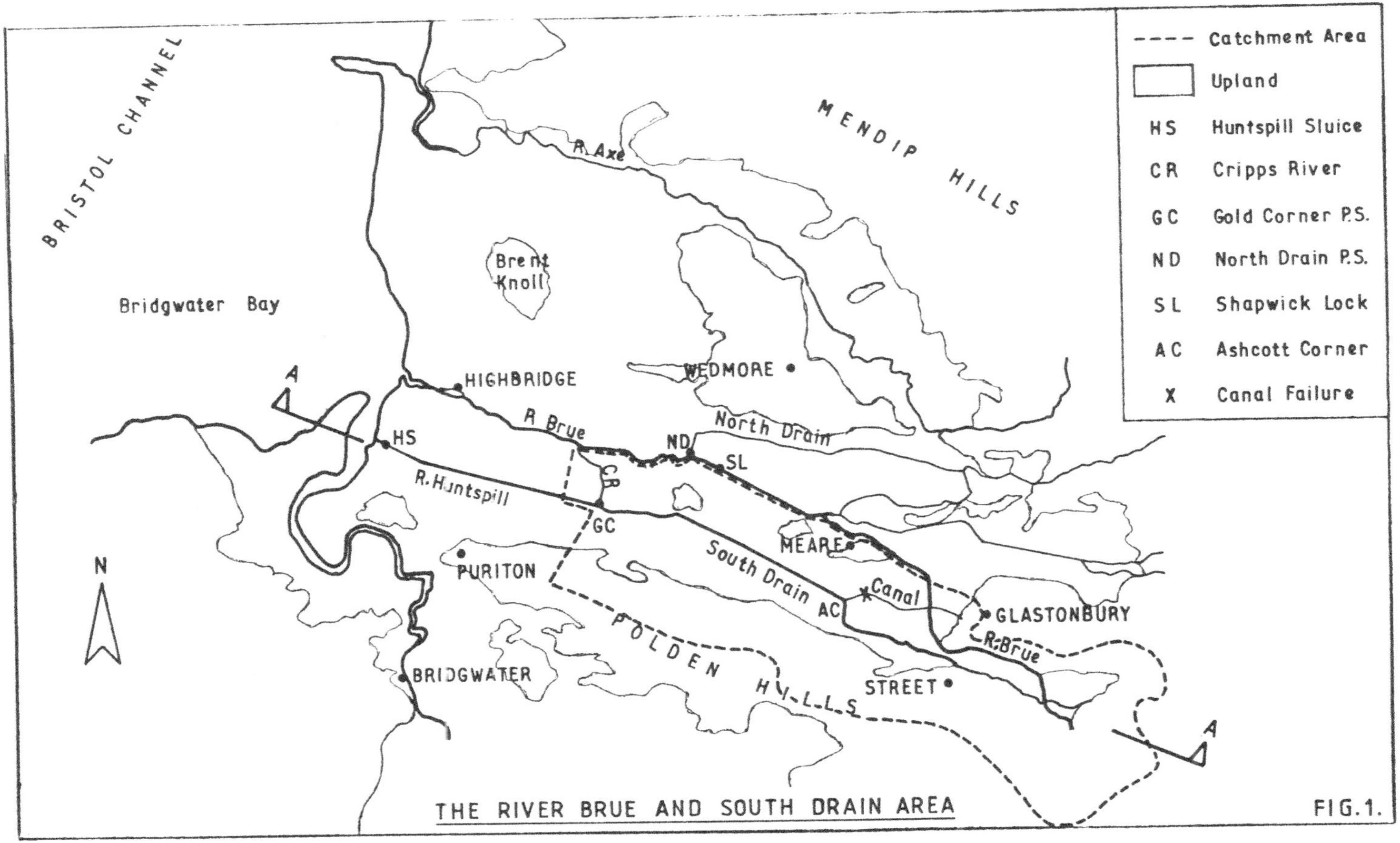

THE RIVER BRUE AND SOUTH DRAIN AREA

FIG.1.

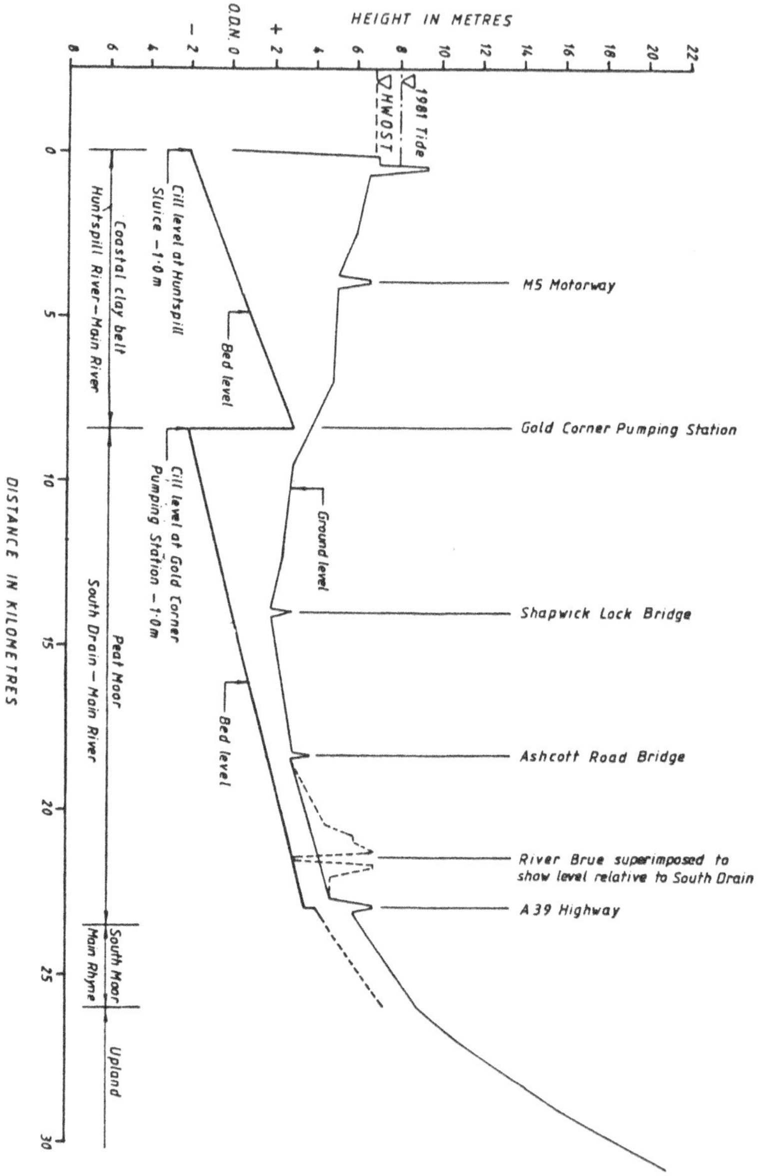

FIG. 2.

SECTION OF SOUTH DRAIN BASIN ON LINE A–A

Integrated Drainage Systems for an Alluvial Wet Land – A Case Study

Govind S. Rajput
J.N.K.V.V., College of Agriculture, Indore M.P. 452 001, India

INTRODUCTION

Waterlogging in alluvial fans or valley bottoms is caused due to very slowly moving body of ground water, which is fed by springs, surface streams or subsurface percolation around the perimeter of the valley; and by surface runoff or infiltrating rainfall on the area itself.

Malagarazi plain (latitude 4°S and longitude $30^{o}10'$E), essentially represents an alluvial wet land, is situated in southern part of the Mosso region in Republic of Burundi. The plain is bound by Muyovozi river in north, Mutsindozi river in south, Kiofi Hills in west and Malagarazi marsh developed along the Malagarazi river. The gently sloping plain originally built by the process of planation but later filled up by colluvial material from the hills forming a narrow zone of piedmont plain and by the alluvium of Malagarazi, Mutsindozi and Muyovozi rivers. The lower end of this alluvial plain is marsh land, which apparently resulted due to flooding and poor infiltration capacity of the surface soils. The study area, an upper part of the plain covering an area about 1200 ha, remains wet during the rainy season (October to May) due to poor surface and subsurface relief, was taken up for its development for extensive cultivation of sugarcane. Location of the area is shown in Figure (1). A project was formulated, and based on detailed investigations, an integrated design of drainage, irrigation and roads was developed for the area. This paper focuses on the integrated drainage systems for the project area.

DATA COLLECTION AND INTERPRETATION

Topography

The project area has a gentle slope (about 0.5%) across the plain in south-west to north-east direction and mild longitudinal slope (0.1%) in the south-north direction (Figure 2). The upper boundary of the study area is at altitudes about 1185 m in south to 1174 m in north while the lower limit of the area is determined by the marsh water varying from 1168 m in south to 1163 m in north.

Climate

The area is located in tropical zone. Mean annual rainfall of about 1200 mm is distributed over 8 months from October to May. Mean daily temperature varies from 20 to 24 °C during a year. The climate of the area is characterized by a dry season of about 4 to 4.5 months from mid-May to September.

As rainfall data for Malagarazi station were scanty, the rainfall frequency analysis was performed with available long term (25 years) data from Musasa, a nearby meteorological station. The analysis revealed that daily and five-days maximum rainfall for 10 years recurrence period are 70 mm and 100 mm, respectively.

Flood levels and river flows

The flood levels in the area are mainly governed by Mutsindozi river in south, by Muyovozi river in north and by marsh water in east. The highest water spread levels along the marsh based on local enquiry and short period gauging were 1168 m in south and 1163 m in north. The results of frequency analysis for maximum and minimum river flows carried by Euroconsult (1977) and ISABU (1978) are summarised in Table 1.

Table 1 Maximum and minimum river flows at different return periods

Return period, years	Maximum flow, m^3/sec		Minimum flow, m^3/sec	
	Mutsindozi	Muyovozi	Mutsindozi	Muyovozi
10	78	95	1.40	1.91
20	95	125	1.20	1.55
50	115	170	0.91	1.03
100	130	205	0.73	0.68

Soils
Detailed soil studies carried out by Frankaart (1974) as reported by Euroconsult (1977) and The Mehta Group International Ltd. (1979) revealed that the eastern part of Malagarazi plain has heavy clay soils with poor drainage only during some part of the year, while the soils of the northern part are lighter in texture with a sandy substratum facilitating better aeration. In the western part near the foot hills, the soils are not saturated with water and are better drained. The sandy soils met within these areas constitute old channel beds. The soils of the levee area of Muyovozi and Mutsindozi have stratified materials of sand, silt and clay. The soils of the area can be grouped mainly into standard soil units 1, 2, 3, 5 and 7 covering area of 235, 360, 225, 135 and 230 ha, respectively. (The Mehta Group International Ltd. 1979). In general the soils can be made suitable for crop production with adequate drainage and nutrient management.

Hydraulic conductivity (K) Auger hole permeability tests were conducted for different soil layers at locations in the vicinity of the observation wells as shown in Fig.(2). The analysis of the test data indicated that the surface soils (0-40 cm) are relatively less permeable than sub-soil (0.5 to 3.0 m) with K values of 0.2 to 2.8 m/day and 3 to 18 m/day, respectively. There was mostly relatively, very low permeability layer encountered below 3 to 3.5 m in the area.

Infiltration rate Double ring infiltration cylinders were used for the test. However, the tests could only be conducted in low ground water table areas. A high basic infiltration rate in the range of 4 to 8 cm/hour was recorded along the western boundary of the plain.

Ground water fluctuation Spatial and temporal fluctuations of ground water levels in the plain were monitored through observation wells sunk up to 5 m depth. The locations of observation wells are shown in Figure(1). Periodic records on ground water fluctuations from this study and previous studies implied that during the dry season the lowest depth of ground water level with respect to ground surface varies between 1.0 to 2.8 m. The ground water level rises with the occurrence of heavy storm and reaches close to the surface almost all along the plain. The general direction of ground water flow is almost same as that of ground surface, however, its gradient is decreased near marsh as indicated by the comparative study of the ground and ground water contours.

Water quality
The quality of ground water and river water based on chemical analysis is classified as excellent to good for irrigation as reported by The Mehta Group International Ltd. (1979).

Irrigation needs
To determine irrigation requirement of sugarcane, the methodology reported by Doorenbos and Pruitt (1977) was followed. An average yearly evapotranspiration deficit worked out to be 440 mm and the peak irrigation need during the dry period can easily be met through available minimum flow in Mutsindozi river.

PLANNING AND DESIGN

Planning
An integrated plan was prepared to facilitate (i) efficient distribution of irrigation to each field, (ii) removal of excess rainfall and/or irrigation water (iii) easy access and use of agricultural machinery for various operations. The layout plan is shown in Figure (3). However, only drainage systems are discussed in this paper.

Fields of 3 to 4 ha size are to be graded with 0.2 to 0.6% and 0.2 to 0.3% slope in longitudinal and crossfield directions, respectively. The average earthwork is estimated to be in the range of 200 to 300 m^3/ha. The excess surface water in the field is to be drained through the furrows into field drains running across the furrows and dropped into the block drain running parallel to the furrows. The block drain, an open ditch, functions as surface-cum-subsurface drain for a block (number of fields in a series). Considering the prevailing topography and soil conditions a herringbone system of parallel block drains was planned. The drainage water from block drains is ultimately disposed of through open collectors and open main drains into the Muyovozi river. The collector drains were laid mainly in the natural micro-relief (depression) lines leading to the marsh as demarcated from the contour map. To avoid seepage from the marshes the drain is planned to be laid at about a metre higher elevation than the highest flood level along the periphery of the marsh land. Interception drains and flood embankments are planned so that superficial water from neighbouring areas & rivers does not enter into the project area.

Design considerations

Surface drainage The drainage coefficient for surface drainage in areas having slope within 1% was based on drainage capacity of about 65% of daily maximum rainfall at 10 years recurrence period and evapotranspiration rate (ET) of 3.6 mm/day. The surface drainage coefficient was, thus, estimated to be 43 mm/day or 5 lit/sec/ha. However, for the areas having land slope greater than 1%, the Rational Method was used for predicting the design discharge.

Subsurface drainage As it is planned to grow sugarcane, which is relatively resistant crop to waterlogging, it was assumed that the crop growth will not be severely affected with the maximum period of waterlogging upto 5 days. Considering 35% of total 5-days maximum rainfall at 10 years frequency as maximum contribution to the ground water re-charge and ET losses at the rate of 3.6 mm/day, the quantity of deep percolation due to designed rainfall is estimated to be 0.0034 m/day or 0.4 lit/sec/ha. For practical reasons it was considered to maintain the depth of ground water level at minimum 1 m below the ground surface. For prevailing soil characteristics in the plain with a block drain depth = 2 m, height of ground water table above the drain level at midpoint between the drain = 1 m, thickness of aquifer below the drain level = 1 m and subsurface drainage coefficient = 0.0034 m/day; the drain spacing for K values of 5, 8 and 12 m/day were estimated to be 133, 168 and 206 m, respectively. The allocation of drains with different spacings were made according to spatial distribution of hydraulic conductivity in the area. Since the block drains are planned parallel to furrow direction, there is scope for further allocation of a drain in between two block drains.

Mannings equation was used for determining the dimensions of various drains at different sections. The value of roughness coefficient (n) was adopted as 0.04 and side slope of drains as 1.5:1. However, the dimensions were modified considering certain constraints and compromises were made for different drains as described below.

<u>Field drain</u> The design discharge for a field drain was calculated to be maximum 20 lit/sec. As these drains are temporary and the shape would be disturbed by the machinery during various field operations, a V-section of 20 cm depth with side slope 8:1 was planned. This facilitates easy access to machinery as practically observed and have carrying capacity of more than 50 lit/sec, which is safe.

<u>Block drain</u> Considering the soil characteristics and construction economy an average depth of 2 m was decided for the block drains.Hooghoudt's steady state design principle as reported by USDA (1971) was used for determining the spacing of block drains. The dimensions of the block drains were determined with the constraints of side slope 1.5:1, uniform bed width and drainage depth in the range 1.5 to 2.2 m. The depth of block drain at its junction with collector drain was kept within the range 1.5 to 1.7 m below the ground.

<u>Collector and main drains</u> The depth of collector drain limited to vary between 1.6 to 2.8 m below ground surface. However, the bed of the collectors at the junction to the main drain could not be lowered arbitrarily as it was guided by water surface profile of the main drain and by the main outfall conditions.

The following important guidelines were kept in view while designing the main drain (i) maximum available slope and depth of flow in the main outlet under the constraint of maximum obtainable drop over the total length of the main drain (ii) estimate of maximum non-erosine velocity for the design runoff as calculated for each junction with collector (iii) water surface profile of the main drain during peak flow conditions should be below the collector exit bed level, otherwise submergence caused due to flooding in Muyovozi river for a shorter period considering the economics of the cross-section.

In order to satisfy the criterion (iii) above, the section of main drain would have to be enormously large. An economic section was possible through a certain degree of sacrifice, i.e., allowing the water surface profile in main drain above the bed level of collectors over a short period of time.

Flood embankment Flood embankments are designed for 50 years flood frequency. Earthen embankments of 2 m height and 2 m top width shall be provided along the north bank of Mutsindozi, the eastern periphery of the Malagarazi plain along the main drain and the southern bank of Muyovozi river as shown in Figure (3). Four outfalls shall be provided in the embankments with control gates to drain the accumulated drainage discharges on the hind side. The outfall channels shall be stonepitched to prevent scour. The entire length of flood embankments works out to be 19400 m. The embankment will be mostly constructed from spoil excavated from drainage ditches.

Although, this paper attempts a special case of integrated drainage systems for an alluvial wet land, the same principles could be applied to any area with similar problems.

ACKNOWLEDGEMENTS

This study forms a part of the turnkey "Moso Sugar Project" awarded to The Mehta Group International (TMGI) by the Government of Burundi, for which the author was deputed as an expert by Agrima Project Engineering & Consultancy (APECS) Bombay. The facilities rendered by TMGI, APECS and Societe Sucriere du Moso (SOSUMO), Burundi are gratefully acknowledged.

REFERENCES

Doorenbos, J. and Pruitt, W.O. (1977) Crop water requirements. Irrigation & Drainage Paper 24, FAO, Rome.

Euroconsult (1977) Moso Sugar Project - A preliminary Feasibility Report. Ministry of Planning, Republic of Burundi.

ISABU (1978) Annual Report, Institute des Sciences Agronomiques du Burundi.

The Mehta Group International Ltd. (1979) A Feasibility Report, Vol.II Soils and Hydrology, Moso Sugar Project. Ministry of Planning, Burundi.

USDA (1971) Drainage of Agricultural Land. In: National Engineering Handbook. Soil Conservation Service, U.S.Dept. of Agric.

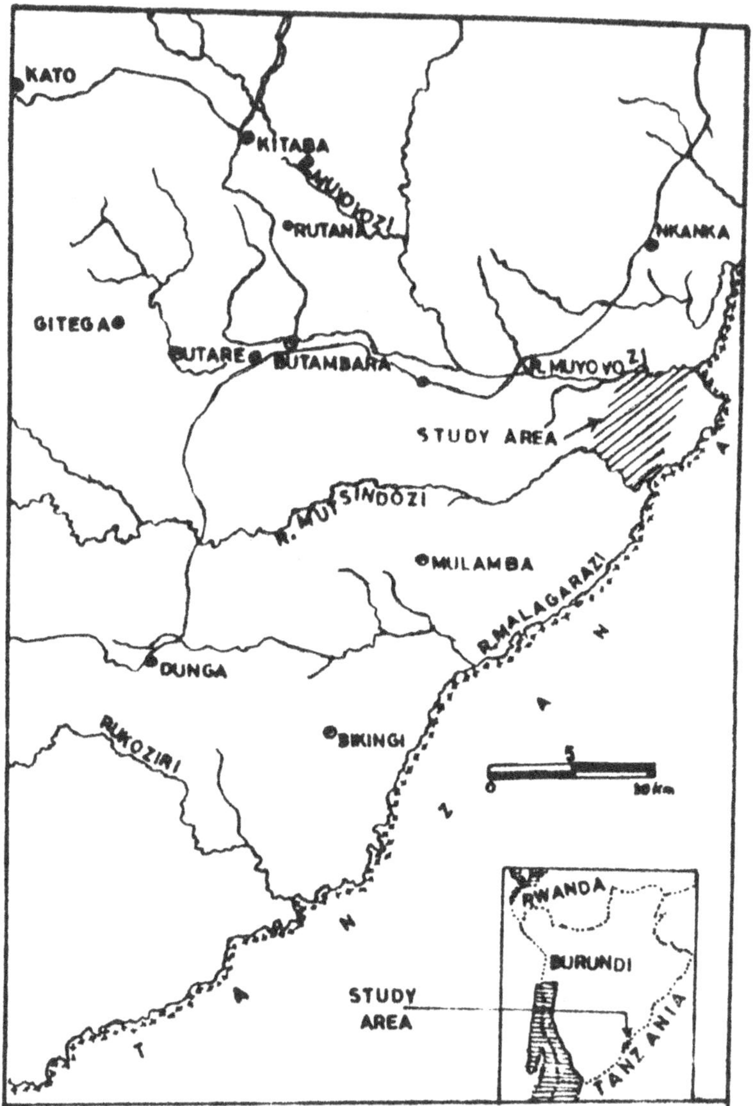

FIGURE: 1-LOCATION MAP OF STUDY AREA

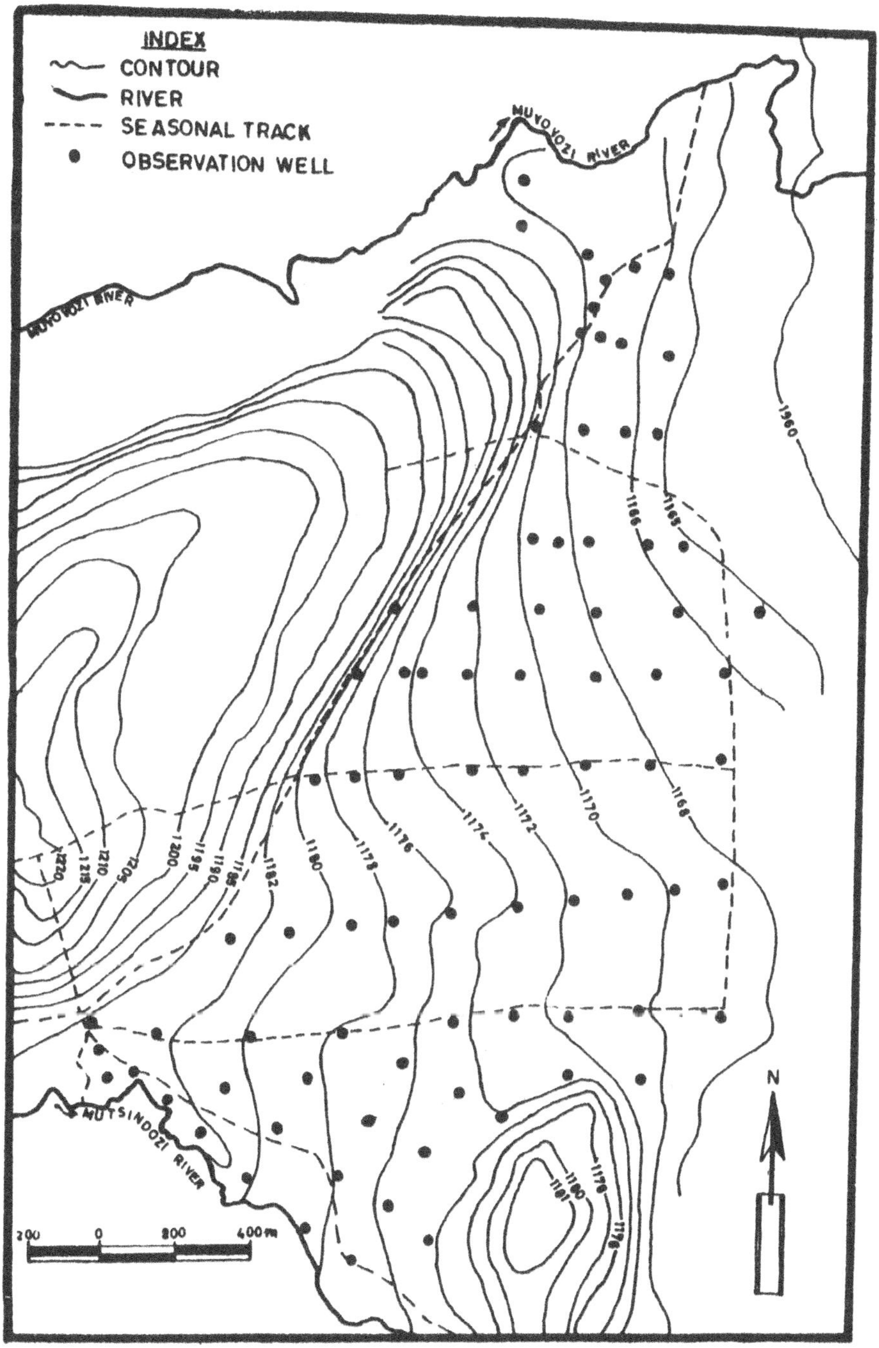

FIGURE: 2 - TOPOGRAPHY AND LOCATION OF OBSERVATION WELLS IN THE STUDY AREA

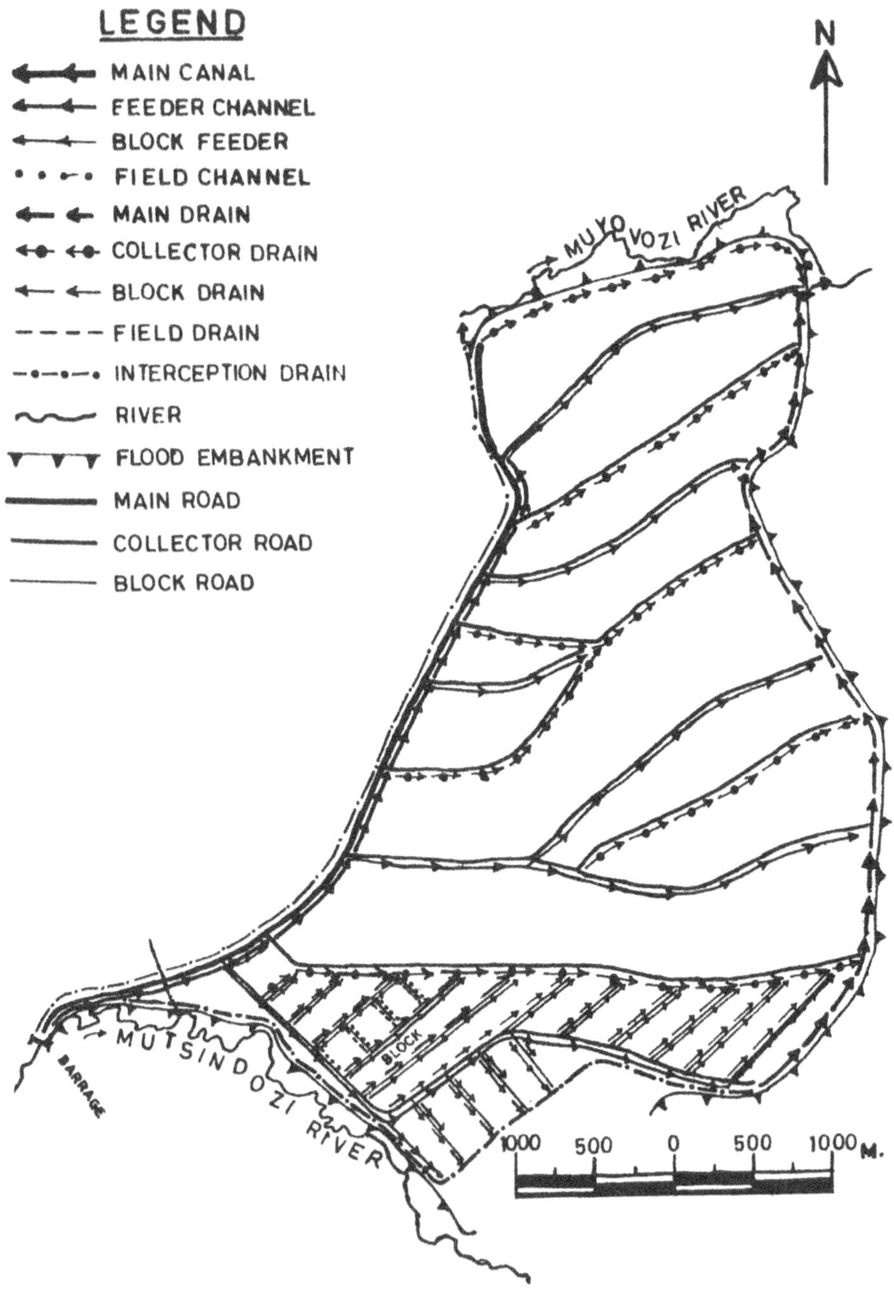

FIGURE: 3 - DRAINAGE, IRRIGATION AND ROADS LAYOUT.

Development of Lahore at West Bank of Ravi

Dr. Javed Yunas Uppal
Engineering Project Development Consultants

SUMMARY

Lahore, second largest city of Pakistan and 30th in the World, situated on the eastern bank of the river Ravi, is in compelling need of expansion across on to the West Bank, due to the pressures of urban growth, national communications grid and proximity to the international borders. The West Bank of the river, however, being a vast low lying flood plain of mud and silt is an unhabitable wasteland. The river Ravi, deprived of its waters most of the year, carries abrupt flood discharges up to 5700 m^3/sec in monsoons. The river therefore has to be trained and the land has to be reclaimed. There are numerous examples in the World, where flood affected low plains have successfully been turned into benefitial urban areas. Method has been suggested in this paper to provide a flood protection dike at the West Bank with remodelling of the bed of the channel, providing of reclamation fill behind the dike and construction of associated drainage works. Alignment of the West Bank dike is a dominant cost/benefit effective parameter. An optimisation exercise has been conducted through dynamic programing to arrive at the best alignment of the dike. A positive feasibility has emerged on account of the benefits by the enhancement of the value of land in relation to the costs.

WEST BANK OF RAVI

Beyond the Western bank of the river Ravi, and around the present Sharakpur Road, are lying promising areas of land. In the North, the land is already connected to Lahore through the existing bridge over Ravi. In the South, it is proposed that a second bridge may here be constructed over Ravi near Niaz Beg Thokar at Multan Road. This second bridge will provide a bypass for the South-north inter-city traffic and a relief for the existing road network. The new area will be located at a strategically safer position in wartimes. Since the land is almost completely barren, there will be plenty of room to plan beautiful environments and skylines on modern levels. The vast area however has to be properly reclaimed and the problem of flooding has to be resolved. The costs of flood protection, reclamation and the new river bridge are off set by the increase in selling price of the developed land. This makes this proposition economically feasible.

THE RIVER RAVI

Today, the waters of the Ravi are mostly used by India for their own irrigation purposes. However, the floods are released. Most of the year, the Ravi remains more or less dry and any time between June to September the influx of high flood is received. The flood discharges through the Ravi at Lahore are of the order of 5700 m^3/sec 50 years return.

The Ravi has its first head works at Modhopur about 32 km up in India. From the borders, down to the confluence with the Chenab, the Ravi has a stretch of about 680 km with a fall of about 100m spread almost uniformly. There are two major head works namely Balloki taking off 230 m^3/sec and Sidhnai 150 m^3/sec.

Flood is the most important aspect not only with respect to direct losses but also with respect to the restrictions provided for general development. Floods are normally generated in the upstream mountainous areas where snowmelt, intense rainfall and flashy stream build up high river discharges. The flood control measures have largely been through dike construction.

EXAMPLES

Methods can be studied, where solutions to similar problems have been successfully sorted out elsewhere in the World. Some typical examples are:

Macan China

Macan, a coastline in China was once a low lying reed marsh on soft and massive mud. The sandy soil available in the adjacent sea bed was dredged, stock piled on the sea coast and then pumped and spread on the mud. Water from mud was drained through a network of horizontal and vertical drains. Filling over the entire area was thus provided in layers ranging 1 to 1.5m in depth. Each layer required 4 to 6 weeks time for consolidation. The land was ultimately developed for a high rise housing and commercial development.

Lagos

In Nigeria, the commercial housing development in the South of Lagos is an interesting example. Hundreds of thousands of acres were reclaimed which had once been low lying sea shores. Sand filling of entire area was done by dredging and pumping. The sandfill varies in thickness, and at places was as much as 4m. Compaction was carried out by watering. At places, settlements by more than 1m in a period of one year or so occured before the area was finally ready for development.

Netherlands

Netherlands consisted of low lying waste lands, mud dumps, swamps and uncultivable marine salt sands. As early as in the 17th century, large areas were reclaimed by constructing dikes and water fronts between Amsterdam and Alkammar in the North. In 1918 and afterwards, over half a million acres of cultivable land was developed by drying the sea of Zuyderzee.

At present under the Delta Plan, colossal stretches of delta are being turned into rich useful agricultural and urban areas.

LAND PROTECTION AND RECLAMATION FOR LAHORE

The West Bank Dike

A dike along the Eastern bank is already existing between the present bridge down to Dholanwal which effectively protects the Western parts of the city of Lahore from floods. A similar dike can be constructed along the Western dike of the river so that the land between this dike and the Sharakpur Road now lying almost waste can be developed. The length of such a dike as shown on the plan is of the order of about 14.5 kilometers. The section between the dikes can be effectively trained through the dredging and revetment so that it can pass the the floods safely.

The Filling

The land beyond this dike can be dried through a network of drainage and a system of pumping. Once the land is dry, the filling can be done in layers using the dredged materials mixing adequate proportion of sand. As the filling proceeds, small waterways and ditches will be necessary in order to facilitate natural drainage. Drainage of the surrounding area, also, has to be remodelled.

ANALYSIS

Description

The problem so identified is (i) to provide a flood protection dike (5) at the West Bank, (ii) remodelling bed of the channel, (iii) reclamation fill behind and (iv) associated drainage works. A dike at the East Bank already exists, so its alignment is fixed. Although the number of interdependent of parameters is so large that it is extremely difficult to isolate the influence of any single one of them, the alignment of the dike at the West Bank is a dominant variable. In the present exercise, the 'alignment' of the West Bank dike is the single parameter studied. In making of the analytical system, references 1,2,3 and 4 have been followed. The depth and width of the regulated channel is so adjusted that the sediment transport capacity is not largely affected. The longitudinal slope is also adjusted to match with the adjacent reaches. The possibility of coincidence of flood waves as well as the possibility of creation of upstream backwater have also been studied.

Behind the dike, the fill material is foreseen.Part of it will be supplied from the dredging of the channel and a part of it will be imported. Drainage of this area at various stages as well as the drainage of the surrounding area has also been included in the system.

The nonlinear programing formulation for optimisation (6) can be conceptually stated as:

minimise $C\ (A,L,Q,H,h,p)$
maximise $P\ (A)$
subject to $G\ (A,L,Q,H,h,p)$

where $C = Cd + Cc + Cf + Cr$ = Sum Cost
Cd = Cost of dike
Cc = Cost of regulating channel
Cf = Cost of fill
Cr = Cost of drainage
P = Sum price of developed area
G = Set of constraint functions
A = Resulted developed area = $f\ (Q)$
L = Level of fill = $f\ (Q,h)$
Q = Vector of dike alignment coordinates
B = Distance between dikes
H = Array of topographical levels

h = Array of Hydraulic parameters
p = Array of unit prices.

Parameters

Parameters used in the study are listed out in Table-1. They have been taken from the available data of last 30 years and from the type of development foreseen. Whereas the natural channel has been catered for floods of 5 years return period, the regulated channel between the dikes has been foreseen to cater for floods of 50 years return period. Top levels of the dike have been taken to cater for floods of 200 years return period.

The soil type found in the bed and in the surrounding area is mainly silt. Topographic characteristics of the flood plain were noted. The contours are generally equi-distant and are parallel to the natural stream.

The unit costs of the material were obtained from local sources and the available cost indices.

TABLE-1: FLOW PARAMETERS

Stream	Ephemereal
Data	30 years
Natural channel, Q5	500 m^3/s 5 years return
Max flood, Q50	5700 m^3/s 50 years return
Dike levels for Q200	7400 m^3/s 200 years return
Regulated reach	14.75 km
Free board	1.00 m
Bed soil	Silt
Side slopes	1:3
Mannings 'n'	0.030
Longitudinal slope, I_o	1:7500

ALDR Computer Program

A computer program 'ALDR' has been developed on the flow chart described in Figure-1. The program takes as input, the flow parameters such as (i) longitudinal slope (ii) flood discharges, (iii) sediment mass and (iv) an array of topographical levels. A trial alignment of the West Bank dike is then fed. The program performs hydraulic calculations and calculates cost involved in regulating the channel. The program works out earthwork quantities both for the dredged material and also for the fill material behind the dike. The program then performs drainage calculation and works out the costs of works provided to cater for seepage of the development area and drainage of the surrounding area. The computer then works out the total developed area so evolved and estimates its selling prices. The program performs dynamic optimisation and works out the costs and benefits by successively choosing different alignments of the West Bank dike.

Results

Different alignments of the West Bank dike, expressed as 'distance between two dikes', are given in Figure-2 and 3. When it is varied, its effect on the system cost is shown by curve-1 in Figure-4. The relationship with the system benefits is shown by curve-2. A third curve-3 gives the ratio of benefits/cost. It can be observed that the optimum alignment (at maximum ratio) of the dike is the one where the distance is 5km. This enables development of an urban area of about 10000 acres. An economic feasibility based on the resulted optimum alignment is can now be evaluated.

ECONOMIC FEASIBILITY

Although an accurate feasibility analysis can only be performed after a detailed study, a rough picture can be made to assess the order of the parameters. The project of this magnitude will naturally be phased out such that the capital expenditures and the returns will be spread to a number of years in an admissible cash flow arrangement.

Capital Expenditures

Capital Expenditures can be assessed as below:

Land:10000 acres	$ 14.00 m
Bridge:	$ 4.50 m
Bund:	$ 4.70 m
Reclamation:	$ 10.30 m
Services:	S 2.00 m
Establishment:	$ 2.00 m
Total capital expenditures:	$ 37.50 m

Recurring expenditures can be listed as:

Interest on Loans:	$ 1.00 m/yr
Taxes:	$ 1.50 m/yr
Repair & Maintenance:	$ 1.50 m/yr
Establishment:	$ 1.00 m/yr
Total recurring expenditure	Rs. 5.00 crores per year

At 15% rate of interest, the discounted value of 15 years recurring expenditures at zero year is:	$ 30.00 m

Total expenditures therefore are: $ 67.50 m

Returns

As far as returns are concerned, out of a total of Say 10,000 acres, 2/3rd might be saleable, 80% land may be residential and 10% may be commercial/industrial. 30% fund may be spent on infrastructure development. On this basis, the returns can be estimated to be of the order of $ 100 m.

REFERENCE

1. Henderson,F.M.(1966) Open Channel Flow,Macmillan, New York, N.Y.
2. U.S. Corps of Engineers, (1961) Manual of Hydraulic Design Criteria, Vicksburg.
3. Chow, V.T. (1959) Open-Channel Hydraulics, McGraw-Hill New York, N.Y.
4. P.Ph.Jansen, Ph et al. (1979) Principles of River Engineering,Pitman,London.
5. Franco, J.J. (1967) Research for river regulation dike design,Jour,Waterways & Harbors Div.,ASCE, 93(WW3).
6. Froise, S. and Burges, S.J. (Nov 1978) Least Cost Control Strategies in Urban Drainage, Journal of the Water Resources Planning and Management Division, ASCE, Vol. 104, No. WKI, PP 75-92.

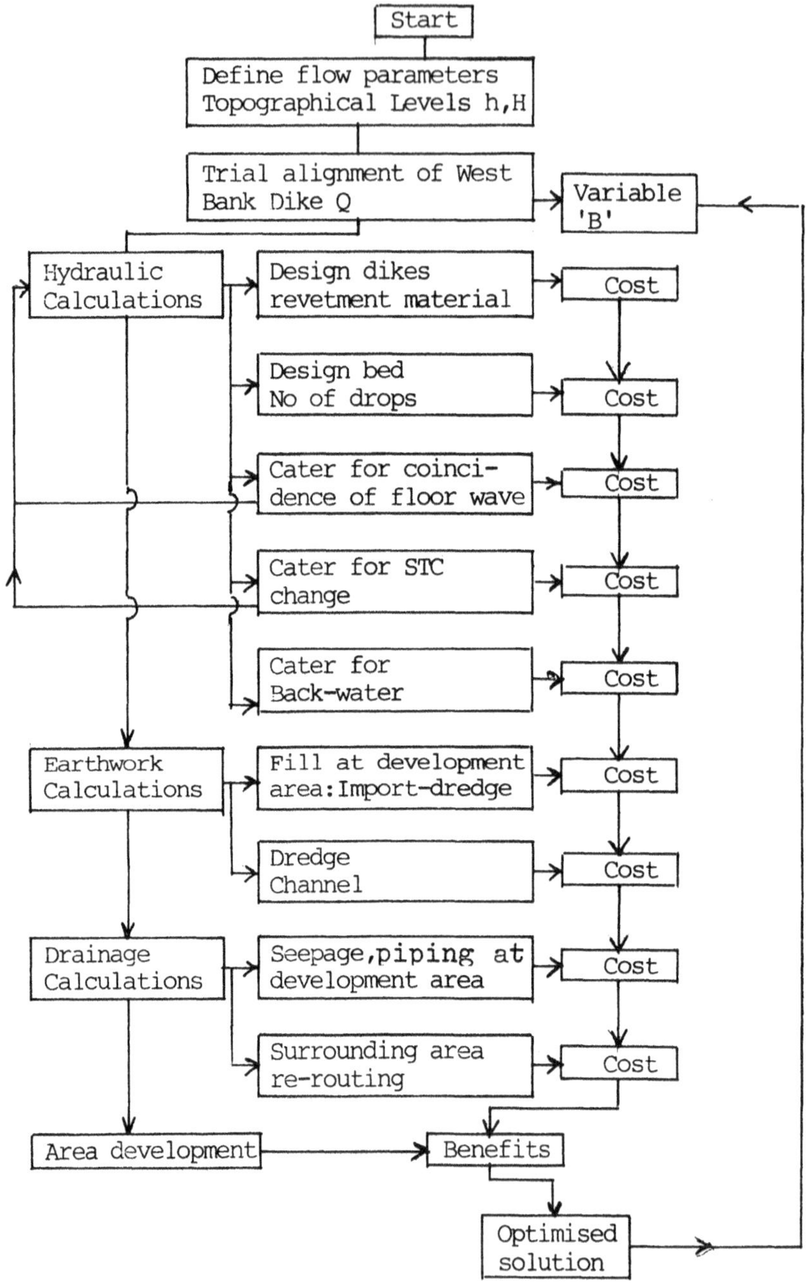

FIGURE - 1: FLOW CHART FOR ALDR

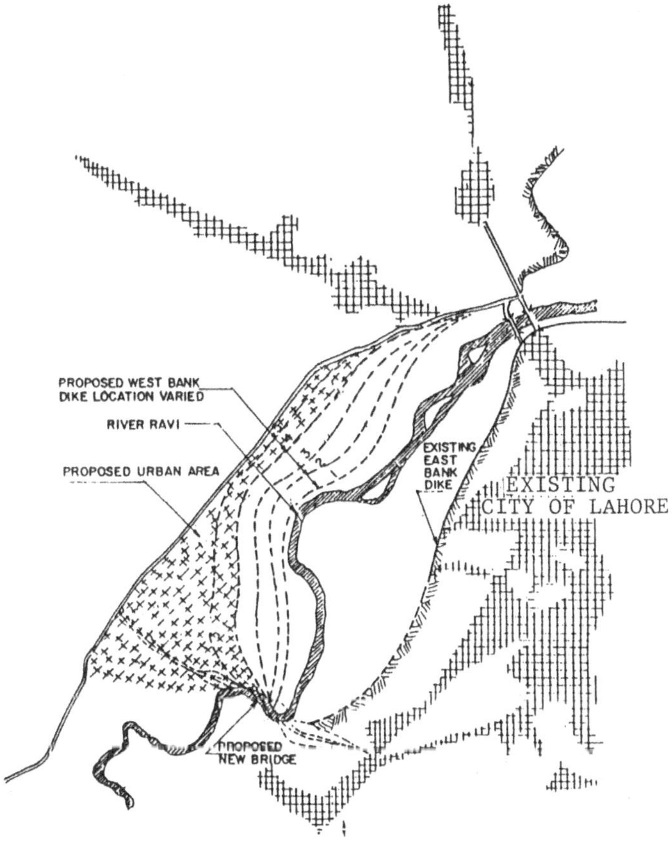

FIGURE-2 PLAN OF RIVER RAVI

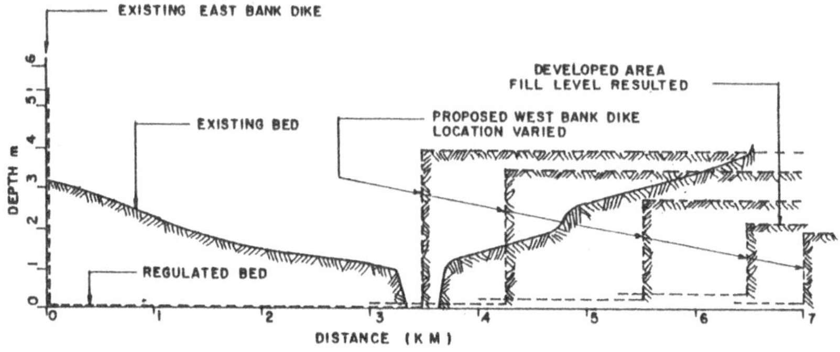

FIGURE-3 CROSS SECTION

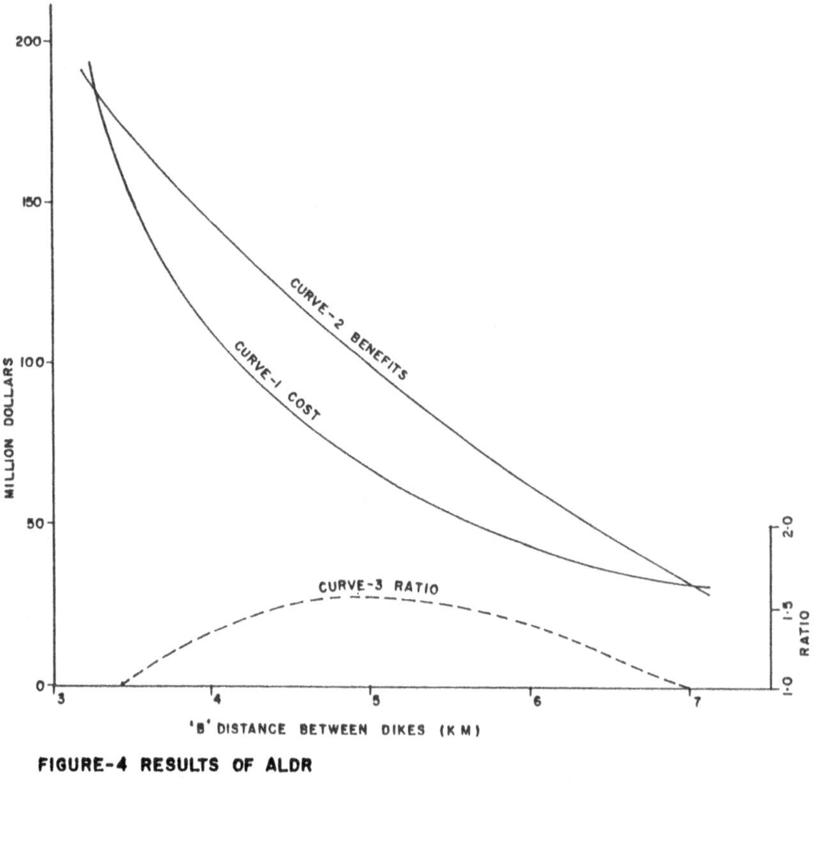

FIGURE-4 RESULTS OF ALDR

The Effects of Artificial Subsurface Drainage on Flood Discharge

Richard W. Harms
Institut für Wasserwirtschaft, Universität Hannover, Callinstr. 32, D-3000 Hannover 1

INTRODUCTION

The positive effects of artificial subsurface drainage to agriculture have never been questioned (e.g. increase in yield, improvement of trafficability). However, every manipulation of a system not only has effects on itself but also on related systems. This study was conducted to investigate the impact of subsurface drainage of agricultural lands on flood discharge. In this connection a number of questions have to be answered or at least to be considered: Will the soil properties be effected by the implementation of a subsurface drainage system? Will there be a change of land use, of what kind, and what are the hydrological consequences? In what way can a series of rainfall events be considered in a study? Especially the last question makes a long term simulation of the rainfall-runoff-process necessary. Therefore a soil water balance model on the basis of a hour-to-hour simulation has been incorporated in a runoff-concentration model to derive catchment related discharge hydrographs.

SIMULATION MODELS

Water Balance Model

The applied soil water balance model is DRAINMOD which has been developed by Skaggs (1978) as a water management model for shallow water table soils.
The basis for the computer model is a water balance for the soil profile (Figure 1). The rates of infiltration, drainage, and evapotranspiration, and the distribution of soil water in the profile can be computed by obtaining numerical solutions to non-linear differential equations. However these methods

would require prohibitive amounts of computer time for long term simulations and thus could not be used in the model.

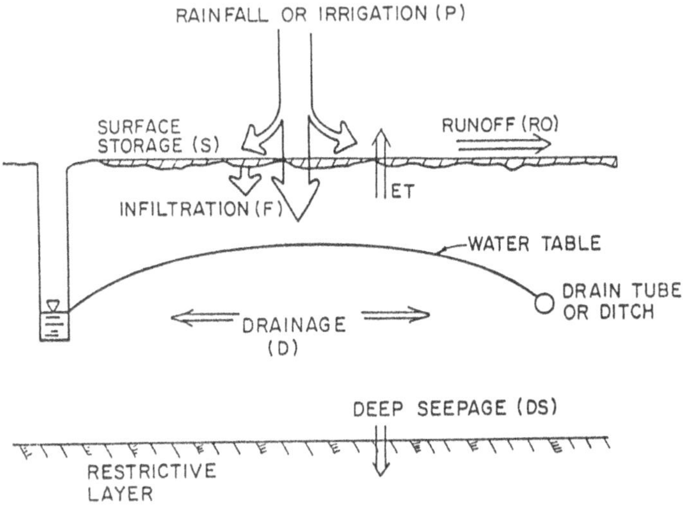

Figure 1: Schematic of water management system with drainage to ditches or drain tubes (Skaggs, 1978)

Instead, approximate methods were used to characterize the water movement processes. In order to insure that the approximate methods provided reliable estimates, Skaggs compared them to exact methods for a range of soils and boundary conditions. Further, the reliability of the total model was tested using field experiments.
The basic relationship in the model is a water balance for a thin section of soil of unit surface area which extends from the impermeable layer to the surface and is located midway between adjacent subsurface drains or ditches. The water balance for a time increment of Δt may be expressed as,

$$\Delta V_a = D + ET + DS - F \qquad (1)$$

where ΔV_a is the change in air volume (cm), D is drainage (cm) from the section, ET is evapotranspiration (cm), DS is deep seepage (cm) and F is infiltration (cm) entering the section in Δt.

The terms on the right-hand side of Equation 1 are computed in terms of the water table elevation, soil water content, soil properties, site and drainage system parameters, crop and stage of growth, and atmospheric conditions. The amount of runoff and storage on the surface is computed from a water balance at the soil surface for each time increment which may be written as,

$$P = F + \Delta S + RO \qquad (2)$$

where P is the precipitation (cm), F is infiltration (cm), ΔS is the change in volume of water stored on the surface (cm), and RO is runoff (cm) during time Δt. The basic time increment used in Equations 1 and 2 is 1 hour. However, when rainfall does not occur and drainage and ET rates are slow such that the water table position moves slowly with time, Equation 1 is based on Δt of 1 day. Conversely, time increments of 3 minutes are used to compute F when rainfall rates exceed the infiltration capacity.

Model Components

Infiltration Of the proposed approximate equations for predicting the infiltration the Green-Ampt equation is used to characterise the infiltration component in DRAINMOD. The equation may be written as,

$$f = k_s + k_s \cdot M_d \cdot S_f/F \qquad (3)$$

where f is the infiltration rate, F is accumulative infiltration, k_s is the hydraulic conductivity of the transmission zone, M_d is the difference between final and initial volumetric water contents ($M_d = \Theta_o - \Theta_i$), and S_f is the effective suction at the wetting front. For a given soil with a given initial water content Equation 3 may be written as,

$$f = A/F + B \qquad (4)$$

where A and B are parameters that depend on the soil properties, initial water content and distribution, and surface conditions.

Surface Drainage Surface drainage is characterised by the average depth of depression storage that must be satisfied before runoff can begin. It is assumed that depression storage is evenly distributed over the field and to be constant during the year.

Subsurface Drainage The rate of subsurface water movement into drain tubes or ditches depends on the hydraulic conductivity of the soil, drain spacing and depth, profile depth and water table elevation. The method used in DRAINMOD to calculate drainage rates is based on the assumption that lateral water movement occurs mainly in the saturated region. The effective horizontal saturated hydraulic conductivity is used and the flux is evaluated in terms of the water table elevation midway between the drains and the water level or hydraulic head in the drains. Hooghoudt's steady state equation, as used by Bouwer and van Schilfgaarde (1963), was selected for use in DRAINMOD.

Evapotranspiration The determination of evapotranspiration (ET) is a two-step process in the model. First the daily potential evapotranspiration (PET) is calculated by the method developed by Thornthwaite in terms of daily maximum and minimum temperature values and is destributed on an hourly basis. In a second step each ET-calculation involves a check to determine if soil water conditions are limiting. When the water table is near the surface or when the upper layers of the soil profile have a high water content ET will be equal to PET. However, for deep water tables and drier conditions, ET will depend on a relationship between maximum rate of upward water movement and water table depth which has to be determined and read as input to the model in tabular form.

Concentration Model

As stated earlier DRAINMOD computes the surface and subsurface discharge for a thin section of a soil of unit surface area. For calculating the discharge of a specific catchment the concentration process has to be described. The study is based on a semi-synthetic catchment of 6 km^2 in size and a hydrological approach (time area method, single linear reservoir model). The discharge values obtained by DRAINMOD were read into the concentration model and catchment based hydrographs were computed by convolution.

SIMULATION RUNS

In most cases artificial subsurface drainage improves the trafficability considerably with consequences to land use. Not seldom grassland turns to arable land when drainage works take place. This has to be considered when studying the impact of land drainage

on flood discharge.
Mainly during the seedbed preparation in spring arable lands are not sheltered by plants and heavy rainfall may cause clogging at the surface which increases surface runoff. This phenomenon has been modeled by using different values for the depression storage; i.e. 6 mm for grassland and 3 mm for arable land.
Changes of soil properties due to subsurface drainage proofed not to be significant to the results.
Simulation runs have been carried out for grassland use without subsurface drainage and arable landuse with subsurface drainage each of these for three different soils: clay, sandy loam, and sand. Model parameters related to the soils have been determined from literature. The main drainage parameters are listed in Table 1. K_s is the saturated hydraulic conductivity.

Table 1   Soil and Drainage Parameter

Parameter		Clay	Sandy Loam	Sand
K_s	(cm/h)	0,108	1,2	125
Depth of Drain	(m)	0,75	0,75	1,20
Drain-spacing	(m)	5	9	30
Depth of Ditch	(m)	0,75	0,75	1,20
Ditch-spacing	(m)	30	200	300

Note, that the depth and spacing of the drains will be used in the model in accordance with the arable land-use, whereas the depth and the spacing of the ditches belong to the grassland use.
Furthermore, results were obtained by varying the degree and location of tube drainage within the catchment.

RESULTS

Based on a 22-year rainfall series as input data Figure 2 shows the change of the discharge pattern due to subsurface drainage. In this Figure 100 % correspond to the total amount of rainfall within the 22-year simulation period. Independent of the soil properties, on the long term the discharge volume is increasing after subsurface drainage works have been carried out. This is due to shorter water-paths to the drains. It can be seen clearly, that only the increase of the subsurface drainage volume is responsible for these results even though the surface runoff as part of the discharge volume is decreasing.
Since the surface runoff appears only during intensive

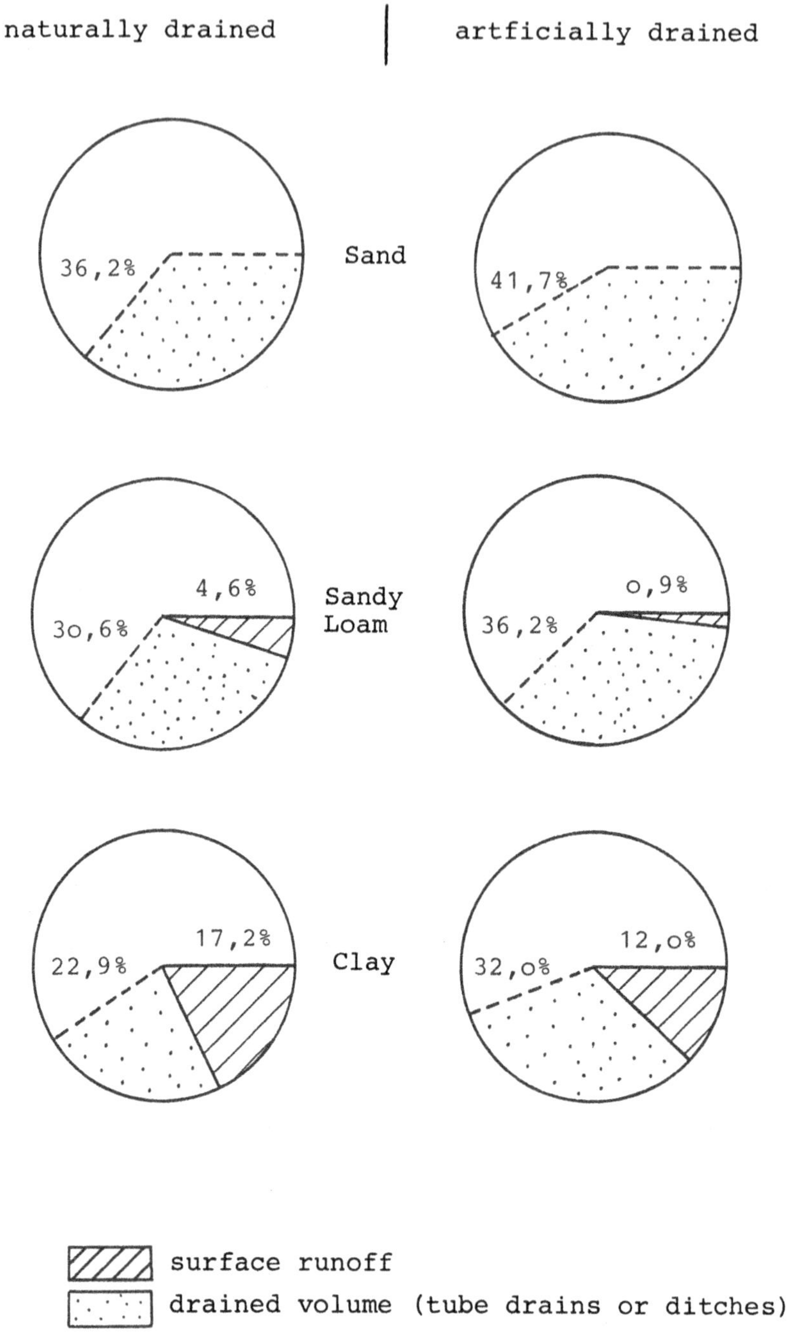

Figure 2 Change of discharge pattern due to subsurface drainage

rainfall events and/or groundwater tables up to the surface it can be stated, that subsurface drainage reduces the volume of quick-respond-runoff. In this investigation no surface runoff at all from the sandy soil has been simulated.
These results have been obtained by applying only DRAINMOD. According to the objective of this study the effects of a subsurface drainage system on flood discharges were of interest in the first place, which means that the simulations have to be catchment-related and the results evaluated statistically. The peak-discharge-frequency-curves are shown in the Figures 3-5. It becomes obvious, in a first place the effects of artificial drainage on peak-discharge depends strongly on the type of the soil. For recurrence intervals more than 2 years for a clay soil (Figure 3) the flood situation will not be effected by subsurface drainage. For a highly permeable sandy soil the flood situation is increasing dramatically, however (Figure 5). The reason is the decrease of retention due to the relative small drain-spacing compared to the spacing of the ditches. These results, however, have to be interpreted: A clay soil might not be considered of being worth to be artificially drained because of only little improvements from agricultural point of view. On the other hand for sandy soils normally there is no need for any tube drainage system. From this, artificial drainage mainly will be employed to loamy soils. It can be seen from Figure 4, that for those soils only floods of recurrence intervals smaller than 3 years will be effected significantly. On top of this, catchments having a proportion of 50 % or more artificially drained are likely to be rather small and therefore the increasing effects are of smaller significance to the receiving waters.
Quite interesting are the effects of artificial drainage on single events. Figure 6 shows the hydrographs of an event of a recurrence interval of about 1 year. The simulation revealed nearly identical hydrographs for the three soils, when these lands are artificially drained. However they differ quite a lot when a natural drainage system by ditches is assumed. While for the sandy soil only a slight increase of the baseflow can be determined, the discharge from the clay even exceeds that one simulated for the artificial drainage system. This result is contrary to the result shown in Figure 3. Looking at another event (Figure 7) of a comparable magnitude again shows the opposite. The explanation for the different behaviour are the different depths of the groundwater tables at the beginning of the

Peak Discharge vs. Recurrence Interval

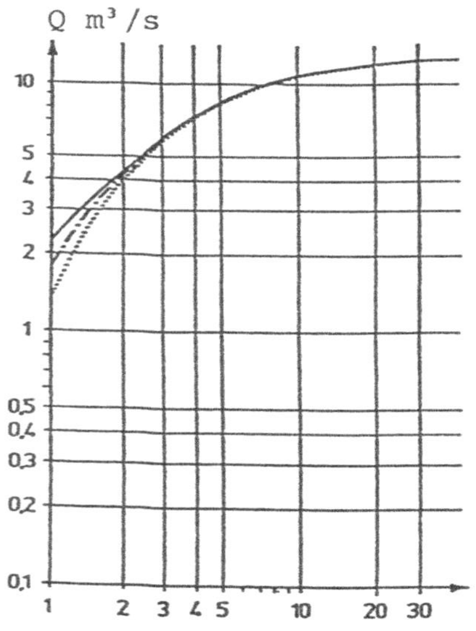

Figure 3 Clay

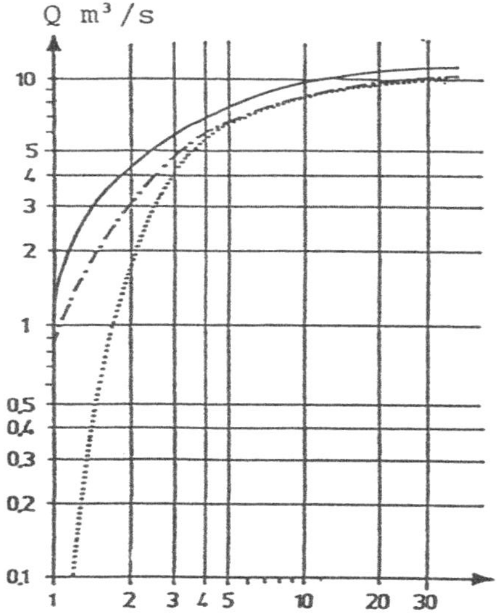

Figure 4 Sandy Loam

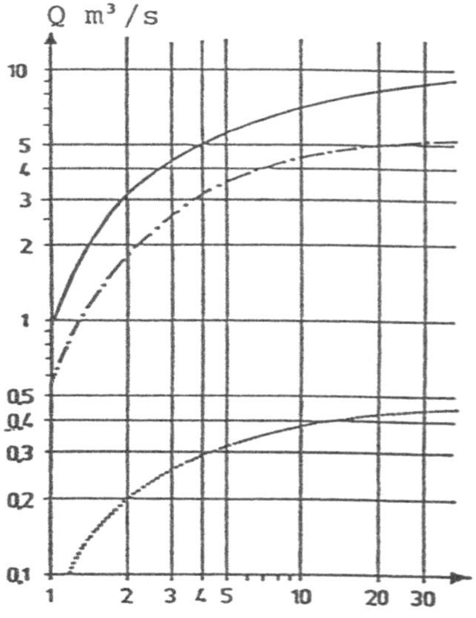

Figure 5 Sand

——— 1oo % artificially drained
—·—·—·— 5o % artificially drained, 5o % naturally drained
·············· 1oo % naturally drained

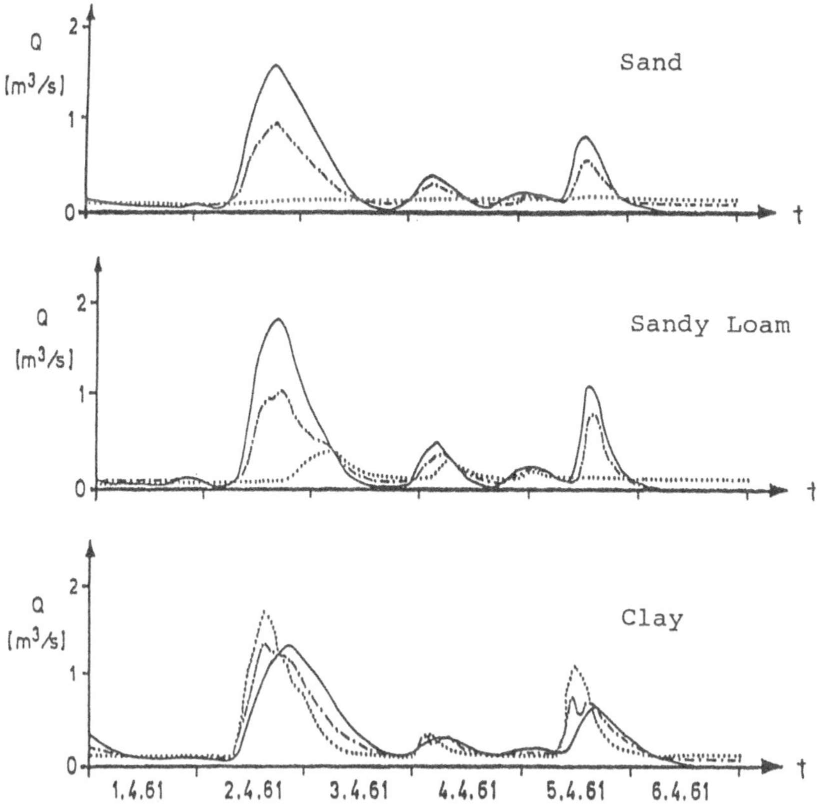

Figure 6   Storm Event of April 1961

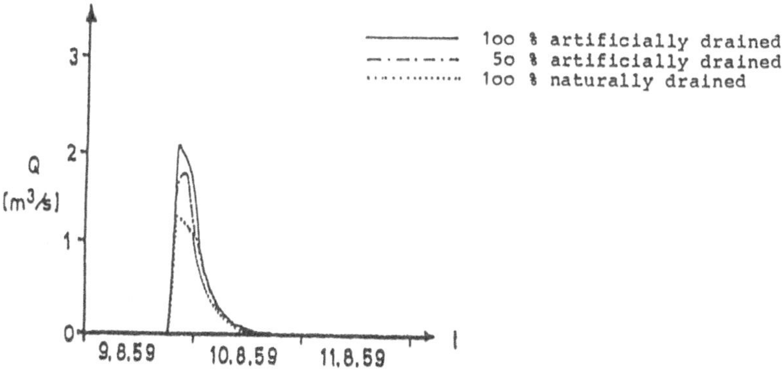

Figure 7   Storm Event of August 1959

events. At April 1, 1961, (Figure 6, clay) the groundwater table was 74 cm below surface in the case of an artificial drainage system and the soil could cope with the relative small event. In the case of the natural drainage system the groundwater table was right at the surface and overland flow occured. At August 9, 1959, (Figure 7) the groundwater tables were 208 cm and 168 cm below surface in case of an artificial and natural drainage system respectively and were not relevant at all. Overland flow took place because of rainfall intensity exceeding the infiltration rate. The differences between the hydrographs for artificial and natural drainage conditions are due to the different land use, i.e. different values for depression storage. By the way, for the sandy loam and the sand neither surface nor subsurface discharges were simulated for the event of August 9, 1958.
The study of these single events reveals very clearly, how important a continous simulation of the whole process and a statistical analysis is to avoid mis-interpretations.

CONCLUSIONS

- The effects of an artificial drainage system on flood discharge is depending on the soil.
- The more permeable a soil is, the more will the flood discharge increase.
- The spatial arrangement of artificial drained areas within a catchment, is not relevant to small catchments up to about 20 km^2 in size.
- For practical purposes, the effects are not of that magnitude to be considered in dimensioning procedures of the receiving stream and rivers.

REFERENCES

Bouwer, H., van Schilfgaarde, J. (1963) Simplified method of predicting the fall of water table in drained land. Transactions of ASAE 6, 4:288-291,296.
Brakensiek, D.L., Onstad, C.A. (1977) Parameter Estimation of the Green and Ampt Infiltration Equation. Water Resources Research, 13, 6:1009-1012.
Eggelsmann, R. (1981) Dränanleitung. 2. Auflage, Verlag Paul Parey, Hamburg und Berlin.
Skaggs, R.W. (1978) A water management model for shallow water table soils. Report no. 134, Water Resources Research Institute, North Carolina State University, USA.

Flood Protection and Drainage of the East and West Fens Hydrology

T.E. Evans and D.G. Thorn
Sir M. MacDonald & Partners, Demeter House, Station Road, Cambridge, CB1 2RS, England

SUMMARY

Improvements to land drainage was required for the East and West Fens covering an area of 40,000 ha. Because of the extensive network and the large number of engineering options available to improve the drainage, it was clear at the outset that simulation models of the two systems were required. Due to the substantial storage in the system and the time dependent transient hydraulic characteristics, it was considered inappropriate to use a steady-state model to predict water levels resulting from a large number of inflow hydrographs. Consequently, a computational transient hydraulic model was developed.

The model takes as its primary input concurrent inflow hydrographs occurring at 51 nodes in the Fenland system and 13 nodes in the Catchwater drains. A major part of the hydrological studies involved the derivation of the inflow hydrographs. For the Fenland area problems arose in modelling the system using inflow for actual flood events estimated by the UK Flood Studies (FSR) procedure (NERC, 1975). This led to modifications in the time to peak of the FSR regional unit hydrograph. Effective rainfall was set equal to the pumped outflow so eliminating the need to estimate this main input parameter. Some difficulties also arose in the selection of design events for evaluating engineering options. As a result a variety of different parameters were investigated to assist in assigning frequencies to recorded events.

The catchwater drain discharges by gravity through the Maud Foster sluice. No records were available of outflow which was calculated using tide levels in the Haven and water levels in the drain. Confirmation of the validity of calculated outflows was obtained by comparison with calculated volume of inflow (NERC, 1975) and modelled water levels. Of particular interest

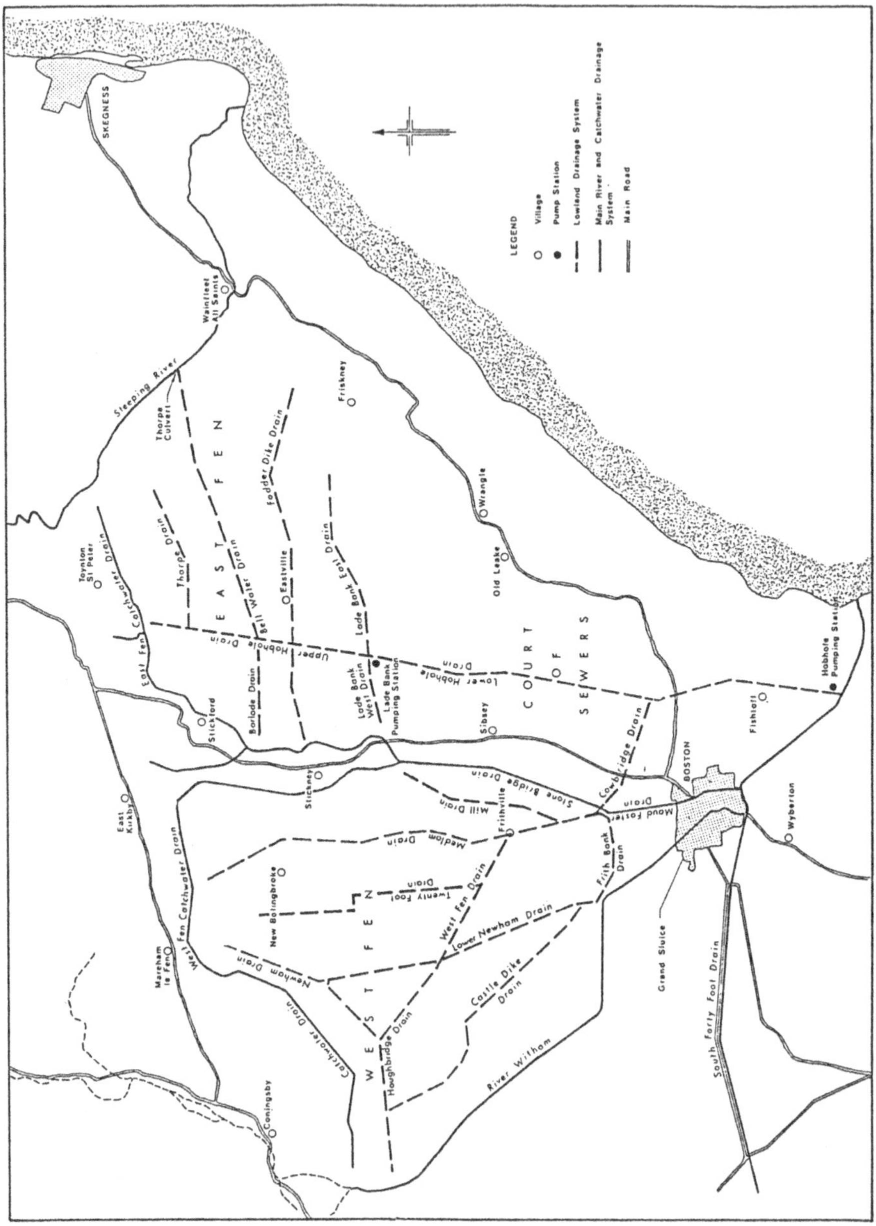

FIGURE 1 PROJECT LOCATION MAP

was a comparison between the FSR unit hydrograph approach to predicting flow from the Wolds and from flat Fenlands. It was also found that the usual design problem of calculating joint probabilities, in this case tide levels and flood runoff, could be resolved quite simply.

1. INTRODUCTION

The East and West Fens cover an area of some 40,000 ha north of Boston in Lincolnshire. Large parts of the area are at or below sea level and are drained by a complex system of channels constructed over many years, dating back to the sixteen century. The Fen drains feed into the Hobhole drain which outfalls at the Hobhole pumping station. All drainage during dry weather can be discharged by gravity into the Haven at times of low tide. Pumping is only required when inflow over the tidal cycle exceeds the gravity discharge capacity. The East Fen, in the north-east of the project area, contains the lowest land levels and its drainage water is pumped into the head of the lower Hobhole drain at Lade Bank.

Runoff from the Wolds, north of the Project Area, is intercepted by catchwater drains which channel water at a higher level through the Fens to a gravity outfall at Maud Foster sluice (first constructed in 1568) and thence into the river Witham estuary some four miles above Hobhole pumping station.

Progressive deterioration in the Lade Bank and Hobhole pumping stations and major floods in 1977 and 1981 led the Witham Fourth I.D.B. to put forward proposals for improvements to the drainage system. During the 1981 flood the high level catchwater system also spilled water into the Fens and in addition there is a danger of the catchwater causing urban flood damage. As a result, Anglian Water (AW) in May 1983 invited consultants to submit proposals for a comprehensive drainage study of both the East and West Fens and the Catchwater systems. The proposal submitted by Sir M MacDonald and Partners (MMP) was accepted and work started in July 1983. The terms of reference for the study are summarised below:

- identify alternative ways of carrying out drainage improvements which will provide a modern standard of drainage, defined as 1 m of freeboard during the one year in ten flood event and the prevention of flooding to the optimum degree;
- carry out hydrological studies;
- carry out hydraulic studies, including modelling studies;
- extend earlier assessments of benefits;
- develop engineering details and costs to allow selection of a preferred project option.

2. LOWLAND DATA AVAILABILITY

2.1 Flow Records

No direct flow records existed within the project area. However, detailed records of pumping hours and water levels, upstream and downstream, exist for the Hobhole pumping station which were used with pump characteristics to calculate hourly pumping rates. The calculations included corrections for pump speeds and wear to impellers and volutes. Gravity discharges were calculated by assuming a range of tidal levels in the Haven and a range of gravity discharges and using the hydraulic characteristics of the discharge culvert in a backwater analysis to construct a table of water levels/tidal levels/discharges.

Cumulative discharge volumes were calculated for nine major flood events between 1958 and 1982. table 1 summarises the results.

Table 1 Comparison between Areal Rainfall and Hobhole Outflow

Event period	Rain Event duration (days)	Areal rain (hm^3)	Outflow (days)	Outflow (hm^3)	Per-cent runoff
Dec '65	15	27.37	15	15.45	46
Feb '66	6	10.82	8	8.46	63
Jan '69	10	20.09	13	16.30	69
Jan '71	11	18.96	12	12.68	54
Jan '71	25	34.43	25	28.56	67
Dec '78	10	24.24	11	15.07	52
Jan '80	13	22.61	16	19.57	74
Mar '81	10	19.48	11	15.67	70
Mar '81	12	26.74	12	18.50	59

2.2 Lowland Areal Rainfall

21 rainfall stations are located within or immediately outside the catchment. Daily data was obtained from the Meteorological Office on magnatic tape and supplemented by hourly recorded charts for Dirby and Tetford which lie to the north of the Project Area. Effective rainfall based on discharge volumes calculated at Hobhole pumping station were compared with areal rainfall calculated using a multi-quadratic surface fitting approach (Lee, 1974). The results (Table 1) suggest runoff ranging from 50% to more than 70%, significantly more than reported by Beran (1982) for Newborough Fen where three events analysed in detail gave 32%. Further data supplied by Beran for Anderby Fen in Lincolnshire gave runoff coefficients exceeding 50% and 70% for 54% and 32% respectively of winter months. Although the runoff coefficients were suspected of being high (Beran, 1983 private communication) they were consistent and deemed not unreasonable for a well drained artificial catchment.

3. LOWLAND HYDROGRAPH ANALYSIS

Calculation of runoff hydrographs was based on the method described in Chapter 6 of the Flood Studies Report (NERC, 1975). The method is essentially a unit hydrograph convolution procedure and gives a number of empirical formulae for estimating parameters which cannot be derived from observed data. The two principal parameters required are effective rainfall and the unit hydrograph. The method employs catchment characteristics such as stream length, slope and average annual rainfall to calculate a synthetic triangular unit hydrograph. Catchment areas are drained by a large number of man-made channels and the normal methods of estimating stream lengths do not apply. Slopes assumed for main drains varied from 1 : 10 000 to 1 : 20 000 whilst sewer slopes ranged from between 1 : 3 000 and 1 : 5 000. The estimated stream slopes were based on the composition of drains and sewers along the main streams of the subcatchments.

The lowland system was divided into 53 subcatchments for the purpose of modelling (see Figure 2).

Effective rainfall based on daily rainfall and the runoff coefficients in Table 1 was distributed throughout each 24 hour period assuming a similar time distribution to that given by analysis of concurrent rainfall recorder charts at Tetford and Driby.

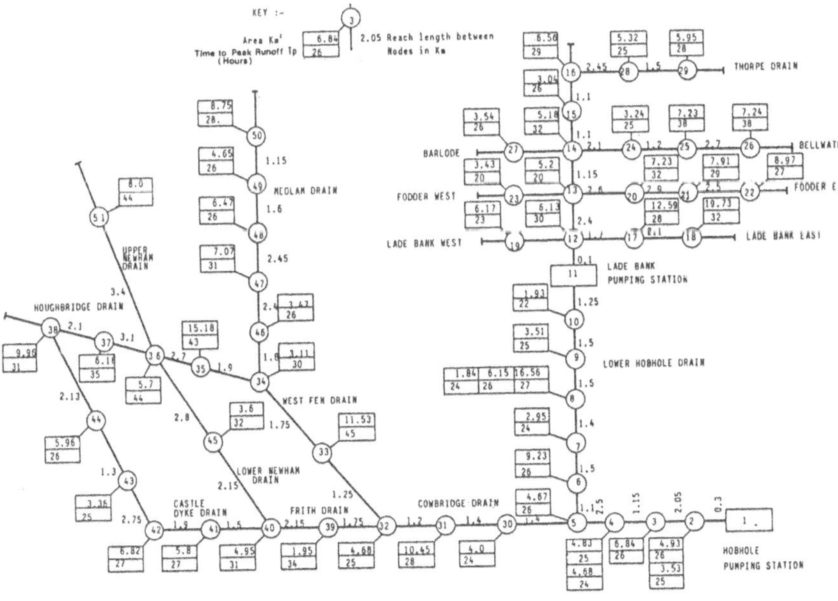

FIGURE 2 NETWORK DIAGRAM FOR FENLAND SYSTEM

Using the standard FSR unit hydrograph it was discovered that the time to maximum storage of the routed hydrographs was considerably longer than indicated by the recorded water levels. An example is shown in Figure 3. Although there are several possible explanations for this anomaly, the most likely is an over-estimate in the time to peak of the unit hydrograph. The FRS equation for estimating time to peak (Tp) was based on recorded data from natural catchments. The intensive field and open drainage systems in the Fens would be expected to reduce the time to peak. In addition, with the majority of the small catchments used in the UK Flood Study located in upland catchments with steep gradients, it is suspected that an inherent bias might exist in the regression analysis used to derive the Tp equation, which would lead to an over-estimate in Tp for small flat gradient catchments.

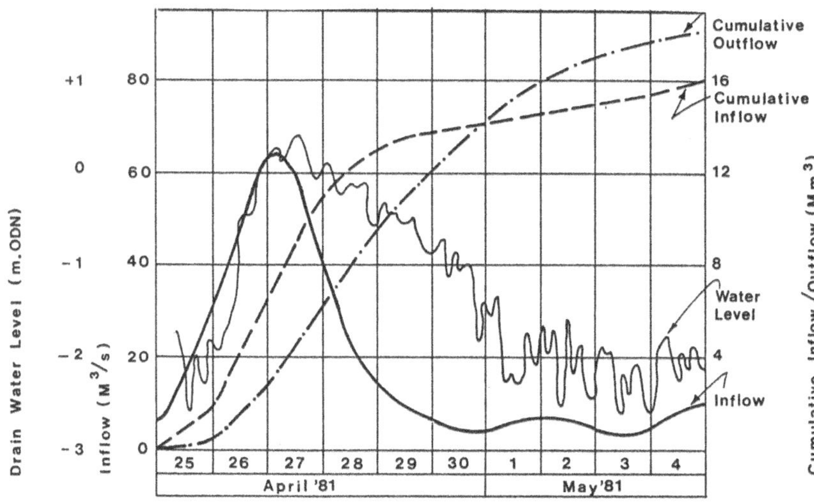

FIGURE 3 LOWLAND SYSTEM STORAGE AND HOBHOLE WATER LEVELS

To remedy this defect in the derived flood hydrographs, modifications were introduced using two criteria:

- recorded maximum instantaneous water levels should occur reasonably close to the calculated time of maximum storage;
- modified hydrographs should produce flood levels at Hobhole and Lade Bank pumping stations similar to recorded levels.

A justification for using the first criterion is that maximum water levels at Hobhole, Lade Bank and West Fen sluice tend to occur concurrently. Estmated FSR time to peak were decreased to reduce the time difference between maximum storage and maximum water levels. Figure 5 illustrates the effect of reducing FSR time to peak by a factor of 0.4 for nine major storms.

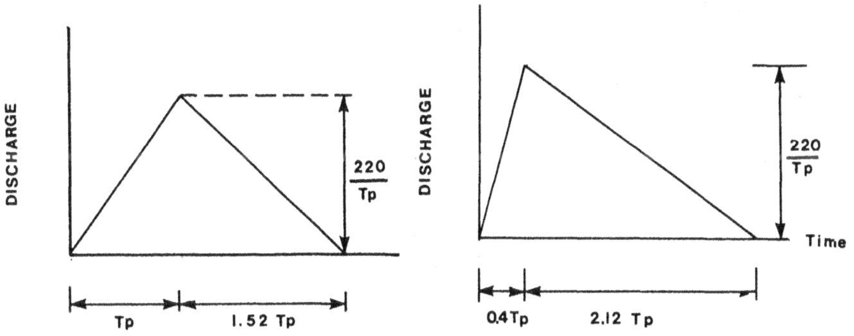

FIGURE 4 MODIFICATION OF FSR UNIT HYDROGRAPH

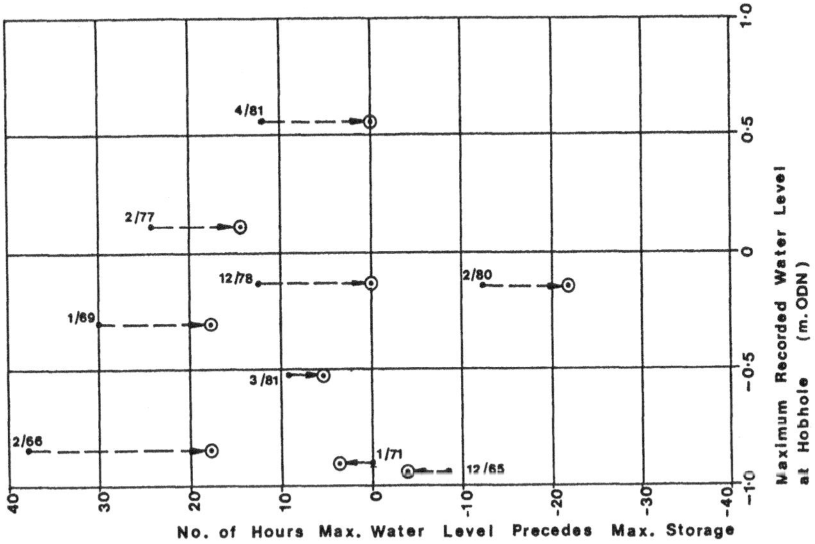

FIGURE 5 EFFECT OF REDUCING TIMES TO PEAK

The application of a constant reduction factor of 0.4 to Tp was accepted and led to a reasonable correlation with recorded levels when fed into the hydraulic model.

4. LOWLAND DESIGN CRITERIA

Fenland drainage systems are often designed to discharge flows expressed as runoff depths in 24 hours. A modern standard in UK is often taken as 12 mm; corresponding to a constant outflow of 1.4 l/s/ha or 20 cusecs/100 acres. Lade Bank has a design capacity of 1.12 l/s/ha whilst Hobhole (which is also equipped with a gravity sluice) has a capacity of 0.83 l/s/ha. The terms of reference for the present study specify a modern standard as:

- a minimum of 1 metre freeboard above the water level of a 1 in 10 year flood event to lowest general land levels;
- the prevention of flooding to the optimum degree of any significant area of land surface.

It was decided to base inflow conditions for a 1 in 10 year flood on recorded data. The main contributary factors in determining the severity of a flood event in the project area are rainfall amount, rainfall duration, soil moisture/antecedent rainfall and water levels in the Haven. As a first requirement it was necessary to estimate the return period of past events. A number of different parameters are available for establishing the probability of occurrence of past events by applying frequency analysis techniques. Those investigated included; number of pumping hours, volume discharged at Hobhole, winter rainfall, effective rainfall, maximum storage in system and maximum recorded water levels. As would be expected the latter proved the best indicator as this parameter integrates all the contributary factors leading to flooding i.e. tide levels, rainfall, runoff etc.

Effective rainfall was calculated using an in-house computer program, BALANCE, which simulates a daily soil moisture balance. Three rainfall stations were used and a continuous daily soil moisture level estimated for the project area. The calculated soil moisture deficits (SMD) were used in the FSR procedures to calculate the catchment wetness index (CWI) and so derive percentage runoff (PR) for extreme events:

$PR = SPR + 0.22\ (CWI - 125) + 0.1\ (P - 10)$.

The annual maximum 2-day effective rainfall so derived was fitted to a normal distribution and gave return periods of 56, 25 and 10.5 years for the 1981, 1977 and 1978 events.

A frequency analysis of maximum water levels is reproduced in Figure 6.

The series of maximum water levels up to zero m ODN can be considered relatively homogeneous, though it is partially controlled by the level of pumping. Above zero m ODN outflow starts to be influenced by two factors: limiting pump capacity and flood attenuation. Flooding of land tends to reduce flood levels whilst the throttling effect of the pump station increases level. The frequency plot should not therefore be extended above a level of zero m ODN. The analysis suggests that the February 1977 flood was in excess of 1 in 10 year event whilst the December 1978 event was somewhat lower.

Rather than selecting a single event on which to base engineering designs a better approach, when a simulation model of a system has been constructed, is to operate the system for a range of different inputs and base the recommended engineering works on their performance under differing conditions. It was therefore decided to modify both the 1977 and 1978 inflow hydrographs to give a peak water level at Hobhole with return periods of 10 years. In addition engineering options were also investigated by routing unmodified 1969, 1977, 1978 and 1981 floods.

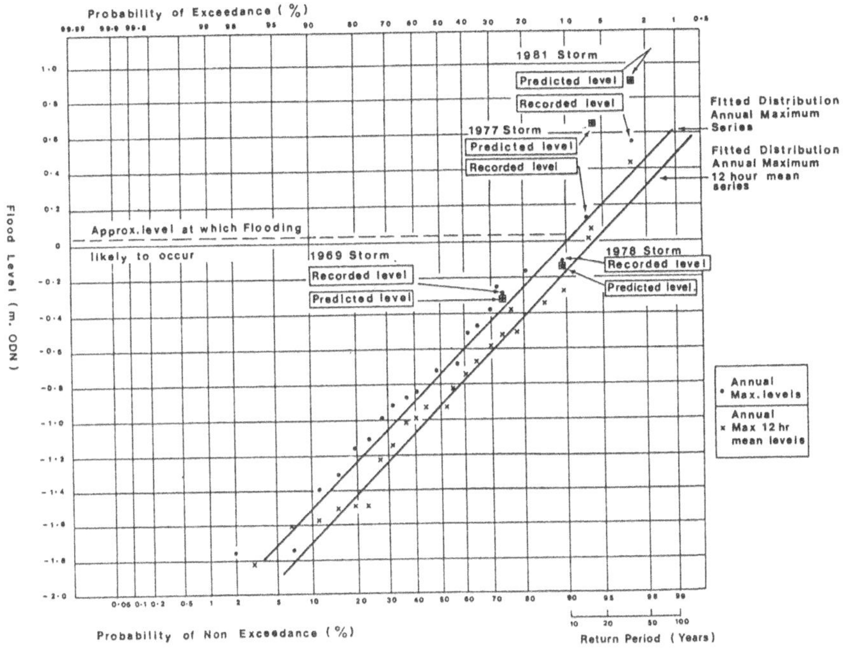

FIGURE 6 FREQUENCY ANALYSIS OF MAXIMUM WATER LEVELS AT HOBHOLE PUMPING STATION

5. CATCHWATER DATA AVAILABILITY

5.1 Maud Foster Sluice Outflow

The northern margin of the Witham Fourth District is protected by a catchwater drainage system which outfalls to the Haven through the Maud Foster sluice. From recorded levels in the Maud Foster drain and in the Haven it has been possible to calculate outflow from the catchwater drain for flood events over the past nine years. It also proved possible to calibrate the discharge equations used in calculating outflow by comparing outflow with calculated inflow based on rainfall.

Discharge equations used to estimate outflow through Maud Foster sluice were derived by HRS, the Hydraulic Research Station, and are quoted in the FLUCOMP program.

Sluice flow - gate partially open

$$Q = Cs \times F_1 \times A \times 2g \sqrt{Hu - w/2}$$

F_1 = Drowned flow function = $0.581 \sin (2.104 \times \emptyset^{0.45})$

$$\emptyset = \left\{\frac{W}{Hd} - \frac{W}{Hu}\right\}$$

where : Hu = upstream head over sill

Hd = downstream head over sill

W = gate opening

A = area of gate opening

Cs = coeficient = 0.7

and two associated discharge equations for weir flow with gate fully open and overflow with gate closed were also used. A typical coefficient of 0.7 was selected which gave inflow values which corresponded closely to the calculated inflow based on the FSR (1975) catchment parameter equation for estimating effective rainfall. A comparison of inflow and outflow estimates is presented in Table 2.

Table 2 Comparison Between Inflow Estimated by FSR Method and Calculated Outflow Volumes for Maud Foster Sluice

Storm	Maud Foster outflow (hm^3)	Volumes of effective rainfall (hm^3)
Feb 1977	1.4	1.5
Dec 1978	2.0	1.7
Feb 1979	0.9	0.7
Mar 1981	1.0	1.2
Apr 1981	3.3	3.1
Apr 1983	0.8	0.7

5.2 Simulated Tidal Conditions in the Haven

The nine years of water levels at Maud Foster sluice are not sufficient to obtain reliable estimates of extreme flood events for return periods of up to 100 years. A different approach was required to that adopted for the Lowland system and involved the routing of synthetic flood hydrographs. Consequently, the routing model developed for the catchwater system requires as input tidal levels. It was necessary therefore to simulate synthetic tidal sequences for use with different design floods.

Ranking of peak flood levels recorded in the drain and highest recorded low tide levels in the Haven showed a high degree of correlation indicating that extreme run-off within the river Witham and the Catchwater catchments tend to be coincident. With the probability distributions for catchwaterpeak flows and Haven low tide levels highly correlated, joint probability solutions were therefore avoided. A partial series of highest low tide levels recorded in the Haven at the time of maximum levels in the catchwater drain as shown in Figure 7 and typical sequences of tidal fluctuations are reproduced in Figure 8.

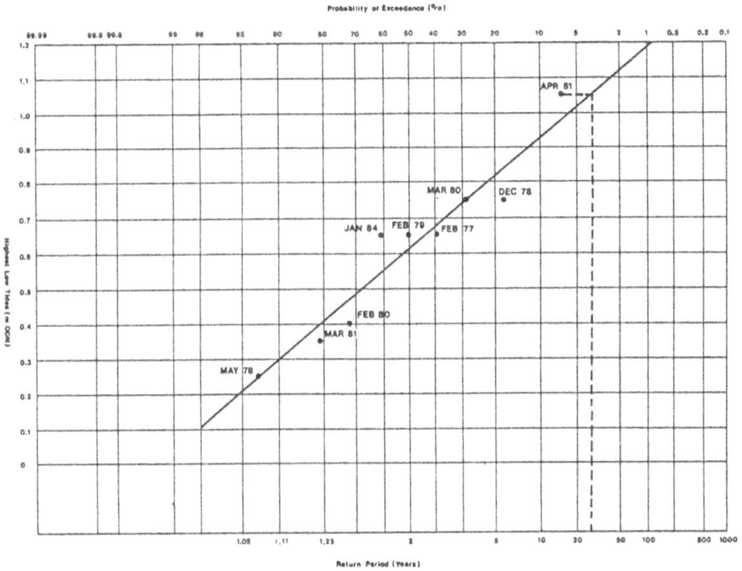

FIGURE 7 PARTIAL SERIES ANALYSIS FOR HIGHEST LOW TIDE LEVELS IN THE HAVEN, MAUD FOSTER SLUICE (1975 - 83)

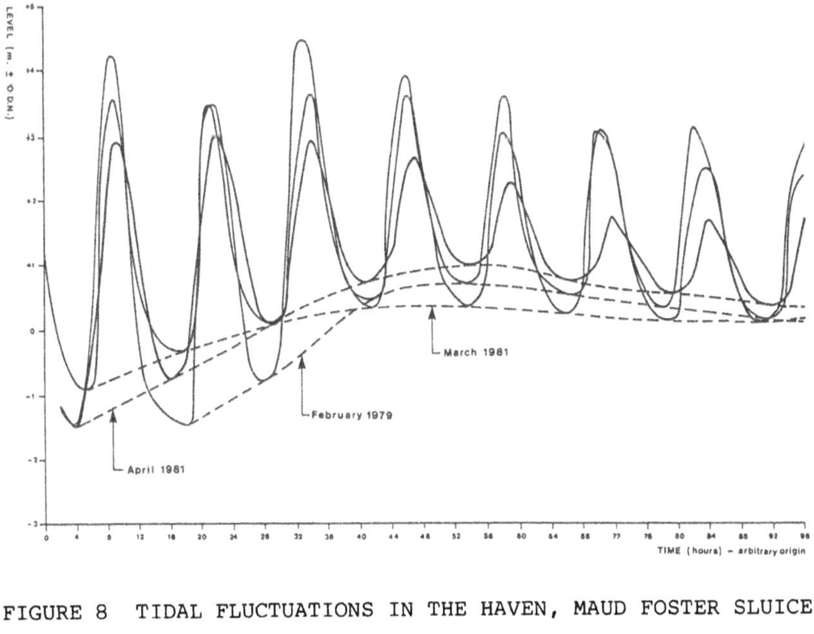

FIGURE 8 TIDAL FLUCTUATIONS IN THE HAVEN, MAUD FOSTER SLUICE

The level sequence recorded during the March 1981 flood was selected as the most critical, with higher levels maintained over a longer duration, and values adjusted linearly to give a maximum high low water level of 4.18 metres for the 100 year return period event. Similarly sequences for 50, 25 and 10 year return period events were obtained.

5.3 Catchwater Areal Rainfall

As with the Fenland system effective rainfall was estimated for the main storm events using the FSR methodology. Volumes of run-off for six storm events are given in Table 2. Daily soil moisture was estimated from rainfall data recorded at Hagworthingham and East Kirby for the period 1886 - 1983 and used to calculate the catchment wetness index. A trend analysis was undertaken with the data which was also used to assess the effect on run-off of changes in farming practices resulting from greater compaction of soils.

6. CATCHWATER HYDROGRAPH ANALYSIS

Unit hydrographs were derived for the river at Partney, 10 Km north-west of the Project Area. Rainfall losses were not distributed according either to the constant loss rate or by constant percentage as recommended by the FSR. A non-linear technique was applied which gave more satisfactory results. The unit hydrographs so derived were compared with estimated regional synthetic FSR hydrographs. The results were extremely satisfactory and gave good confidence in applying the FSR approach to the Catchwater catchment.

The area draining to the Catchwater system was divided into 13 subcatchments and the drainage system simplified into a node and reach network, as shown in Figure 9.

The Flood Studies method was therefore used to estimate synthetic design hydrographs with return periods of 10, 25, 50 and 100 years. The steps involved are similar to those applied in the lowland system. Hagworthingham rainfall data were used to evaluate effective rainfall for use with standard FSR rainfall profiles before convoluting with standard FSR unit hydrographs. A rainfall duration of 14 hours was found to produce the highest peak flow.

As a check on the FSR derived flow into the catchwater system, a comparison was made between recorded April 1981 water levels in Maud Foster drain and storage as defined by inflow less outflow. The results as shown in Figure 10 demonstrate that time of maximum and minimum storage based on estimated outflow and inflow is compatible with the times of maximum and minimum water levels and therefore the design inputs should enable the catchwater model to produce realistic results.

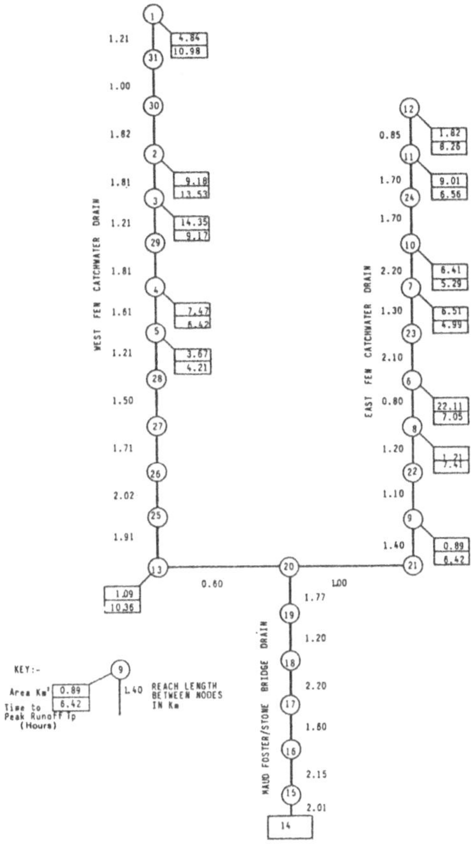

FIGURE 9 NETWORK DIAGRAM FOR CATCHWATER SYSTEM

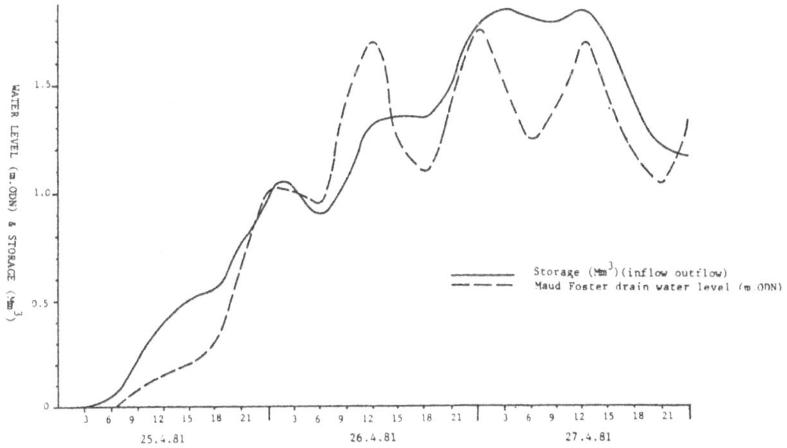

FIGURE 10 COMPARISON BETWEEN WATER LEVELS AT MAUD FOSTER AND STORAGE IN CATCHWATER SYSTEM, APRIL 1981

For model simulation, inflow was introduced into the system at each of the 13 nodes and routed to Maud Foster sluice using an unsteady state dynamic model. Details of the model, its calibration and application,are provided in an accompanying paper. As in the Lowland system a range of inflows was modelled including both synthetic and recorded floods, details of which are given in Figure 11.

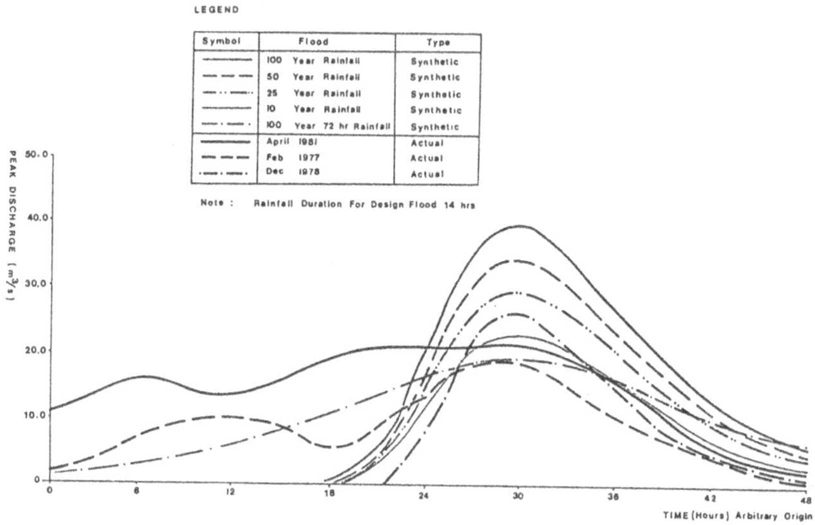

FIGURE 11 COMPARISON BETWEEN SYNTHETIC AND ACTUAL INFLOW TO THE CATCHWATER SYSTEM.

7. ACKNOWLEDGEMENTS

This paper is presented with the permission of Anglian Water and the Witham Fourth Internal Drainage Board and the Consultants wish to acknowledge the help, assistance and adivce given by both authorities.

8. REFERENCES

Guganesharajah K and Thorn, DG, (1986), Flood Protection and Drainage of the East and West Fens: Unsteady State Modelling Studies. 2nd Int. Conf. on Hydraulic Design in Water Resources Engineering: Land Drainage. University of Southampton, April

Thorn, DG and Davidson, W (1986), Flood Protection and Drainage of the East and West Fens: Engineering and Economics. 2nd Int. Conf. on Hydraulic Design in Water Resources Engineering: Land Drainage. University of Southampton, April.

NERC (1975) Flood Studies Report.

Beran, MA (1982) The Drainage of low lying Flatlands. Conference of River Engineers, Cranfield, U.K.

Beran, HA (1983) Personal Communication.

Flood Protection and Drainage of the East and West Fens Unsteady Flow Modelling Studies

D.G. Thorn and K. Guganesharajah
Sir M. MacDonald & Partners, Demeter House, Station Road, Cambridge, CB1 2RS, England

1. INTRODUCTION

The East and West Fens cover about 40 000 ha. with large parts of the area at or below sea level, and are drained by a complex system of interconnected channels. Companion papers describe hydrological studies (Evans, 1986) and studies of engineering and economics (Thorn and Davidson, 1986). This paper is confined to unsteady flow modelling studies aimed at evaluating the hydraulic performance of alternative development options.

Some of the development options could be compared using simple backwater analyses and cost estimates. Others required use of an unsteady state computational model to account for the effect of channel storage in attenuating flows in the drainage system, in conjunction with alternative outfall capacities and a possible offstream storage site. The model described is flexible in use and allows closed loops in the drainage system, where water can drain via more than one route to a common point. Boundary conditions include the options of pumping or gravity discharge to a tidal river estuary.

2. MODEL FORMULATION

The model uses conservation of mass at points of interest (nodes) and within discrete channel reaches and conservation of energy in channels. Node locations are determined by considering channel capacities and their catchment areas and by the requirement that channel reaches between nodes are not excessively long. Inflows to the drainage system are considered only at nodes. Evans (1986) gives the resulting node/reach network diagrams for the low level drainage system, which directly drains the East and West Fens, and for the catchwater system which prevents runoff from the Lincolnshire Wolds from entering the project area. The existing low level system was represented by 51 nodes and 52 channel reaches, and the catchwater model by 31 nodes and 30 channel reaches. Channel geometries were defined by surveyed cross-sections covering some 180 km of channels with spacings ranging from 100 m to 3 km.

The following notation is defined:

h_i^t = water level at node i, time t

I_i^t = surface water inflow to node i, time t

$Q_{i,k}^t$ = outflow from node i to channel k, time t

x_k = length of channel reach k

$S_{i,k}^t$ = energy slope in reach k, time t, evaluated near node i

V_k^t = water stored in channel k at time t

$A_{i,k}^t$ = cross-sectional area of flow in reach k at time t, evaluated near node i

$B_{i,k}^t$ = water surface width in reach k at time t, evaluated near node i

N = number of nodes

M = number of channels

Mass conservation at node i:

$$\Sigma Q_{i,k}^t - I_i^t = 0 \tag{1}$$

Mass conservation within channel k:

$$\left(\frac{dv}{dt}\right)_k^t = Q_{i,k}^t + Q_{j,k}^t \tag{2}$$

where i and j refer to upstream and downstream nodes respectively

Energy conservation in channel k

$$h_i - h_j - \frac{x_k}{2}(S_{i,k}^t + S_{j,k}^t) = 0 \tag{3}$$

Equation 3 is a simplified form of the dynamic St. Venant equation in which inertia terms are neglected. Variations in water velocity with both time and distance are small in the artificial drainage networks considered, which have comparatively large storage capacities and a correspondingly slow response to inflows. The dominant terms in the dynamic equation are therefore the water surface slope and the friction slope.

Finite differences are used to approximate the rates of change of variables with time. Only equation 2 is expressed in finite differences, other equations are considered as applying to a particular time instant. Approximating channel storage by water area times channel length:

$$V_k^t = \frac{x_k}{2}(A_{i,k}^t + A_{j,k}^t)$$

The total change in channel storage over the time period (t-1) to t equals the average net inflow

$$Q_{i,k}^t + Q_{j,k}^t - Q_{i,k}^{t-1} - Q_{j,k}^{t-1} = \frac{x_k}{\Delta t}(A_{i,k}^t + A_{j,k}^t - A_{i,k}^{t-1} - A_{j,k}^{t-1})$$

where Δt = time period

A solution for time (t-1) has already been found by the model or represents known initial conditions. Collecting known terms gives

$$f_k^t = Q_{i,k}^{t-1} + Q_{j,k}^{t-1} + \frac{x_k}{\Delta t}(A_{i,k}^{t-1} + A_{j,k}^{t-1}) \qquad (4)$$

$$Q_{i,k}^t + Q_{j,k}^{t-1} - \frac{x_k}{t}(A_{i,k}^t + A_{j,k}^t) + f_k^t = 0 \qquad (5)$$

Equations 1, 3 and 5 define a set of (N + 2M) non-linear equations in (N + 2M) unknowns. These equations are solved by successive approximations using a generalised form of the Newton-Raphson method. A description of the method as applied to channel routing is given by Williams and Chidley (1976).

3. BOUNDARY CONDITIONS

Nodes on the boundary of a network often have specific characteristics which must be included in the model if realistic results are to be obtained. For example, discharge rates at Hobhole pumping station depend on relative upstream and downstream water levels and either the number of pumps in operation or the setting of the gravity discharge sluice gate. The model includes a set of critical levels specifying 'switch-on' and 'switch-off' pumps for the pumps, checking that downstream water levels are higher than upstream drain water levels: if this is not so then gravity discharge is assumed. Hydraulic characteristics of the pumps or gravity sluice are incorporated by introducing additional terms into the finite difference representation of the basic equations. These terms give the variation in discharge with water level.

Boundary nodes proved difficult to model, mainly because of the large variations in discharge that can occur with small changes in water level, which can result in a pump being switched on or off (at Hobhole) or a sluice gate being closed or opened (at Maud Foster). In extreme cases it was found that the model continuously oscillated between two approximate solutions and a smoothing procedure was used. However, this is not as serious as it sounds: water level fluctuations at boundary nodes are rapidly attenuated by channel storage with increasing distance upstream and the fluctuations have very little effect on most interior nodes.

4. INITIAL WATER LEVELS AND FLOWS

The model requires an initial consistent set of water levels and reach flows in order to begin computation. These values are calculated by assuming steady-state conditions, simulated by omitting all time-dependent terms from the model, which thus degenerates to calculation of back-water surface profiles. In practice, it proved sufficient to always start the model from the same set of initial conditions, irrespective of the actual inflow hydrographs and tidal fluctuations to be simulated.

5. COMPUTATIONAL ASPECTS

It was required that the developed model be flexible in use, allowing rapid reconfiguration to model any proposed development. The model therefore incorporates a system of indexing, where all nodes, channel reaches and cross-sections are assigned working reference numbers at the start of model execution. Thus all model components can be allocated any meaningful reference numbers by the user, these numbers being included in a data file, defining for each channel reach the user-given reach number, upstream and downstream cross-sections and upstream and downstream node numbers. Data contained in other files but not referenced by the reach file is ignored by the model. Simple changes to the reach file thus allow easy reconfiguration of the drainage network and substitution of proposed for surveyed cross-sections.

The heaviest computational load is imposed by the requirement of the Newton-Raphson method to successively solve a number of linear equations. In the case of the low-level system there are more than 150 equations to be solved at each iteration. A typical run of the model covered 24 time steps, each with an average of 5 iterations, thus requiring 120 solution sets. The matrix representing the simultaneous solutions is sparse, most rows having not more than 3 non-zero entries, and the model incorporates a sparse matrix routine which greatly reduces computation times.

6. MODEL CALIBRATION

Calibration of the model consisted of varying Manning's roughness coefficients and control rules used for pump and sluice operation until reasonable agreement was obtained between simulated and recorded water levels. In practice, calibration was a compromise, with better results being obtained for some storm events at the expense of others. Furthermore, calibration data was limited and available information for storm events causing flooding was of restricted use because the model did not include overbank spills: predicted levels for these events were thus higher than recorded. Tables 1 and 2 compare predicted and observed water levels for the low-level and catchwater systems respectively and Figure 1 gives a sample comparison at Hobhole pumping station. Given the limited data for calibration and the degree of empiricism inherent in estimation of inflow hydrographs, the model was accepted as adequate to assess the effect of alternative proposals for flood prevention and drainage improvements.

7. MODEL USAGE

The model was used to design a number of alternative projects such that a common standard of flood protection and land drainage was achieved between projects and throughout the project area. The types of project component included:

- increasing the capacity of outfalls and pumping stations
- raising channel banks at places susceptible to overtopping
- channel dredging to improve conveyance and increase storage
- install small pumping stations to maintain low water levels in local catchments.

Each alternative project involved running the model a number of times to investigate the effect of changing project components. Each run of the model required changes to be made to input data, particularly that relating to channel cross-sections and network details. These changes were easily made by modifying the channel reach data file as described in Section 6. Changes to other input data (inflow hydrographs, tide levels etc.) are accommodated by the user specifying the various data files to be used when prompted by the model. The end result is a very flexible arrangement which allows the user to specify new input data with a minimum of effort.

Model output is principally a set of water levels and flows at selected points in the drainage network. These data are used to draw up surface water profiles showing the effectiveness of each alternative project.

Table 1: Performance of the Low-Level Model

Locations	Peak water levels predicted by model (m ODN) January 1969	December 1978
1. Hobhole pumping station	-0.27 (-0.29)	-0.17 (-0.15)
2. Lower Hobhole drain (ch 6 000 m)	0.22	0.42
3. Lower Hobhole drain (ch 11 800 m)	0.43 (0.48)	0.59 (0.40)
4. Lade Bank pumping station	-0.94 (-0.96)	-0.77 (-0.88)
5. Upper Hobhole drain (ch 5 850 m)	-0.86	-0.53
6. Thorpe drain (ch 3 950 m)	-0.71	-0.46
7. Cowbridge drain (ch 4 000 m)	0.37 (0.30)	0.56
8. Frith Bank drain (ch 3 900 m)	0.47	0.64
9. Medlam drain (ch 9 400 m)	0.48	0.70
10. Upper Newham drain (ch 3 400 m)	0.49	0.71
11. Castle Dike drain (ch 6 150 m)	0.54	0.70
12. West Fen drain (ch 4 900 m)	0.43	0.64

Note: observed water levels shown in brackets.

Table 2: Performance of the catchwater model

Location	Peak water levels predicted by model (ODN) Feb* 1977	Dec 1978	Apr* 1981	Apr 1983
A Maud Foster sluice	1.65 (1.57)	1.67 (1.57)	2.20 (1.80)	0.95 (0.88)
B Cowbridge lock	1.66	1.67	2.21 (1.85)	0.96 (0.99)
C Confluence of East and West Fen catchwater drains	1.71	1.67	2.26 (1.90)	0.98
D Hagnaby beck East Fen catchwater (ch 5 600 m)	1.92	2.17	2.40 (2.40)	1.16
E Halton East Fen catchwater (ch 14 500 m)	2.16	2.44	2.57 (2.40)	1.36
F Hagnaby lock West Fen catchwater (ch 6 600 m)	1.81	1.85	2.37 (2.05)	1.05
G New York West Fen catchwater (ch 19 300 m)	2.25	2.42	2.73 (2.80)	1.71

Note: observed water levels shown in brackets.

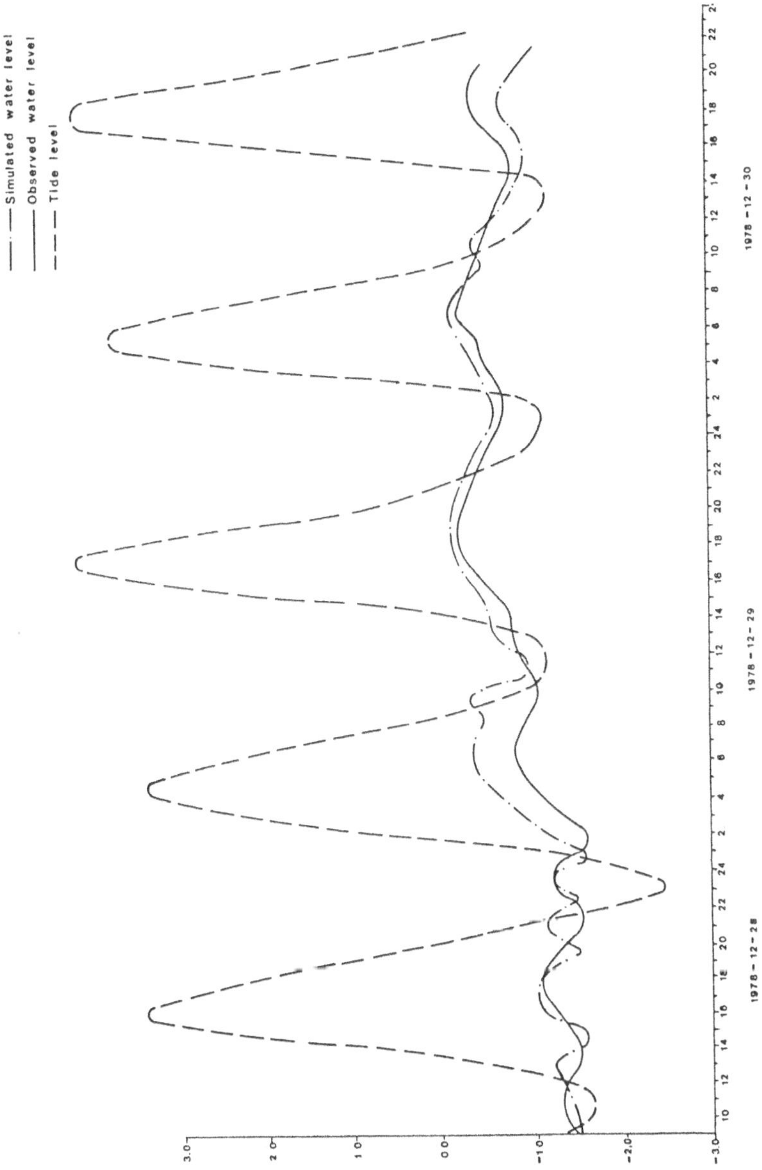

FIGURE 1

Simulated and observed water levels at
Hobhole pumping station

8. ACKNOWLEDGEMENTS

This paper is presented with the permission of Anglian Water and the Witham Fourth Internal Drainage Board.

References

Evans, T.E. and Thorn, D.G. (1986) Flood Protection and Drainage of the East and West Fens : Hydrology. 2nd Int. Conf. on Hydraulic Design in Water Resources Engineering : Land Drainage. University of Southampton. April.

Thorn, D.G. and Davidson, W.G. (1986) Flood Protection and Drainage of the East and West Fens : Engineering and Economics. 2nd Int. Conf. on Hydraulic Design in Water Resources Engineering : Land Drainage. University of Southampton. April.

Williams, J.J.R. and Chidley, T.R.E. (1976) Nonlinear Analysis of Unsteady Flow in Open Channel Networks. Int. Symposium on Unsteady Flow in Open Channels. BHRA, Bedford. Paper F1.

Flood Protection and Drainage of the East and West Fens Engineering and Economics

D.G. Thorn and W.G. Davidson
Sir M. MacDonald & Partners, Demeter House, Station Road, Cambridge, CB1 2RS, England

A large number of options were identified to reduce flood risk and improve drainage in the East and West Fens. Selection between these options was by a three-stage process involving increasing detail. The first stage used criteria of comparative costs or practicality and the second stage involved backwater analyses to give a sound basis for comparing costs of channel improvements and options requiring pumping. The last stage covered 9 alternative projects for the lowland system and a range of options for the catchwater system, each of which was designed using an unsteady state numerical model to give common standards of flood protection and drainage. Final project selection was made on grounds of cost.

Economics studies included a farm survey covering about 40% of the project area to identify drainage problems and agricultural practices. Project benefits were calculated separately for reduced flood risk and improved drainage. The latter benefits were 90% of total benefits.

1. INTRODUCTION

The East and West fens cover an area of about 40 000 ha north of Boston, Lincolnshire, extending from near the coast for about 22 km inland. Large parts of the area are at or below sea level. A major flood in 1981, progressive obsolescence of pumping stations and increased awareness of the growing value of agricultural production have been instrumental in prompting studies aimed at improving standards of flood protection and drainage.

The lowlying area forming the bulk of the project area is drained by a complex system of channels (Figure 1). The largest of these is the Hobhole drain which outfalls to the

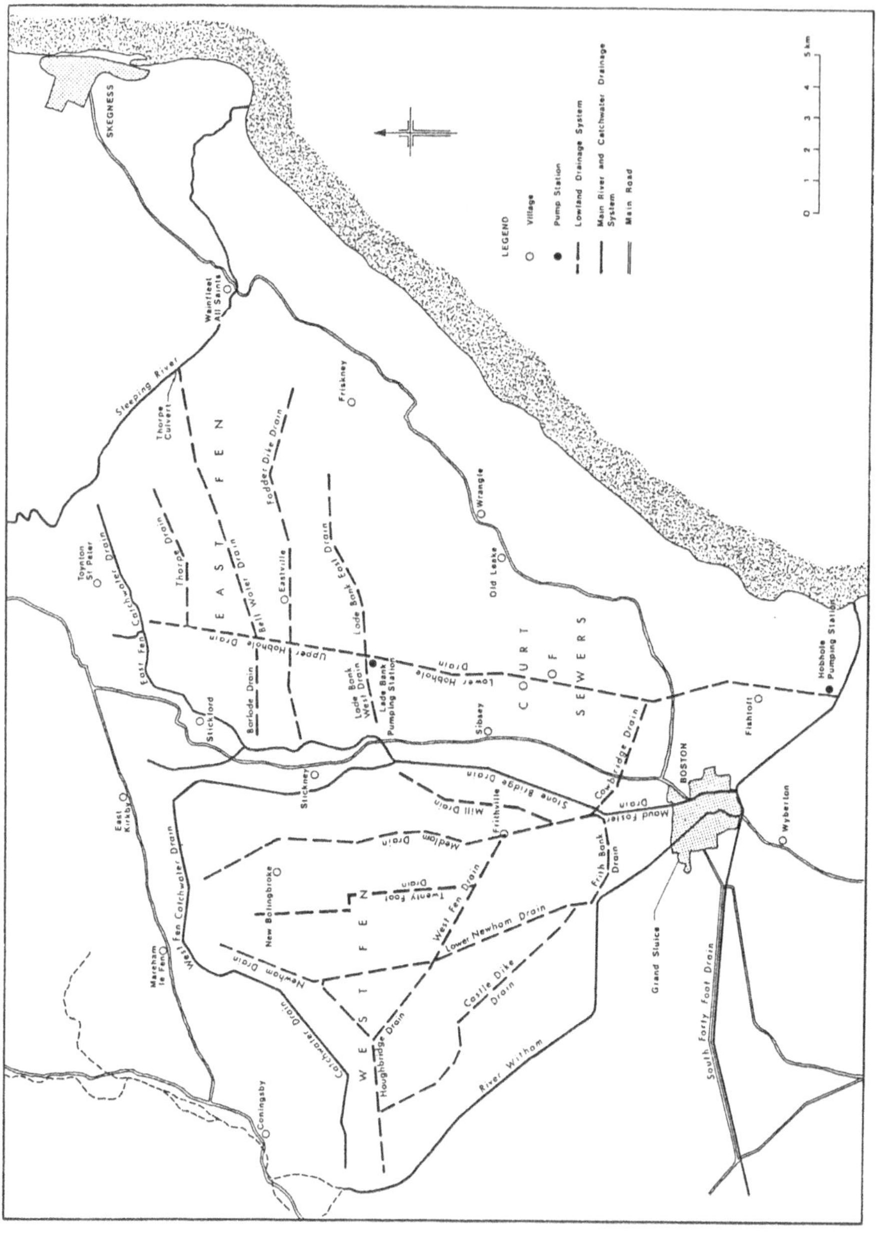

FIGURE 1

THE PROJECT AREA

tidal R. Witham at Hobhole pumping station, commissioned in 1957. Drainage water is discharged by gravity or pumping depending on relative water levels upstream and downstream of the outfall. The East Fen forms the northern catchment of the Hobhole drain and contains the lowest land levels. It is pumped by Lade Bank pumping station, commissioned in 1941 and upgraded in 1964. The western part of the project area comprises the West and Wildmore Fens which drain via the Cowbridge drain into Hobhole drain below Lade Bank.

Runoff from the Wolds, north of the project area, is intercepted by the East and West Fen catchwater drains and routed through the high level Stone Bridge and Maud Foster drains to outfall at Maud Foster sluice by gravity. Discharge rates depend on gate openings and relative water levels upstream and downstream of the sluice which are subject to tidal variations.

A companion paper (Evans, 1986) describes hydrological studies and the formulation and calibration of an unsteady state numerical model of the drainage systems. This paper is concerned with the identification, evaluation and selection of alternative schemes to improve flood protection and drainage, together with concurrent studies of project economics to assess benefits.

2. PROJECT COMPONENTS

The combination of a very flat topography and the location of the lowest areas of land nearest to the head of the drainage network results in difficulties in maintaining sufficiently low water levels to ensure adequate freeboard during normal winter conditions and in achieving swift drainage during severe storms. This causes frequent waterlogging and occasional flooding. The two existing pumping stations are considered to be of inadequate capacity to provide efficient drainage and one of these, at Lade Bank, is near the end of its useful life. The other at the outfall of the Hobhole drain is not automated and is difficult and uneconomic to operate especially when handling duty flows. To overcome these problems a large number of project components were considered and combined into a number of alternative developments designed to provide a common standard of flood protection and drainage. This standard was defined as a minimum freeboard of 1.0 m to be maintained during a 1 year in 10 storm. The types of project components can be grouped under the following headings:-

- Channel improvements to improve conveyance and storage capacities.
- Small pumping stations to maintain low levels within localised areas.
- Improvement or replacement of the existing Lade Bank pumping station, either directly or by new pumping stations at more than one location.
- Augmenting capacity at Hobhole pumping station, thereby extending the life of existing diesel driven pump sets,

reducing operating costs and providing increased security.
- Offstream storage at Lade Bank to reduce peak flows into the lower Hobhole drain.
- Improvements to Maud Foster sluice.
- Structural changes to the drainage system, for example by pumping water from the low level into the high level catchwater system.

Each of the above component types, particularly the first two, comprise a number of options for consideration when building up a complete development alternative. Some of the components were mutually exclusive, for example the same objective could sometimes be obtained by either localised channel improvements or installation of a small booster pump: selection of either component would automatically exclude the other. Nevertheless, there were a very large number of alternative development projects which could be identified. In order to simplify matters to some degree, advantage was taken of the continued requirement for pumping in the vicinity of Lade Bank thus allowing the drainage systems above and below this point to be considered independently.

3. LOW LEVEL SYSTEM PROJECT SELECTION

Project selection for the low level drainage system was carried out by a 3 stage procedure to reduce the number of alternatives requiring detailed study. The first stage allowed some project alternatives to be eliminated for practical reasons or after simple cost comparisons.

The second stage of the selection procedure utilised backwater analyses to predict main drain water levels for different options at two sets of steady state runoff rates which were selected as being reasonable given the catchment areas drained.

Water levels throughout the drainage network were predicted from the backwater analysis and used to identify areas with less than 1.0 m freeboard. The number of small booster pumping stations that would be required to drain the localised catchment areas not achieving 1.0 m freeboard were estimated and detailed cost comparisons made. The second stage selection procedure did not consider options for a storage reservoir, varying the capacities of Hobhole and/or Lade Bank pumping stations or assessing the capacity of a possible new pumping station at Midville as these could not be adequately studied under steady state conditions.

The third stage of the selection procedure utilised an unsteady state numerical model to study the most beneficial options identified by the backwater analysis along with those not studied during that stage. Seven alternative project options were identified for detailed study; two of these had a subsidiary option including a 2.7 m^3/s pumping station on the Medlam drain to improve drainage in its upper reaches. There were thus 9 alternative projects for detailed study as summarised in Table 1.

Two design storms were adopted to give inflow hydrographs to the model. These were the storms of February 1977 and December 1978, both modified to represent a 1 in 10 year event as described in the companion paper (Evans, 1986). The peak water levels predicted by the model were used to estimate the freeboard between the lowest ground levels for individual small catchments and the main drain water level draining each catchment. In order to make cost comparisons between the options the cost required to bring each option up to the same standard of drainage was determined by estimating the number of small booster pumping stations required to give 1.0 m freeboard throughout the project area.

Final project selection was based on total present value costs and on operational considerations, and option 4A was chosen accordingly. The performance of the recommended option is shown in Figure 2 for the 1977 1 in 10 year design storm and areas of reduced freeboard summarised in Table 2.

Table 2: Areas of Reduced Freeboard for Recommended Option at 1977 1 in 10 Year Design Storm

Condition	Approximate area (ha) for freeboards of different values				
	Less than 0.6m	Less than 0.7m	Less than 0.8m	Less than 0.9m	Less than 1.0m
Existing	6300	8200	10100	11700	15100
Recommended Improvement	0	0	400	1100	2100

The proposed scheme provides for the construction of small booster pumping stations to serve isolated areas of land where the desired freeboard of 1.0m during the design storm is not achieved. Most of these pumping stations will be located at the head of the main drains where some of the lowest lying land is situated. Their capacity will range from 0.1m^3/s to 0.9m^3/s and each will comprise a submersible or axial flow pump, weedscreen and gravity bypass as shown in Figure 3.

Table 1: Options Considered for Final Project Selection: Low Level Drainage System

Option 0	Existing condition.
Option 1	Increase Hobhole pump capacity by 30%. New pump stations at Midville (11.2 m^3/s), Lade East (4.7 m^3/s) and Lade West (1.1 m^3/s). Channel improvements to Upper Hobhole (2 500 to 5 850), Thorpe, Fodder Dike East, Cowbridge and West Fen (0 to 3 000) drains.
Option 2	Increase Hobhole pump capacity by 30% and increase Lade Bank pump station by 30%. Channel improvements to Upper Hobhole (0 to 5 850), Thorpe, Fodder Dike East, Lower Hobhole (0 to 6 000), Cowbridge, West Fen (0 to 7 500) and Medlam drains.
Option 3	Increase Lade Bank pump capacity by 30%. Storage reservoir (1 Mm^3) at Lade Bank. Channel improvements to Upper Hobhole (0 to 5 850), Thorpe, Fodder Dike East, Lower Hobhole (0 to 6 000), Cowbridge, West Fen (0 to 7 500) and Medlam drains.
Option 4	Increase Hobhole pump capacity by 30%. New pump stations at Midville (11.2 m^3/s), Lade East (4.7 m^3/s) and Lade West (1.1 m^3/s). Channel improvements to Upper Hobhole (2 500 to 5 850), Thorpe, Fodder Dike East, Lower Hobhole (0 to 6 000), Cowbridge, West Fen (0 to 7 500) and Medlam drains.
Option 4A	As Option 4 but additional new pump station (2.7 m^3/s) at ch. 5 900 m on Medlam drain.
Option 5	As Option 4 but no channel improvements to Lower Hobhole drain.
Option 6	Increase Hobhole pump capacity by 15%. New pump stations at Midville (11.2 m^3/s), Lade East (4.7 m^3/s) and Lade West (1.1 m^3/s). Channel improvements to Upper Hobhole (2 500 to 5 850), Thorpe, Fodder Dike East, Lower Hobhole (0 to 6 000), Cowbridge, West Fen (0 to 7 500) and Medlam drains.
Option 6A	As Option 6 but additional new pump station (2.7 m^3/s) at ch. 5 900 m on Medlam drain.
Option 7	As Option 4 but capacity of Midville pump station reduced to 9.0 m^3/s.

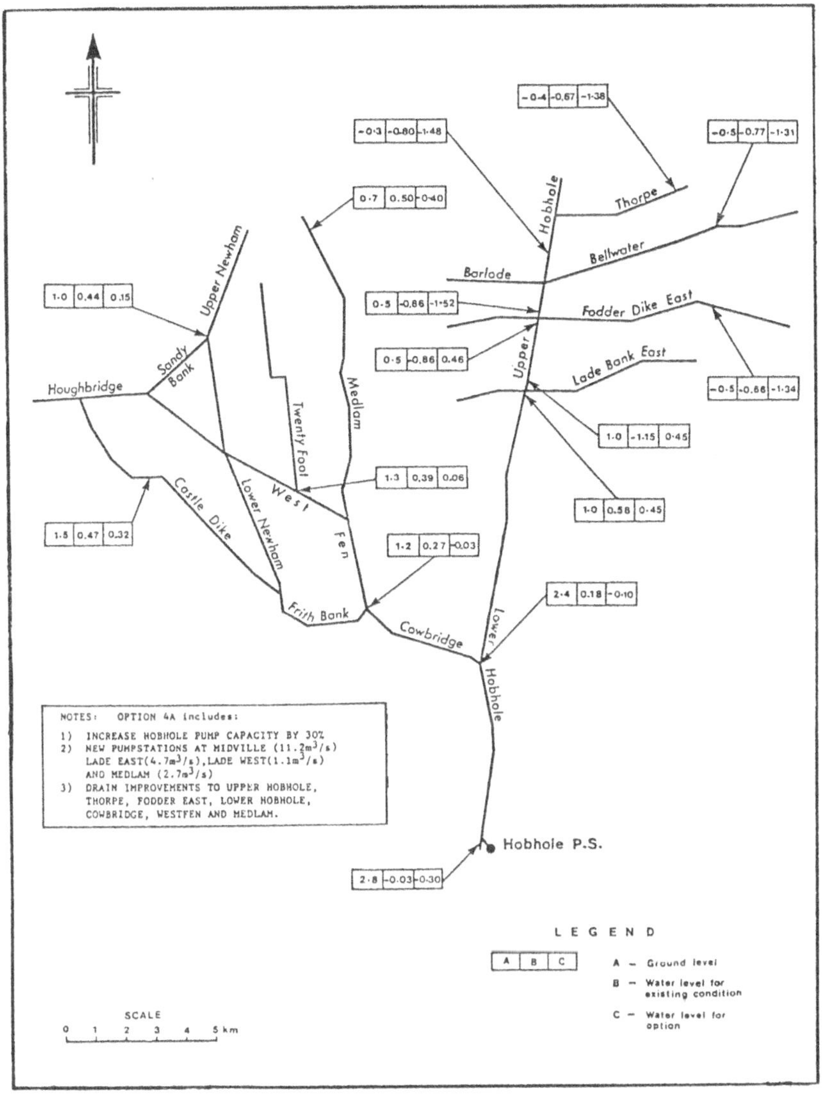

FIGURE 2

PREDICTED PEAK WATER LEVELS FOR EXISTING SYSTEM AND OPTION 4A

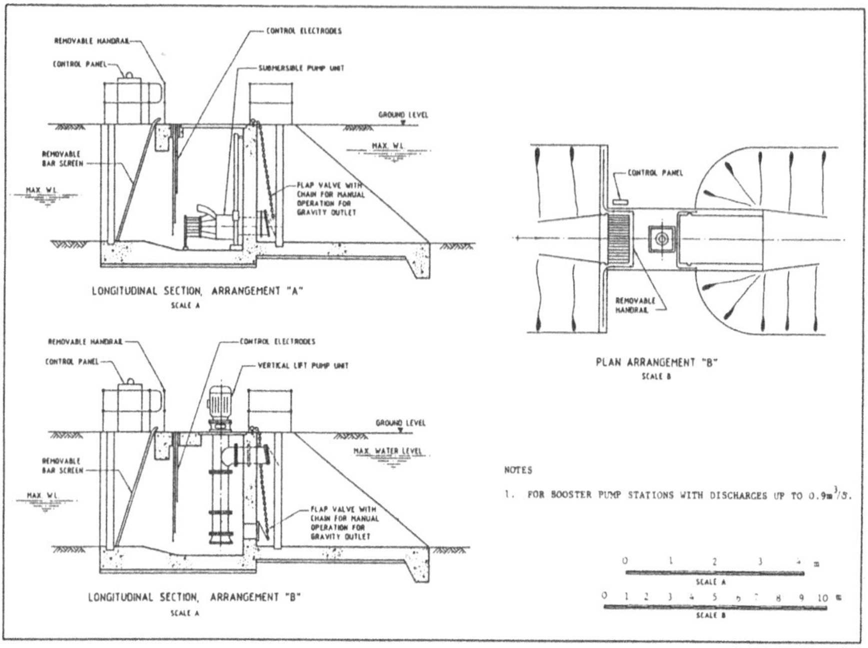

FIGURE 3

DRAIN BOOSTER PUMPING STATION

The new pumping station at the outfall of the Hobhole drain will contain electrically driven axial flow pumps with a total duty capacity of 8.5 m^3/s and will be fully automatic. The existing pumping station at Hobhole contains three diesel driven pumps together with a large gravity bypass and was commissioned in 1957. By handling all the normal low flow pumping the new station will effectively increase the life of the existing station which would then only be required for gravity discharge and for storm discharge pumping.

4. CATCHWATER SYSTEM PROJECT SELECTION

The unsteady state model was used with a number of design storms to simulate the effect of possible improvement options, singly and in combination, on peak water levels. A study of these results and consideration of component costs and other factors allowed selection of the project summarised below:

- install new timber pointing doors and motor driven gates in all three bays of Maud Foster sluice. The third bay contributes little in terms of reduced water levels but was included to guard against failure of the others during times of flood runoff and high tide.

- raise banks along Maud Foster drain to 2.40 m ODN where required to provide flood protection in urban areas against a 1 in 100 year event.
- raise banks along Stone Bridge and East and West Fen catchwater drains to provide flood protection against a 1 in 10 year flood for adjacent low-lying areas.
- install six small booster pumps to drain adjacent areas of low-lying land directly into the catchwater system.
- carry out minor channel improvements to a small length of the West Fen catchwater drain.

5. PROJECT COSTS

Table 4 summarises project costs, expressed as net present values of capital costs, operation and maintenance costs over the life of the project, and engineering design and supervision costs. The recommended project includes some works to replace existing installations which are at or near the end of their useful life. The costs of these works have therefore been excluded from improvement costs shown in Table 4.

Table 4: Project Costs (£ million)

	Capital Costs		Operation and Maintenance Costs	
	Total	Improvement	Total	Improvement
Lowland System				
Hobhole pump station	0.91	0.91	0.38	0.38
Midville pump station	0.83	} 0.60	0.44	} 0.26
Lade Bank E pump station	0.49		0.27	
Lade Bank W pump station	0.12		0.04	
Medlam pump station	0.20	0.20	0.17	0.17
Booster pump stations	0.13	0.13	0.22	0.22
Main drains	2.22	2.22	-	-
Minor (sewer) drains	1.36	1.36	-	-
Telemetry & control	0.25	0.25	-	-
Design & supervision	0.65	0.58	-	-
Catchwater System				
Maud Foster sluice	0.32	0.01	0.02	0.02
Bank raising	0.13	0.13	-	-
Booster pump stations	0.11	0.11	0.23	0.23
Channel improvements	0.01	0.01	-	-
Design & supervision	0.05	0.02	-	-
Totals	7.78	6.53	1.77	1.28

Total present value of improvement costs = £7.81 million.

Notes:

1. 1983 prices
2. Improvement costs do not include works required without the project.
3. Design and Supervision costs are 10% of capital costs
4. Operation and Maintenance costs discounted at 5% over 30 years.

6. PROJECT BENEFITS

Benefits are obtained by reduction in agricultural losses due to improved flood protection and better drainage. These losses include loss of crops, fertilisers and other inputs, reductions in crop yields, extra costs for repeated cultivations etc. The two types of benefits were estimated separately.

6.1 Benefits due to reduced flooding

A comparison of Lade Bank rainfall and recorded incidences of flooding showed no significant correlation. Indeed, only two recorded floods (1977 and 1981) have occurred since Hobhole pumping station was commissioned in 1957 and only two other floods are remembered since 1941, when Lade Bank pumping station was constructed. Flooding to any extent is therefore a rare event.

Flood probabilities were estimated directly from observed floods, regarding the data points as the areas flooded (1 500 ha in 1977, 11 800 ha in 1981) with corresponding probabilities given by their Gringorten plotting positions (0.062 and 0.022). An estimated probability function was drawn through these points and a third hypothetical event with negligible flooded area. The area under the curve gave an expected flooded area of 393 ha/year. A similar analysis for the project case (which will not prevent flooding under extreme conditions) showed an expected flooded area of 16 ha/year.

The benefits of reduced flooding per hectare were estimated by considering the effect of winter and spring floods on the crops grown in the area. Floods in summer and autumn are extremely rare and were not considered. The probabilities of winter and spring floods were assumed equal and seasonal losses were averaged to give an effective loss per crop which were weighted according to the cropping pattern to give an overall loss per hectare of £199.5.

Total benefits for flood protection were found by discounting annual benefits (199.5 x 393 - 199.5 x 16 = £75 211) at 5% over 30 years to give a present value of £1.16 million.

6.2 Benefits due to improved drainage

The drainage system was viewed as being comprised of primary and secondary subsystems, respectively covering the areas above and below field drain outlets into open drains. If these outlets are not submerged then drainage problems in the field are due to deficiencies in the primary system (low soil permeabilities and/or poor field drainage) and cannot be alleviated by improvements to the secondary sub-system. A farm survey covering 40% of the project area indicated extensive field drain submergence, which was subdivided into a number of classes. The most severe of these were:

- Persistently submerged for most of the winter : 8150 ha
- Frequently submerged for 1 to 7 days : 2800 ha

Only the above categories were considered to have a significant effect on agricultural losses, other categories having a lower incidence of field drain submergence.

Calculations of losses per hectare were prepared for each crop at both levels of waterlogging for each month, November, December, January .... June. These were combined with relative monthly probabilities of high drain water levels as estimated by an analysis of pumping records at Hobhole pumping station. Months July to October inclusive were found to have negligible probabilities of high water levels. The final outcome was a set of annual average losses per hectare for each crop and each level of waterlogging. These were used with areas and cropping patterns falling into each category (ascertained from the farm survey) to obtain a value of £100.7 per hectare as the annual average loss, equivalent to a net present value of £16.91 million at 5% over 30 years.

The above calculations assume wrongly that field drain outlet submergence causes waterlogging throughout the 10 950 ha and that the project would completely eliminate waterlogging. Neither assumption is true and the benefits attributable to the project were assessed as £13.5 million, representing 80% of the calculated value.

Two other estimates of waterlogging losses were made based on a consideration of soil/rainfall relationships and on increased yields suggested by the farm survey. These estimates were substantially higher than that described above. An estimate of the area affected by waterlogging was made from results of the unsteady state model which confirmed the 10 950 ha obtained from the farm survey.

6.3 Project evaluation

The economic cost-benefit analysis of the improvement works, at 1983 market prices, is:

Discounted costs	£7.81 million
Discounted benefits	£14.66 million
Net present value of benefits	£6.85 million

Benefit/cost ratio = 1.88.

7. ACKNOWLEDGEMENTS

This paper is presented with the permission of Anglian Water and the Witham Fourth Internal Drainage Board.

Reference

Evans, T.E. and Guganesharajah, K. (1986) Flood Protection and Drainage of the East and West Fens: Flood Hydrology and Modelling. 2nd Int. Conf. on Hydraulic Design in Water Resources Engineering: Land Drainage. University of Southampton, April.

Effect of Subsurface Drain Discharge on Streamflow

Ross W. Irwin and Graham Bryant
School of Engineering, University of Guelph, Guelph, Ontario, Canada, N1G 2W1.

INTRODUCTION

Environmentalists in Canada attack the practice of land drainage on the premise the works destroy the habitat for waterfowl, buries trout spawning beds in silt, adds to erosion problems, and puts unneeded land into cultivation adding to an already high agricultural surplus. They state, usually without documentation, that projects are seldom cost effective and environmental appraisals, when done, are often flawed.

In addition to the above are those who view agricultural land drainage as a major contributor to downstream flooding. This is of concern in Ontario where the annual installation of subsurface drains is about 5,000 km. About 500 km is larger in diameter than 100 mm.

In recent years design engineers have been asked to determine the downstream effect that existing drainage measures have on a watercourse or that proposed measures would have if installed. Similiar information can be used for assessing damage claims, determining adequacy of an outlet, and determining the cost of main drains.

Work at the University of Guelph (Serrano et al 1985) have addressed a number of the above problems with the object that documented information is needed to support or reject allegations concerning agricultural water management practices. The study reported here is one phase of a project to determine the effects of agricultural land drainage on the hydrology of streamflow.

LITERATURE REVIEW

The selection of the diameter of a subsurface main drain not only involves the spacing and depth of drains, but also the drainage coefficient. The design of agricultural drains in Ontario was based on a single drainage coefficient for 50 years, primarily because the work was concentrated in a small area (Day 1909).

Influenced by a new standard, Design of Subsurface Drains in Humid Areas (ASAE 1952), a range of drainage coefficients was presented in the first Drainage Guide for Ontario (Hore 1963). These values were not based on local experimental experience. The revised Drainage Guide (Irwin 1985) specifies a drainage coefficient of 9 mm in 24-hours for improved pasture and general grain crops, 12 mm for cash grain crops, and 20 mm for specialized vegetable crops.

Normally subsurface drain design is based on a 12 mm in 24 hours drainage coefficient, but without a clear concept of how often such a quantity can be expected. Practice uses a single coefficient for all soils in an area.

Main drains are designed using a drainage coefficient; however, lateral drains are seldom designed for maximum capacity. Recurrence intervals of less than 0.5 years are not significant in design, often a 2-year return period is adopted.

The drainage rate is influenced by the degree of protection warranted, the climatic regime, drainable porosity, available soil water-holding capacity, and the minimum water table depth needed for field operations. It should be no larger than necessary to remove excess water as quickly as needed.

Hore (1963) determined the probability of tile drain discharge for Brookston clay. Data were limited to a 5-year period. Maximum 24-hour average flow rates of 12 mm were exceeded once in 5 years. Maximum 12-hour values of 8 mm, and 48-hour maximum values of 14 mm, were exceeded annually.

Sharma and Irwin (1975) modelled subsurface drainage as depletion from two linear reservoirs. A fast-reacting reservoir described the plow layer and a slow-reacting reservoir, the subsoil. Data from the model were used to calculate a drainage coefficient. A coefficient of 12.5 mm in 24-hours was expected once in 4 years. The same rate of discharge could be expected to occur in 48-hours once in 8 years.

EXPERIMENTAL SITE

Location

The experimental site is at the Woodslee substation of Harrow Research Station, Agricultural Canada, in Essex county, near Windsor, Ontario (Figure 1).

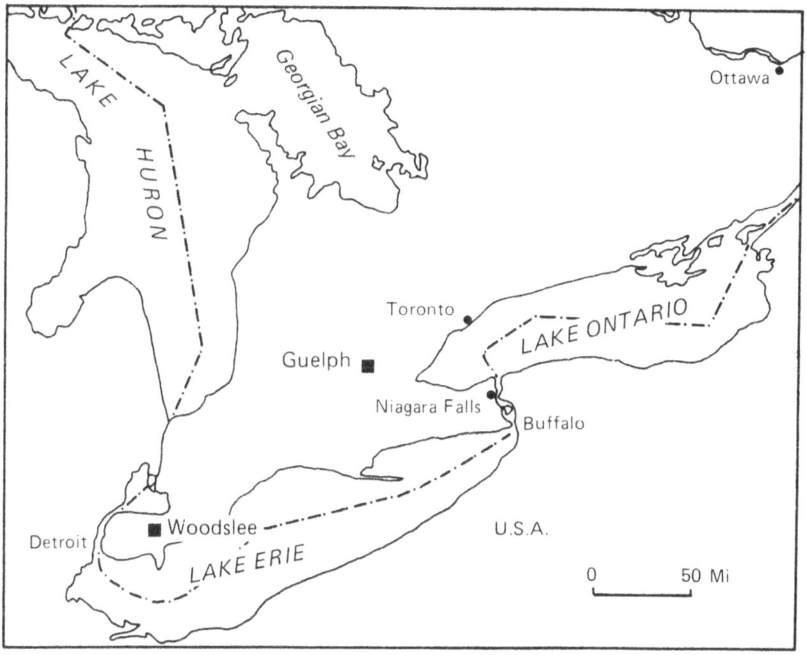

Figure 1.- Location of Experimental Site

<u>Soil</u>

The agricultural area of southern Ontario is 5.26 million hectares. The Brookston soil series is found in 14 counties and represents 1.1 million ha of which one-third is found in the counties adjacent to the experimental site. Brookston is an important agricultural soil in Ontario.

The Brookston soil is a naturally poorly drained orthic humic gleysol developed under wet forest conditions. Surface slope is 0.05%. Physical properties of the soil are in Table 1.

Table 1. Physical Properties of Brookston Series.

Horizon	Depth (cm)	Grain Size (%) Sand	Silt	Clay	Dry Density $g\ cm^{-3}$	O.M. (%)
Ap	0-16	25	38	32	1.40	2.5
Bg1	16-45	21	34	45	1.58	1.0
Bg2	45-116	21	34	45	1.58	0.4
Ckg	116+	21	39	40	1.52	0.6

The surface horizon is described as medium texture having platy or granular structure. The B horizon is fine textured with medium to course sub-angular to angular blocky structure. The C horizon is fine textured, medium to course blocky, prismatic or massive.

The soil often cracks to drain depth in late summer and hydraulic conductivity is difficult to measure with consistent results. Field tests range from 0.05 to 5 m/d; laboratory tests range from 0.12 to 1.6 m/d, with a median value of 0.85 m/d.

Plot Description

Twelve plots, 12.2 m by 76.2 m, having an area of 929.64 m^2 were established in 1957 as a drainage experiment concerned with the loss of nutrients in drainage water. These results were reported by Bolton (1970). A 100 mm tile drain was installed in the centre of each plot, having a spacing of 12.2 m. The plot layout, drain tile gradient (%), corner surface elevations (m), and collection system is shown in Figure 2. The average depth of drains is 70 cm which is existing practice in the area.

Each drain discharges into an access well which contains a Casella flow meter. The Casella meters are equipped with linear head-discharge weir plates. The access wells were heated for year around operation.

Cover on the experiment is randomized duplicate plots of continuous corn, continuous bluegrass sod, and a four year rotation of corn-oats-alfalfa-alfalfa. Superimposed on the cover experiment are two levels of fertility. Plots 7 to 12

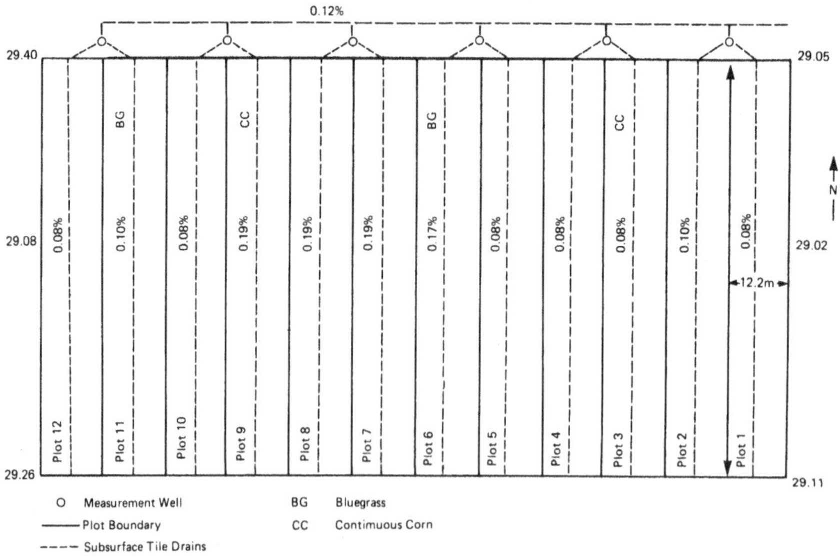

Figure 2.- Experimental Plot Design

receive 336 kg/ha/yr of 5-20-10 fertilizer, and corn receives an additional side-dressing of 112 kg/ha/yr of N. Plots 1 to 6 are unfertilized. Each crop is grown each year.

Data

Climatological data is collected at the site as part of the Environment Canada meterological observation network. Primarily this includes a continuous record of precipitation from 1957, twice daily standard rain gauge depth, air and soil temperature and wind movement.

Surface runoff was collected from four plots for the years 1980 to 1982. Water table elevation was not measured.

Tile flow records from the 12 plots are available from 1961 to 1982, a continuous record of 22 years. Data from the strip charts were digitized and stored on a microcomputer hard disk. The recession time was often too long due to the meter float sticking and similar operational problems. A small correction was applied to the rising and recession limbs of the event hydrograph. The total volume was reduced, but not significantly for the period of flow. The excessively long recession times were eliminated.

The digitized data was used in a computer program to produce time to peak, total flow duration, total volume, peak flow, and maximum flow depth for 12, 24, 36 and 48-hours. These data were then analysed on a seasonal and annual basis.

ANALYSIS OF DATA

Summary of the Data

There were 1,927 tile flow events during the 14-year period 1969 to 1982 of which 1,641 were over 12-hours duration. The record was divided into three seasons. Planting (P) for the months of April, May and June. Growing (G) was July, August, September and October. The balance of the year was termed a Dormant (D) season. Table 2 shows the distribution of events by duration and season.

Table 2. Number of Tile Flow Events, 1969-1982.

Duration, h	12	24	36	48
Planting season	530	435	343	295
Growing season	233	171	138	116
Dormant season	878	765	678	612
Total	1641	1371	1159	1023

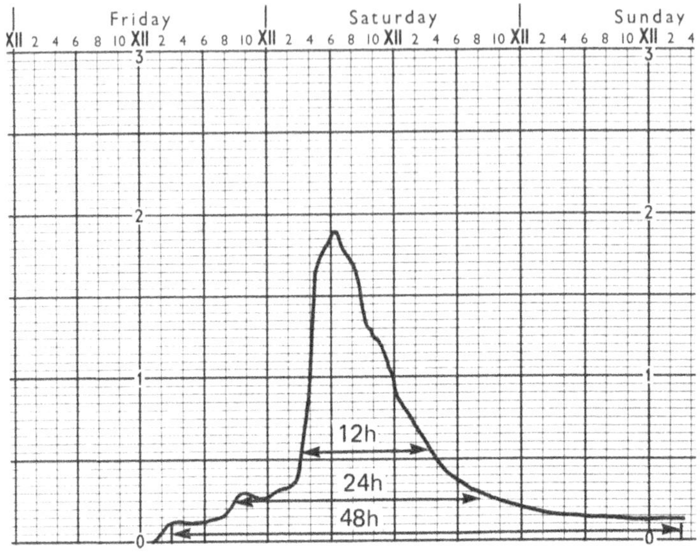

Figure 3.- Tile Discharge Hydrograph Showing Durations

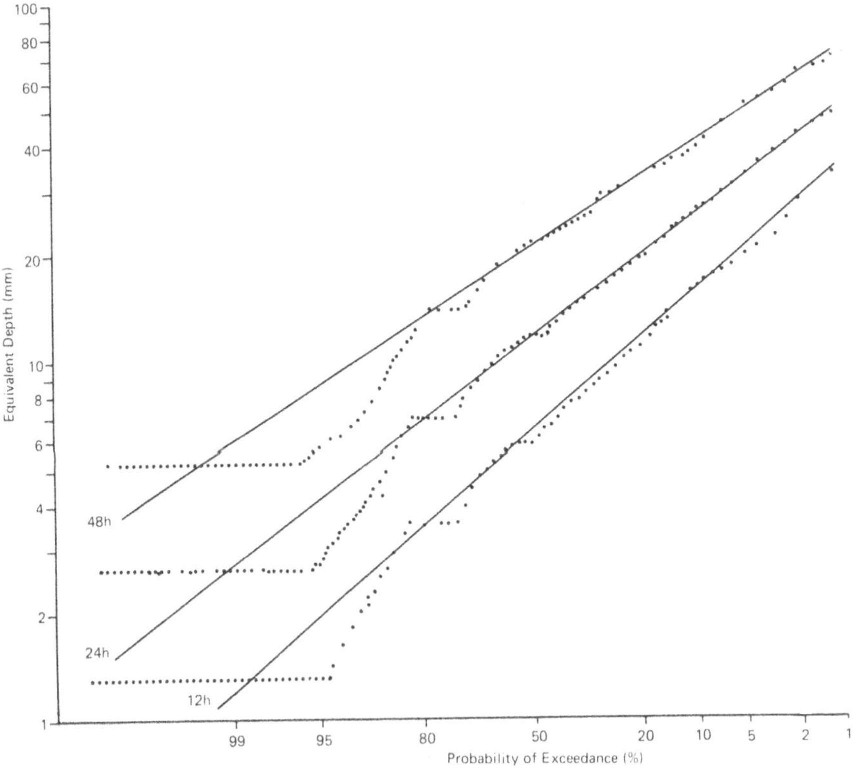

Figure 4.- Depth Frequency Analysis - Total Record

Frequency Analysis

Maximum equivalent depths for periods of 12, 24, 36 and 48 hours were extracted from the discharge hydrograph for each event, as shown in Figure 3. The digitizing technique created some preference for certain values at low differences in stage and appeared to round values creating single values at probabilities less than 30 percent. Analytical tests showed the data best fitted a log-normal theoretical probability distribution.

Figure 4 shows a complete duration series for the total data for three durations of 12, 24 and 48-hours. Deviations are due to seasonal effects.

The large number of events during the Dormant season tends to influence the character of the distribution but all seasons are similiar in the medium range of probability. Storm pattern in the Planting and Growing seasons cause a higher discharge at probabilities of exceedance less than 20%.

If there is a 24-hour tile flow event, the probability of a contribution to streamflow greater than 11.5 mm is 50%. The contribution is 19 mm at a probability of exceedance of 20% and 7 mm at a probability of exceedance of 80%. The 12-hour value is 6 mm and the 48-hour value is 21 mm at a 50% probability of exceedance.

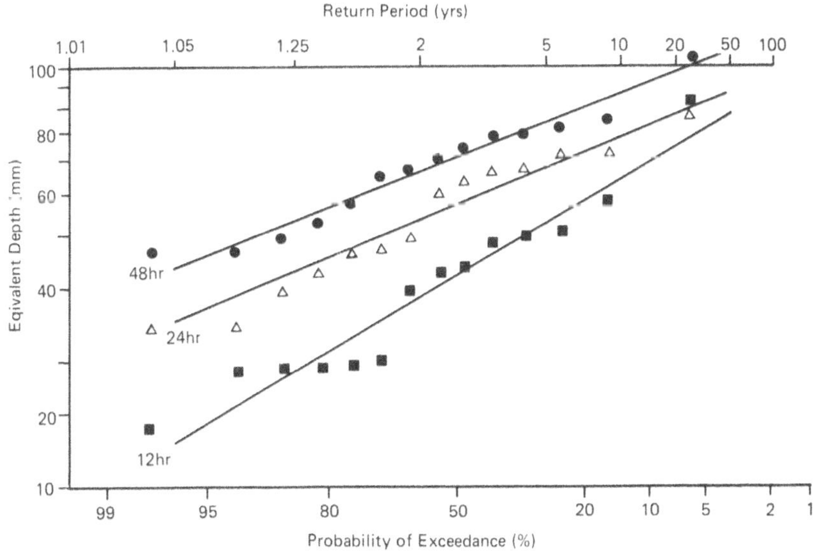

Figure 5. - Depth Frequency Analysis - Annual Series

Figure 5 is a frequency analysis of the extreme annual maximum series consisting of 14 events, including all seasons. These values are shown relative to the theoretical log-normal distribution. There is substantial scatter due to the short record.

The 24-hour and 48-hour theoretical distributions are parallel; however, the 12-hour distribution converges at low probabilities of exceedance. This is due to the number of intense storms of substantial depth which occur for this short duration due to convective type storms in the summer months and cracked soil. Flows over 36 and 48-hours are usually the result of rainfall on saturated soil.

Comparing Figures 4 and 5, the ratio of the annual maximum distribution to the 24-hour total duration at a 50% probability of exceedance is about 4.

Figure 6 shows the theoretical log-normal distribution for the extreme annual maximum 24-hour series for the Dormant and Planting seasons, with values noted. Equivalent depths for the Growing season are also shown.

The Dormant season causes higher depths of flow due to March snowmelt. Soil is wet to saturated during this season and a large event produces a high discharge. Storms tend to be due to frontal conditions, are longer in duration, lower in intensity and are more frequent. About 70% of the 12-hour events shown in Table 2 continue to 48-hours for this season compared to about 50% for the other seasons.

The Planting season converges with the Dormant season at low probabilities of exceedance due to the frontal storms which occur on wet soil in April creating higher depths of discharge.

The Growing season is bounded on the left by four zero values in the 14 year record. Figure 6 shows the probability distribution has two slopes. Equivalent depths greater than 28 mm have a return period of 3 years while the return period for zero flow is 1.25 years. Growing season convective storms are of high intensity, usually less than 24-hour duration, and often of substantial depth. If there is drain discharge, it will be high.

Figure 6 also shows that a given annual drainage coefficient has three exceedence probabilities associated with it, one for each season.

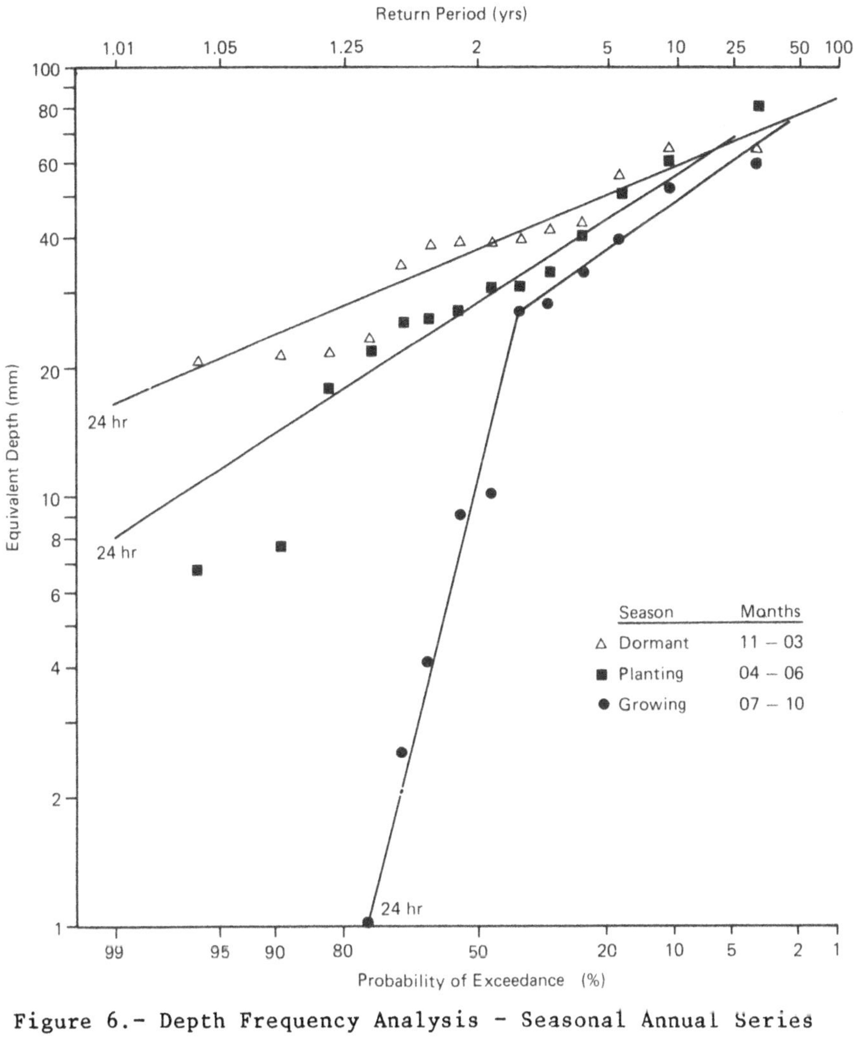

Figure 6.- Depth Frequency Analysis - Seasonal Annual Series

CONTRIBUTION TO STREAMFLOW

Subsurface drainage systems use a main drain to collect water from lateral drains. The diameter of the main drain is based on a drainage coefficient and the area contributing water to the drain. The pipe is presumed to run full for a 24-hour period, but not under pressure, and discharges this volume of water to a watercourse.

The foregoing is conservative inasmuch as lateral drains frequently deliver water to a main drain at different rates. The design value is a maximum flow condition; however, the rate of flow is not uniform over the period.

The contribution of subsurface drains to streamflow is greatest in the Dormant season although the Planting and Growing seasons are more important for agricultural drain design. Flows are highest in March when there are no crops on fields, but farmers still do not want the soil flooded.

In summer the watertable is low creating considerable storage in the soil profile. Substantial rainfall depths are needed for drain discharge. Late summer is characterized by convective storms and cracked clay soils which create high discharge volumes with short time to peak and duration. These events are few in number and are highly variable.

The 14 year flow record from Brookston soil shows the 50% probability of contribution to streamflow can be estimated from the product of the drainage area and a drainage coefficient of 11.5 mm.

REFERENCES

American Society of Agricultural Engineers.(1954) Design and Construction of Tile Drains in Humid Areas. Tenative ASAE Recommendation. Agr. Engr. Yearbook. 1st ed. 89-98.

Bolton, E.F., J.W. Aylesworth, F.R. Hore. (1970) Nutrient Losses through Tile Drainage under Three Cropping Systems and two Fertility Levels on a Brookston Clay Soil. Can. J. Soil Sci. 50: 272-279.

Day, W.H. (1909) Farm Drainage Operations. Ont. Dept. of Agric. Bulletin 175.

Hore, F.R. (1963) Flow Characteristics from Tile Drains in Fine Textured Soils. Can. J. Agric. Engrg. 5:17-20.

Hore, F.R., R.W. Irwin, B.C. Matthews, F.J. Parrish. (1963) Drainage Guide for Ontario. Ont. Dept. of Agric. Publication 29.

Irwin, R.W. (1985) Drainage Guide for Ontario. Ont. Min. of Agric. and Food. Publication 29.

Serrano, S.E., H.R. Whiteley, R.W. Irwin. (1985) Effects of Agricultural Drainage on Streamflow in the Middle Thames River, Ontario, 1949-1980. Accepted for publication. Can. J. Civil Engrg. December.

Sharma, T.C., R.W. Irwin. (1977) Drain Discharge from a Level Agricultural Watershed. J. Irrig. & Drainage Div. ASCE, 103(IR4): 429-443.

Development of the Lower Mono River Valley, West Africa: A Case Study

P.N. Lewis
ELC-Electroconsult

ABSTRACT

The potential of developing the agricultural sector in the lower Mono River valley in Togo and Benin was studied at prefeasibility level in 1983 and 1984. The measures proposed to overcome the frequent flooding of the valley and their estimated cost are presented. It was concluded that the substantial costs involved in protecting the area against flooding were not warranted without further investment in irrigation to raise the expected benefits in the "with project" situation.

INTRODUCTION

The Prefeasibility Study into the agricultural development of the Lower Mono River valley encompassing areas in both Togo and Benin in West Africa was undertaken during 1983 and 1984 by ELC-Electroconsult of Milan in joint venture with Louis Berger International Inc.

The binational study, financed by the African Development Bank, was given impetus by the construction of the Nangbeto Reservoir upstream of the Project area.

The reservoir is principally for electric power generation with the associated benefit of provision of a secure supply of irrigation water; the reservoir does not, however, have a major flood protection role.

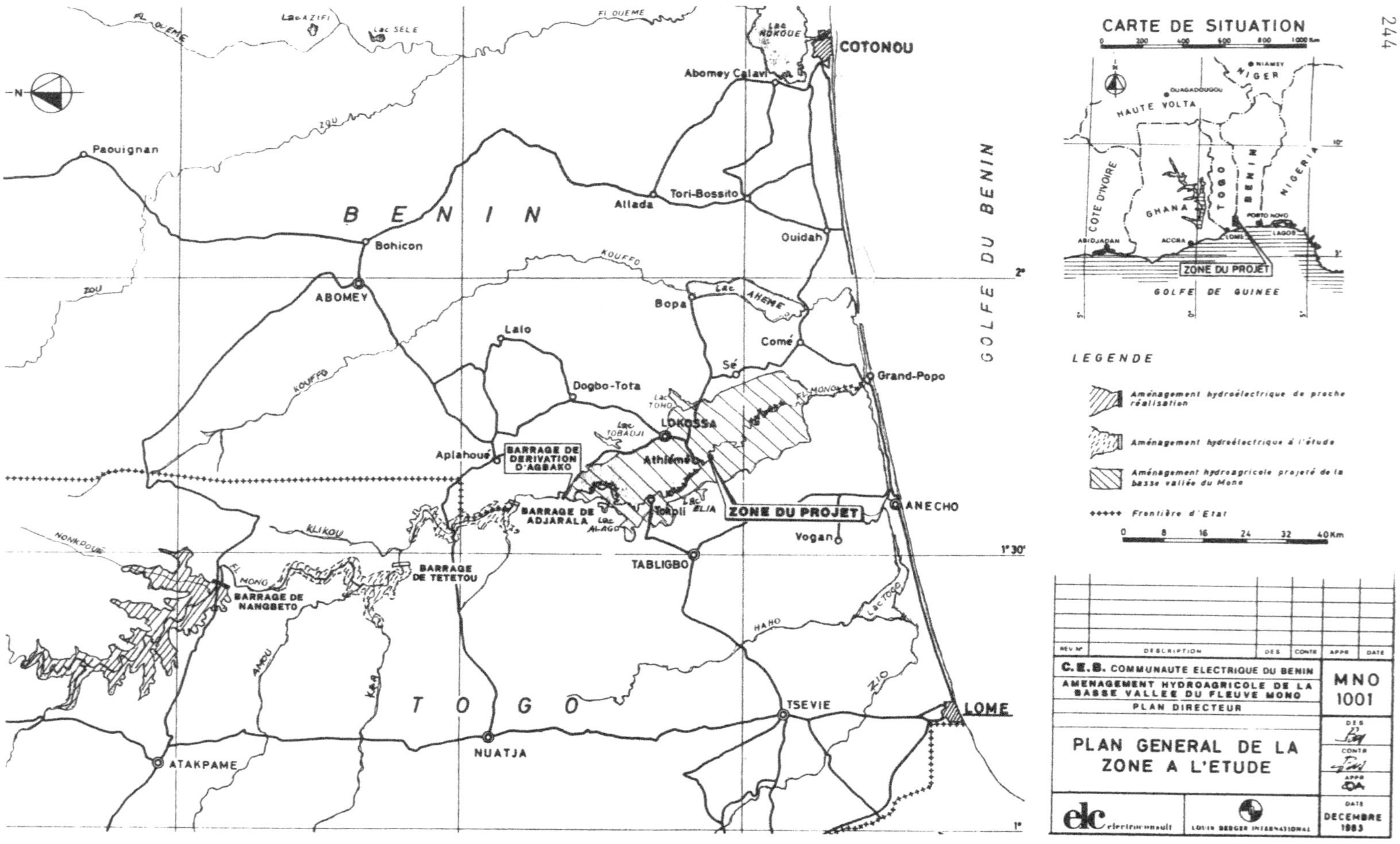
CARTE DE SITUATION
0 200 400 600 800 1000 Km
NIAMEY
NIGER
OUAGADOUGOU
HAUTE VOLTA
COTE D'IVOIRE
GHANA
TOGO
BENIN
NIGERIA
PORTO NOVO
LAGOS
LOME
ACCRA
ABIDJAN
ZONE DU PROJET
GOLFE DE GUINEE
LEGENDE
Aménagement hydroélectrique de proche réalisation
Aménagement hydroélectrique à l'étude
Aménagement hydroagricole projeté de la basse vallée du Mono
Frontière d'Etat
0 8 16 24 32 40Km
REV Nº
DESCRIPTION
DES
CONTR
APPR
DATE
C.E.B. COMMUNAUTE ELECTRIQUE DU BENIN
AMENAGEMENT HYDROAGRICOLE DE LA BASSE VALLEE DU FLEUVE MONO
PLAN DIRECTEUR
MNO 1001
PLAN GENERAL DE LA ZONE A L'ETUDE
DECEMBRE 1983
elc electroconsult
LOUIS BERGER INTERNATIONAL
N
Fl. OUEME
Lac AZIFI
Lac SELE
ZOU
Paouignan
BENIN
Bohicon
ABOMEY
KOUFFO
Lalo
Dogbo-Tota
Aplahoué
BARRAGE DE DERIVATION D'AGBAKO
Lac NOKOUE
COTONOU
Abomey Calavi
Tori-Bossito
Allada
Ouidah
Bopa
Lac AHEME
Comé
Sé
Grand-Popo
Lac TOHO
Lac TOBADJI
LOKOSSA
Athiémé
Fl. MONO
Tokpli
Lac ELIA
Lac ALAGO
BARRAGE DE ADJARALA
ZONE DU PROJET
Vogan
ANECHO
TABLIGBO
GOLFE DU BENIN
2°
1°30'
1°
NONKPOUE
KLIKOU
BARRAGE DE TETETOU
BARRAGE DE NANGBETO
AMOU
KRA
TOGO
HAHO
ZIO
Lac TOGO
TSEVIE
NUATJA
ATAKPAME
LOME

The study was commissioned to establish the optimum agricultural development strategy for the lower Mono River valley, with the knowledge that any large-scale development of the area would involve extensive flood protection works, and to capitalize on the availability of a source of water for irrigation.

This paper outlines the problems involved in the flood protection and drainage of the study area and the proposed possible and adopted solutions.

THE AREA

The project area flanks the Mono River along 70 km of its course as it forms the border between Togo and Benin. The lower extreme of the area is some 8 km from where the river debouches into the Gulf of Benin.

The Mono River, which forms the principal drainage path for the 10 to 30 km wide plain forming the project area, has a catchment of some 40 000 sq km above the project area. Within the project area the river is flanked on both sides by a network of secondary drainage channels, probably prior river beds of the Mono, which run parallel to the river. These channels are interconnected by numerous smaller drainage paths crossing the gently undulating plain which is dotted with shallow lakes. The mouth of the river is more or less permanently open although it forms part of an extremely complex hydrographic system constantly changing due to the effects of the littoral drift. The mean gradient of the streams draining into the valley is about 4 m/km whereas the mean gradient of the Mono River itself is about 0.1 m/km within the project area.

The climate is equatorial with high sustained relative humidity and temperatures. The zone is characterized by the abundance and irregularity of the rainfall.

The soils of the project area are typically of recent alluvial origin. About 80% of the area is currently exploited for agriculture with low quality oil palms or single annual crops being grown on holdings of 2 ha average size. The agriculturally based economy supports a population of some 110 persons/sq km mainly concentrated in villages situated on isolated areas of high ground.

DEFINITION OF THE PROBLEMS CONSTRAINING DEVELOPMENT

The object of the study was to define the necessary action to realize the full potential of the Lower Mono River valley. The principle obstacle to development is the risk of inundation which discourages farmers from cultivating crops and utilizing costly fertilizers and modern farming techniques. A major objective was to ascertain whether it is feasible to protect the valley against flooding from the Mono River, runoff from the surrounding plateaux, and rainfall falling directly on the area. A further objective of the project was to establish the feasibility of irrigation.

Although some 30 years of rainfall and flow records were available, little information was available on the extent of historic floods. Unfortunately, no aerial photography or satellite imagery was available showing the historic floods which, although extremely damaging, were of fairly short duration. 1 in 20 000 scale mapping with one metre contour interval was commissioned within the project and produced from aerial photography. Based on this mapping, information on 1963 and 1964 historic floods, and modelling of flows within the Mono River, the main flow paths and likely areas flooded were delineated. Of the total area of 65 000 ha the flooded areas were estimated as shown in Table 1 below.

Table 1

ESTIMATED INUNDATED AREAS FOR VARIOUS FLOWS IN THE MONO

Flow in Mono River (m^3/s)	Inundation Areas (ha)		
	Certain	Probable	Total
1 200	6 300	12 300	18 600
2 000	14 900	10 600	25 500
3 000	30 000	9 300	39 300

With the estimated bank full capacity of the Mono between 600 and 700 m^3/s it can be seen from the proposed operating criteria (Table 2) of Nangbeto Reservoir that the reservoir does not have a significant flood mitigation effect. In fact, at lower river flows it actually marginally worsens the situation; a condition that will have to be reviewed in the event of a protection scheme downstream not being realized.

Table 2

FLOOD RETURN PERIODS AND CHARACTERISTICS FOR MONO RIVER BELOW NANGBETO RESERVOIR

Return Period (years)	Maximum Flow (m^3/s) In to Nangbeto	Maximum Flow (m^3/s) Out of Nangbeto	Flood Duration (days)	Volume of Flood Superior to 600 m^3/s Flow ($10^6 \times m^3$)
Annual	1 000	1 200	26	1 800
10	1 230	1 200	26	1 950
20	1 430	1 200	27	2 100
50	1 750	1 200	28	2 500
100	2 100	2 070	29	3 000
1 000	3 000	2 995	30	4 150
10 000	4 500	3 700	40	5 500

As regards runoff from the surrounding hills, this was established based on the statistical analysis of measured flow in two tributaries of the Mono. These flows were than checked for consistency with flows according to the "Curve Number" method (S.C.S. 1972) with application of the "Cypress Creek" (S.C.S. 1971) reduction factor for the ten delineated catchments as shown in Plan 2. These results and those of a previous study (SOFRELEC) which utilized the Rational Method are shown in Table 3.

The 1 in 5 years runoff from the surrounding hills as shown in Table 3, contributes about 2 m^3/s to 4 m^3/s per kilometer length of the project perimeter.

Table 3

COMPARISON OF PREDICTED PEAK FLOWS FROM SURROUNDING HILL CATCHMENTS

Basin Concerned	Flows Evaluated for Different Return Period (m^3/s)					
	Statistical Method		Curve Number Method		SOFRELEC	
	1 in 5 yrs	1 in 10 yrs	1 in 5 yrs	1 in 10 yrs	1 in 5 yrs	1 in 10 yrs
D1	58	78	49	55	n.c.	n.c.
D2	32	48	26	29	24	27
D3	23	31	18	20	45	51
D4	16	22	12	13	n.c.	n.c.
D5	29	40	23	26	n.c.	n.c.
G1	33	44	26	29	n.c.	n.c.
G2	44	59	36	41	23	26
G3	8	11	14	16	9	11
G4	70	93	60	68	105	123
G5	31	41	24	27	40	48

n.c. = not considered

Without systematic investigations following flood events as often take place in an urban situation, the estimation of damage is at best very subjective. In this specific case damage was considered as:

- Net product of lost crops.
- Damage to roads and agricultural infrastructure which was taken as 10% of their capital cost based on existing and proposed roads types and type of agricultural infrastructure foreseen.
- Damage to irrigation and internal drainage network taking into account the following considerations:
  - most primary and secondary canals are on high ground;
  - major canals act as flood barriers, except where there is a high cross drainage capacity;
  - high damage will be sustained by low-lying drainage structures;

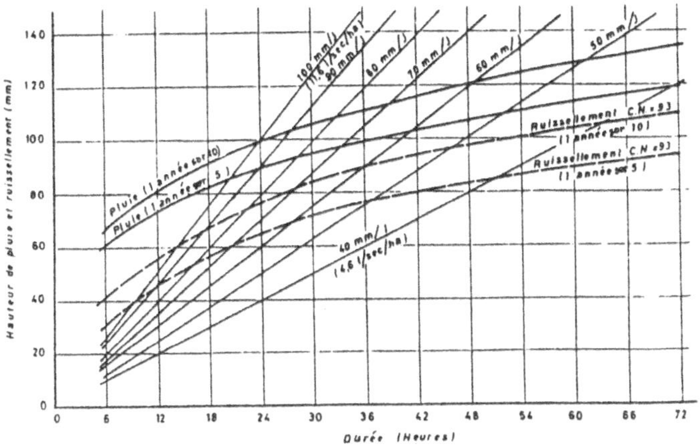

Fig. 1 - Height/Depth/Duration Curves for Rainfall (Pluie) and Runoff (Ruissellement)

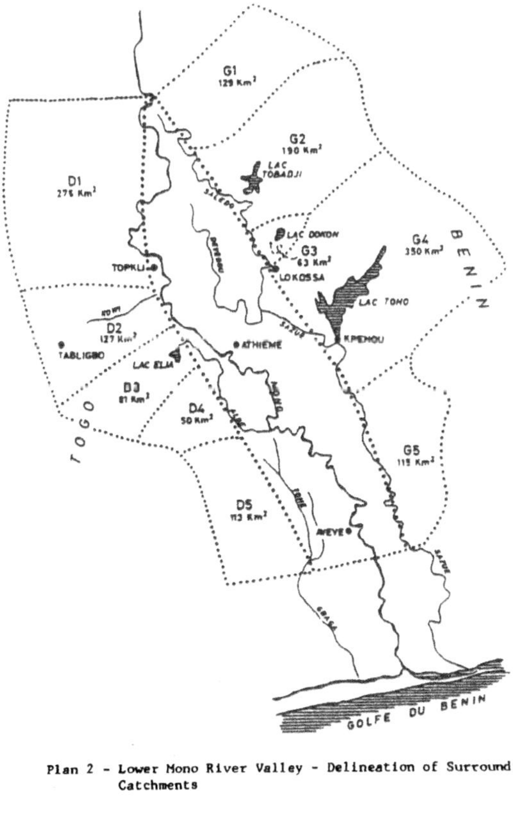

Plan 2 - Lower Mono River Valley - Delineation of Surrounding Catchments

- quaternary canals might be extensively damaged;
- canal structures are rarely damaged;
- silt deposition can be a problem.

The following damage as percentages of capital costs was adopted:

- headworks, pump station, estimated for each case;
- primary/secondary irrigation canals and structures - 5%;
- tertiary/quaternary irrigation canals and structures - 10%;
- drains and structures - 20%.

. Damage to housing in the "without project" situation would be minimal as habitation is restricted to isolated high spots.

To prevent this damage a flood protection system would be required against external inundation from the Mono River and runoff from the surrounding hills. In addition, an internal drainage network would be required to evacuate the excess rainfall falling directly on the project area. A rainfall depth-duration frequency analysis showed the 24 hour one in five year rainfall as 95 mm and the corresponding 7 day rainfall as 185 mm.

ALTERNATIVES FOR FLOOD PROTECTION

After a preliminary study it became obvious that the only means of economically protecting the project area from flooding from the Mono River was by improving the existing course of the river and flanking the river with dikes. The improvement of the river course which basically involved the cutting of eleven major meanders and the enlargment of the river bed reduced the length of the river in the project area from 85 km to 65 km, steepening the bed slope by some 20%. The regularisation also provided suitable material for lateral dike construction.

The selection of the optimum width and height of the dikes was undertaken utilising standard economic techniques. The sum of the present cost of the dikes and their maintenance plus the estimated present cost of damage in spite of the existence of dikes was minimised. The cost of damage in spite of protection (ie.

by overtopping) was estimated by summing the relevant area under the cost versus estimated return period curve.

The study, with estimated construction costs of US$ 28, 35 and 43 million for protection against a 1 200, 2 000 and 3 000 m^3/s floods respectively, revealed an optimum solution providing protection for the area against a 2 000 m^3/s flood. The works required for the selected system included a total length of some 125 km of dikes of 1 m to 4 m height (average 2.2 m) with a spacing of 1 000 m between dikes (see Plan MNO 1023).

The protection of the project area against flooding from the surrounding plateaux was based on the improvement of existing watercourses including regularisation of the bed, erection of protection dikes, and the excavation of short stretches of new drain. The major problem to be resolved was, however, to establish the best means of outfalling the collected flows into the Mono River during high flows in the river once it had been flanked by protection dikes.

After numerous alternatives were examined three (depicted in Plans MNO 1017, 1018 and 1019), comprising varying magnitudes of pumping, were studied in detail and costed. With the usual requirement of massive pump stations (50 to 90 m^3/s) of very high capital cost, the analysis showed clearly that the gravity solution (Plan MNO 1017) was economically the most attractive. The estimated capitalised costs of the solutions in Plan 1017, 1018 and 1019 were US$ 12.6, 16.4, and 24.3 million, respectively for the right bank. For the left bank the capitalised costs of solutions in Plan 1017 and 1018 were US$ 18.7 and 29.0 million respectively. The accepted solution, with an estimated cost per hectare of about US$ 800, also eliminated the maintenance problems inherent in large pumping stations used for very short periods.

INTERNAL DRAINAGE

The internal drainage system to protect the project area against inundation originating from rainfall directly on the project area was designed for the si-

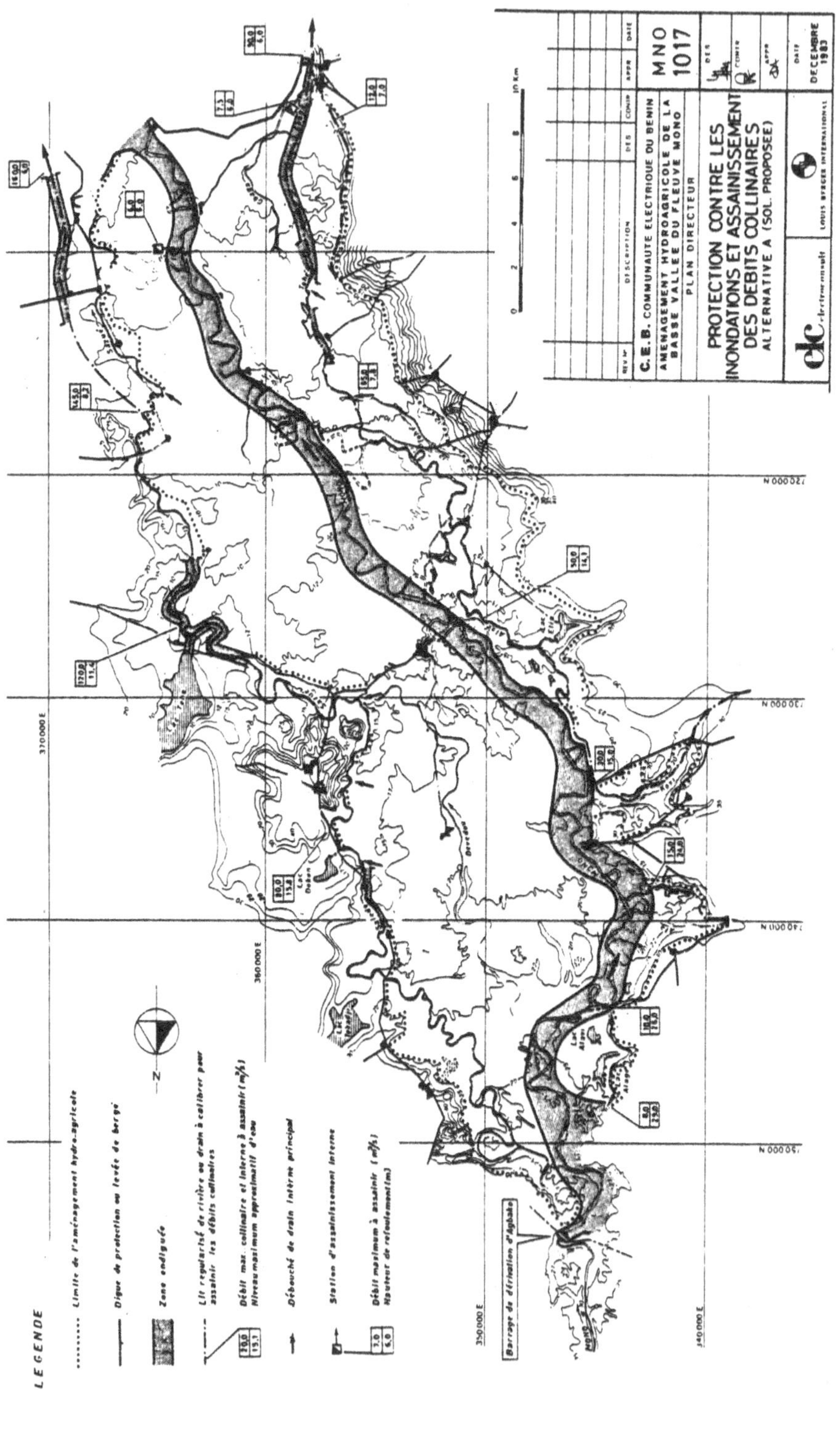
LEGENDE
Limite de l'aménagement hydro-agricole
Digue de protection ou levée de berge
Zone endiguée
Lit regularisé de rivière ou drain à calibrer pour assainir les débits collinaires
Débit max. collinaire et interne à assainir (m3/s)
Niveau maximum approximatif d'eau
Débouché de drain interne principal
Station d'assainissement interne
Débit maximum à assainir (m3/s)
Hauteur de refoulement (m)
C.E.B. COMMUNAUTE ELECTRIQUE DU BENIN
AMENAGEMENT HYDROAGRICOLE DE LA BASSE VALLEE DU FLEUVE MONO
PLAN DIRECTEUR
PROTECTION CONTRE LES INONDATIONS ET ASSAINISSEMENT DES DEBITS COLLINAIRES
ALTERNATIVE A (SOL. PROPOSEE)
MNO 1017
DECEMBRE 1983
LOUIS BERGER INTERNATIONAL

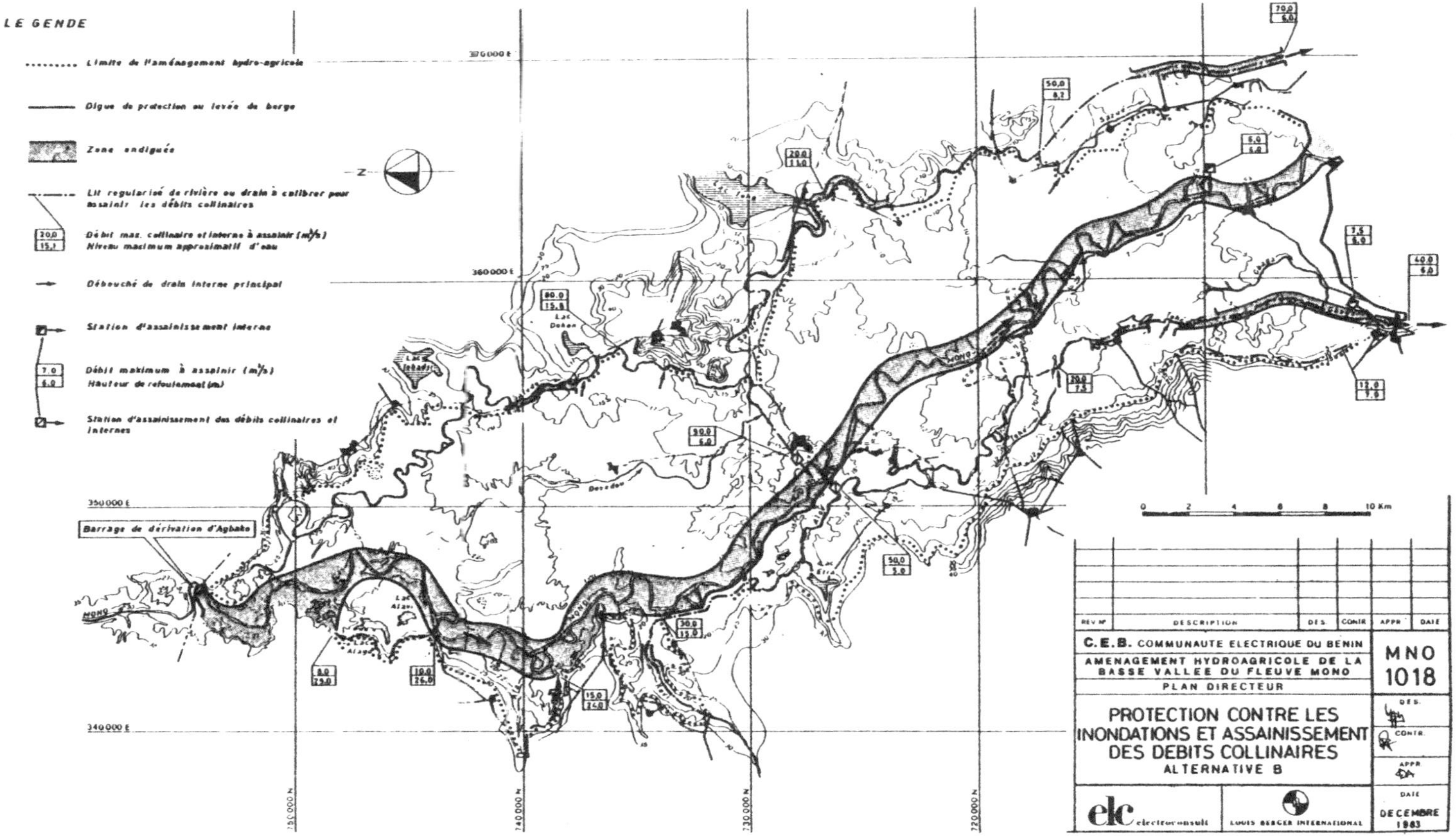
LE GENDE
Limite de l'aménagement hydro-agricole
Digue de protection ou levée de berge
Zone endiguée
Lit regularisé de rivière ou drain à calibrer pour assainir les débits collinaires
Débit max. collinaire et interne à assainir (m³/s)
Niveau maximum approximatif d'eau
Débouché de drain interne principal
Station d'assainissement interne
Débit maximum à assainir (m³/s)
Hauteur de refoulement (m)
Station d'assainissement des débits collinaires et internes
Barrage de dérivation d'Agbaho
0 2 4 6 8 10 Km
REV Nº
DESCRIPTION
DES.
CONTR.
APPR.
DATE
C.E.B. COMMUNAUTE ELECTRIQUE DU BENIN
AMENAGEMENT HYDROAGRICOLE DE LA BASSE VALLEE DU FLEUVE MONO
PLAN DIRECTEUR
PROTECTION CONTRE LES INONDATIONS ET ASSAINISSEMENT DES DEBITS COLLINAIRES
ALTERNATIVE B
MNO 1018
DATE
DECEMBRE 1983
elc electroconsult
LOUIS BERGER INTERNATIONAL

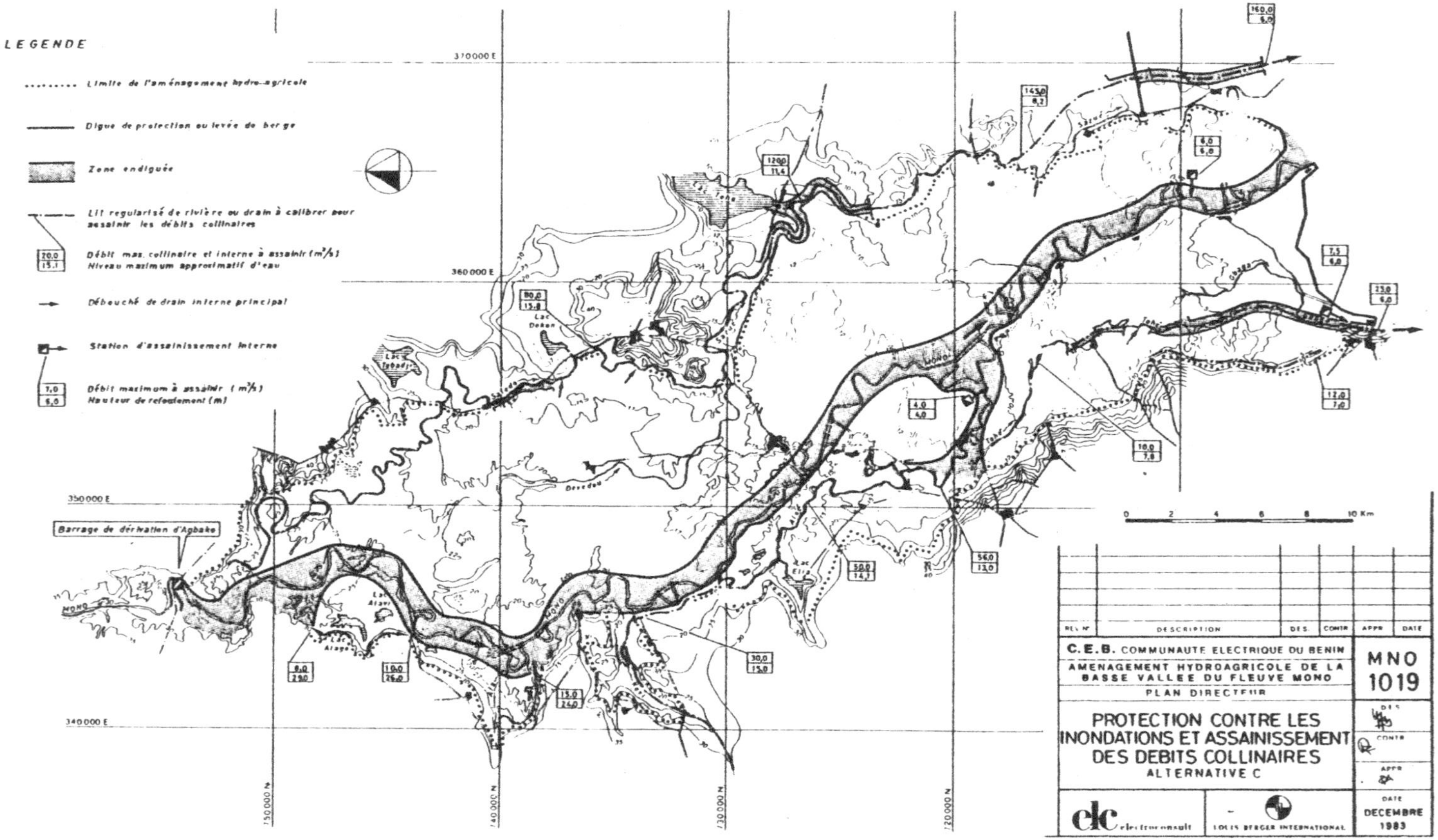

LEGENDE
Limite de l'aménagement hydro-agricole
Digue de protection ou levée de berge
Zone endiguée
Lit regularisé de rivière ou drain à calibrer pour assainir les débits collinaires
20,0
15,1
Débit max. collinaire et interne à assainir (m³/s)
Niveau maximum approximatif d'eau
Débouché de drain interne principal
Station d'assainissement interne
7,0
6,0
Débit maximum à assainir (m³/s)
Hauteur de refoulement (m)
370000 E
360000 E
350000 E
340000 E
150000 N
140000 N
130000 N
120000 N
Barrage de dérivation d'Agbako
MONO
Lac Dokon
Lac Toho
0 2 4 6 8 10 Km
REV. N° DESCRIPTION DES. CONTR. APPR. DATE
C.E.B. COMMUNAUTE ELECTRIQUE DU BENIN
AMENAGEMENT HYDROAGRICOLE DE LA BASSE VALLEE DU FLEUVE MONO
PLAN DIRECTEUR
PROTECTION CONTRE LES INONDATIONS ET ASSAINISSEMENT DES DEBITS COLLINAIRES
ALTERNATIVE C
MNO 1019
DES.
CONTR.
APPR.
DATE DECEMBRE 1983
electroconsult
LOUIS BERGER INTERNATIONAL

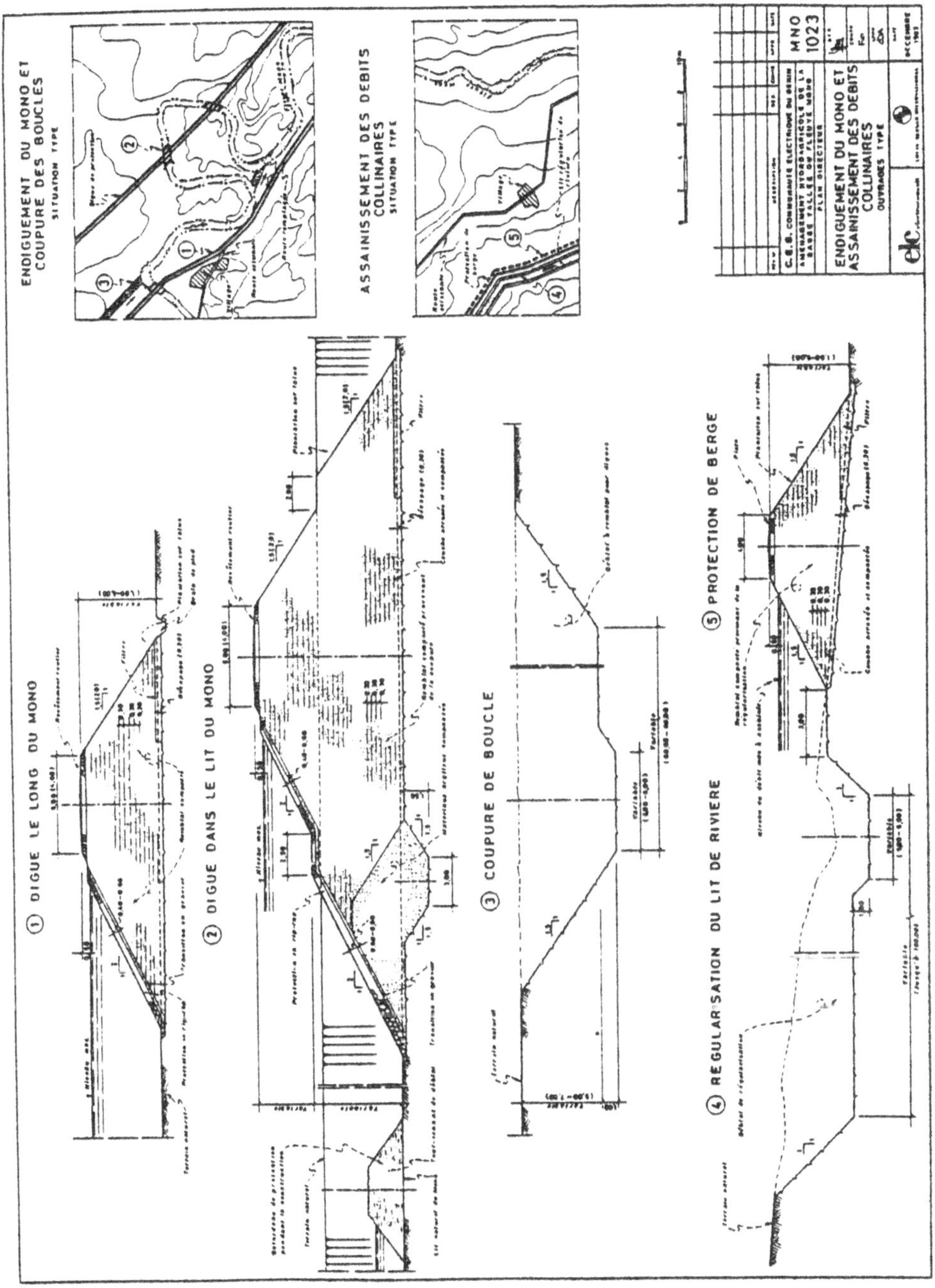
ENDIGUEMENT DU MONO ET COUPURE DES BOUCLES
SITUATION TYPE
ASSAINISSEMENT DES DEBITS COLLINAIRES
SITUATION TYPE
① DIGUE LE LONG DU MONO
② DIGUE DANS LE LIT DU MONO
③ COUPURE DE BOUCLE
④ REGULARISATION DU LIT DE RIVIERE
⑤ PROTECTION DE BERGE
C.E.B. COMMUNAUTE ELECTRIQUE DU BENIN
AMENAGEMENT HYDROAGRICOLE DE LA BASSE VALLEE DU FLEUVE MONO
PLAN DIRECTEUR
ENDIGUEMENT DU MONO ET ASSAINISSEMENT DES DEBITS COLLINAIRES
OUVRAGES TYPE
MNO 1023

tuations involving development both with and without irrigation.

Within the irrigation development plan protection was required for flooded rice growing areas covering about 35% of the area and other crops irrigated by furrows. In the case of rice, three criteria as shown in Table 4 were examined.

Table 4

DRAINAGE MODULE FOR PROTECTION OF RICE GROWING AREAS (l/s/ha)

Criteria	For 1 in 5 year Return Rainfall	For 1 in 10 year Return Rainfall
To maintain the increase in height of water in paddies below 75 mm	2.1	3.8
To evacuate all rainfall within three days	4.6	5.1
To drain 100 mm in 3 days for agronomic reasons	3.8	

For the furrow areas the time necessary to evacuate all run off based on the Curve Number Method is shown as in Figure 1. Final drainage modules of 5 l/s/ha for the rice growing areas, and 8 l/s/ha corresponding to 70 mm/24 hours were accepted for the furrow areas both for the irrigation situation and the rainfed cropping situations. Drainage canals were dimensioned utilizing the Cypress Creck reduction method.

CONCLUSIONS

The project involved a comprehensive study of an area with very limited development options without very substantial costs. The various criteria accepted and the resulting development costs involved a pooling of substantial experience in the field of drainage design and extensive research into flood protection and drainage projects undertaken in similar conditions.

The key results of the economic analysis which show the project to be viable (but only after further investment in irrigation with consequent higher agricultural returns) are reproduced in Table 5.

Table 5

SUMMARY OF BASIC ECONOMIC EVALUATION CRITERIA

Description	Rainfed Development		Irrigated Development	
	Total	per ha	Total	per ha
Net Area (10^3 ha)	45.2		40.0	
Flood Protection Cost (*)				
Total	68.3	1 511	68.3	1 707
Annual	0.3	6.6	0.3	7.5
Total Investment (*)	96.6	2 137	244.0	6 100
Annual Cost (*)	4.6	103	8.0	203
Maximum Annual Benefit (*)				
Gross(incl. labour costs)	20.9	463	69.7	1 740
Net (excl. labour costs)	15.1	337	60.0	1 500
I.R.R. (%) Gross	11.4		17.2	
Net	7.3		15.0	
B/C (i = 10%) Gross	1.1		1.7	
Net	0.8		1.5	
B-C (*) (i = 10%) Gross	12.0		157.7	
Net	-22.6		105.7	

(*) Total costs: 10^6 US$
Cost per ha: US$

The high capital cost of US$ 68 million for flood protection resulting in a cost of over US$ 1 500 per ha protected may lead to serious consideration of alternative operation procedures for Nangbeto Reservoir. It should be pointed out, however, that a substantial portion (US$ 31.3 million) of the total cost was required to protect the area from runoff from the surrounding hills. Although this cost would be lowered in the absence of dikes, it would never be eliminated in the event of a flood protection reservoir being feasible. The detailed study of another reservoir on the Mono River was outside the scope of this study but is currently being examined in the light of further studies commissioned by the Client.

A further aspect of overall project design highlighted within this study was the inadvisability of incorporating costly high capacity pump stations to be used only during events of long return periods yet requiring regular maintenance.

The prefeasibility study has been succeeded by the design of pilot irrigation schemes in both Togo and Benin to be implemented as the first stage of development of the project area.

It is hoped that the results of the study as outlined above will be of assistance to those involved at a preliminary planning level in similar projects.

ACKNOWLEDGEMENTS

The cooperation and the support provided by the client, Communauté Electrique du Bénin, during the study is gratefully acknowledged.

The author would also like to thank ELC-Electroconsult for permission to reproduce various illustrations and the individuals of the company who aided in proof reading and typing.

REFERENCES

Soil Conservation Service (1972) Hydrology, National Engineering Handbook, Section 4, U.S. Department of Agriculture, Washington D.C.

Soil Conservation Service (1971) Drainage of Agriculture Land, National Engineering Handbook, Section 16, U.S. Department of Agriculture, Washington D.C.

Moisture Excess Removal from the Saucers in the North-Bucharest Area

Iulian Mihnea
Institute of Land Reclamation Studies and Design, Bucharest, Romania
Horaţiu Ioniţoaia
Institute for Research & Technological Engineering on Irrigation & Drainage, Bucharest, Romania

ABSTRACT

The saucers, microdepression forms, represent a particular case of the loess plain relief in the South and Eastern areas of the Romanian Plain. Here they are affected periodically by moisture excess due to the conditions of a semi-dry climate and high phreatic leveller in the conditions of a semihumid climate in the presence of a low permeable clayey horizon.

The moisture excess removal works achieved in the North - Bucharest area in the conditions of a low humidity climate constitute a positive example of an intensive reclamation both of the agricultural lands and the environment.

DEFINITIONS AND PROJECT SOLUTIONS

The saucers, small depressions of circular or oval form are frequently met with on flat loess plains or on those presenting a reduced slope in the semi-dry areas. Another category of small depressions are the dells that have different dimensions but with smaller depths than the saucers. They are characteristic of the sub-humid climate.

The appearance of saucers and dells is due to an initial existence of reduced land levelling that allows for the accumulation of waters from rains or snow thawing, thus increasing local infiltrations. The infiltrations in excess have accelerated the soluble salt dissolving process and according to this the loess or the loess deposit settling. In the semi-humid climate the lesser deposit settlings have been accompanied by the clay migration from the soil top horizon. The saucers and dells being located at a level lower than the surrounding land level have generally higher natural humidity and during the periods with a surplus hydrometeorological balance, the water is stagnating and ponding and logging phenomena appear.

Soil logging and the appearance during the last 20 years of some small lakes especially in the semi-humid area of the Romanian Plain have called for the implementation of a row of measures for moisture excess removal.

The project solutions proposed for water excess removal from saucers and dells have been differently carried on according to specific microrelief, soil and excess source conditions of each of the geographical zone.

The project solutions applied within the North Bucharest city zone, located in sub-humid climate conditions, which is the first unit where an integral achievement of moisture excess removal works from saucers (dells) have been carried on, are further presented.

Natural conditions

The North Bucharest zone is situated between the Ilfov rivulet at West, Ialomiţa river at North and the Mostiştea Valley at East, covering land of about 130,000 ha out of which in the South-Eastern part there are about 20,000 ha of microdepressions-saucers and on the rest of the territory are to be met with microdells. The saucer maximum depth is 1.5 m and the microdell frequent depth is of 0.5 - 1.0 m.

Their surface: frequent 2 - 5 ha; maximum about 10.0 ha.

An average of 20 % of the saucer occupied surface may locally reach in the central part of the inter-rivers of the area down-trem stream sector up to 50 %. (Fig.1 Tunari-Dămieni location plan). The general land slope is oriented on the West-Eastern direction and its average value is: 1.1 - 1.5 ‰.

The loessoid deposits in the microdepression areas are characterized by a high content of colloidal clayey (fractions of Ø 0.002 mm, over 25 - 30 %) on large layer thicknesses (over 5 - 6 m) weakly - very weakly permeable. There are also present impervious illuvial - clayey horizons.

The ground water is accumulated at the basis of the weakly permeable loessoid deposits fact that allows for a rapid and important ground water level rise up to 1 - 1.5 m. The ground water level depth is mostly larger than 3 - 5 m in the sub-zone that is upstream from the Bucharest-Buftea-Bujoreanca high road, and between 1 - 3 m in its downstream sub-zone - with the exception of the drained lands in the immediate neighbourhood of the valleys. On the largest part of the territory the moisture excess supplying source are the precipitations of the cold period and those of the spring (March to May) and as a consequence pseudo-gleyied soils. Only in the downstream sub-zones the water excess is due to the high ground water levels (amphy-gleyic soils).

Table 1. Hydrometeorological balance on characteristic years (weather-forecast București - Băneasa station)

Balance (mm)	Period X - III	Period IV - IX	Yearly total T (mm)
Rainy year 1980-81			
Precipitations	375.9	302.2	678.1
Evapotranspiration	89.5	617.2	706.7
Differences	+286.4	-315.0	- 28.6
Average year 1975-76			
Precipitations	213.3	366.8	580.1
Evapotranspiration	57.5	573.2	630.7
Differences	+155.8	-206.4	-50.6
Dry year 1973-74			
Precipitations	121.1	351.0	472.1
Evapotranspiration	67.3	587.4	654.7
Differences	+53.8	-236.4	-182.6

The moisture excess duration is in accordance with the hydrometeorological regime: of a short duration (1-3 days) as a consequence especially of summer torrential precipitations, of a lasting duration (of up to 2-3 months) as a consequence of the abundant winter and spring precipitations, and quasi permanent-continuing from one year to the other - as a consequence of the sequence of some rainy years.

The moisture excess becomes manifest by stagnations, by pondings within the microdepression zone and by "furrow excess" on flat lands.

The moisture excess in that zone has a cyclic character according to the precipitation regime. Such characteristic periods have been 1954 - 1955, 1969 - 1972, 1979 - 1981 years.

ACHIEVED SURFACE AND SUBSURFACE DRAINAGE WORKS

More important moisture excess removal works began to be executed in this zone during 1972-1975 and they comprised a main and secondary canal network for the removal of the stagnant superficial waters on a surface of about 33,000 ha. The superficial water evacuation works - that represent the first working stage for the moisture excess removal and which aimed to ensure the evacuation conditions - were resumed in 1978 when they were intensified. In this second stage the water moisture excess removal has been achieved by modeling-leveling and by surface and subsurface drainage works on plots.

On basis of the research achieved in the experimental and pilot projects within the Otopeni - Tunari subsystem and from the accumulated design activity experience it resulted that :

Moisture excess removal works from the saucer have to be designed and achieved according to the relief - microrelief and excess source conditions.

Following work categories can be individualised according to the moisture excess source :

- surface drainage for moisture excess removal from soil surface, excess mainly due to precipitations;
- subsurface drainage for soil profile moisture excess removal, excess due to infiltrations from precipitations and/or a high ground water level;
- combined drainage, surface associated with subsurface drainage in the conditions when the moisture excess is due to a strong or very strong intensity of both sources, mainly predominant being from precipitations.

The above work technical solutions are chosen according to local conditions i.e. micro depression surface and depth; its position as against the collecting-evacuating canal; soil permeability as shown in fig.nr. 1,2 entitled "Technical solutions".

As to the main technical tile drainage parameters, they are differently applied namely :

- distance between the absorbant drain lines :
  30 m for clayey texture soils;
  35 m for loamy texture soils;
  40 m on sandy texture soils.
- absorbant drain average installing depth:
  0.7 - 0.8 m on lands with moisture excess

mainly from precipitations;

0.9 - 1.0 m on lands with moisture excess mainly or exclusively from ground water.

- filtering materials :

in conditions of moisture excess due to precipitations, ballast of Ø 7 - 15 mm.

filtering layer height will be :

30 - 35 cm for the active zone of the absorbant drains located at the bottom of the saucers;

10 - 15 cm for saucer marginal zone;

5 cm in the drained inactive (transport) zone.

In conditions of moisture excess due to ground water geotextiles are put around the tubes.

No filter is being used when soil conditions allow for these technical solutions and the moisture excess is due to ground water.

Saucer water collecting and evacuation is achieved through a canal network which led the waters to the area natural emissaries. To the moisture excess removal works from the saucers have been associated pedomeliorative works - soil scarifying, chemical amendments - to improve the hydrophysical soil characteristics and its acidity, as well as agrotechnical works such as: fertilizers (especially organic ones) adequate crop structure, deep ploughing etc.

Inside saucer project scheme for moisture excess removal is that shown in fig.nr. 3 (saucer scheme).

OBTAINED RESULTS

The above technical solutions have been included in the execution documentation for such works on agricultural lands covering a surface of about 100,000 ha. These works are at present on the way to being completely achieved.

The first results obtained in the Tunari - Demieni zone confirm the efficiency of the designed adopted solutions.

In order to illustrate with examples here further on is presented the land reclaimed situation, main crop obtained yields in an agricultural unit (CAP Tunari, surface = 1,680 ha) in the pilot perimeter zone.

Table 2. Evolution of areas during 1979-1983 and yields obtained (1970-1980; 1981-1984).

	1979	1980	1981	1982	1983
areas affected by moisture excess (ha)	360	305	-	-	-

obtained yields (kg/ha)	wheat	barley	maize	sunflower
1970 - 1980	2,695	3,388	2,855	1,500
1981 - 1984	4,210	4,226	5,047	2,350

Note: Surface and subsurface drainage works were finished in 1980.

From the above given figures the efficiency of the surface-subsurface works is obvious.

In the conditions of a continental-temperate climate in order to obtain large and stable agricultural yields, it is also necessary to achieve - besides the above mentioned works - irrigation works for drought control (such works began to be executed in 1984).

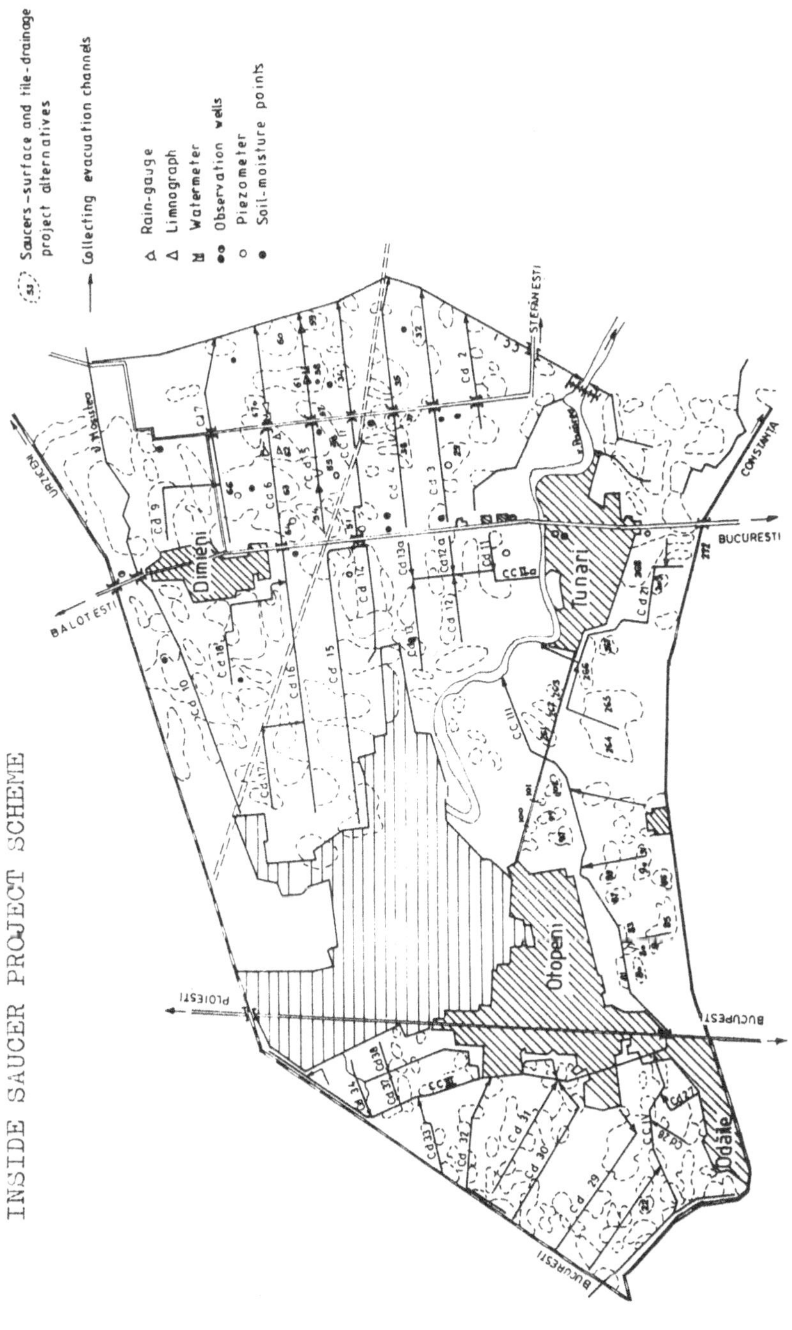
INSIDE SAUCER PROJECT SCHEME
Saucers-surface and tile-drainage project alternatives
Collecting evacuation channels
Rain-gauge
Limnagraph
Watermeter
Observation wells
Piezometer
Soil-moisture points
Dimieni
Tunari
Otopeni
Odăile
BALOTESTI
PLOIESTI
BUCURESTI
STEFANESTI
CONSTANTA
URZICENI

Figure 1

North Bucharest zone

WATER DRAINING
TEHNOLOGY FROM SAUCERS

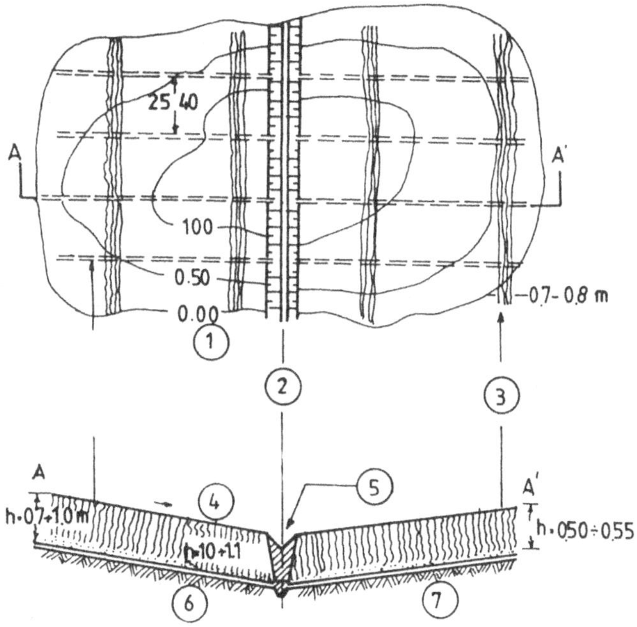

Note: For saucers whose h>0.3-0.4m, their slope levelling has been foreseen according to the above section.
For saucers whose h<0.3-0.4m, their complete filling (levelling) has been foreseen.

Figure 1,2

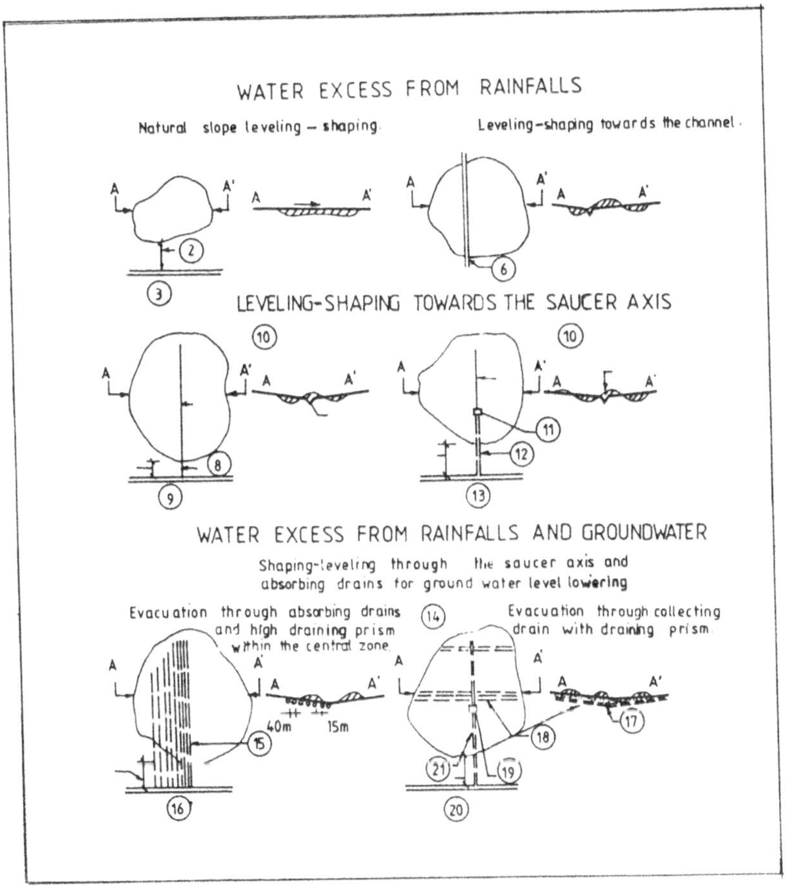
WATER EXCESS FROM RAINFALLS
Natural slope leveling – shaping.
Leveling–shaping towards the channel.
LEVELING-SHAPING TOWARDS THE SAUCER AXIS
WATER EXCESS FROM RAINFALLS AND GROUNDWATER
Shaping-leveling through the saucer axis and absorbing drains for ground water level lowering
Evacuation through absorbing drains and high draining prism within the central zone
Evacuation through collecting drain with draining prism
40m
15m

Figure 2

SECTION 3 HYDRAULIC ASPECTS OF DRAINAGE DESIGN

The Hydraulic Characteristics of Hinged Flap Gates

R. Burrows
University of Liverpool

SUMMARY

The paper reports an experimental programme to investigate the hydraulic energy losses associated with simple hinged flap gates, sometimes referred to as 'tide-flaps'. The graphical results presented, for models of a circular gate, illustrate in particular the influence of the weight of the gate on the energy losses created by the flow constriction.

INTRODUCTION

Flap gates are often installed on the outfalls of sewerage and drainage systems where protection against flood influx from high receiving stream levels, or from tidal levels in estuarial and coastal sites, is required for low lying areas. In these circumstances the gate closes under the action of its self weight and hydrostatic pressure differences, sealing the upstream system against the tailwater. It reopens only when a positive head differential is developed between the pipe system (or the landward detention basin) and the receiving water which is sufficient to overcome the gate's self weight.

Under most sewer flow and tailwater combinations the installation will minimise potential surcharge within the drainage system although the incorporation of the gate introduces a constriction on the flow creating an increased upstream head requirement. Under extreme conditions when high tailwater levels coincide with large drainage flows it is inevitable that the inclusion of the gates will increase the consequential sewer surcharge or flooding unless detention storage in the drainage system is large.

This form of sewer protection has been incorporated extensively in the past and reference to 'tide gate' appurtenances appears in many texts including, for example, Babbitt (1922), Escritt (1962) and Linsley Franzini (1979). In some instances it would appear that the gate's purpose has been primarily to restrict access of vermin to the sewers, with a secondary advantage in restricting potential influx of sediments in circumstances where this may be a potential problem. Certainly many old gates appear to be in a poor state of repair and clearly cannot function efficiently as protectors against tidal influx.

Linsley and Franzini (1979) show a photograph of a bank of tide gates operating with a tailwater surcharge close to springing level of the outfall pipes. The gates appear to be open to an angle of about 45 degrees under these conditions. This is the only evidence of the operation characteristics of such facilities that has been found from an extensive literature search, although it is appreciated that this exercise may not have been exhaustive. Certainly, no texts have been found to provide information on potential head losses. A recent survey of flap gate manufacturers revealed little further information with the exception of the Armco Company whose gates were cited above. This manufacturer gives, (Armco catalogue, 1978), an expression for the head loss through its 'Calco' gates as follows:-

$$h_L = \frac{4V^2}{g} \exp\{- \frac{1.15V}{\sqrt{D}}\} \qquad (1)$$

where h_L = head loss
V = velocity of flow through the gate
g = gravitational acceleration
D = diameter of outfall pipe
exp = base of natural logarithms

Charts of head loss as a function of discharge are also presented for these light flap gates of various size. The empirical results were obtained from full scale tests on gates of 18, 24 and 30 inch diameter conducted at the State University of Iowa. Similar graphical relationships are also presented by the U.S. Soil Conservation Service (1973) which references an earlier handbook dated back to 1936.

Typical figures for the maximum head losses associated with these light gates are 0.05 feet for a 12 inch gate ranging to 0.36 feet for an 84 inch gate and these are considered to have "... but little effect on the discharged capacity of the drainage outlets. Medium and heavy duty flap gates have heavier flaps or covers than the gate model tested. As a result, head losses may be slightly more than those indicated by the test" (Armco, 1978).

These comments may indeed be valid in the majority of applications as evidenced by the widespread application of such facilities with little apparent demand for a more complete understanding of the hydraulic characteristics.

The author's attention was first directed towards flap gate installations in a study of storm overflow provision for the proposed interceptor sewer scheme for the City of

Liverpool, where protection of the low level interceptor sewer against influx would be required for each cycle of the tide. In this particular situation many of the outfall sewers were located at shallow depth and head losses introduced by the new structures (overflow chambers and 'tide gates') were, of necessity, to be kept to a minimum to prevent worsening of local flooding problems. The available information, summarised above, was considered inadequate for these purposes, (since robust heavy duty gates would probably be required) and an exploratory model testing programme was completed (Doyle, 1984; Burrows et al, 1984), the results of which are reported herein. In the event, an alternative solution was adopted for influx protection on this scheme involving costly automatically controlled mechanical penstocks. Both potential head losses and risk of interference of gate seatings by debris were considered as major disadvantages of the flap gate facility.

The experimental work described here relates to the behaviour of a rigid single-hinged circular flap fabricated in-house. No attempt has been made to model in detail any particular commercial arrangement but future tests will be made on a small proprietary device. Notwithstanding this, the relationships developed between head loss and gate weight should be generally representative of behaviour, even for double-hinged systems often incorporated to improve gate seating. However, in view of the limited extent of the data collected the results can only be regarded as provisional pending a more extensive testing programme.

EXPERIMENTATION

The experimental apparatus utilized in the tests consisted of a 9m length of 114.5mm internal diameter acrylic pipe connected between two tanks (Doyle, 1984). This equipment had been used previously for various pipe flow studies and attachment of a flap gate, shown in figure 1, at the outfall into the downstream tank completed the test facility. The tank was sufficiently wide so as not to interfere with the discharging flow and incorporation of an adjustable weir downstream enabled the variation of tailwater surcharge conditions. The overflowing water was returned directly to the upstream header tank using a pair of submersible pumps providing a maximum discharge of approximately 9 litres per second. Flow rates were established by the incorporation of a calibrated V-notch within the header tank, the notch and associated depth gauge being separated from the pumped inflow by a baffle system.

Both the header tank and pipe sections were mounted on adjustable supports which permitted the study of different pipe flow conditions. A slope of 1:100 producing supercritical flow was first established enabling the study of the hydraulic jumps created in front of the gate. Incorporation of a flexible coupling between the header tank and the pipe enabled the study of subcritical conditions by the lowering of the pipe sections and a slope of 1:400 was subsequently considered.

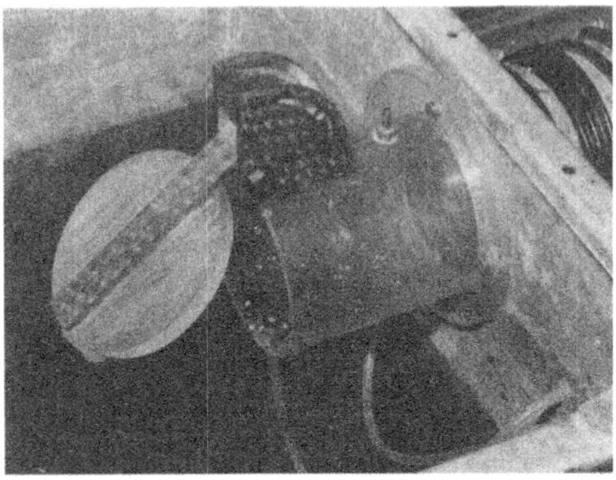

FIGURE 1 FLAP-GATE INSTALLATION

FIGURE 2 FLAP-GATE OPERATIONAL AT AN ANGLE OF 30 DEGREES AND DISCHARGING WITH NO TAILWATER SURCHARGE.

For partially full pipe flow conditions a system of pointer gauges from the pipe soffit could be used to establish water depth profiles and hence velocities of flow at the approach to the gate. With pipe full conditions a bank of 7 piezometer tubes, drawn from the pipe inverts spaced over a 900mm distance in front of the gate, monitored the 'hydraulic grade line' from which the total head could be computed by making allowance for the appropriate velocity head. Two of these tappings can be seen in figure 1 (as can the soffit depth gauge tappings) and the average of the 6 upstream piezometer readings (excluding that at the exit where conditions rapidly vary) was used to establish the total head upstream of the gate. By deducting the corresponding head estimate obtained with the gate removed, the additional head, H_A, required to drive the flow through the constriction was determined. Note, that with this definition, if working upstream from the tailwater level an allowance, in addition to H_A, should be included to account for the minor losses associated with the pipe exit ($\sim V^2/2g$).

The form of construction of the gate and hinge mechanism can be seen clearly from figure 1. The first series of tests investigated head losses for various flow conditions with the gate restrained to a fixed opening. Figure 2 shows such a situation with the gate set at an angle (of opening) $\theta = 30°$ using a graduated scale inserted on the restraint system, which consisted of a wing-nut assembly with travel along curved slots. In later 'gate weight' tests these restraints were removed and the opening angle, θ, was recorded for different flow rates and tailwater conditions.

Tailwater surcharge was defined by the distance, z, of the water surface above outfall pipe invert level and settings of z/D = 0, 0.5, 1.0 and 1.5 were considered.

NOTATION AND DIMENSIONAL ANALYSIS

The first test series involved the measurement of additional head loss, H_A, created by insertion of the gate as flow rate, Q, and gate opening angle, θ, were varied systematically for different pipe slopes, i. The second series involved the measurement of θ as Q was varied for two different gate weights, W. In each case different levels of tailwater surcharge, z, were considered. Application of dimensional analysis produces the following relationships.

For 'head loss' tests: the dimensionless additional head (equivalent to the 'head loss coefficient') may be expressed in terms of the kinetic energy head of the approaching flow,

$$\frac{H_A}{\alpha V^2/2g} = \phi'\left[\frac{Q}{\sqrt{g}\,D^{5/2}},\ \theta,\ i,\ \frac{z}{D}\right] \tag{2}$$

where ϕ is 'an unknown function'
and α is a kinetic energy correction factor.

A value of $\alpha = 1.15$ has been assumed herein. Pao (1961) suggests a range of values for turbulent flows in open channels between 1.03 and 1.36 and for pipe flow between 1.01 and 1.10. It is appreciated that often in analysis α is taken as unity, in which case dimensionless head losses presented later should be adjusted accordingly.

For 'gate weight' tests: the angle to which the gate opens may be expressed as,

$$\theta = \phi''\left[\frac{Q}{\sqrt{g}\,D^{5/2}},\ \frac{z}{D},\ \frac{WD^2}{\rho Q^2},\ i\right] \tag{3}$$

In subsequent figures the gate weight has also been referred to in the form of its mass equivalent, W*(kg), which is of course equal to W/g.

HEAD LOSSES CREATED BY THE FLOW CONSTRICTION

The results of this study are presented in figure 3. In these tests the gate was restrained at set values of θ and the influence of the constriction at small angles is immediately apparent. In the unlikely event that the gate was to become 'seized' at an angle of about 7 degrees the upstream surcharge, amounting to about 10 times the velocity head, may even exceed the potential degree of tailwater surcharge to be protected against.

With the form of presentation adopted in figure 3 it at first sight appears that the longitudinal slope of the

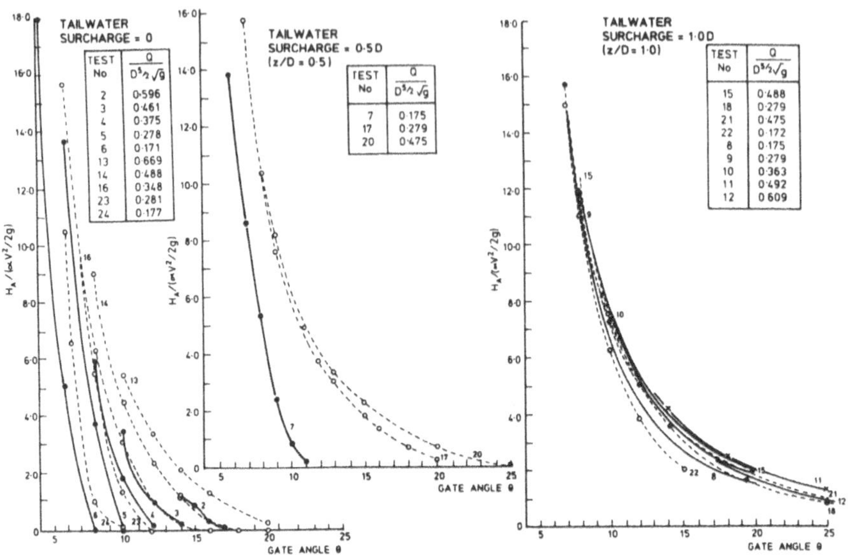

FIGURE 3 RESULTS FROM 'HEAD LOSS' TESTS
(solid lines correspond to 1:100 pipe slope, dotted lines correspond to 1:400 slope)

outfall pipe, i, is a significant factor. This is, in fact, not the case, for this range of slopes at least, as can be seen from figure 4 where the energy head loss is non-dimensionalised in terms of the pipe diameter. The influence of the slope is minimal. The discrepancies in figure 3 between the curves for different slope under a zero surcharge condition ($z/D = 0$) and the same 'Froude' number, $Q/\sqrt{g}\, D^{5/2}$, arise because the natural upstream flow depths (in the outfall with the flap gate removed), and hence $\alpha V^2/2g$, differ. For the 1:100 slope the depth condition is

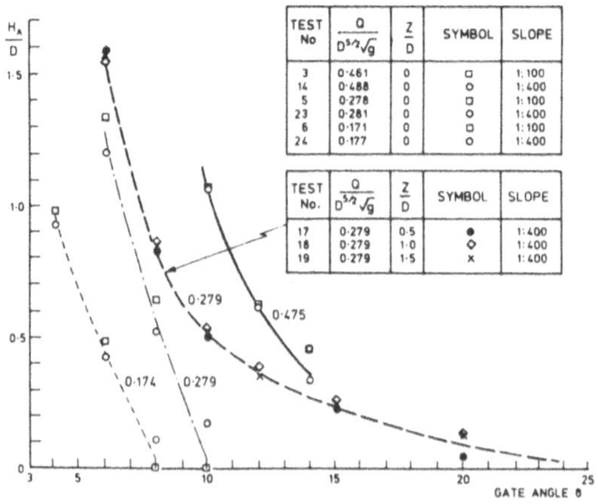

FIGURE 4 HEAD LOSS THROUGH THE FLAP-GATE AS A FRACTION OF PIPE DIAMETER

supercritical and hence the velocity, V, is much larger than that for the 1:400 subcritical flow condition. The curves for the latter, therefore, lie above those for the former in figure 3 ($z/D = 0$).

These discrepancies associated with upstream velocity head reduce to zero as the tailwater surcharge approaches (and subsequently exceeds) 1.0 as can be seen by the graph for this condition in figure 3.

The other interesting observation from this plot is that for conditions where the outfall pipe is full and the jet is fully submerged, the influence of 'Froude' number, $Q/\sqrt{g}\ D^{5/2}$, diminishes to zero as might be anticipated. For engineering purposes, under these conditions ($z/D \geq 1.0$), which are likely to be representative of the 'design' situation, the dimensionless head loss (or equivalent 'minor loss' coefficient) is, therefore, dependent only upon the gate angle, θ, since the tailwater surcharge also has negligible effect as can be seen from figure 4.

Unfortunately, limitations in the accuracy of the piezometer system and the relatively small scale of the model prevented the establishment of the smaller energy losses associated with gate angles above 25°, which are most likely to represent real operation conditions for light flap gates.

THE INFLUENCE OF GATE WEIGHT

Figure 5 shows the relationships between the dimensionless gate weight, $WD^2/\rho Q^2$ and the gate angle, θ, for different tailwater surcharge conditions obtained with gates of two different weights. For clarity most data points have been omitted from the curves which show that the tailwater surcharge is only a significant factor until z/D approaches unity.

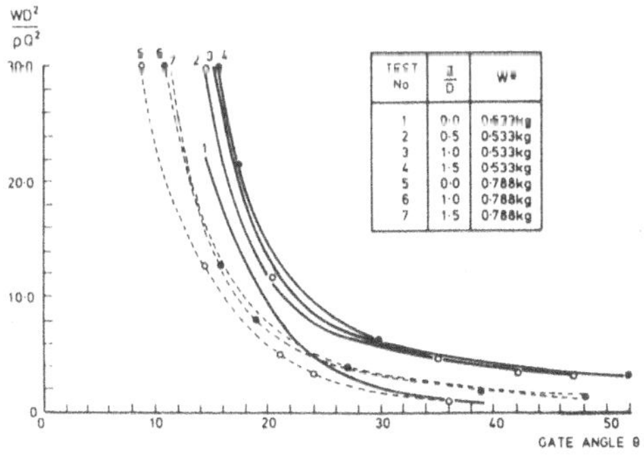

FIGURE 5 RESULTS FROM 'GATE WEIGHT' TESTS

The modelling represented in figure 5 is incomplete as the influence of the flow 'Froude' number, $Q/\sqrt{g}\ D^{5/2}$, identified in equation (3), is not accounted for. Furthermore, the influence of pipe slope, i, has not been investigated although for 'typical' outfall gradients under design (surcharged tailwater) conditions the effect is likely to be negligible. From earlier discussion it might be anticipated that the effect of a 'Froude' number will be greatest for $z/D < 1$ and that it may increase as the angle of gate opening increases permitting the existence of partially full pipe conditions.

The problem can be addressed theoretically by application of 'conservation of momentum' principles to the jet discharge or, alternatively, by a simplified treatment which reduces to a gate equilibrium computation involving appropriate hydrostatic thrusts. For example, considering the 'drowned' gate situation ($z/D \geq 1.0$), moments can be taken about the position of the hinge which reduces to requirement for equality of forces between both sides of the gate to yield:-

$$\{W - \rho g \frac{\pi D^3}{4} (\frac{t}{D})\} \sin \theta = \frac{\pi D^2}{4} . \Delta p \qquad (4)$$

where t is the equivalent uniform thickness of the gate, the term to which it is associated representing a buoyancy term, and

$$\Delta p = \rho g H_A + \rho g (\frac{\alpha V^2}{2g}) \qquad (5)$$

It is assumed, therefore, that Δp is the pressure resulting from the 'additional' head required to drive flow through the constriction plus allowance for upstream velocity head and 'exit' losses. This implies that the velocity head for the flow in contact with the gate itself is small.

Figure 6, which has been prepared from Md. Said (1985), illustrates from the idealised two dimensional flow net that velocities over much of the gate surface might be sufficiently small to permit the incorporation of the velocity head term in equation (5). Figure 6 also shows the pressure variation on the back of the square gate tested to be near hydrostatic as assumed above.

Equations (4) and (5) can be rearranged in terms of the non-dimensional parameter groupings used previously, i.e.

$$\frac{WD^2}{\rho Q^2} = \frac{\pi}{4} (\frac{t}{D}) (\frac{Q}{\sqrt{g}\, D^{5/2}})^{-2} + \frac{2\alpha}{\pi \sin \theta} [\frac{H_A}{(\alpha V^2/2g)} + 1] \qquad (6)$$

This relationship is plotted out in figure 7 using head losses from figure 3 and $(t/D) = 0.1$ which matches the experimental conditions. Superimposed on the plots are the data points from the experiments conducted with the different gate weights and conditions associated with the Froude numbers displayed are joined. These are seen to follow the theoretical distributions reasonably well,

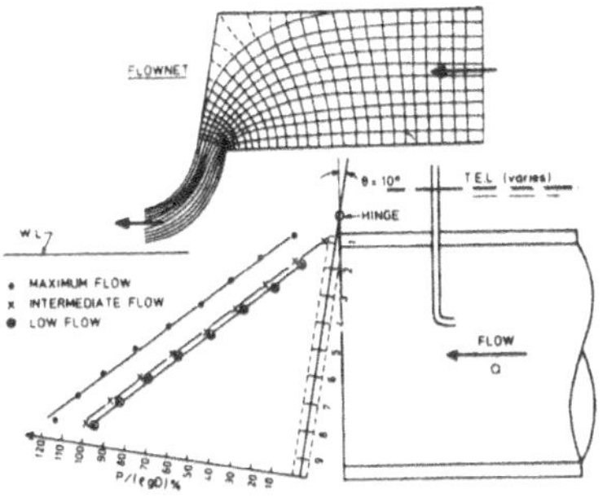

FIGURE 6 PRESSURE VARIATION ON UPSTREAM FACE OF A SQUARE SECTION FLAP GATE (SIDE D = 200mm; GATE ANGLE 10°) AND SIMPLE TWO DIMENSIONAL FLOWNET

although clearly much more data would be required to adequately verify the above simplistic analysis. Indeed, a more rigorous analysis would account for the precise distribution of gate mass (herein assumed to act at the gate's centre) and the small difference between the pipe and gate diameters.

Notwithstanding these potential refinements, it would appear that equation (6), or figure 7, coupled with figure 3 can provide a basis for assessment of the hydraulic characteristics of flap gate operation. For a given gate weight and stipulated flow conditions, figure 7 will indicate the approximate angle to which the gate should open and figure 3 will then predict the corresponding head loss. (Note that if $(t/D) \neq 0.1$ then a revised plot of figure 7 will be required using equation (6) and figure 3).

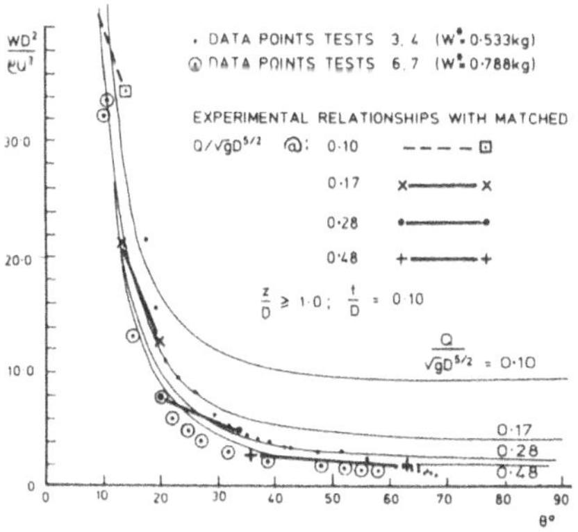

FIGURE 7 THEORETICAL RELATIONSHIPS BETWEEN GATE WEIGHT AND OPENING ANGLE COMPARED WITH EXPERIMENTAL RESULTS

A preliminary attempt at verification of the procedure has been made using equation (1). This was achieved by considering a range of flow 'Froude' numbers for a gate of notional size and computing from equation (1) the head losses. These were then used to obtain from figure 3 the angles θ and from figure 7 the associated values of $WD^2/\rho Q^2$. By checking that these dimensionless numbers produced gate weight estimates that were not grossly different, nor unrealistic, a crude test of the procedures was completed.

CONCLUSIONS

An exploratory study has produced relationships which can potentially be used to assess the hydraulic characteristics of simple hinged flap gates. These findings do, however, require further verification and should, therefore, be treated accordingly.

ACKNOWLEDGEMENTS

The author wishes to acknowledge the contribution of former students Mr. K.J. Doyle and Mr. M.A.B. Md. Said in respect of the experimental programme completed. He also wishes to express thanks to Mr. D.L. Hockin of the North West Water Authority for the kind contribution of reference literature and addresses of gate manufacturers.

REFERENCES

ARMCO Canada Ltd., (1978), 'Armco Water Control Gates -Catalogue', P.O. Box 3000, Guelph, Ontario, Canada.

Babbitt, H.E., (1922), 'Sewerage and Sewage Treatment', pub. Wiley, New York.

Burrows, R., Ali, K.H.M. and Pateman, D.R., (1984), 'Storm Sewage Overflow Model Testing for Outfalls L1-L5', Report to Liverpool City Engineer, Dept. Civ. Eng., U. of Liverpool.

Doyle, K.J., (1984), 'An Assessment of the Hydraulic Characteristics of Hinged Flap Gates', M.Eng. dissertation, Dept. Civ. Eng., U. of Liverpool, May.

Escritt, L.B., (1962), 'Sewerage and Sewage Disposal', pub. Contractors Record Ltd., London.

Linsley, R.K. and Franzini, J.B., (1979), 'Water-Resources Engineering', pub. McGraw-Hill.

Md. Said, M.A.B., (1985), 'An Assessment of the Hydraulic Characteristics of Hinged Flap Gates', Student dissertation, Dept. of Civ. Eng., U. of Liverpool.

Pao, R.H.F., (1961), 'Fluid Mechanics', pub. John Wiley and Sons Inc., p458.

Soil Conservation Service, (1973), 'Drainage of Agricultural Land', U.S. Department of Agriculture, Pub. Water Information Center Inc., New York.

Laboratory Experiments on the Flow Resistance of Aquatic Weeds

H.J. van Ieperen and M.S. Herfst
Department of Hydraulics & Catchment Hydrology, Agricultural University, Wageningen, The Netherlands

1. INTRODUCTION

In the early seventies interest was created for other methods of maintenance of ditches than traditional ones in order to decrease costs and to fulfil requirements with respect to environmental and recreational demands. Investigations were carried out to explore the possibility of restricting or preventing the growth of weeds by means of light interception effectuated by floating leaves of certain species (Pitlo, 1978 and 1982). The design and management of vegetated channels require a procedure for predicting flow resistance due to aquatic weeds. A survey of the various research done in the past to accomplish such a procedure is very well embodied by Lindner (1982). A relatively simple flow model based on the drag force on vegetation in flow was described by Petryk & Bosmajian (1975). For a further development of this model flow resistance due to aquatic weeds was studied on monocultures of white water-lily (Nymphaea alba), broad pond-weed (Potamogeton natans) and reed (Phragmites australis) at the Agricultural University of Wageningen during the period 1983-1985. The aquatic weeds were also investigated, using a new technique in order to make a more profound estimation of plant biomass.
This paper describes the measurements and discusses the results.

2. BASIC EQUATIONS

In general flow resistance in an open channel is described by using Manning's coefficient as a friction factor:

$$\bar{u} = \frac{Q}{A} = \frac{1}{n} R^{2/3} S_H^{1/2} \qquad (1)$$

where $\bar{u}$ = mean velocity; Q = discharge; A = cross-sectional area; n = Manning's coefficient; R = hydraulic radius; S_H = energy gradient.

The coefficient of Manning represents the influence of the boundary roughness. In case of vegetated channels the model of Petryk & Bosmajian (1975) gives a description of the flow resistance through a vegetated channel in which both the influence of boundary roughness and aquatic weed are represented. From momentum considerations they obtained the following result:

$$S_H = \left\{\frac{C_d \Sigma A_i}{2g\ AL} + \frac{n^2}{R^{4/3}}\right\} \bar{u}^2 \qquad (2)$$

where C_d = the drag coefficient for the vegetation; A_i = the projected area of the *i* th plant in the streamwise direction; g = the gravitational constant; L = the length of the channel reach being considered.

Several steps can be taken to refine this model and to make it more suitable for the analysis of actual measurements involving aquatic weed and flow resistance. The numerator of the first term between brackets refers, in case of the monocultures mentioned above, to the total projected area of the stems and its drag coefficient. The length of the stems can be related to the initial or the actual waterdepth in case of water-lily and pond-weed or reed respectively. Then $C_d \Sigma A_i$ in Equation 2 can be written as:

$$C_d \Sigma A_i = \mu\ dD \qquad (3)$$

where μ = aquatic weed resistance; d = initial or actual waterdepth; D = diameter of the stems.

Note that the number of stems is unknown in the right hand side of Equation 3, and that the sum is omitted. In case detailed information about the stems does exist, then it is possible to calculate the sum of the projected area. However, it seems more realistic to take the mean values of stem length and diameter allowing a better reproduction in case of estimations. In this case the numerator can be written as:

$$C_d \Sigma A_i = \xi\ m\ \bar{l}\ \bar{D} \qquad (4)$$

where ξ = drag coefficient of stems; m = number of stems; $\bar{l}$ = mean length of stems; D = mean diameter of stems. Both the coefficients μ and ξ depend on the Reynolds number and represent not only a drag coefficient but also the deficiency due to the simplicity of the model.

The influence of boundary roughness represented by the second term between brackets of Equation 2, can also be described with a non-dimensional roughness parameter. Analogous to the flow resistance in closed conduits the energy gradient without aquatic weed becomes:

$$S_H = \frac{\Delta H}{L} = \frac{\lambda}{R}\frac{\bar{u}^2}{2g} \qquad (5)$$

where ΔH = difference in total head; λ = wall roughness coefficient. Now, from Equation 2, 3 and 4 it can be derived that:

$$S_H = \left\{\frac{\mu\ dD}{A\ L} + \frac{\lambda}{R}\right\}\frac{\bar{u}^2}{2g} \qquad (6)$$

Using Equation 4 instead of Equation 3, one obtains:

$$S_H = \left\{\frac{\xi\ m\ \bar{l}\ \bar{D}}{A\ L} + \frac{\lambda}{R}\right\}\frac{\bar{u}^2}{2g} \qquad (7)$$

This formula uses only dimensionless coefficients and consistently describes the influence of aquatic weed and wall roughness on the flow resistance. These coefficients can be related to the Reynolds number (μ or ξ) or to the relative wall roughness (λ) i.e. the ratio between the hydraulic radius and the equivalent roughness (R/k)

3. EXPERIMENTAL DESIGN

3.1 General

The model described above can be evaluated by taking measurements with the objective (i) to study the resistance at different flow conditions but at equal amount of aquatic weed, and (ii) to determine the relation between the resistance and the time dependent biomass. The measurements were taken in the period 1983-1985. Because of the fact that the three variables discharge, waterdepth and aquatic weed are independent variables (corresponding to Equation 7), in each year the experiment focussed on one of these variables.

All measurements were carried out in an outdoor laboratory from which a plan view is shown in Figure 3.1.

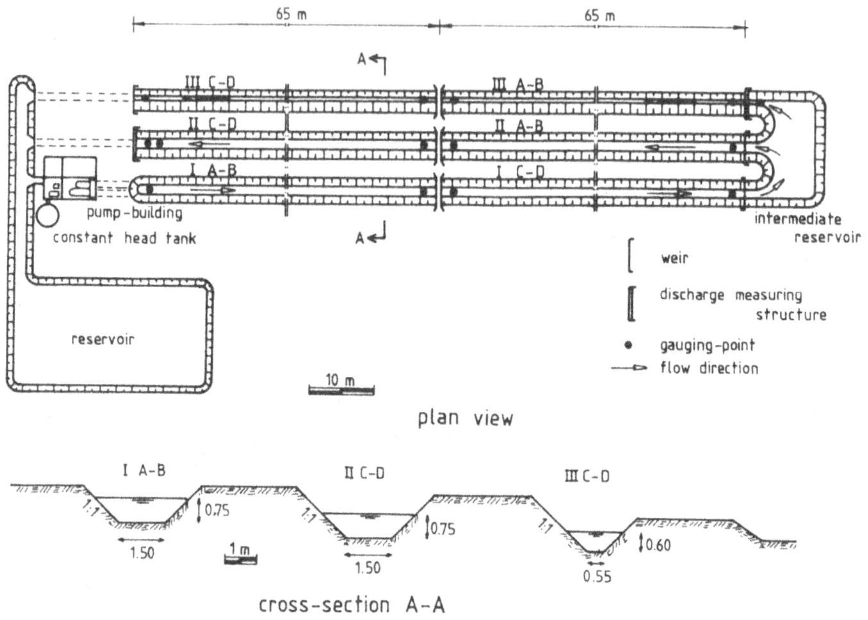

Figure 3.1 Diagram of the experimental ditches

Ditch section I-CD was planted with white water-lily, ditch section II-AB with broad pond-weed and ditch section III-AB with reed.

3.2 Flow resistance measurements

The cross-sectional area and wetted perimeter of the ditch section were obtained by levelling cross-sections every 5 meter with a 0.25 meter distance between measuring points. After calculating the cross-sectional area and wetted perimeter for each cross-section their averages were calculated. The levelling has been carried out in 1983 and was repeated in 1985.
The discharge was measured with two volumetrically calibrated V-notches at the end of intake flumes (Van Ieperen & Herfst, 1985). The waterheight was obtained using point gauges above a stilling well at the side of the flume.
A pressure transducer attached to an amplifier, data-logger and mini-computer was used to registrate the difference in waterheight between the gauging points in the ditch sections. This system was calibrated each year and regularly checked during the measurements by means of additional point gauge readings at stilling wells.

Concerning the flow resistance measurements, in 1983 the water depth was taken at a low, medium or high level with the discharge increasing from 0.030 to 0.130 m^3/s, with intervals of 0.020 m^3/s. The measurements were repeated 6 times during the season and each measurement lasted approximately 6 days. In 1984 also 6 repetitions took place but only at medium waterdepth and discharge increasing with an interval of 0.005 m^3/s, each measurement lasting 2 days. In 1985 measurements were performed weekly. In one day the discharge was increased from 0.040 to 0.115 m^3/s with intervals of 0.015 m^3/s at medium waterdepth. In this year ditch section III-AB was left out of consideration.

3.3 Estimation of plant biomass

Being a standard measure for plant biomass, the estimation of the percentage of cover by a vegetation when projected perpendicular on a horizontal plane has been used (Hoogers, 1963). In case of a high cover of floating leaves, it is hardly possible to make an estimation of the submerged parts of the vegetation. This and subjectivity of the observer make this method less suitable for research on dense vegetation in ditches.

In 1983 a start was made with collecting additional data on the vegetation next to the estimation of cover. Length and diameter of stems and area of leaves were measured once or twice.

In 1984 reed stems in ditch section III-AB have been counted monthly. On behalf of the vegetation in the other two ditches several experiments have been carried out to develop a better method for estimating biomass. Sampling the vegetation by harvesting a 0.50 m strip of the cross-section was very labour intensive but gave the best results. It was further developed to the method described below which has been used for ditch section II-AB in 1985.

During the growing season (April-October) a weekly sample of the vegetation was taken by cutting of the stems at the bottom of the ditch within a 0.30 x 0.40 m area. For this purpose in early spring two rows of 0.30 x 0.40 m rectangular frames were placed at the bottom of the ditch at a 0.40 m distance (Figure 3.2). These frames existed of 0.15 m long parts of a PVC ventilation shaft, around which polythene bags were folded, and were pushed into the mud until they leveled the ditch bottom. Strings were attached to the polythene bags to pull them up in order to prevent the vegetation from floating away while cutting.

From the obtained vegetation samples the composition was noted. This consisted of broad pond-weed, water-thyme and other species (remainder). The number, length and diameter of stems and the number and area of leaves of pond-weed were obtained. Also the volume, fresh weight and dry weight of stems and leaves of pond-weed, water-thyme and remainder were measured.

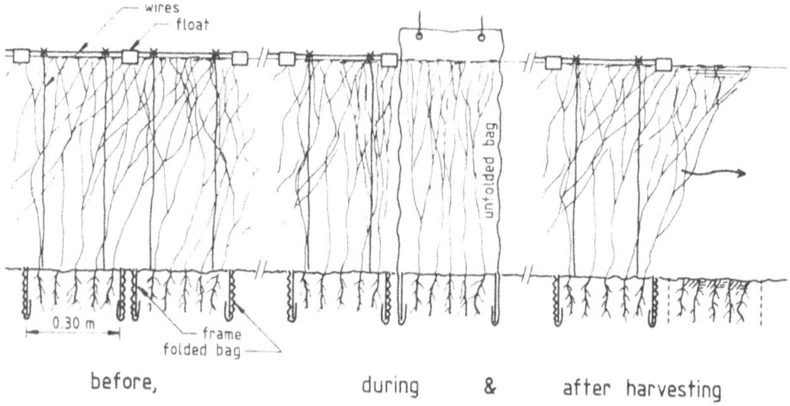

Figure 3.2 Sampling of the vegetation in ditch section II-AB

Vegetation sampling in ditch section I-CD existed of cutting off one water-lily plant. Number, length and diameter of leaf- and flowerstems, leaf area and fresh- and dry-weight were noted. In this section leaves of water-lily were counted on two fixed areas of 3 m^2 each.
From the hydraulic point of view number, length and diameter of the stems are important. The other parameters, as dry weights and leaf area, will be examined on their relation to the aquatic weed resistance coefficient and their suitability for working in the field in order to replace the method of estimating cover. They also can be used by the development of simulation models for the growth of water-lily and pond-weed.

4. RESULTS AND ANALYSIS

The measurements of 1983, when the variation of the waterdepth was emphasized, are to be considered as a first exploration of the influence of different flow conditions and the relation between the aquatic weed resistance μ and the Reynolds number ($Re = u.D/\nu$, with ν = kinematic viscosity). A respresentative example of the results is given in Figure 4.1. There were no significant differences in aquatic weed resistance at different waterlevels.

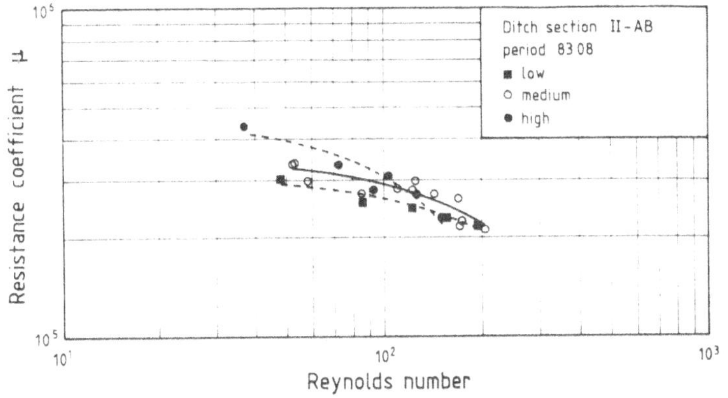

Figure 4.1 μ as function of Re, Ditch section II-AB, August 1983

In 1984, emphasizing the discharge, the relation between μ and Re was determined by a third degree polynomial regression on the approximately twenty data of each run. Table 4.1 shows the correlation coefficient r of the regression, the estimated standard deviation σ of the μ values, mean value of μ and the ratio between σ and μ.

Ditch section	Month	r	σ 10^3	$\bar{\mu}$ 10^4	$\sigma/\bar{\mu}$ 10^{-2}
I-CD	04	0.975	0.84	1.03	8.1
	05	0.954	0.80	1.50	5.3
	06	0.983	1.02	2.89	3.5
	07	0.962	2.05	3.58	5.7
	08	0.988	6.11	2.73	2.2
	09	0.980	5.71	2.30	2.5
II-AB	04	0.987	4.56	7.01	6.5
	05	0.973	18.50	17.64	10.6
	06	0.985	28.32	35.29	8.0
	07	0.988	19.30	28.45	6.0
	08	0.996	14.05	37.13	3.8
	09	0.993	22.98	42.75	5.4
III-AB	05	0.942	6.00	5.09	11.8
	06	0.995	1.71	6.52	2.6
	07	0.968	10.63	13.73	7.7
	08	0.953	5.68	12.77	4.5
	09	0.967	11.49	17.64	6.5

Table 4.1 Parameters from regression analysis

Notice that the mean value of μ gives no information about the actual resistance coefficient during the season because of change in range of the Reynolds number. It can be concluded from Table 4.1 that the regression is quite satisfactory. A remarkable feature however is the rather high ratio of σ/μ, despite the high correlation coefficient and high accuracy of the measurements. An error analysis showed that the relative error in μ is about 3 to 4% and in Re about 2%. So the phenomena itself causes the deviations to be rather high. The relation between μ and Re itself showed a decreasing value of μ at increasing Reynolds number. The order of magnitude of the negative gradient at the mean value of μ was 250, 3000 and 400 for ditch I, II and III respectively.

As expected nearly all the 1985 harvesting data showed a large scattering, especially those of pond-weed. For this species Table 4.2 shows the correlation coefficient of a fifth degree polynomial regression on the various plant parameters, estimated standard deviation σ, the mean values y and the ratio between σ and y.

	r	$\sigma/\bar{y}$ 10^{-2}
Stems		
number	0.499	53.7
mean length	0.704	15.3
mean diameter	0.855	53.8
fresh weight	0.411	51.8
dry weight	0.429	54.0
Leaves		
number	0.730	45.7
area	0.728	52.8
fresh weight	0.747	52.2
dry weight	0.699	57.7
Total		
fresh weight	0.461	43.0
dry weight	0.494	52.1
volume	0.505	49.0
$\Sigma A_i/L$	0.436	52.8

Table 4.2 Parameters from regression analysis

The method of taking one sample each week together with the very small sample area explains the high ratio between σ and y. The low ratios for stem length and diameter promise a better result when sufficient repetitions can be done.

The calculated values of the area of stems are shown in Figure 4.2. The drag coefficient ξ can be calculated by Equation 7 using the predicted values of $\Sigma A_i/L$. A value of 0.045 was taken for λ, this being the wall resistance coefficient at the begin of the season. Without this the value of ξ would have been extremely high in case of small areas of stems. In comparison to the drag coefficient of a cylinder the values of ξ are rather high. However, the decrease of ξ at increasing Reynolds number is similar to the decrease of the drag coeffi-

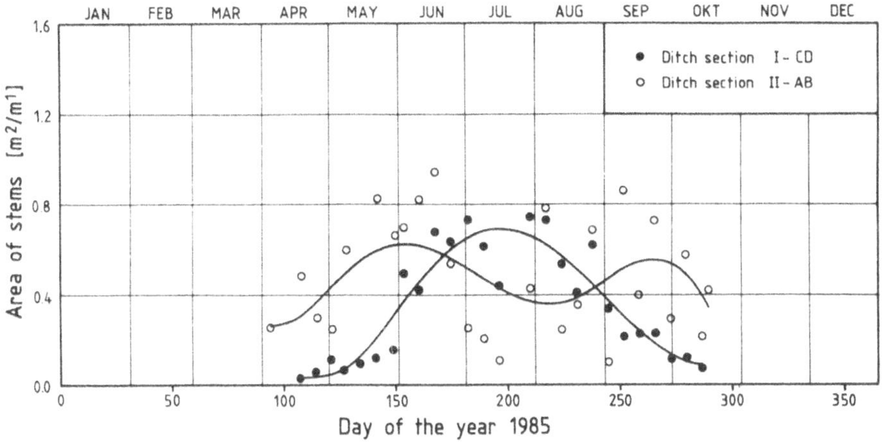

Figure 4.2 Area of stems per unit length $\Sigma A_i/L$

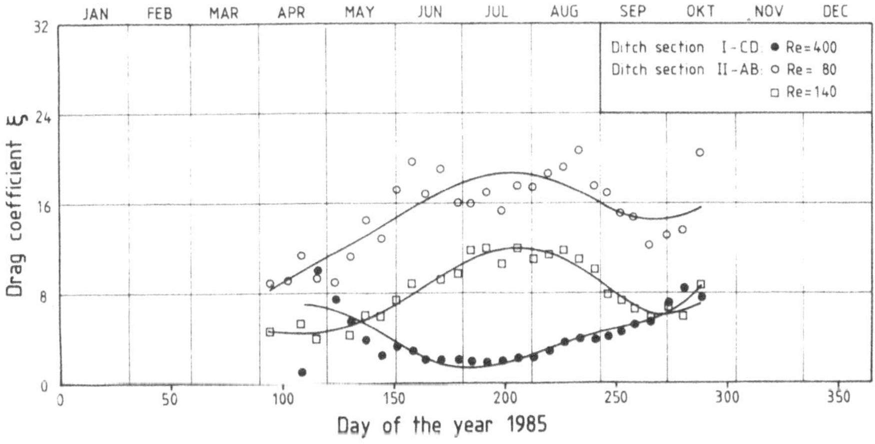

Figure 4.3 Drag coefficient of stems ξ

cient of a cylinder at increasing Reynolds number (Figure 4.3). Despite the difference in structure of the aquatic weed in ditch section I-CD (single stems with a 10 mm diameter) and II-AB (branched stems with a 2.5 mm diameter) the drag coefficients of stems link up nicely with each other. The regression-analysis resulted in a value for σ/ξ of 30% in case of ditch I-CD and 10% in case of ditch II-AB.

5. CONCLUDING REMARKS

An attempt has been made to model the flow resistance of aquatic weeds using plant biomass related parameters. The

number, diameter and length of the stems are the most relevant from a hydraulic point of view. Three species of aquatic weeds were investigated within a limited range of Reynolds numbers. The results might be influenced by scatter in the data of the aquatic weeds. This scatter would not occur when the sample area could be enlarged. In this project the limited range of homogeneous vegetation restricted the sample size. Instead of harvesting a technique which allows sampling data more accurate would be preferable.
The drag coefficient used in the model shows a time dependent variation. This should not occur in case of the model being perfect. The question arises whether this is caused by neglecting the energy correction factor in the model. It is quite obvious that the velocity distribution in a vegetated ditch is different from a (nearly) clean ditch. Nevertheless the results are encouraging and a further development of the model could be accomplished by increasing the range of Reynolds number at one species and by observing different species with flow conditions causing an overlap in Reynolds number.

REFERENCES

Hoogers, B.J. (1963) Het botanische onderzoek van slootgegevens. Jaarboek I.B.S. 1963, pp. 65-69.

Ieperen, H.J. van & M.S. Herfst (1985) Stromingsweerstand van slootvegetaties: metingen 1983, Department of Hydraulics & Catchment Hydrology, Agricultural University Wageningen.

Lindner, K. (1982) Der Strömungswiderstand der Pflanzenbeständen, Leichtweiss-Institut für Wasserbau der Technischen Universität Braunschweig, Mitteilungen Heft 75/1982.

Petryk, S. & G. Bosmajian (1975) Analysis of Flow through Vegetation, Journal of the Hydraulics Division, Proc. ASCE, Vol. 101, No. HY7, July, 1975.

Pitlo, R.H. (1978) Regulation of aquatic vegetation by interception of daylight, Proc. EWRS, 5th Symposium on Aquatic Weeds.

Pitlo, R.H. (1982) Flow resistance of aquatic vegetation, Proc. EWRS, 6th Symposium on Aquatic Weeds.

LIST OF SYMBOLS

A	area	$[L^2]$
C_d	drag coefficient	$[1]$
D	diameter of stems	$[L]$
d	waterdepth	$[L]$
g	gravitational constant	$[L \cdot T^{-2}]$
H	total head	$[L]$
k	equivalent wall-roughness	$[L]$
L	length of channel reach	$[L]$
l	length of stems	$[L]$
m	number	$[1]$
n	Manning's coefficient	$[L^{-1/3} \cdot T]$
Q	discharge	$[L^3 \cdot T^{-1}]$
R	hydraulic radius	$[L]$
Re	Reynolds number	$[1]$
r	correlation coefficient	$[1]$
S_H	energy gradient	$[1]$
u	velocity	$[L \cdot T^{-1}]$
Δ	difference operator	
λ	wall roughness coefficient	$[L]$
μ	aquatic weed resistance	$[1]$
ν	kinematic viscosity	$[L^2 \cdot T^{-1}]$
ξ	drag coefficient of stems	$[1]$
σ	standard deviation	

Diffusion of the Sluice Way Jet

S.K. Al Naib
North East London Polytechnic

INTRODUCTION

When a stream passes under a submerged sluice gate into water on the downstream side, it forms a jet which goes far downstream before it is dispersed [1]. A knowledge of the behaviour of such a jet is of engineering importance because it allows designers to predict the rate of deceleration of the jet which is necessary in the prevention of scour downstream of hydraulic structures. Furthermore, it leads to a better understanding of the mechanism of turbulent diffusion of jets in restricted spaces. Although it was one of the classical problems in hydraulics it has been little studied in the past. Early publications mainly describe model tests for guiding design and predicting the depths and heads associated with the discharge characteristics. Consequently the correlations were given in terms of one-dimensional forms of continuity, momentum and energy relationships [2].

The approach followed by these investigators has been, in general, empirical in character concerning an overall study of the flow variables. While information of this type is essential for industrial design, it does not lead far towards a fundamental understanding of the flow mechanism. However, when as in the present investigation the details of flow are examined a definite picture emerges of the jet dispersing in the deeper water downstream, in the manner occurring in other diffusion problems and conforming with similar statistical picture.

Only a little relevant work has yet been published. Liu [3] investigated by pitot-cylinder the mean patterns with different depths of water downstream. He found that the distance required for the jet to diffuse in a given depth is independent of the Froude number. By using a pitot-tube operating an electrical circuit, Henry [3] measured the longitudinal component of turbulence in certain of the patterns previously studied by Liu. He discovered that the energy of the turbulence is less than one per cent of the energy of the mean flow.

Their results contained discrepancies such as the one between the measured discharge and that calculated through the integration of the velocity profiles. Rajaratnam [4] attempted to analyse the jet as the case of a plane turbulent wall jet under an adverse pressure gradient over which a backward flow has been placed. He presented the forward flow simply as a plane wall jet and evaluated the backward flow using the results of Liu and Henry. The two parts were joined to predict the characteristics of the jet. Unfortunately, the theoretical profiles did not compare at all with the measured surface profiles, either in shape or magnitude [5]. The above three works, while containing some unexplained discrepancies, nevertheless clearly contribute to a better understanding of the general flow pattern.

As a further contribution, the present investigation was undertaken with a view to overcoming the difficulties of the above investigators. To this end a preliminary series of instrumentation tests were undertaken, followed by extensive measurements to study the flow which proved very complicated and unsteady [6,7], requiring the development of a photographic technique to measure velocities [8]. When this was achieved, distributions of mean velocity were determined in various sections of the jet for differing ratios of the limiting downstream depth to sluice opening for the values 2,3,4,5.5 and 8 at a fixed Froude number of the flow.

EXPERIMENTAL APPARATUS AND METHOD

The experiments were carried out in a perspex flume, 300mm wide by 900mm deep and 6m long, fitted with a sharp edged sluice gate 12.5mm thick. To damp the surging and turbulence, the water entered the inlet reservoir through a central perforated pipe which was shielded by a curved sheet and additional baffles were inserted across the tank. Irregularities in the discharge were reduced by a cylindrical overflow weir. An adjustable weir near the end of the flume regulated the downstream level. The discharge was measured in a calibrated tank. Gate openings, d, of 25mm, 50mm and 75mm were used. The Froude number of the flow $F = \sqrt{qd_o{}^3}$ where q is the unit discharge and $d_o = 0.61d$, was kept constant at 2.3 for all the flow patterns. Because of an inherent fluctuation of the flow pattern, the velocity distributions in the jet were measured by a photographic technique developed by the author [8]. A mercury-vapour discharge lamp was used to produce a sheet of light 10mm thick by 200mm long along the centreline of the flume. By using a stroboscopic light, the paths of the illuminated oil tracers over successive periods of one-hundredth of a second were obtained as a series of dashed streaks. By comparing the length between the centres of adjacent dashes with a linear scale included in the photograph, the magnitude of the instantaneous vector and hence the horizontal and vertical components of velocity parallel to the photographic plane were obtained. When successive values of the velocity at a point were recorded, then with S as the number of measurement and N the total number of intermittents measurements, the mean value of the velocity was found by

$$u = \lim_{N \to} \sum_{s=1} u_s$$

The final velocity profile was then drawn for the best representation of the plotted points.

DESCRIPTION OF FLOW

Although the physical aspects of the dispersion of a sluice jet have some similarity with the simpler problem of the dispersion of a jet issuing into infinite fluid, it has several fundamental properties which distinguish it. Apart from the contraction due to the sharp edge of the gate, the differences are those described below.

The simple concept of a jet issuing from a point source into an infinite surrounding is no longer appropriate on account of the much larger size of the sluice opening in comparison with the dimensions of flow; the width of opening may be half of the downstream depth. A schematic section of the flow is shown in Fig.1. The jet issues into a slow moving medium and is guided by a lower boundary. The mixing process between the jet and the surrounding fluid causes part of the latter to be carried forward with the jet under conditions in which not only the forward momentum but also the total discharge is conserved. The process of entrainment leads to recirculation to replace the fluid entrained. Expressed in another way, the action of the jet is to produce a low pressure region around it which is then fed from regions of higher pressures downstream. Thus a zone of circulatory motion is established which is usually termed a "roller". Meanwhile, the mixing process causes the jet to expand vertically until the upper mean flow line reaches the free surface. This point is the stagnation point, which defines the downstream end of the roller.

In comparison with a free jet, the chief difference is that in the sluice jet the upper stream or limit of the flow is a free surface which slopes upward in the downstream direction. Since this surface is nearly free from shear stress, its slope is evidence of positive pressure rise in the direction of the flow downstream of the gate. The amount of the resulting pressure difference depends on the sluice openings, the discharge and the downstream depth of the flow, as indicated in in Figs. 2 and 3. The former concerns variation in downstream depth, the upstream head being constant; whilst the latter shows a number of photographs selected to describe the kinds of patterns which occur under the particular conditions marked a, b, c, d and e in Fig.2.

WATER SURFACE PROFILES

Figure 4 shows the mean depth of water plotted non-dimensionally against distance for five patterns of flow. It is seen that the depth h at first decreases from its value at the gate to a minimum value h_o at some distance downstream. This is partly due to the stagnation at the gate together with effects due to contraction of the jet. Further downstream, the depth gradually increases until it reaches the final downstream level. For the semi-submerged patterns, the rate of increase in depth is seen to diminish with distance downstream and since the shear stress at the surface is negligible its slope indicates considerable conversion of kinetic head into potential head and this is seen to occur mostly in the first half of the circulation zone.

In the analysis of the half-jet boundary, Tollmien [9] found that along the streamline Ψ=0, forming the upper boundary of the jet, the velocity is 0.566 U_o. Since this streamline terminates at the stagnation point at the downstream end of the circulation zone, the velocity head 0.32 $U_o^2/2g$ must be partly converted into pressure head which is (h_2-h_o), and the rest is dissipated in turbulence. The proportion of pressure recovery varies with the degree of submergence of the jet as indicated by the comparison of the calculated and measured heads given in Table 1 below:

Table 1 - Recovery of Pressure Head

h_2/d	2.0	3.0	4.0	5.5	8.0
$0.32U_o^2/2gd_o$	0.293	0.262	0.159	0.111	0.056
$(h_2-h_o)/d_o$	0.228	0.259	0.129	0.070	0.030

In the semi-submerged patterns it is surprising to find that at the boundary of the jet with the overlying circulation, where the shear stresses are maximum, there is relatively little head lost in turbulence and most of the velocity head is converted into pressure head. For the highly submerged patterns Tollmien's theory is less applicable as the length of the potential core is small compared to the length of the circulation zone.

SPREADING OF JET BOUNDARIES

The curves of Figs. 5 and 6, show the spreading of the jet boundaries for various ratios of submergence. The angles of spread are compared in Table 2 with those of other confined jets with recirculation zones and also with the classical results of Tollmien for the plane free jet. The following conclusions may be derived:

(i) The rate of growth of the boundary layer δ near the bed of the channel is almost linear with an inclination of 0.5° in the entrainment region. Over most part of the wall layer, the Reynolds number of the flow is greater than $5x10^5$, indicating that the flow there is turbulent.

(ii) The inner boundary 01 of the potential core expands at an angle of 4.8° or about 1 in 12, which is the same as that for a half-jet boundary.

(iii) The mean spreading angle of the outer boundary 02 is 15.3° which closely conforms to Tollmien's solution for a plane free air jet with an expansion angle of 14° or about 1 in 4. The section where 02 reaches the water surface marks the end of the entrainment region and the position of maximum reverse velocity in the circulation zone above the jet.

(iv) The boundary of the jet 03 defined by Ψ=0 increases gradually at an angle of 1.3° until near the end of the circulation zone it curves sharply upward to reach the free surface, whereby the stagnation point is reached.

(v) As the submergence increases, the distance of the stagnation point from the gate increases, the locus being approximately a straight line with a slope of the boundary 04 being 9.8° or about 1 in 5.83 which is practically the same as predicted by Tollmien for a half jet boundary. Additional experiments with different Froude numbers of the flow gave the same results indicating that the length of the circulation zone is only dependent on the degree of submergence.

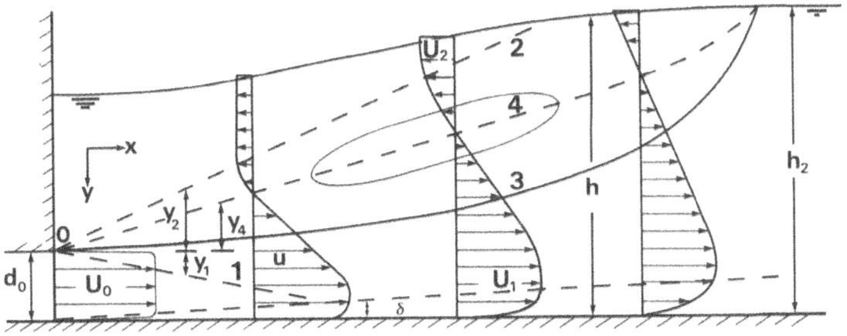

Fig. 1

Table 2 - Spreading of Jets with Circulation Zones

FLOW	BOUNDARY			
	01	02	03	04
Diffusion of Sluiceway Jet Naib (Present Data)	4.8°	15.3°	1.3°	9.8°
Surface Motion of Plane Liquid Jet Naib [10]	6.4°	15.3°	1.3°	9.4°
Rectangular Channel Expansion Naib [11]	5°	13°	1.2°	7.5°
Confined Wall Jet Forthmann [12]	5°	15.2°	1.5°	8.9°
Flow Behind Bluff Body Abramovich [13]	6.3°	14°	1.3°	9.4°
Half Jet Boundary Tollmien [9]	5°	-	0.91°	9.8°
Plane Free Jet Tollmien [9]	5°	15°	-	-

DECAY OF MAXIMUM VELOCITY

The longitudinal distributions of the maximum velocity U_1 divided by the maximum jet velocity U_o are plotted against $\sqrt{x/d_o}$ in Fig.7. The curves, each of which represents the results of one pattern, show that the maximum velocity remains constant for a certain region downstream of the gate. The length of this potential core region, x_c, as given by the horizontal part of each curve, is variable depending on the submergence as shown in Table 3.

Table 3 - Length of Potential Core

h_2/d	2	3	4	5.5	8
x_c/d_o	3.3	4.4	5.6	6.5	8.2

Following the core region is a transitional length, which is characterized by a gradual decrease of the maximum velocity. This zone opens into the fully developed turbulent region where surprisingly the jet decays linearly with the square root of the distance x, as compared to an inverse variation for the plane free jet. The straight lines indicate that for both the entrained and recirculated regions of the jet, the decay of maximum velocity is governed by the same power law, as given by the following general equation:

$$U_1/U_o = c - m\sqrt{x/d_o} \qquad \text{.........} \; [1]$$

The variations of c and m with h_2/d are given in Fig.8, where it is seen that the value of c increases to a maximum of 1.77 for $h_2/d = 4$ and then decreases to 1.52 for $h_2/d = 8$. The straight line corresponding to $h_2/d = 8$ lies highest and the rate of decay is least, indicating that mixing takes place most slowly in the highly submerged case. For the present experimental conditions the following equations have been derived:

$$h_2/d = 2.0: \quad U_1/U_o = 1.68 - 0.30\sqrt{x/d_o} \qquad \text{.....} \; [2]$$

$$h_2/d = 3.0: \quad U_1/U_o = 1.73 - 0.28\sqrt{x/d_o} \qquad \text{.....} \; [3]$$

$$h_2/d = 4.0: \quad U_1/U_o = 1.77 - 0.26\sqrt{x/d_o} \qquad \text{.....} \; [4]$$

$$h_2/d = 5.5: \quad U_1/U_o = 1.67 - 0.22\sqrt{x/d_o} \qquad \text{.....} \; [5]$$

$$h_2/d = 8.0: \quad U_1/U_o = 1.52 - 0.16\sqrt{x/d_o} \qquad \text{.....} \; [6]$$

MEAN VELOCITY PROFILES

Distributions of mean velocity for one pattern $h_2/d = 5.5$, non-dimensionalized by the maximum jet velocity, are plotted in Fig.9. The variation of the velocity distribution with respect to the distance downstream was found to be substantially the same for all the flow patterns and is described typically as follows.

At the gate, the distribution is characterized by a constant velocity up to the middle of the sluice opening, decreasing above this, apparently in a parabolic manner, to zero at the tip of the gate. Downstream of the gate, the jet undergoes a contraction, while entraining the surrounding fluid. As a result the velocity distribution near the vena contracta section is marked by an increased maximum velocity which remains constant up to a depth of 0.4d. Further downstream, the flow is fully developed. In the entrainment region, Fig.10, the profiles lie very nearly on one curve and therefore are similar despite the different rates of pressure rise under various degrees of submergence. The profiles are also seen to be in good agreement with the theoretical profile for a plane wall jet [14] with a superimposed reverse velocity of $U_2 = 0.25\,U_1$. In the recirculation region, Fig.11, the profiles vary from section to section and therefore are no longer similar. This lack of similarity is due to different characteristics of the flow in this region, such as the increasing curvature of the streamlines as the stagnation point is reached and the excess turbulence generated due to large time variations in the length of the circulation zone up to ±15 per cent. Photographs of the flow in this region show a distinctive pattern of large eddies which because of their size, convective motion and random formation cause the jet to oscillate both vertically and horizontally. These changes are often accompanied by separation of the boundary layer along the bed of the channel. It appears that such a large scale motion carries a mixture of smaller eddies distributed across the flow and so intensifies the mixing. This in turn causes the jet to decay much faster than in the entrainment region.

REFERENCES

1. Naib, S K A: Flow patterns in a submerged liquid jet diffusing under gravity. Nature, Vol.210, p694, May 14 1966
2. Addison, A: Hydraulic Measurements Chapman Hall Ltd 1948
3. Henry, H R: Discussion on diffusion of submerged jets. Proc.ASCE, Vol.75, p1541, 1949.
4. Rajaratnam, N: Submerged hydraulic jump. J.Hyd.Div, ASCE, Vol.91, July 1965.
5. Ramaprasad and Ramanoorthy, M V: Discussions on submerged hydraulic jump. J.Hyd.Div.,ASCE, Vol.92, Feb.1966.
6. Naib, S K A: Unsteadiness of the circulation pattern in a confined liquid jet. Nature, Vol.212,p753 November 12, 1966.
7. Naib, S K A: Abnormal grouping of large eddies in a submerged liquid jet. La Houille Blanche, No.3, p282, 1967.
8. Naib, S K A: Photographic method for measuring velocity profiles in a liquid jet. The Engineer, Vol.221, p961, June 24 1966.
9. Tollmien, W: Calculation of turbulent expansion processes NACA T.M No.1085, 1945.
10. Naib, S K A: Surface motion of a plane liquid jet. La Houille Blanche, No.6, p377, 1980.
11. Naib, S K A: Mixing of a subcritical stream in a rectangular channel expansion. J.I.Wat.E, p199, May 1966.
12. Forthman, E: Uber Turbulente Strahlausbreitung. Ing.Arch. Bd 5, p42, 1934.
13. Abramovich, G K: The theory of turbulent jets. M.I.T Press, Boston, USA, 1963.
14. Glauert, M B: The wall jet. J.Fluid Mechanics Vol.1, p625, 1956.

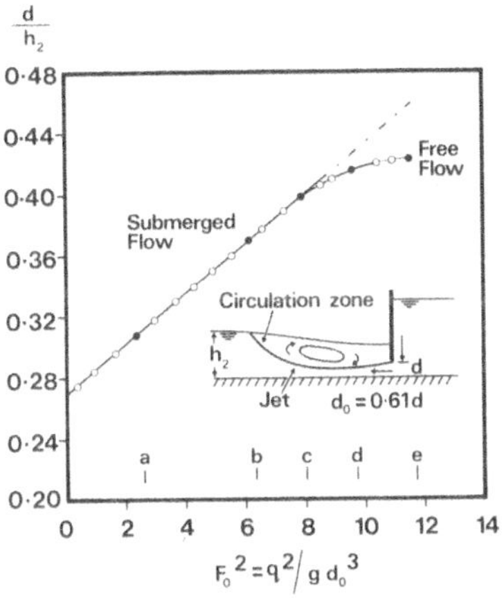

Fig. 2

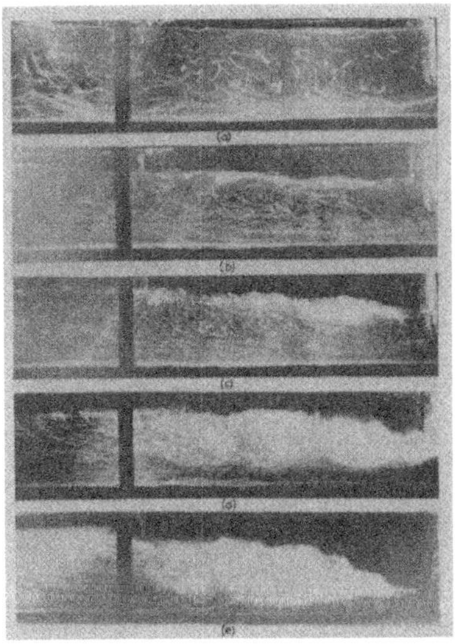

Fig. 3

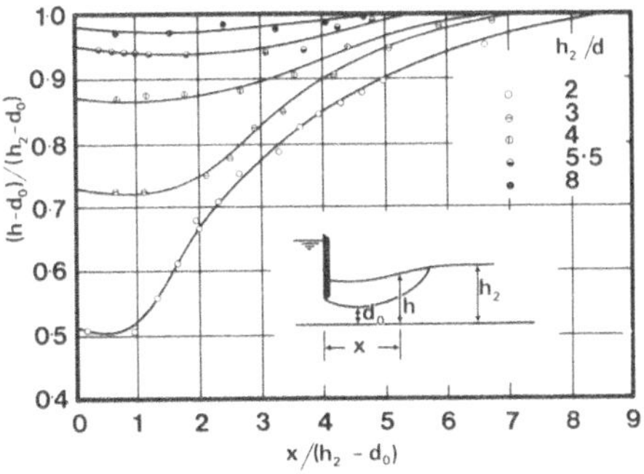

Fig. 4

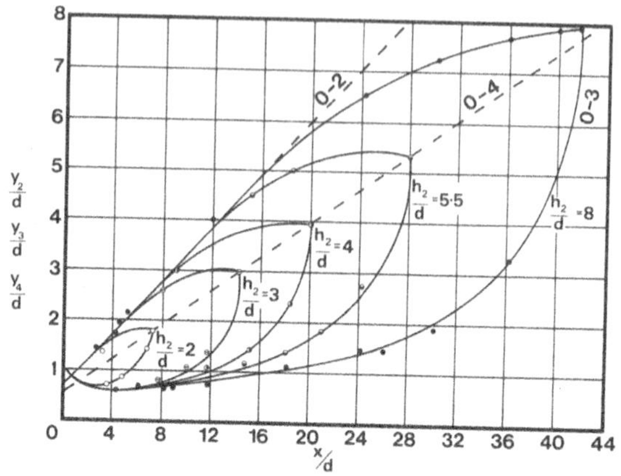

Fig. 5

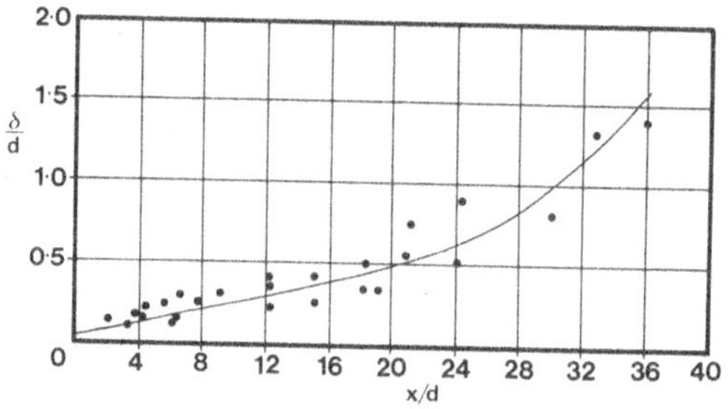

Fig. 6

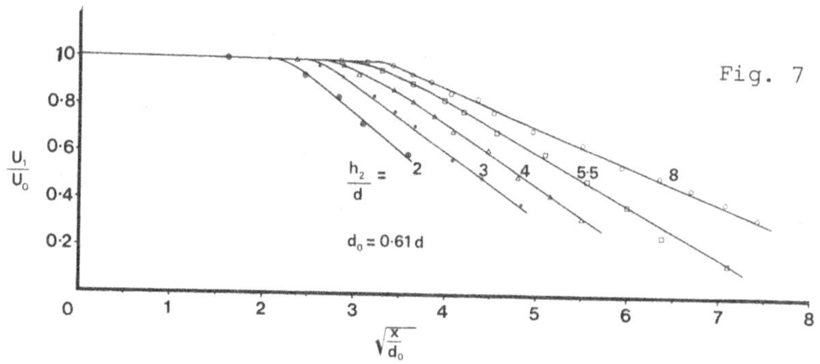

Fig. 7

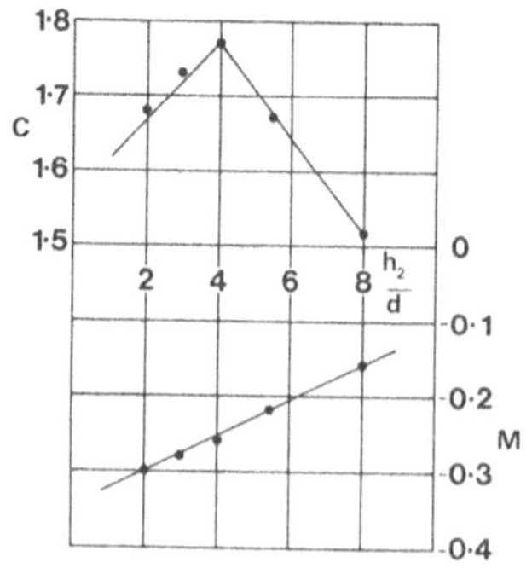

Fig. 8

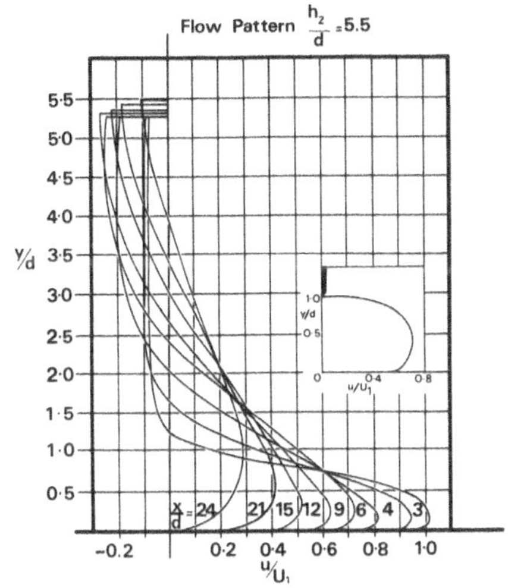

Fig.9

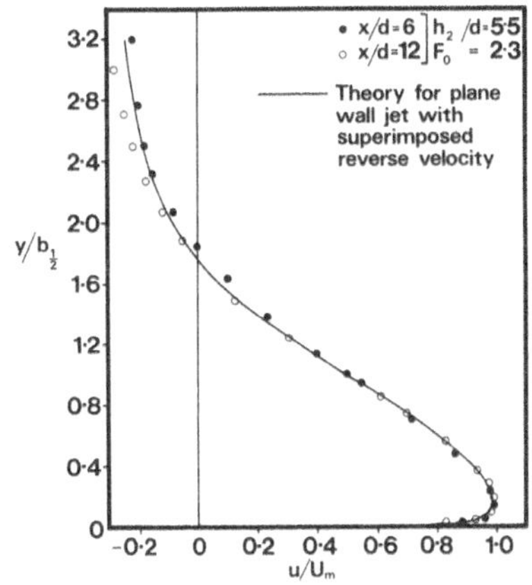

Fig. 10

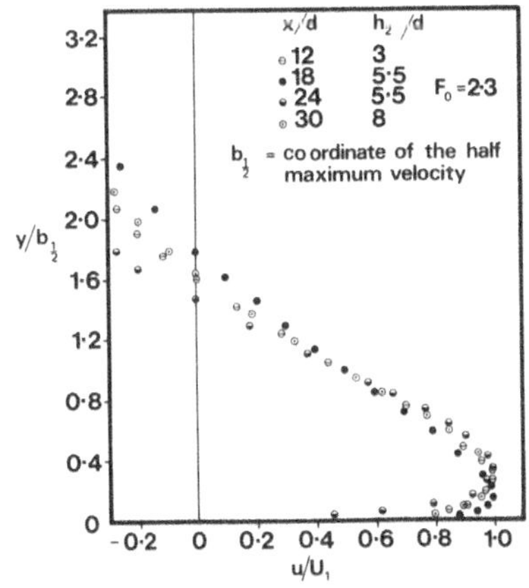

Fig. 11

Regime Approach to the Design of Drainage Channels

Kenneth V.H. Smith
Civil Engineering Department, University of Southampton, U.K.

INTRODUCTION

In designing any water conveyance channel the engineer is very much concerned that the cross-section and slope of the channel will remain stable, thereby reducing maintenance costs and avoiding disruption to adjacent land by bank erosion and shifting of the channel course. The Regime method was originally developed in the Indo-Gangetic plain for the design of large irrigation canal systems so that they remained reasonably stable and free of operational problems. It must be remembered however that, as compared with drainage channels and rivers, the discharge in an irrigation canal during the operating season is relatively constant. Accordingly it is not possible simply to translate the Regime method equations for the design of irrigation canals into equations for the design of drainage channels. Some thought has to be given to selecting a characteristic discharge from amongst the whole range of flows which occur in the channel. This paper is not specifically concerned with the straightforward application of Regime method equations, but rather makes use of the regime concept of a channel which is in balance with the water and sediment flowing in it.

THE REGIME METHOD - LACEY EQUATIONS

The regime method equations were concerned with the stable values of the channel cross-section and longitudinal slope as a function of discharge and also of a factor related to the sediment moving with the flow. There are many forms of the regime method equations, but in metric units they can be expressed as:

$$P = 4.83\ Q^{1/2} \qquad (1)$$

$$R = 0.473\ (Q/f)^{1/3} \tag{2}$$

$$S = f^{5/3}\ Q^{-1/6} / 3340 \tag{3}$$

from which it also follows that

$$V = 0.63\sqrt{fR} \tag{4}$$

In these expressions P = wetted perimeter (often approximated as the water-surface width), Q = discharge, R = hydraulic radius, S = channel slope for uniform flow and V = mean velocity in the channel. f is the silt factor, which should represent both the sediment size and concentration. The equations were derived from a study of the cross-sectional properties, slopes and discharges in canal systems which had become stable, or reached "regime" (Lacey 1929, 1933). Many other investigators have deduced systems of Regime method equations, notably Blench (1957). In practice it is found that the values of f in Equations 2, 3 and 4 may not be the same, although the proportionality relationships hold. The coefficient 4.83 in Equation 1 is known to apply to canals with silty-clay banks such as existed in the areas from which these equations were derived. The approximation $f = 1.76\sqrt{D_m}$ is often used to derive a silt factor, where D_m is the average size of the bed material in mm. This value for the silt factor neglects some relevant factors, and for any major canal development proposal a research programme would probably be undertaken to deduce appropriate values for the parameters involved in Equations 1 to 3.

CHANNEL DESIGN WHEN DISCHARGE IS VARIABLE

In the case of drainage channels or channels carrying diverted flows around drained areas, the flow may vary quite substantially depending on seasonal and other factors. If the channel is designed for the largest discharge then for a large part of time the water will move slowly through the channel and sediment will deposit on the bed and in berms; the channel cross-section will accordingly be reduced. On the other hand, if the channel is designed for a low discharge, the cross-section will be small and for most of the time the channel will scour until the cross-section is increased. It is a fairly common concept that there is a discharge most closely related to the stable cross-section for the channel (i.e. when deposition and scour roughly balance). This discharge has been called the formative discharge, or sometimes the dominant discharge, and it would seem to correspond to the magnitude of the discharge which moves most sediment in the channel. For a given section and slope of channel and a given hydrograph of flow (or representative hydrograph of flow) it is possible to calculate the dominant discharge from this definition.

Dominant discharge calculation
From the available or anticipated information about discharge it is possible to construct a distribution diagram for discharge, as shown by the dotted line in Figure 1 (see footnote), such that the area under the graph between two discharges Q_1 and Q_2 represents the percentage of time that the discharge in the channel lies between Q_1 and Q_2. The sediment transport rate at each of a series of discharges must then be calculated and this is similarly plotted as a distribution graph of sediment transported (solid line in Figure 1). Figures 1 and 2 have been plotted from canal data and quite clearly for the canal in Figure 1 the dominant discharge corresponds to the peak discharge rate which is between 300 and 400 m^3/s. Figure 2 shows a case in which there are two possible peaks of sediment transported and probably it would be difficult to maintain a stable regime channel in this particular case. Figures 1 and 2 are reproduced from Smith and Yates (1973).

Channels transporting mostly fine sediment For channels transporting fine sediment there is no direct relationship between the rate of sediment transport and appropriate hydraulic parameters of the flow. The amount of sediment in the flow is only dependent on the availability of fine sediment which can be eroded from the catchment area. At the same time, the amount of fine sediment available is usually dependent on the season of the year and quantity and intensity of rainfall. Accordingly it correlates quite well with channel discharge, and a relationship is often plotted of fine sediment transport against channel discharge. Alternatively the sediment entering the channel can be estimated directly from rainfall and soil data for the catchment or drainage area from soil conservation calculation techniques, such as those of the U.S. Department of Agriculture. Milos Holy (1966) has carried out investigations of soil erosion from drained areas in Czechoslovakia.

Channels transporting sand-size material For channels transporting sand-size material there are quite well-established methods for calculating sediment transport from hydraulic parameters of the flow. A technique which has been used by the writer is the graphical method developed by Colby (1964) which gives a relationship between sediment transport and dominant parameters which include mean sediment size, depth of flow and velocity of flow. Figure 3 illustrates this relationship. A relationship such as this can be used to calculate the "sediment transported" curve of Figures 1 and 2 directly from the distribution curve of discharge.

Note: Figures are printed near the end of the paper.

USE OF BANK-FULL CAPACITY AS A FORMATIVE DISCHARGE

The concept of a relationship between the discharge which just causes the channel to flow bank-full, and parameters of the channel cross-section such as the water-surface width, has an obvious attraction to engineers. The concept has been investigated by several researchers. Bank-full discharge occurs when the channel flow is just about to overflow the banks over a fairly substantial length of channel. It could be evaluated by using a large number of observers, or making observations from a vantage point or a helicopter. The corresponding discharge can be read from gauging station records.

Wolman and Leopold (1957) considered the frequency of occurrence of bank-full discharge, mainly for American rivers. Although they were of the opinion that the frequency of occurrence of bank-full discharge should be affected by geographical location and historical factors, their study suggested that for most rivers the recurrence interval of bank-full flow lies closer to 1 year than to 2 years. Dury (1959) as a result of investigations of the Nene and Great Ouse considered that the bank-full discharge has a recurrence interval between 1 and 2 years.

Marshall Nixon (1959) made a study of the occurrence of bank-full discharge for rivers in England and Wales. He concluded that the frequency of bank-full discharges of rivers in regime is of the order of 0.6%, equivalent to a little over two days/year. He considered this to be the standard bank-full discharge and demonstrated that this discharge could be used in regime-type equations to relate channel width, depth etc. to discharge.

Figure 4 shows the scatter of the observations of frequency of occurrence of bank-full discharge about the selected value of 0.6% used by Marshall Nixon. Some of the higher values of frequency of occurrence can probably be attributed to special characteristics of the site considered. Sites were normally located close to flow gauging stations. On the basis of the bank-full discharge as defined, Q_b, Marshall Nixon investigated the generality of Regime type equations to the English and Welsh rivers included in his study i.e.

$$W = a\ Q_b^{1/2} \tag{5}$$

$$d = c\ Q_b^{1/3} \tag{6}$$

$$V = K\ Q_b^{1/6} \tag{7}$$

In these equations, W = water surface width, d = depth of flow,

V = mean velocity of flow, and a, c, K are coefficients. Marshall Nixon evaluated the coefficients, which were fairly universal for English and Welsh conditions. However, for these areas, fine sediment probably predominates in suspension and in the bed so that it would be unwise to use these values in other countries without a local investigation. In fact, one of the main purposes of writing this paper has been to draw attention to the possibilities of using observations from channels in a particular part of the world which are stable in order to design new channels in the same part of the world which will operate under similar conditions.

EXPLICIT METHODS FOR DESIGNING REGIME CHANNELS

Channels with relatively constant discharge

According to the definition of a regime channel it should be possible to design a channel for a specified water and sediment flow provided reliable water discharge and sediment transport relationships are available. The object of the design is to devise a channel which is capable of transporting the maximum discharge of water and is also able to carry through the channel all the sediment entering it. It should be added that there are variables which can be adjusted in order to achieve the balance between sediment entering the channel and the sediment transporting capacity of the flow; these variables are essentially the water surface width and the longitudinal slope of the channel. Smith (1977) carried out an investigation on a sand-bed irrigation canal system using (a) a modification of the Einstein-Barbarossa prediction technique to produce a computerised method for discharge computation, Einstein and Barbarossa (1951), and (b) the relationship for sediment transport due to Colby (1964). The investigation showed that there was a range of widths, close to the widths given by Equation 1, for which the head loss between inlet and outlet to the canal system was a minimum.

Drainage channels with varying discharge

For channels in which the discharge varies considerably it is obvious that the stable (regime) section will be one which is overflowed occasionally. Accordingly river and drainage engineers have adopted a cross-section which comprises a deep-water section having overflow berms contained within banks. A typical section as proposed for the Black Bush Polder (Scott, Camacho and Sharman, 1960) is sketched in Figure 5.

The writer has encountered the problem of achieving a stable channel with highly variable discharge in flood diversion channels located in northern India and carrying flood flows across the plain of the River Sutlej and hence discharging into the Sutlej. A typical diversion channel would collect the flows from four or five mountain torrents. Land on either side of the diversion channels was being levelled and drained. These channels were mainly sand-bed as a result of the characteristics of the sandstone hills they were eroding. Data for the flows entering these diversion channels was difficult to obtain and in order to devise a procedure for designing such channels so that they would remain essentially stable, Castello-Branco (1983) together with the author used data from some U.S.A. sand-bed rivers for which regular water and sediment gauging had been carried out over a considerable period of time. The slopes of these rivers were comparable with the Indian diversion channels. In this investigation, similar techniques to those used by Smith (1977) were used, it being assumed that the flows could be considered as steady over short periods of time.

The selected channel sections used a deep-water channel which could carry the dominant discharge as previously defined in this paper, together with overflow berms. A major requirement of the design (apart from having adequate discharge capacity) was that over a period of time there was a sediment balance in the channel (i.e. apart from small periods of scour and deposition, the sediment transporting capacity of the channel equalled the quantity of sediment entering the channel. This is demonstrated in Figure 6, which shows a comparison between the cumulative value of the observed sediment load which would enter the diversion channel (1) and the cumulative value of the sediment transport capability of the diversion channel (2). The particular design which was examined in Figure 6 was for a diversion channel having a deep water channel bed width of 18 m and a slope of 0.0009. The comparison in Figure 6 is for a period covering a number of years and it will be seen that for about the last year the channel would be scouring. However an alternative design with a slightly smaller slope might have provided a satisfactory solution with no overall deposition or scouring at the end of the period studied. It was in fact found generally that diversion channel slopes less steep than the ground slope gave the most satisfactory results. This study was computerised so that it was possible to investigate a number of alternative designs.

CONCLUSIONS

Some concepts have been discussed relating to the selection of stable channels to transport flows which are varying. There is inadequate knowledge about flow in composite channels (deep water channel and overflow berms) which suggests that field observations of such channels would be useful.

REFERENCES

Blench, Thomas (1957) "Regime behaviour of canals and rivers", Butterworths, London.

Castello-Branch, V. (1983) "Investigations related to the design of flood diversion channels", Ph.D. thesis, University of Southampton.

Colby, B.R. (1964) "Discharge of sands and mean-velocity relationships in sand-bed streams". U.S. Geological Survey, Professional Paper 462-A.

Dury, G.H. (1959) "Analysis of regional flood frequency on the Nene and Great Ouse". Geogr. J vol 125.

Einstein, H.A. and Barbarossa, N.L. (1951) "River channel roughness", Jnl Hydraulics Division ASCE vol 77.

Holy, M. (1966) "Protection against sedimentation in drainage channels", International Commission on Irrigation and Drainage, Sixth Congress, Question 20, R11.

Lacey, G. (1929) "Stable channels in alluvium". Proceedings Institution of Civil Engineers, Proc. paper 4736 vol 229.

Lacey, G. (1933) "Uniform flow in alluvial rivers and canals", Proceedings, Institution of Civil Engineers, Proc. paper 4893, vol 237.

Nixon, Marshall (1959) "A study of the bank-full discharges on rivers in England and Wales", Proceedings, Institution of Civil Engineers, Proc. Paper no. 6322.

Scott, P.A., Camacho, R.F. and Sharman, F.A. (1960) "The reclamation and irrigation of the Black Bush Polder in British Guiana". Conference on Civil Engineering Problems Overseas, ICE, paper no. 4.

Smith, K.V.H. and Yates, T.J. (1973) Civil Engineering Department, University of Southampton, Departmental Report CE/7/73.

Wolman, M.G. and Leopold, L.B. (1957) "River flood plains : some observations on their formation", U.S. Geological Survey Prof. Paper 282 C.

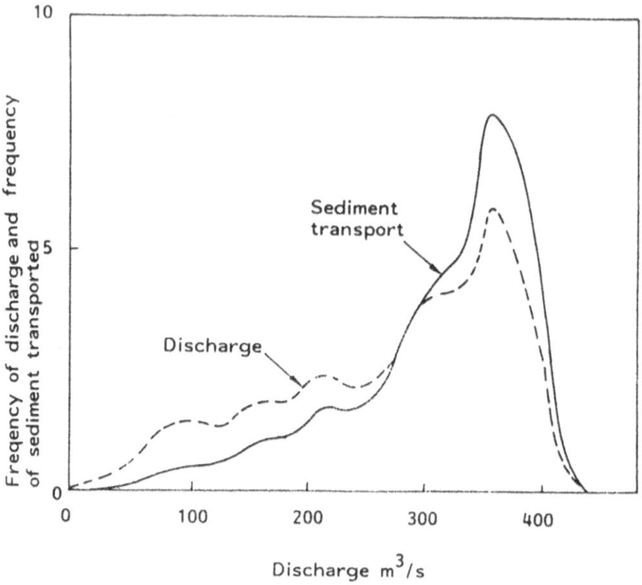

Fig.1 Baloki-Sulleimanki Canal RD53

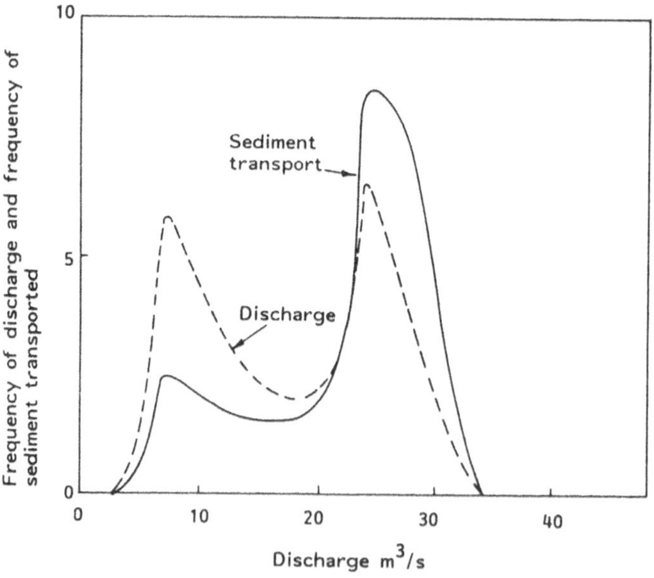

Fig.2 Abbasia Canal RD11

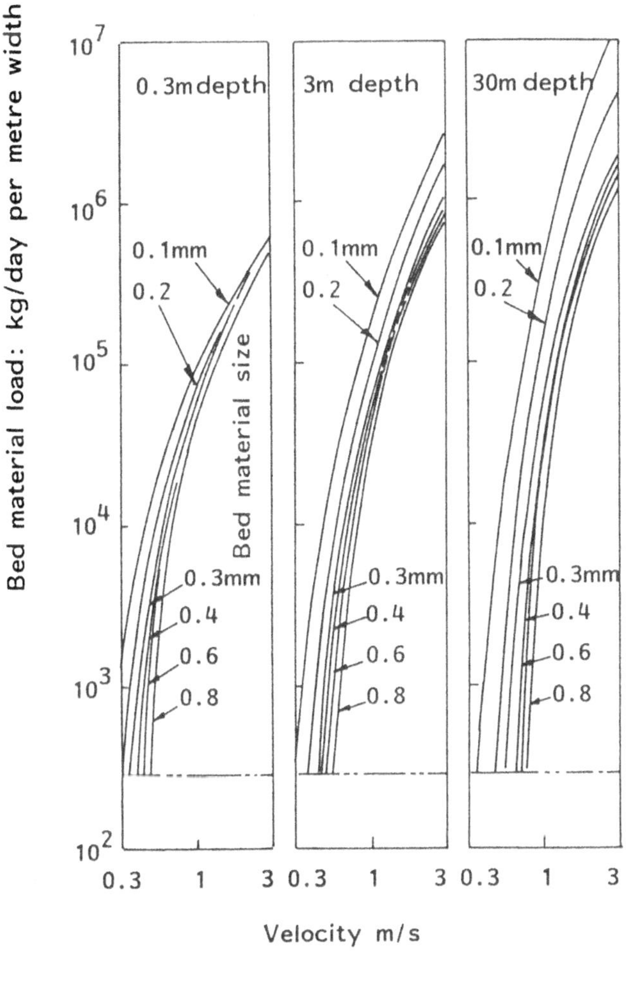

Fig. 3 Colby's correlation of sediment transport with mean velocity and depth

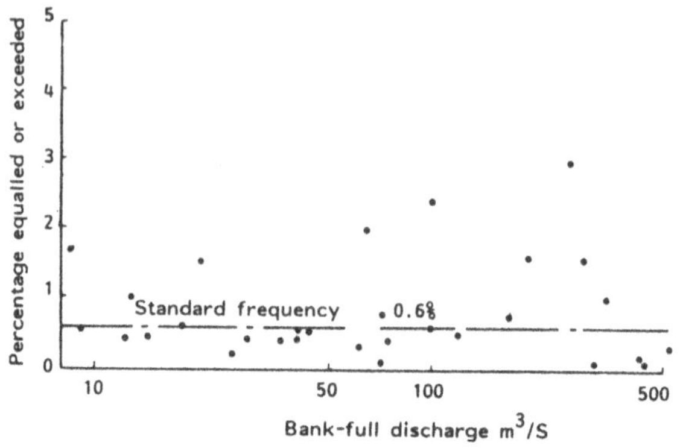

Fig. 4 Frequency of bank-full discharge after Marshall Nixon (1959)

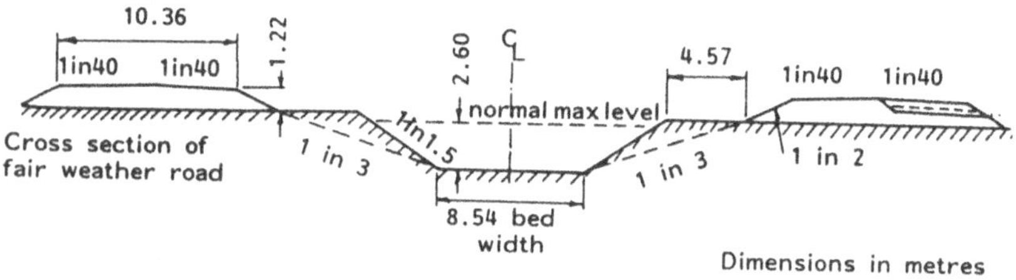

Fig. 5 Typical channel section with berms, after Scott et al (1960)

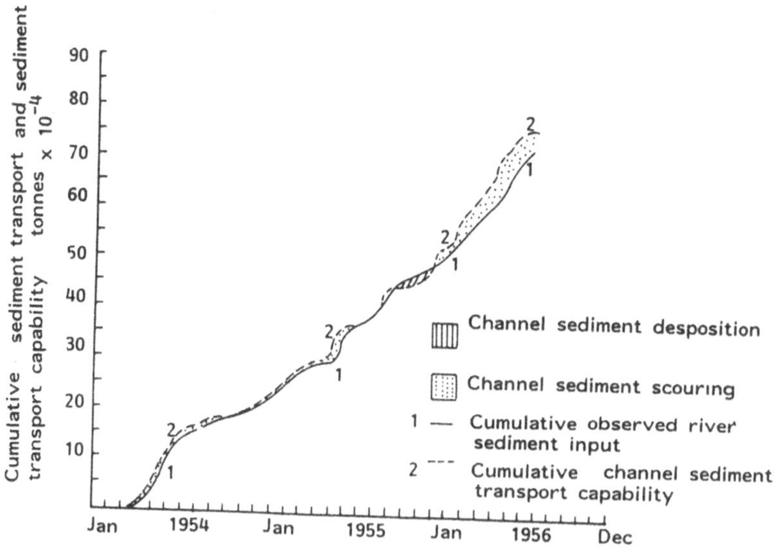

Fig. 6 Areas showing periods of sediment desposition and scouring for the Rio Grande diversionchannel combined cross-section.
Bed width of deep water channel=18m. S=0.0009

Sediment Transport in Circular and Non-Circular Conduits

J.H. Loveless
King's College London

INTRODUCTION

The entry of sediment into a drainage system is an undesirable but almost unavoidable circumstance. Once there, the sediment must be transported right through the system without giving rise to deposition in appreciable quantities at any point. This requirement has, until recently, been allowed for in the design office using the Crimp and Bruges condition, which requires that the mean velocity shall be at least 0.76 m/s, when the conduit is half full.

In a subject as complex as sediment transport however, it would be remarkable if all the possible variations of flow, conduit and sediment could be ignored in determining the mean velocity necessary to prevent sediment deposition. Furthermore, since this arbitrary condition often determines the choice of pipe size, or gradient, any information as to its validity would lead to more precise, reliable and economic designs.

The situation where the sediment in passage through a conduit is just about to form a stationary bed, is known as the **condition of incipient deposition**. It has been noted by several researchers that the condition is always very well defined; the slightest change in gradient or velocity being sufficient to produce the effect at the limit. Also, at this limit, the nature of the sediment transport changes from that of **flume traction**, where the particles are sliding over a rigid boundary, to **stream traction** where they are part of a continuously moving and possibly duned bed. All the researchers into the subject of sediment transport in closed conduits have therefore endeavoured to define this condition because it is both readily observable and, of vital importance to the efficient performance of the conduit.

This paper presents new ideas on the condition of incipient deposition which follow, in part, the treatment of May (1982) and also attempt to clarify the question of the effective width for sediment transport (at the limit of flume traction) raised by Ackers (1984).

The results of a limited series of tests on rectangular, oval and circular shaped conduits are also presented. The results of these tests are then compared with theory, using a number of different predictions.

THE THEORY OF FLUME TRACTION

In order to see the effect of different conduit cross-sections on the mechanisms of sediment transport it is necessary to consider the basic theory. Figure 1 shows the forces acting upon a sediment grain in flume traction. The velocity of the grain is u and the velocity of the fluid adjacent to the grain is v. The effective local velocity v is equal to βV where β is some constant and V is the mean velocity of the flow. The relevant forces shown are the friction, lift, drag and weight forces.

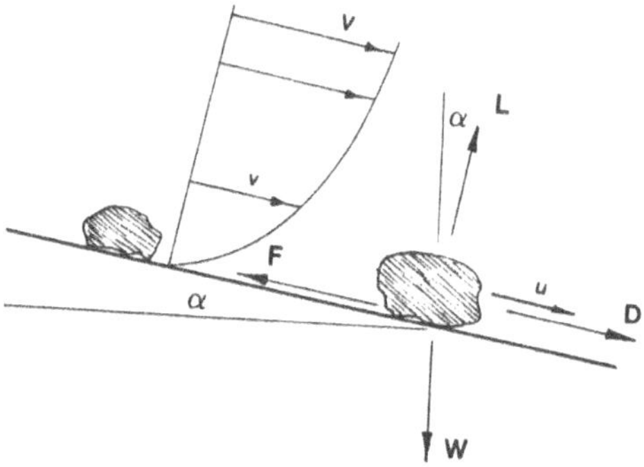

Figure 1: Forces on a sediment particle in flume traction

Resolving these forces in the direction of the flow gives:

$$D = F - W\sin\alpha \qquad (1)$$

where

$$\left.\begin{aligned} D &= \tfrac{1}{2}\,\rho C_D A_p(\beta V - u)^2 \\ F &= \tan\phi(W\cos\alpha - L) \\ \text{and} \quad L &= \tfrac{1}{2}\,\rho C_L A_p(\beta V - u)^2 \end{aligned}\right\} \qquad (2)$$

In these equations ϕ is the friction angle between the surface and the grain; A_p is the area projected to the flow by the grain and C_D and C_L are the drag and lift coefficients for the grain.

Since the grains are not usually spherical both the projected area,

A_P and the weight, W require additional shape coefficients in order to be defined, thus,

$$\left.\begin{aligned} A_P &= \alpha_1 d^2 \\ W &= \alpha_2 \rho g (S_S - 1) d^3 \end{aligned}\right\} \qquad (3)$$

Here, d is the nominal diameter of the sediment and α_1 and α_2 the shape coefficients. For a sphere, $\alpha_1 = \pi/4$ and $\alpha_2 = \pi/6$.

Substituting Equations 2 and 3 into Equation 1 gives the following result:

$$\frac{(\beta V - u)^2}{g(S_S - 1)d} = \frac{2\alpha_2}{\alpha_1} \cdot \frac{(\tan\phi\cos\alpha - \sin\alpha)}{(C_D + \tan\phi C_L)} \qquad (4)$$

This is a general equation for flume traction, which can be used to determine V, the necessary mean velocity to transport grains with a nominal diameter d. The principal difficulty in its application, which will be apparent to the reader, is the correct definition of the coefficients α_1, α_2, β, ϕ, C_D and C_L. However, even so, it does provide a suitable foundation for the consideration of those factors which are most relevant to the sediment transport process. Each of the variables in the equation will now therefore be considered in turn.

Gradient of the conduit, α It may easily be shown that, since drainage conduits are always laid to shallow gradients, the effect of the variation of α upon Equation 4 is always negligible. Provided the slope of the conduit is less than 1:50 the effect of α may be ignored. The gradient does of course have an indirect effect in that it helps to determine V.

Friction angle ϕ It was assumed in the development of Equation 4 that the grain moved by sliding but, in fact, the grain can also roll. Indeed where they are almost spherical in shape this type of motion is the more likely. However, close observation of the two sediments used in the present experiments showed that the majority of the grains were sliding when conditions were close to incipient deposition.

	Fine Grains d_{50}=0.45mm			Coarse Grains d_{50}=1.5mm		
% REMAINING AT ANGLE	90%	50%	10%	90%	50%	10%
SMOOTH ks=0.003mm	30°	37°	46°	25°	28°	36°
INTERMEDIATE ks=0.04mm	48°	50°	52°	36°	40°	42°
COARSE ks=0.2mm	51°	55°	65°	42°	44°	49°

Table 1: Values of the friction angle ϕ submerged for different grains and surfaces

Table 1 shows the results of a simple test to determine the value of ϕ for the grains used in the experiments. Three different surfaces having k_s values of 0.2, 0.04 and 0.003 mm, were used for the tests.

A wide variation in the mean values of ϕ was obtained ranging from 28° for the coarse particles on the smooth surface, to 55° for the fine particles on the rough surface. It is concluded therefore, that the friction angle is strongly dependent on d/k_s and the shape of the particles. The exact relationship is not known but, it is an easily determined quantity for any combination of surface and sediment.

<u>The shape coefficients α_1 and α_2</u> A practising engineer might (I suspect) consider the determination of the above shape coefficients to be a fruitless exercise and, that their presence in Equation 4 severely restricts the value of this equation as a tool in sediment transport studies. In fact both α_1 and α_2 can be estimated quite reliably by comparing the sediment shape to the nearest regular shapes. Also, when the ratio of α_2 and α_1 is taken the degree of variation is not as large as might be expected.

Table 2 gives the values of α_1, α_2 and α_2/α_1 for the various regular shapes which resemble typical sediments.

SHAPE	α_1	α_2	α_2/α_1		
			n = 0·3	n = 1·0	n = 3·0
sphere  d	$\pi/4$	$\pi/6$		0·67	
cylinder  nd	n or $\frac{\pi}{4}$	$n\pi/4$		0·79	
			0·3	1	3
cube 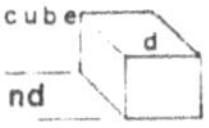 d, nd	n or 1	n	0·3	1	3
			1	1	1
lozenge  d, nd	$\pi/4$ or	$\pi/6 + n\pi/4$	0·97	1·67	3·67
	$n + \pi/4$		0·7	0·73	0·76

Table 2: Values of the shape coefficients α_1 & α_2

It may be seen, from Table 2, that extremely small values of α_2/α_1 represent plate like sediments while extremely large values of α_2/α_1 represent "pencil shaped" sediments. Neither of these extremes commonly occur with sediment sizes greater than 0.1 mm and, if they did, they would be quickly rounded by abrasion. In conclusion, it seems that for most sediments, the value of α_2/α_1 will vary only slightly between a value of about 0.5 and 1.5. Assuming a mean value of 1.0 is not unreasonable.

<u>Drag and Lift Coefficients</u> Most researchers seem to ignore the lift force when developing their theories of sediment transport. It would appear that because White (1940) in his original paper on sediment transport ignored it, other researchers have chosen to do

so too. There is little doubt however, that a substantial lift force does exist and, experiments on cylinders close to a flat boundary show that, at least for high Reynolds numbers, it is typically about 30% of the drag force. The coefficients of drag and lift will also vary with,

(i) The grain Reynolds number, $\frac{\beta V d}{\nu}$

and (ii) The spacing of the grains.

Raudkivi (1967) gives various values of C_D for spheres and also for angular sand grains. For the experiments in this study it is estimated that the grain Reynolds Number was generally about 100 at which value the drag coefficient would be about 1.7. (The water temperature in all the experiments was between 19 and 21°C).

The second factor which has a significant influence on the coefficient of drag is the particle spacing. This may be defined by a particle spacing coefficient, η as follows:

$$\eta = \frac{Q_s}{u\ d\ w_e\ \alpha_2} \qquad \text{-------- (5)}$$

where u is the mean particle velocity, w_e is the effective width for sediment transport by flume traction and Q_s is the volumetric sediment load. When η equals one the sediment particles are closely packed with each occupying an area of d^2. As η decreases the particles become more widely spaced. The concept is best explained by the "chessboard analogy" in which when $\eta=1$ every square is occupied by a piece but, when $\eta=0.2$ only every fifth square has a piece on it.

In order to determine at what spacing the drag coefficient begins to be influenced by the wakes of the particles around it reference has been made to a paper by Morris (1959) on wake interference flows. From this it would appear that for η values less than 0.15 wake interference should not occur.

<u>Velocity distribution coefficient, β</u> May (1982), calculated values of v/V for both smooth turbulent and rough turbulent flow. Working from the normally used equations he found that:

$$\frac{v}{V} = 1 + 2.033\sqrt{\lambda}\ \log_{10}(2.113\ y/R) \qquad \text{-------- (6)}$$

for smooth turbulent flow and,

$$\frac{v}{V} = 1 + 2.033\sqrt{\lambda}\ \log_{10}(2.245\ y/R) \qquad \text{-------- (7)}$$

for rough turbulent flow. (R being the hydraulic radius and λ the friction factor). He then went on to derive equations for β, to determine the effective velocity around the grain. Since however,

Equations 6 and 7 are so nearly the same (with respect to the velocity distribution) a mean value equation has been used here to define β. Thus,

$$\beta = 1 + 2.033 \sqrt{\lambda} \left(\log_{10}(2.18y/R) - 0.434\right) \qquad \text{--------- (8)}$$

for both rough and smooth turbulent flow. The depth y is chosen as the nominal grain diameter, d.

In the present experiments, values of λ for the different flows and conduits varied from 0.018 to 0.026. The D_{50} size of the fine sand used was 0.45mm and a typical value of R was 14 mm. Substitution of these values into Equation 8 gives β values of 0.57 and 0.48 respectively.

Outcome of the theoretical analysis

Taking Equations 4 and 5 together a prediction of the mean velocity necessary to prevent sediment deposition can be made for any conduit, any sediment load, any conduit roughness and any sediment type. This velocity, multiplied by a suitable factor of safety to allow for the inaccuracies of pipe laying etc., can then be used to find necessary conduit sizes and/or gradients using the Ackers (1969) charts.

The effective width w_e, for sediment transport by flume traction in a circular pipe must be less than D, the pipe diameter. Indeed, it must be considerably less even than ϕD. Application of Equation 5 then shows that the largest pipe that can be expected to convey fine sediments (≤ 0.4 mm) by flume traction (at C_V = 100 ppm) is about 600 mm in diameter. Above this size, for the stated conditions, some sediment bed must form and thus any flume traction equation will no longer apply. Consequently, for pipes larger than this, a stream traction theory (e.g. the modified Ackers and White op. cit.) is more appropriate. It is interesting to note, that in a rectangular conduit, or one with a slightly dished floor, flume traction can persist beyond this size.

For an isolated grain, where there is no sediment load, the velocity of the grain, u will indeed be zero when deposition is about to occur. However, Equation 5 shows that whenever Q_s has some finite value, the value of u must also be significant, simply to prevent a "log jam" of particles at some point. This is probably the reason why the deposition condition is so well defined, since, the moment u falls below some critical value the majority of the sediment will suddenly stop moving. The critical value of u may be determined from Equation 5 by assuming η=0.5; it may then be substituted into Equation 4 to find V.

OTHER THEORIES FOR INCIPIENT DEPOSITION

One of the first people to study the condition of incipient deposition was Ambrose (1953) whose work was summarised by Laursen (1956). He plotted an empirically derived function for sediment transport against y/D, the relative depth of flow in a pipe of diameter, D and found that values of the function at the limit of flume traction all lay on a single curve. This curve, which is shown in Figure 2(a), appears to show a strong relationship with y/D. Unfortunately, on inspection, it is found to be compounded with the function $(y/D)^2$ viz:

$$\frac{Q}{\sqrt{g(S_S-1)}\, D^{2.5} C_V^{1/3}} \simeq \frac{V}{\sqrt{g(S_S-1).y}\; C_V^{1/3}} \cdot \left(\frac{y}{D}\right)^2 \qquad (9)$$

where C_V is the concentration by volume of sediment.

When the limit function $V/\sqrt{g(S_S-1)y}.C_V^{1/3}$ is plotted against y/D (as in Figure 2(b)) it is found to be virtually constant. (This result is also to be expected from the form of Equation 4).

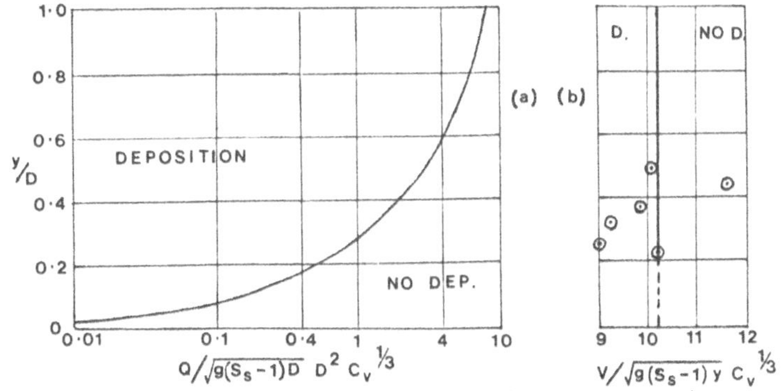

Fig. 2: Limit of flume traction (after Laursen)

Vickers et. al. (1968) also found that, for single particles at least, the conditions for them just not to move were constant. A wide range of particles with different densities and shapes all settled at $\sqrt{\tau_0/\rho}=1.9$ cms/sec. Using the equation,

$$V = V^* \sqrt{\frac{8}{\lambda}} \qquad (10)$$

This may be converted to a critical deposit velocity of 0.31 m/s. (for $\lambda=0.03$).

Novak and Nalluri (1978) also carried out experiments on incipient deposition and derived the following equation for discrete particles,

$$\frac{V}{\sqrt{g\,(S_S-1)d}} = 0.61\left(\frac{d}{R}\right)^{-0.27} \qquad (11)$$

They also derived equations to determine the critical deposit velocity for non-discrete particles but, space does not permit their inclusion here.

EXPERIMENTAL APPARATUS

The conduits Three conduit shapes were used in the experiments and one of these was utilised for two tests by rotating it through 90°. The specifications are shown in Figure 3. All were fabricated from perspex throughout and assembled in 1.0 m or 0.9 m lengths. The circular pipe was 8.0 m in length while the other conduits were 7.2 m long. For the non-circular conduits special fibreglass transition sections were constructed to ensure smooth entry conditions; all joint connections were made with care to ensure the smoothest possible flow conditions.

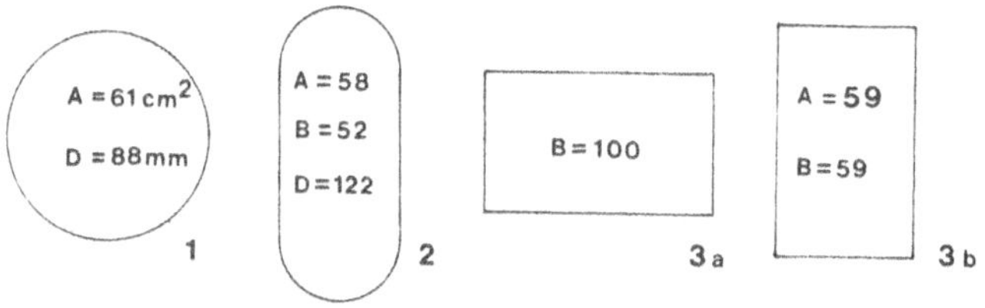

Fig. 3: Conduit shapes tested.

Support Structure The conduits were supported on a specially fabricated tilting frame constructed from rigid steel channel sections. The slope of the frame was true to within 2 mm over 8.5 m and it could be varied with a fine adjustment. Measurement of the slope was made using two interconnecting water columns set at 6.63 metres apart.

Sediment properties The sediments used were non-cohesive fine and coarse sand (although only the fine sand results are reported here). Both sands were nearly uniform in size having D_{50}'s of 0.45 mm and 1.5 mm respectively. The sediments are shown, enlarged, in Figure 4.

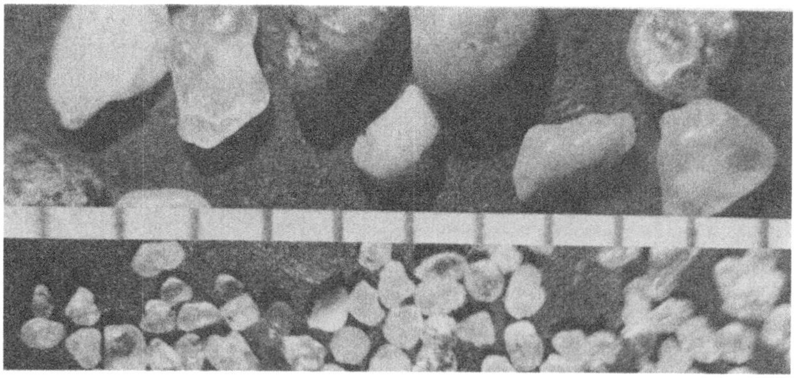

Fig. 4: Fine and Coarse sand sediment shown against a millimetre scale.

Procedure adopted in the tests For each series of tests the procedure used was as follows: First the gradient, sediment discharge and flow were set at values producing non-deposit flume traction flow with the conduit part full. Next the gradient was gradually reduced until deposition began to occur. The flow was then increased to clear the settled deposit and this procedure was repeated several times. Sediment discharge was measured by weighing and flow discharge using a V-notch weir. Water levels in the conduits were recorded at regular intervals.

RESULTS OF INCIPIENT DEPOSITION TESTS

A selection of the results obtained in the tests are shown in Table 3 and those for the circular pipe are also plotted on Figure 2(b). It may be seen that the results (for the circular pipe) agree very well with the Ambrose/Laursen equation (as modified at Equation 9.RHS).

The most interesting finding is that the rectangular conduit placed with its longer axis horizontal proved to be the most efficient for sediment transport; for cases tested, the maximum value of V_c, the critical, no-deposit velocity was less than 0.5 m/s. For the circular pipe, however, the maximum was 0.63 m/s and for the oval pipe it was 0.71 m/s. These results, although surprising at first, are in conformity with the contention that the effective width for sediment transport is a key determinant in the process.

Comparison of the model test results with the values of V_c predicted by Equations 4 and 5 also gives fairly good agreement as is shown in Table 3.

SHAPE	FLOW cm^3/s	C_v ppm	R mm	SLOPE	y mm	CRITICAL VELOCITY EXPTS.	EQ. 9	EQ.4&5
▯	536	1120	12	0·0094	19	0·47		0·53
	761	823	15	0·0080	28	0·45		0·54
	2751	228	21	0·0056	74	0·62		0·60
▭	486	1400	10	0·0116	12	0·41		0·37
	825	800	14	0·0083	19	0·43		0·40
	1546	427	19	0·0056	31	0·50		0·44
0	730	904	13	0·0152	28	0·64		0·77
	825	800	14	0·0136	30	0·66		0·79
	962	686	15	0·0125	32	0·71		0·77
○	282	1914	13	0·0148	14	0·45	0·58	0·53
	561	962	16	0·0098	21	0·52	0·57	0·54
	1112	486	20	0·0075	32	0·56	0·56	0·58
	1546	850	21	0·0066	37	0·63	0·54	0·60

Table 3: A selection of test results

CONCLUSIONS

1. A new theory for sediment transport in closed conduits has been derived following, in part, the treatment of May (1982).

2. The theory shows that, for moderate concentrations of fine sediments, the maximum pipe size for which flume traction can persist is about 600 mm diameter. Above this size flume traction cannot occur and the sediment must either be fully entrained or allowed to form a bed.

3. The experimental results confirm the importance of the effective width (for sediment transport) as one determinant of the critical velocity necessary to achieve non-deposit conditions.

4. The most efficient conduit shape for the conveyance of high concentrations of non-cohesive sediments was found to be a rectangular conduit with its longer axis horizontal - the shape of a typical box culvert!

REFERENCES

Ackers, P. (1969) Charts for the Hydraulic Design of Channels and Pipes. Paper No.2. HRS

Ackers, P. (1984) Sediment Transport in Sewers and the Design Implications. Conference on Sewerage Systems. Reading, England, BHRA.

Ambrose, H.H. (1953) The transportation of sand in pipes. Proc. 5th Hydraulics Conf. Bull. 24, Iowa Univ., U.S.A.

Laursen, E.M. The hydraulics of a storm drain system for sediment transporting flow. Bull. No.5, Iowa Highway Research Board.

May, R.W.P. (1982) Sediment Transport in Sewers. HRS Report No. IT.222.

Morris, H.M. (1959) Design methods for flow in rough conduits. P ASCE HY7.

Novak, P. and Nalluri, C. (1978) Sewer design for no sediment deposition. Proc. Inst. Civil Engrs., Vol.65, Sept.

Raudkivi, A.J. (1967) Loose Boundary Hydraulics. Pergamon Press.

Vickers, J.A., Calvert, J.T. and Francis, J.R.D. (1968). Erosion of sewer and drains. CIRIA Research Report 14.

White, C.M. (1940) The equilibrium of grains on the bed of a stream. Proc. Roy. Soc. (A) 174, No.958.

On the Hydraulics of Corrugated PVC Drains with Different Clogging Degrees

I. Cojocaru, V. Blidaru, I. Bartha, H. Leibu, A. Nicolau
Polytechnic Institute of Iași-România

INTRODUCTION

The corrugated PVC drains have extended and will continue to extend their use range both in the case of classical drainage and particularly so in that of the drainage of sloping land with sliding potential, due to their multiple economic and technical advantages: minimum cost, deformation at land displacement or settling, facility of installation.
Studies on the drain tube hydraulics have been carried out by different researchers (Morris, H-1955, Wesseling, J.-1967, Beken, A.-1972, Stanciu, I.-1973, etc) but they referred especially to the transport capacity of the unclogged drains operating with full section.
The danger of drain clogging is much higher on sloping land with sliding potential than on other types of land, even when the drains are protected by filters.
The high content in fine and very fine particles, their inadequate texture and structure, as well as the extremely low cohesion of these lands result in the partial or total clogging of drains. On the other hand, the research works carried out in Romania (Cojocaru, I.-1985) have shown that the drainage on these lands works under unsteady state conditions, fact which makes the water within the drain flow, most of the time, at filling ratios smaller than the unit.
Consequently, we have intended to determine such hydraulic operational characteristics as roughness coefficient, transport capacity and the relative velocity for a clean corrugated PVC drain with two degrees of clogging. The effect of lateral water inflow in the drain on the hydraulic state of flow was neglected as suggested by some earlier research con-

clusions (Beken - 1972).

EXPERIMENTAL INSTALLATION

The experimental installation made within the Irrigation and Drainage Laboratory of the Polytechnic Institute Iaşi-România enabled the study of the afore-

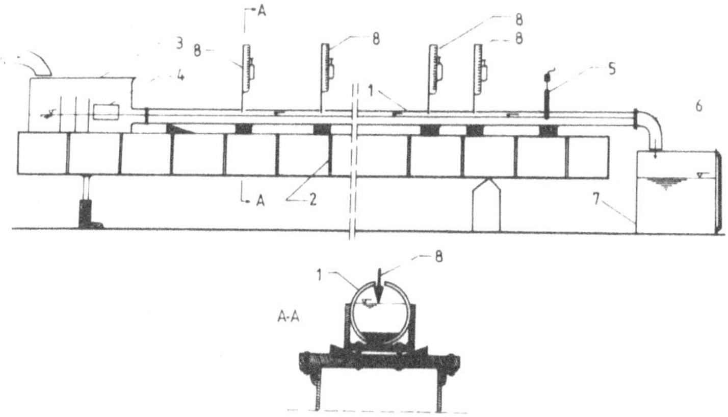

Fig.1. The experimental installation site.

mentioned characteristics for variable slopes and clogging degrees in the conditions of a steady state water flow in the drain.
The corrugated PVC drain (1) Ø 1oo mm (the exact value of the inner diameter: Dn=98,4 mm) was placed in a U shaped metal profile held into position by a support (2) with variable tilt. The drain small openings were sealed with a special solution.
The change in the supply parameters was brought about by means of a reservoir (3) fitted with an adjustable lateral spillway (4).
The steady,uniform state of water flow in the drain was obtained by an adequate operation of the downstream spillway (5).
The water outflow was effected through the oscillating bend (6) within the graduated reservoir (7) or in the discharge channel.
The measurement of water depth in the drain was made by means of measuring devices (8) after the upper part of the drain had been pierced in different places (3x5 cm).
Drain clogging simulation was achieved by a two stage pouring of cement mortar in the horizontally located drain. Two clogging degrees (β) were obtained: β =o,14 and β =o,27.
Sand granulation in the mortar was the same with the

one found in some clogged drains in the field (Cojocaru - 1985).

PROCESSING AND INTERPRETATION OF EXPERIMENTAL RESULTS

Using the present-day computation technique for the drain tubes, the experimental results were processed with the aim to find out the value of Chezy's roughness coefficient "n" able to satisfy Manning's relation.

Unclogged drain case

For the calculation of roughness coefficient (n) the geometric and hydraulic elements were expressed function of the filling degree ($\alpha = \frac{h}{D}$)

$$S = \frac{D^2}{4}\left[\arccos(1-2\alpha)-(1-2\alpha)\sqrt{1-(1-2\alpha)^2}\right] \quad (1)$$

$$P = D \arccos(1-2\alpha) \quad (2)$$

$$R = \frac{D}{4}\left[\frac{\arccos(1-2\alpha)-(1-2\alpha)\sqrt{1-(1-2\alpha)^2}}{\arccos(1-2\alpha)}\right] \quad (3)$$

where: S-section; P-perimeter; R-hydraulic radius; D-diameter; h-depth of water in the drain.
By substituting equation (1),(2),and (3) in Manning's equation it gives:

$$n = \frac{D^{8/3}\, I^{1/2}\left[\arccos(1-2\alpha)-(1-2\alpha)\sqrt{1-(1-2\alpha)^2}\right]^{5/3}}{10{,}079\, Q\left[\arccos(1-2\alpha)\right]^{2/3}} \quad (4)$$

where: I-drain slope, Q-transported flow.
The processed experimental data (fig.2) have shown the following correlation between "n" and "α":

$$n = 0{,}015\,(0{,}763\alpha^2 - 0{,}878\alpha + 1{,}282) \quad (5)$$

By applying the relation of steady, uniform state of motion for the experimental values it was possible to calculate the discharge and the relative velocity ($Q_r = \frac{Q}{Q_o}$; $V_r = \frac{V}{v_o}$) respectively, where Q_o and v_o conespond to full section functioning and Q and V to various filling degrees) for different values of "α". The graphical correlation obtained are shown in Fig.3 and 4.
It can be noted that:
-the maximum relative flow (Q_r) is obtained for $\alpha = 0{,}90 \div 0{,}92$, but not for $\alpha = 0{,}95$ as given in literature (Mateescu,C.-1983) for closed channels of annular section;
-the maximum relative velocity (V_r) is obtained for a lower filling degree ($\alpha = 0{,}75$) than the one found in literature for closed channels of annular section(α=0,81)

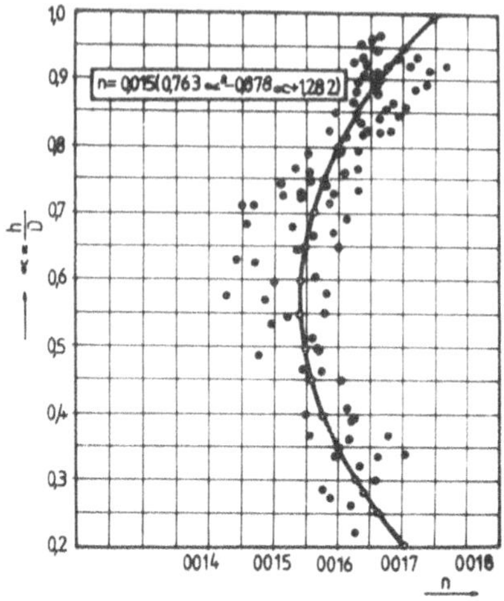

Fig.2. Variation of roughness function of filling degree for a clean,corrugated drain Ø loo mm.

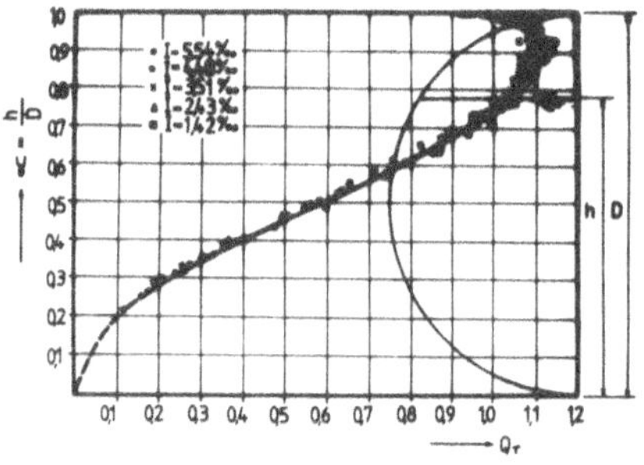

Fig.3. Variation of relativ flow function of filling degree for a clean corrugated drain Ø loo mm.

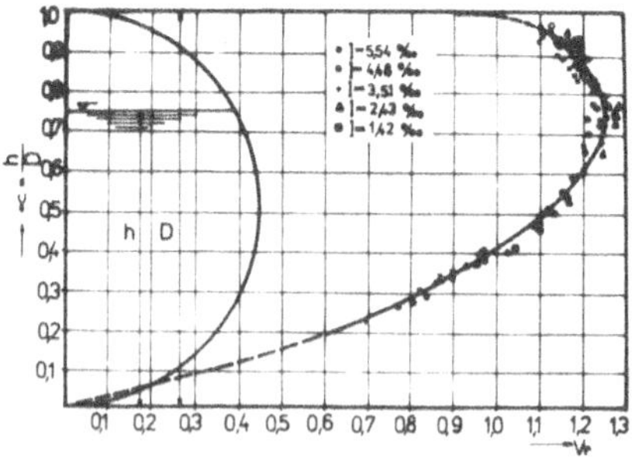

Fig.4. Variation of relative velocity function of felling degree for a clean corrugated drain Ø loo mm.

The clogged drain case

For the clogged drain the geometric and hydraulic elements were expressed function of the clogging degree:(β =c/D)

$$S= \frac{D^2}{4}\left[\arccos(1-2\alpha)-(1-2\alpha)\sqrt{1-(1-2\alpha)^2}-k_1\right] \quad (6)$$

$$S_c= \frac{D^2}{4}\left[\arccos(1-2\beta)-(1-2\beta)\sqrt{1-(1-2\beta)^2}\right] \quad (7)$$

$$h_c= D\sqrt{1-(1-2\beta)^2} \quad (8)$$

$$P = D\left[\arccos(1-2\alpha)+k_2\right] \quad (9)$$

$$R = \frac{D}{4}\,\frac{\arccos(1-2\alpha)-(1-2\alpha)\sqrt{1-(1-2\alpha)^2}-k_1}{\arccos(1-2\alpha)+k_2} \quad (1o)$$

S_c-clogged section; c-maximum hight of the clogging layer and $\alpha = \frac{h+c}{D}$.

If Manning's equation is taken into account, the use of equation 6-1o leads to:

$$n= \frac{D^{8/3}\, I^{1/2}\left[\arccos(1-2\alpha)-(1-2\alpha)\sqrt{1-(1-2\alpha)^2}-k_1\right]^{5/3}}{1o,o791\, Q\left[\arccos(1-2\alpha)+k_2\right]^{2/3}} \quad (11)$$

Fig.5 a and b shows graphically the values of "n" obtained on the basis of experimental data with respect to the filling degree "α" for both clogging degrees.

The experiments have also shown that, by clogging, there takes place a reduction of the average flow carried by the drain, of approx.12 % for β = o,14 and of 32 % for β =o,27.

Applying the values experimentally obtained in the relations 6÷11 the relative flows could be obtained (Fig.6) and the relative velocities (Fig.7) with respect to the filling degree for both filling degrees. It can be seen that the reduction of the section due to clogging, the reduction of the hydraulic radius and the alteration of the roughness value causes a reduction of 12 % in the relative flow and of 4 % in the relative velocity in comparison with the clean drain (for β=o,14) of 32 % and 12 % respectively (for β =o,27).

The processing and interpretation of the experimentally obtained data enabled the plotting of a graph (Fig.8) for estimating the maximum length of a PVC, corrugated drain, Ø 1oo mm, function of the specific flow (q_{sp}), slope (I), distance between the drains (L) and clogging degree (β).

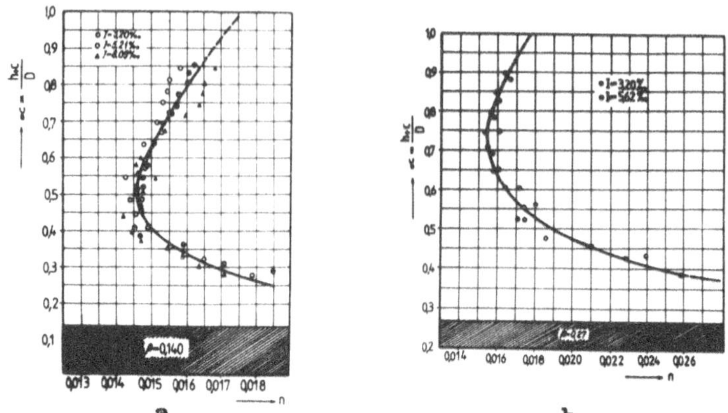

a b

Fig.5. Variation of clogged corrugated drain roughness function of filling degree.a) β =o,14;b) $\beta = 0{,}27$.

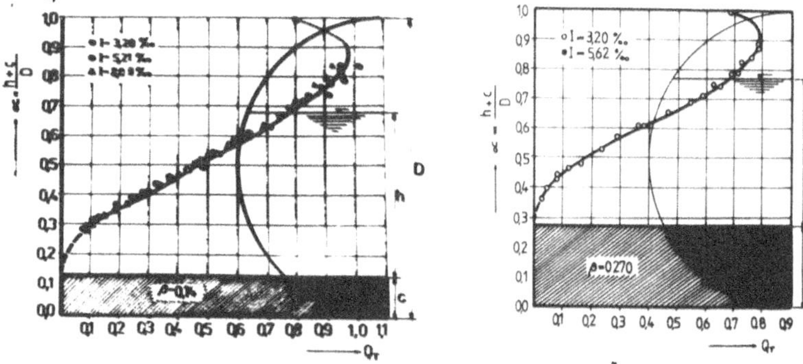

a b

Fig.6. Variation of relative flow function of filling degree in a clogged corrugated drain. a) β =o,14; b) β =o,27.

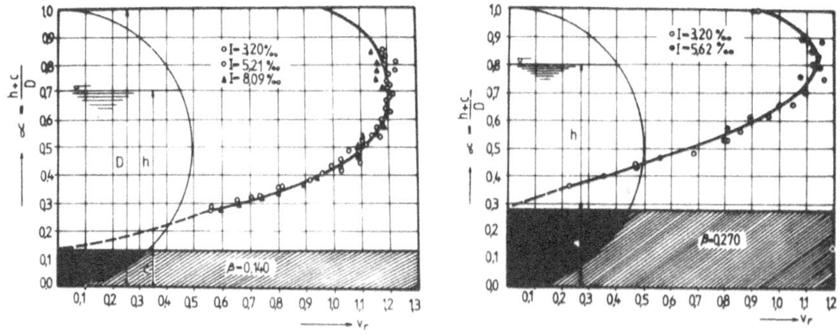

a b

Fig.7. Variation of relative velocity function of filling degree in a clogged corrugated drain. a) β = o,14; b) β = o,27.

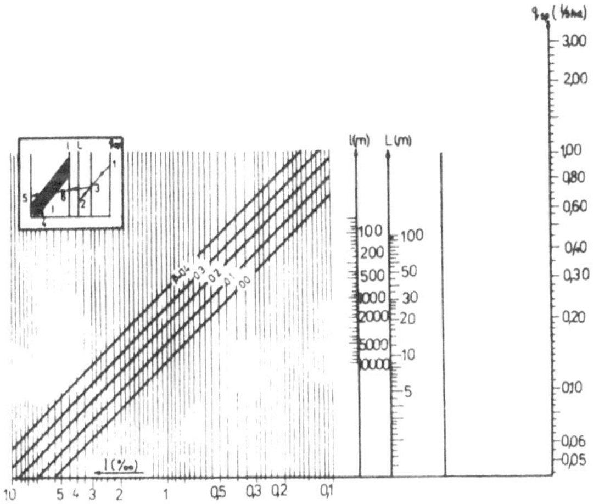

Fig.8.Graph for estimating maximum length of an unclogged PVC corrugated drain(β =o) or of one with various degrees of clogging (β).

CONCLUSIONS

After processing the results concerning the hydraulics of clean or clogged PVC corrugated drains the following conclusions can be drawn:
-the transport capacity of a Ø loo mm drain is reduced by clogging with 12 % for β =o,14 and with 32 % for β =o,27;
-water velocity in the drain decreases due to clogging with about 5 % for β =o,14 and 12 % respectively for β =o,27 ;
-maximum relative flow and velocity in the clean PVC corrugated drain Ø loo mm, are obtained for lower filling degrees than in closed channels. This is due to superposing of macro-roughness effect on micro-roughness effect;
-dimensioning of PVC corrugated drains Ø loo mm, clean or with different clogging degrees can be effected using the graph in fig.8.

BIBLIOGRAPHY

1. Beken,A.(1972) -Research on the Hydraulics of Drain Pipes. Eigth Congress on Irrigation and Drainage, Question 27, Ref.27,Varna.
2. Cojocaru,I.(1985) - Contribution to Estimation of Sliding Slope Drainage Functional Parameters

with References to Experimental Platform in Dîmbovița Watershed Area.Ph.D.Thesis.Polytechnic Institute of Iaşi-România.

3.Morris H.M.(1955) - Flow in Rough Conduits. Trans. A.S.C.E. vol.12o.

4.Stanciu,I. (1973) - Nomograms for Hydraulic Dimensioning of Tubes used in Absorbent Drain Construction, Imb.Funciare, Documentare curentă nr.7-8, C.I.D.H-Bucureşti.

5.Wesseling,J; Homma,F. (1967) - Hydraulic Resistence of Drain Tubes.Tech.Bul.5o,Inst.for Land and Water Managements Research, Wageningen.

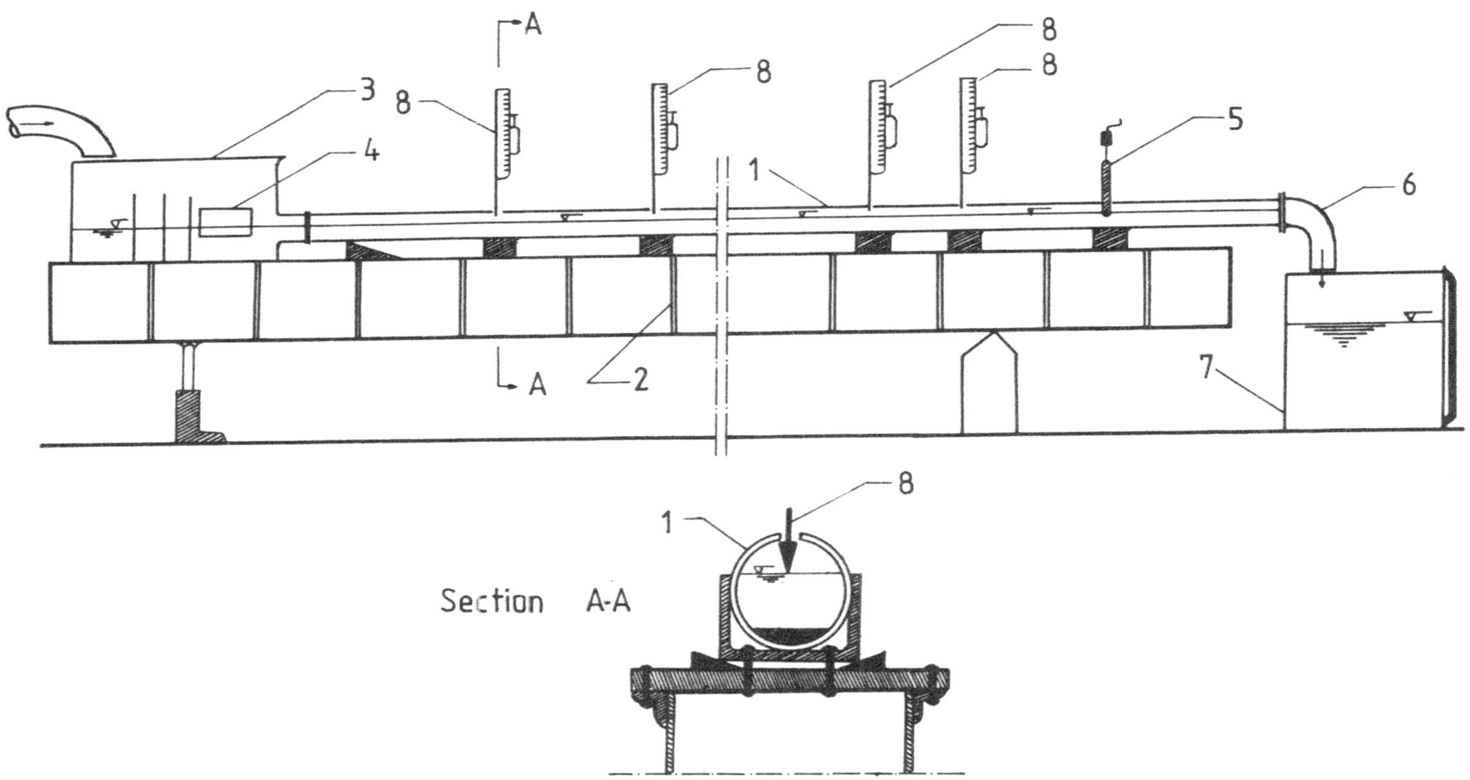

Fig. 1

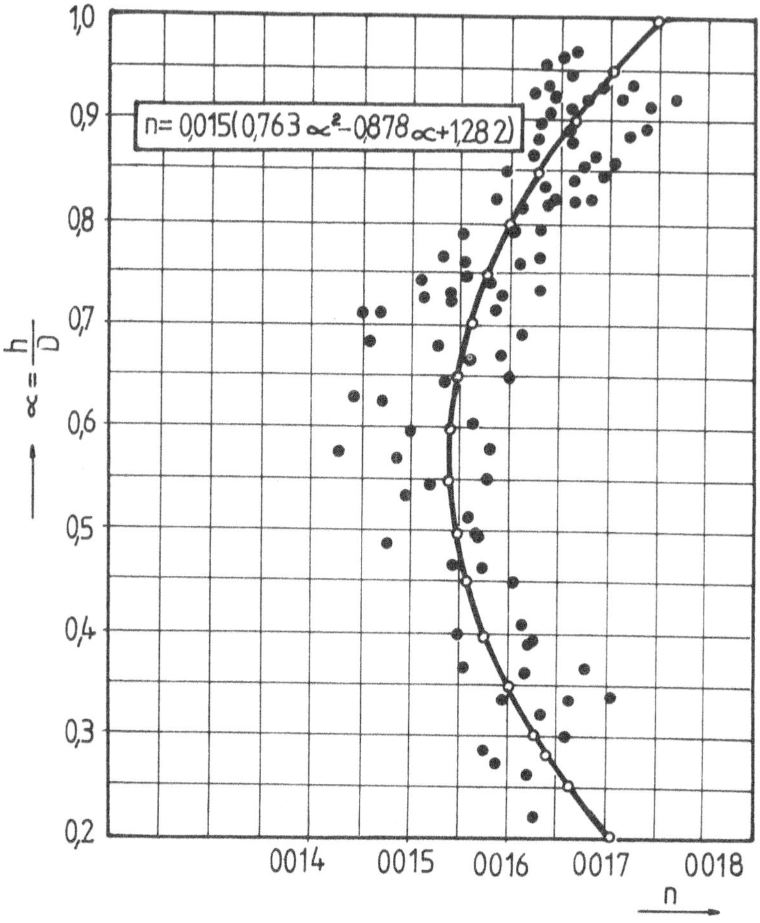

n= 0,015(0,763 α²-0,878 α+1,282)
1,0
0,9
0,8
0,7
0,6
0,5
0,4
0,3
0,2
α = h/D
0014
0015
0016
0017
0018
n

Fig. 2

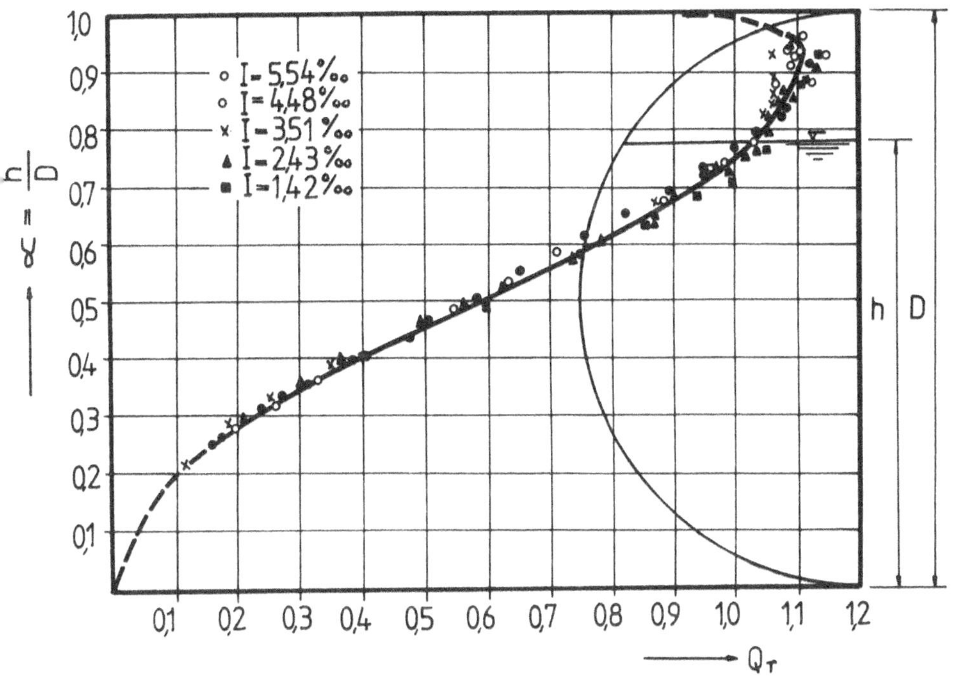

1,0
0,9
0,8
0,7
0,6
0,5
0,4
0,3
0,2
0,1
α = h/D
I = 5,54‰
I = 4,48‰
I = 3,51‰
I = 2,43‰
I = 1,42‰
h
D
0,1
0,2
0,3
0,4
0,5
0,6
0,7
0,8
0,9
1,0
1,1
1,2
Q_r

Fig. 3

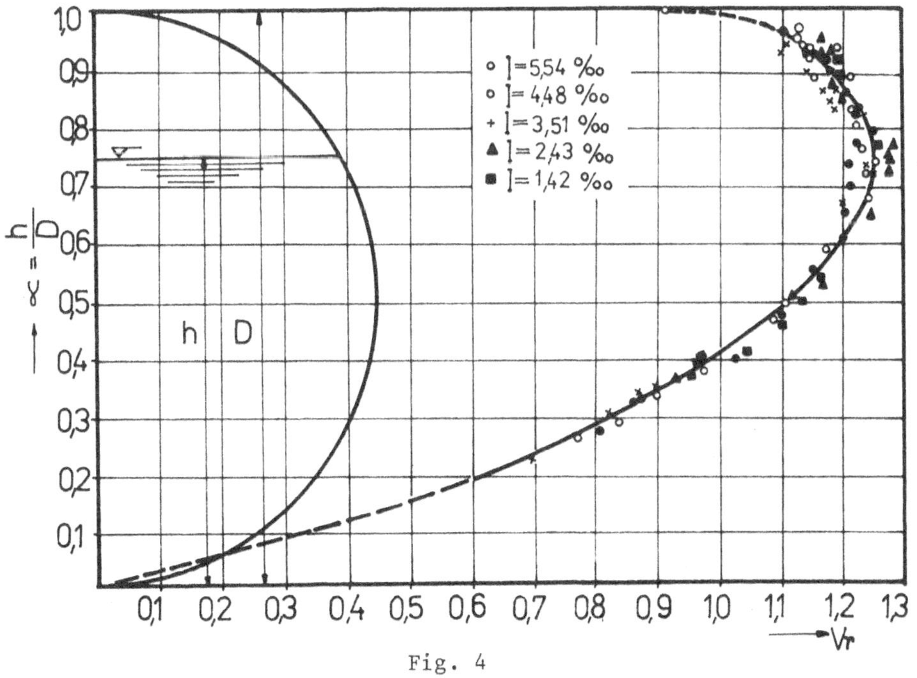

1,0
0,9
0,8
0,7
0,6
0,5
0,4
0,3
0,2
0,1
α = h/D
I = 5,54 ‰
I = 4,48 ‰
I = 3,51 ‰
I = 2,43 ‰
I = 1,42 ‰
h
D
0,1
0,2
0,3
0,4
0,5
0,6
0,7
0,8
0,9
1,0
1,1
1,2
1,3
V_r

Fig. 4

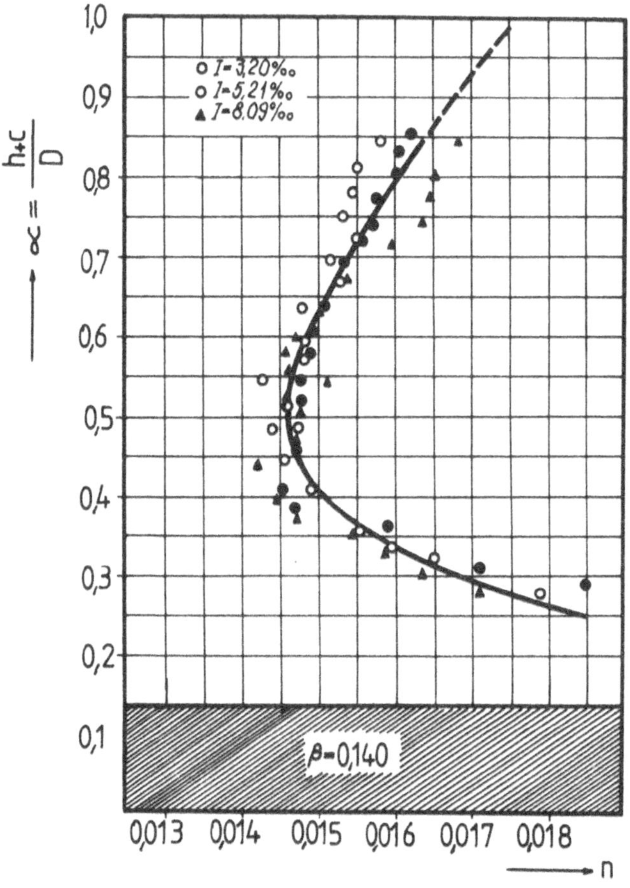
I=3,20‰
I=5,21‰
I=8.09‰
1,0
0,9
0,8
0,7
0,6
0,5
0,4
0,3
0,2
0,1
α = (h+c)/D
β=0,140
0,013
0,014
0,015
0,016
0,017
0,018
n

Fig. 5a

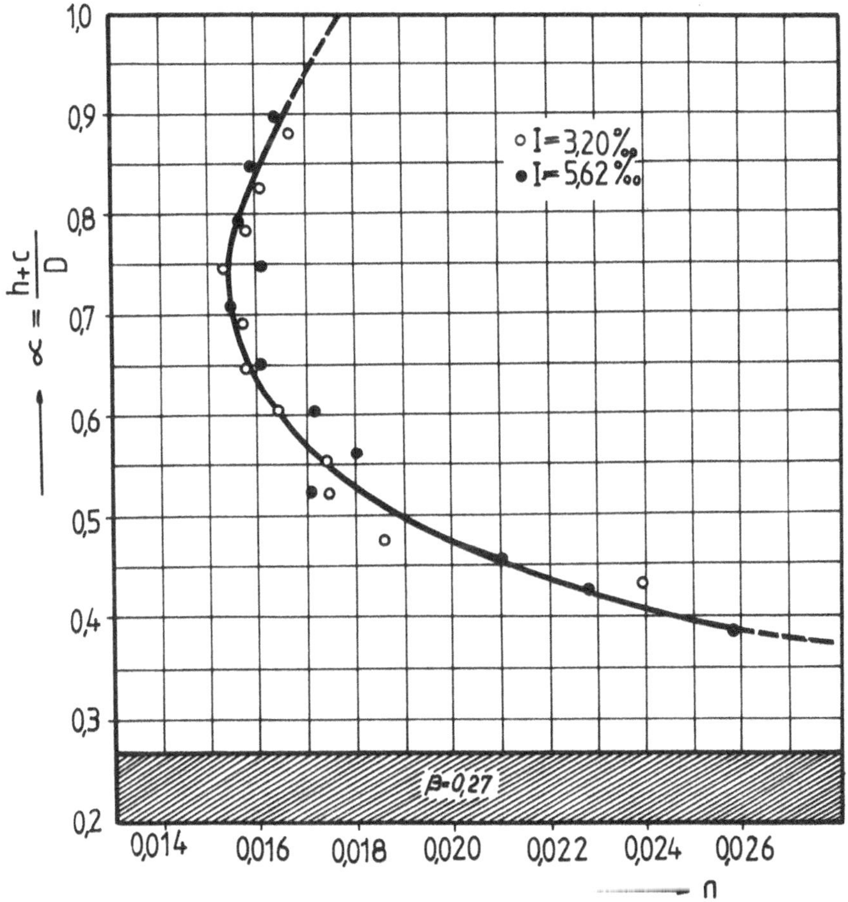

1,0
0,9
0,8
0,7
0,6
0,5
0,4
0,3
0,2
$\alpha = \frac{h+c}{D}$
○ I=3,20‰
● I=5,62‰
β=0,27
0,014
0,016
0,018
0,020
0,022
0,024
0,026
n

Fig. 5b

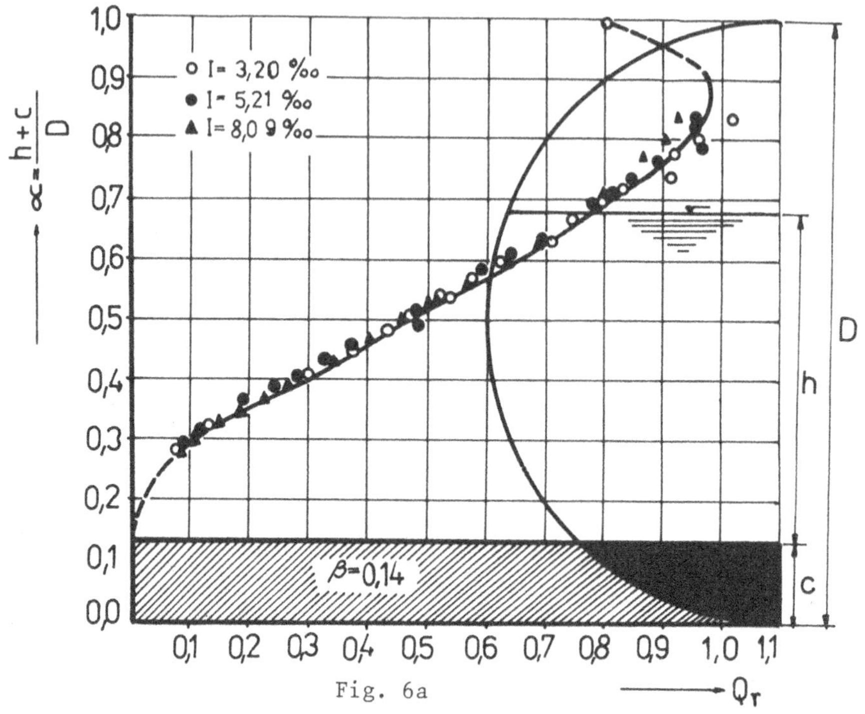

1,0
0,9
0,8
0,7
0,6
0,5
0,4
0,3
0,2
0,1
0,0
α = (h + c)/D
I = 3,20 ‰
I = 5,21 ‰
I = 8,0 9 ‰
β=0,14
0,1
0,2
0,3
0,4
0,5
0,6
0,7
0,8
0,9
1,0
1,1
Q_r
D
h
c

Fig. 6a

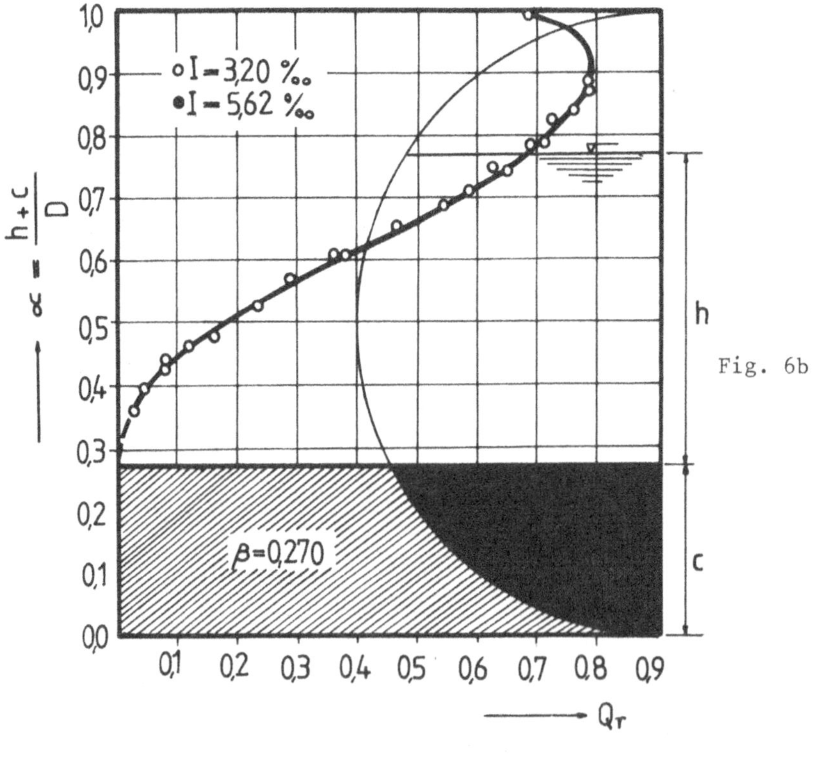

1,0
0,9
0,8
0,7
0,6
0,5
0,4
0,3
0,2
0,1
0,0
α = (h + c)/D
I = 3,20 ‰
I = 5,62 ‰
β=0,270
0,1
0,2
0,3
0,4
0,5
0,6
0,7
0,8
0,9
Q_r
h
c

Fig. 6b

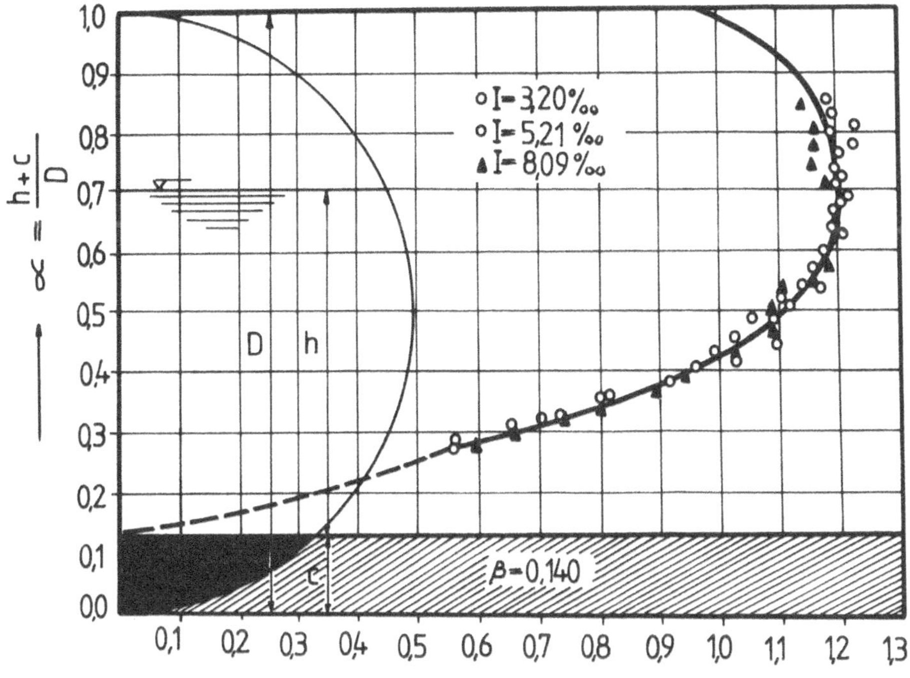

1,0
0,9
0,8
0,7
0,6
0,5
0,4
0,3
0,2
0,1
0,0
α = (h+c)/D
∘ I=3,20‰
∘ I=5,21‰
▲ I=8,09‰
D
h
c
β=0,140
0,1
0,2
0,3
0,4
0,5
0,6
0,7
0,8
0,9
1,0
1,1
1,2
1,3
Vr

Fig. 7a

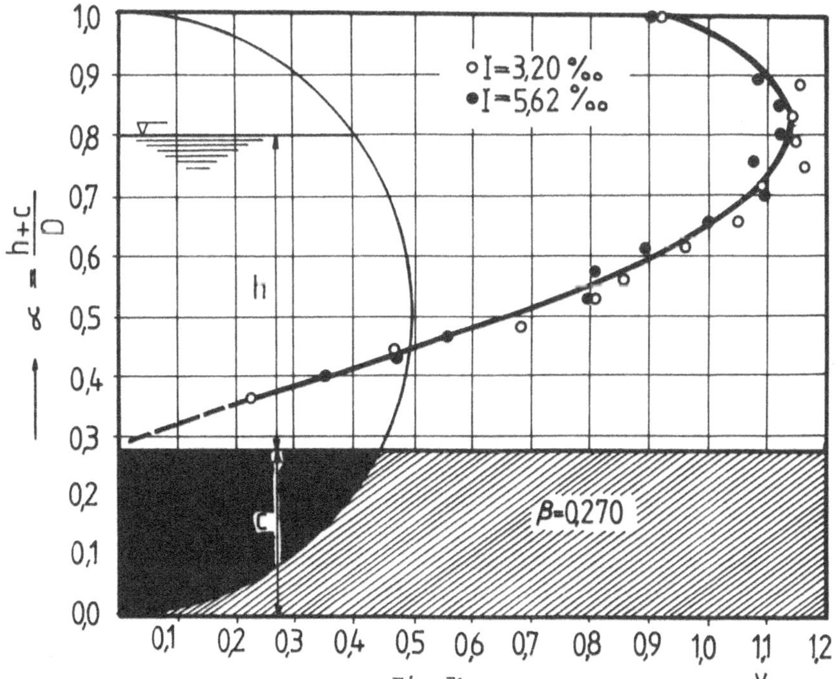

1,0
0,9
0,8
0,7
0,6
0,5
0,4
0,3
0,2
0,1
0,0
α = (h+c)/D
∘ I=3,20 ‰
• I=5,62 ‰
h
c
β=0,270
0,1
0,2
0,3
0,4
0,5
0,6
0,7
0,8
0,9
1,0
1,1
1,2
Vr

Fig.7b

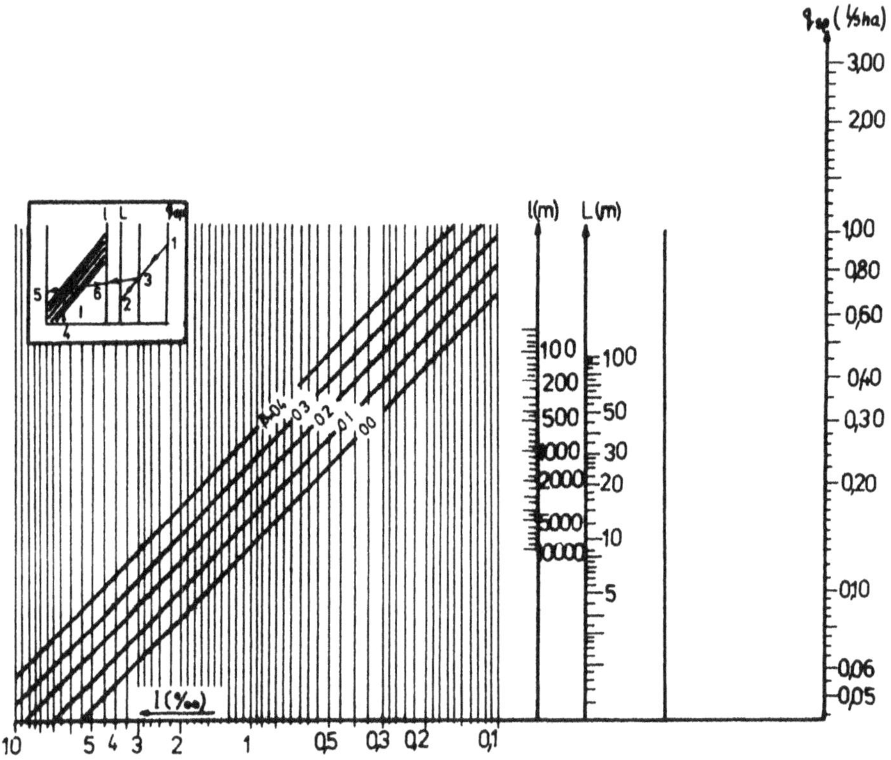

q_sp (l/s ha)
3,00
2,00
1,00
0,80
0,60
0,40
0,30
0,20
0,10
0,06
0,05
l(m)
L(m)
100
200
500
1000
2000
5000
10000
100
50
30
20
10
5
I (‰)
10
5
4
3
2
1
0,5
0,3
0,2
0,1

Fig. 8

Open Drain Bank Stabilization Before and After Construction

L.S. Willardson and Mohammed Al-Adhami
Utah State University

ABSTRACT

Bank instability and erosion during and after construction of open agricultural drains in cohesionless soil is a serious problem in many parts of the world.

A sand tank was used as a two-dimensional physical model to investigate drain bank stability problems. The model was run without an interceptor drain and with an interceptor drain parallel and close to the open drain. The placement of an interceptor drain lowered the water table and reduced the hydraulic gradient near the drain bank. A normal slice slope stability analysis showed that an interceptor drain increases the factor of safety for stability of a 2:1 side slope from (0.5) to (1.5).

A computational procedure was developed to find the position of the water table between the interceptor and the open drain and to find the proper location for the interceptor drain.

INTRODUCTION

The permanence of an open drain depends on the continuous stability of the side slopes of the drain. The characteristics of the soil have a great deal of influence on drain bank stability. Sandy and silty soils possess little cohesive strength and are relatively unstable. When open drains are dug in these materials, the side slopes readily slough because of the pressure of the water entering the side slope of the drains (11). This water pressure tends to lift the soil particles away from the side of the drain and causes them to move to the bottom of the drain. The saturated soil mass has reduced structural strength and may also fail by sloughing.

When an earth slope intersects a permanent or intermittent ground water table, it is very important to provide some type of treatment to prevent slope sloughing. The usual treatment is the placement of a heavy porous rock fill on the slope, installation of toe-drains, a flattening of the side slopes or a combination of these treatments.

The purpose of this study was to show the effect of placement of an interceptor drain at the hydraulic gradeline of the open drain, parallel, and close to the open drain, on the stability of an open agricultural drain bank. Such an interceptor drain accomplishes bank stabilization during and after excavation, minimizes maintenance and excavation costs, allows efficient farming operations, reduces waste of land due to flat drain side slopes, and increases the effectiveness of the outlet drain.

PROCEDURE

The physical model

A sand tank was used as a two-dimensional physical model, Figure 1.

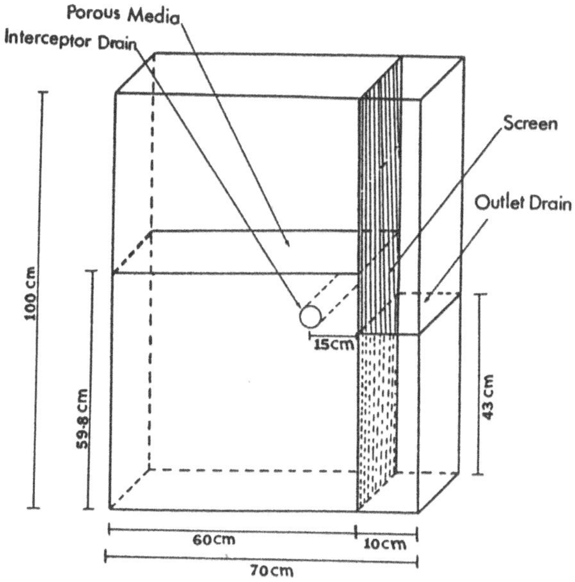

Fig. 1--Sand Tank Model

The tank was 10 cm thick, 100 cm high, and 70 cm wide. An open drain was initially simulated by a rectangular section approximately 10 cm wide on one side of the tank. The outlet drain portion of the model was separated from the other part of the model by a layer of synthetic envelope fabric supported by a layer of window screen which was supported by a perforated heavy aluminum plate. The soil used was a fine sand from dunes near St. Anthony, Idaho. Micro tensiometers were installed in the model to measure hydraulic potentials. Figure 2 shows the complete experimental installation.

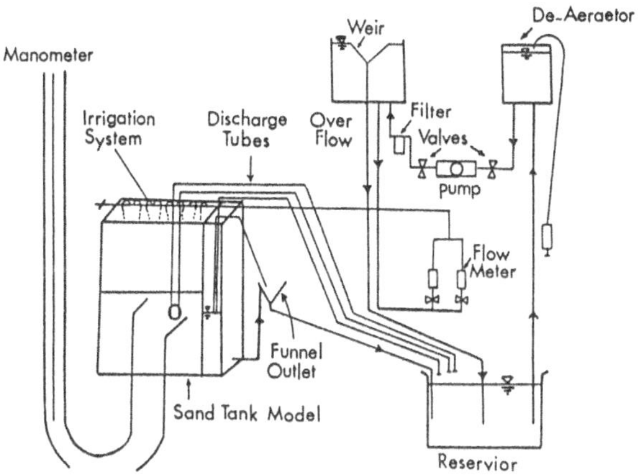

Fig. 2.--Water Supply System. Sand Tank and Manometers

A simulated irrigation system was placed at the top of the sand tank. The irrigation system represented recharge from irrigation or rainfall.

EXPERIMENT RESULTS

Using a rectangular outlet drain shape supported by the screen, the equipotential lines for the highest recharge (240 ml/min) were drawn, which gives the highest water table close to the soil surface with and without the interceptor drain functioning as shown in Figures 3 and 4, respectively.

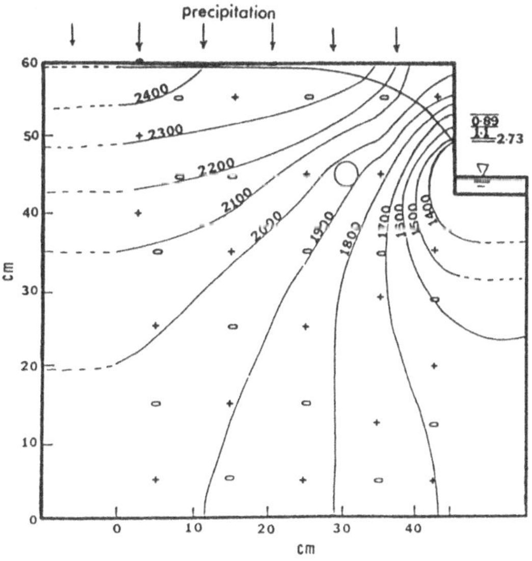

Fig. 3.--Equipotential Lines, Exit Gradients, and Water Table for Inflow-Outflow of 240 mL/min and Interceptor Drain Not Functioning

The water table was very near the soil surface when the interceptor drain was not functioning, Figure 3, and a considerable surface of seepage developed at the vertical drain face. No failure occurred because of the mechanical support of the soil with the screen system.

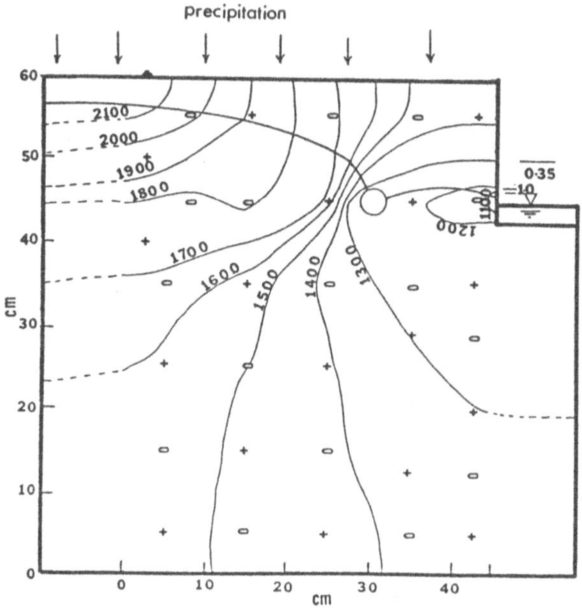

Fig. 4.--Equipotential Lines, Exit Gradients, and Water Table with Inflow-Outflow of 240 mL/min with 163 mL/min Interceptor Drain Discharge

When the interceptor drain was allowed to function, Figure 4, with the potential inside the drain being equal to the potential in the open drain, the water level in the soil decreased and the surface of seepage at the open drain face decreased dramatically. The exit gradients in the soil also decreased, due to reduced flow through the open drain bank.

Following the tests with a mechanically supported vertical drain bank, a simulated open drain with a 1:1 side slope was formed in the model. The length of the rainfall simulator was adjusted to prevent having water fall directly on the slope. The tests were repeated with and without the interceptor drain functioning. The recharge rate was increased until bank failure occurred. Figure 5 shows the position of the water table for a relatively low recharge rate when the interceptor drain is not functioning.

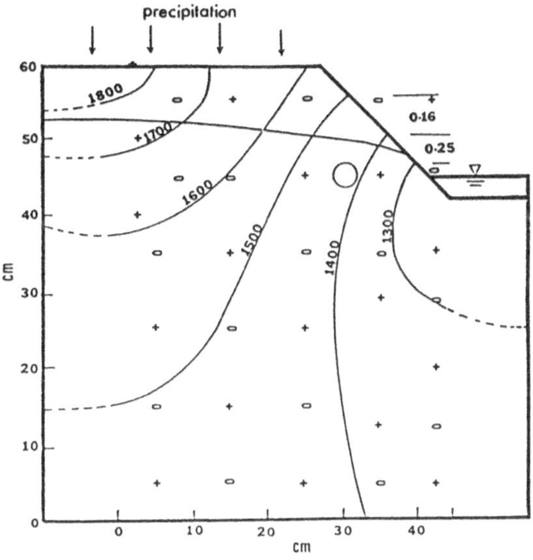

Fig. 5.--Equipotential Lines, Exit Gradients, and Water Table with Inflow-Outflow of 75 mL/min, 1:1 Side Slope, Without Interceptor Drain, and no Indication of Failure

Figure 6 shows the water table position with the interceptor drain functioning and a high recharge rate.

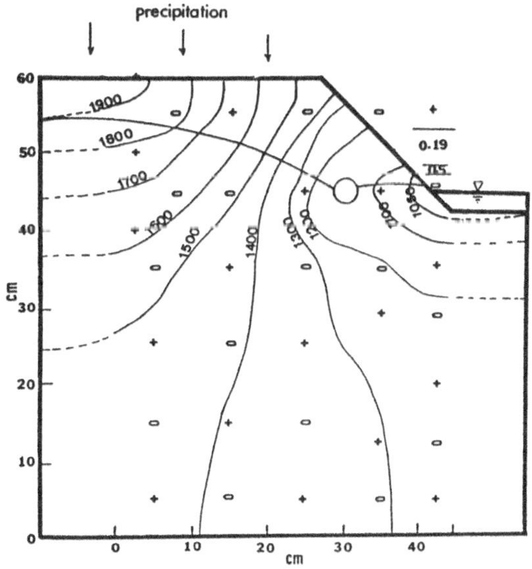

Fig. 6.--Equipotential Lines, Exit Gradients, & Water Table for Inflow-Outflow of 130 mL/min (0.37 mL/min/cm^2), 1:1 Side Slope, Interceptor Drain Functioning, and no Indication of Failure

A normal slice stability analysis was used to determine the stability of open drain slopes and their factors of safety (13). A long term stability analysis with steady state seepage was used as a reference. In order to find the minimum factor of safety, which represents the lowest resisting force, several failure circles were tried and the most critical circle was chosen. Figure 7 shows the critical circle analysis geometry for a 1:1 slope. The sliding mass was divided into 5 slices.

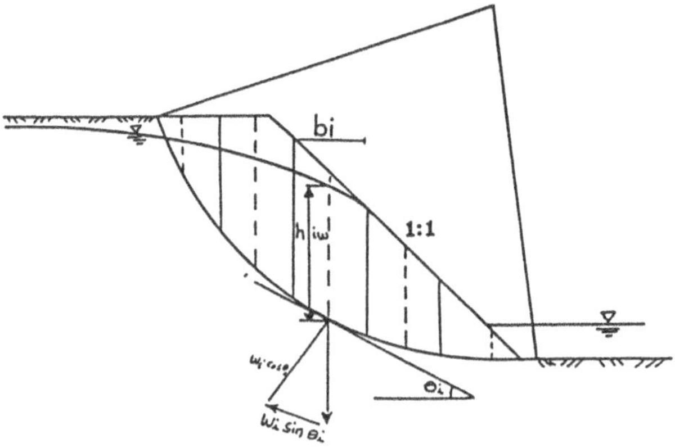

Fig. 7.--Stability Analysis Parameters

The cohesion force for the sand used was found to be equal to zero. The friction angle was found to be 39.5 degrees.

The factor of safety for the slope is defined as the ratio between the effective resisting force and the driving force and can be represented mathematically by

$$FS = \frac{\sum_{i=1}^{n} CL_i + N_i \tan \emptyset}{\sum_{i=1}^{n} W_1 \sin \theta_1} \qquad [1]$$

RESULTS AND DISCUSSION

The slope stability analyses for the conditions in physical model produced the data on slope stability shown in Table 1.

TABLE 1.--Factor of Safety for Slope Stability

Slope	Factor of Safety: Without Interceptor Drain	Factor of Safety: With Interceptor Drain
1:1	0.414	1.07
2:1	0.768	1.505
3:1	1.504	1.96

The data given in Table 1 show that the stability of the side sloping is significantly increased by using an interceptor drain close to and parallel to the open drain.

The fact that the stability of the side slope is increased by using an interceptor drain was verified using the physical model for a drain with a 1:1 side slope. The 1:1 side slope failed with a water application of 0.29 ml/min/cm^2 in the absence of an interceptor drain. The factor of safety for the side slope that failed was 0.541. The corresponding water application rate just prior to sloughing was 0.25 ml/min/cm^2.

With the interceptor drain operating, the factor of safety for the slope was 1.10 and the slope did not fail at a recharge rate of 0.37 ml/min/cm^2 (the maximum pump discharge).

When the interceptor drain was functioning in the model, the hydraulic gradient near the open outlet drain is decreased to 1.0 which is almost 1/3 of the first case, but it is still higher than the theoretical vertical flow critical gradient and so sloughing would still be expected in a drain constructed in a cohesionless soil having a steep side slope with the existence of a very high water table.

In the case of a 1:1 side slope, the hydraulic gradient with the interceptor drain operating is 0.25 which is less than critical gradient. A higher discharge was used, 0.37 ml/min/cm^2 with the interceptor drain functioning, and the hydraulic gradient was 0.50 which is less than the theoretical critical gradient. Placement of an interceptor drain close and parallel to an open outlet drain effectively stabilizes the side slope.

Where initial side slope sloughing occurred, the hydraulic gradient was 0.95 which is higher than the theoretical critical gradient.

EXIT GRADIENTS

The theoretical critical vertical hydraulic gradient to establish a quick condition in the porous medium used in the model corresponds to that for a uniform loose sand when saturated:

$$i_c = \frac{117.7 - 62.4}{62.4} = 0.886$$

The actual critical gradient for soil particle movement in the model is less than the theoretical vertical critical gradient since the direction of flow in soil has a slight downward direction and the soil particle instability therefore becomes greater. The hydraulic gradient near the seepage surface in the open drain when the interceptor drain is not functioning is 2.73 as shown in Figure 3. The actual hydraulic gradient is much higher than the vertical critical gradient (0.886) near the outlet drain vertical bank in the model and verifies the sloughing potential for banks of drains constructed in cohesionless soil. The vertical bank of the outlet drain in the model could not fail because it was mechanically confined by the screen system described earlier.

FIELD APPLICATION

A computational procedure was developed for drain location design in the field and to find the optimum distance of the interceptor drain from an open drain. Table 2 shows the results of the computational procedure for the conditions of the laboratory physical model.

TABLE 2.--Factor of Safety for Different Drain Positions and 2:1 Side Slope

Drain Location (cm)	Factor of Safety
15	1.505
30	1.549
45	1.047

The closer the interceptor is located to the open drain, the higher will be the factor of safety for the side slope; however, it appears that the exact positioning of the interceptor near the open drain is not critical.

The procedure for field application is as follows:

1. Determine a steady state recharge rate that would be used for relief drains in the area.
2. Fix a practical location for the interceptor drain. The depth of the interceptor should be the same as the hydraulic gradeline in the proposed open drain. Assume that the open drain is a closed subsurface drain for purposes of computation, to get a trial drain spacing.
3. Determine the steady state midpoint water table height from a steady state equation such as the Hooghoudt equation (4).
4. Multiply the midpoint water table height by 1.37, an empirical factor developed from the model results.

5. Put the new midpoint water table height into the Hooghoudt equation and calculate a new value for V/K (the recharge rate divided by the hydraulic conductivity of the soil).
6. Use the new V/K value in the hydraulic head form of the Hooghoudt equation:

$$y = (((1 - (4x^2/S^2))\ (S^2/4)\ (V/K))^{1/2}$$

To find the approximate water table position y in the soil for various distances x from the interceptor drain toward the proposed open drain.
7. Use the computed water table position to make normal slice slope stability analyses.
8. Repeat the process for different interceptor drain locations and select the drain location and side slope that gives a maximum factor of safety.

The field practice to be followed to implement the findings of this study are relatively simple. Before the construction of the open drain, subsurface plastic drains are laid with a trencher at the design distance from the open drain centerline. The drains are laid at the level of the design hydraulic gradeline of the open drain to be constructed. The drains begin at the centerline and curve outward until they are parallel to the centerline of the open drain at the design distance. The subsurface drains proceed upstream until their hydraulic capacity is reached, then a new subsurface drain is started. As the open drain is constructed, the downstream ends of the subsurface drains are exposed, providing an outlet and allowing them to dewater the soil ahead of the construction of the open drain. The subsurface drains are left in place as permanent parts of the system. Figure 8 is a representative sketch of the system.

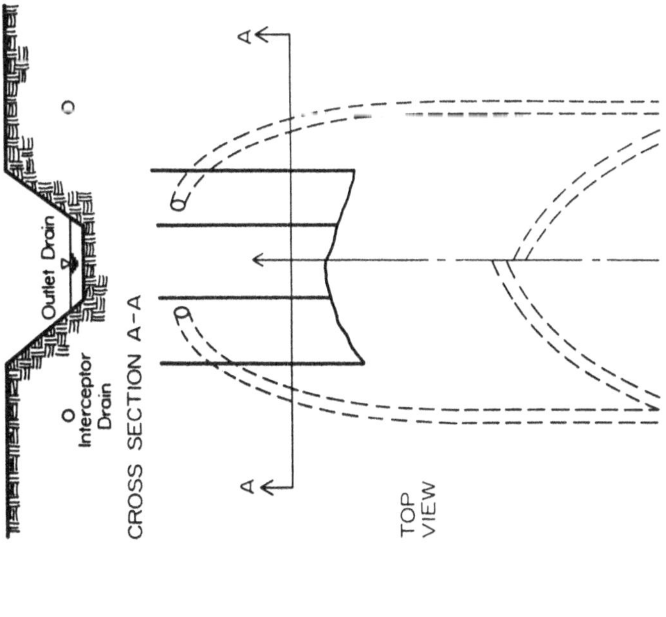

Fig. 8.--Interceptor Drain Installation

SUMMARY AND CONCLUSIONS

The objective of the work reported here was to determine the effect of a closed interceptor drain, near an open agricultural drain, on hydraulic gradient and water content of the soil near an open agricultural drain, and to develop tentative criteria for design of a closed interceptor drain system to stabilize drain banks.

A two dimensional sand tank was used to evaluate flow in the simplest possible system, i.e., a system of steady state flow between the simulated irrigation system and the interceptor and open drain.

The normal slice method was used to check the stability of the outlet drain bank. The interceptor drain changed the factor of safety for slope stability from (0.5) to (1.5).

Critical hydraulic gradient was also used to check the soil stability by comparing it to the gradient developed near the open outlet drain with and without an interceptor drain. The exit gradient was reduced 64 percent by the interceptor drain.

A tentative design procedure was developed for the placement of interceptor drains close to and parallel to an open outlet drain.

The following conclusions can be drawn.

1. Slope stability analyses for the soil condition of the physical model shows that a 2:1 side slope could be used with an interceptor drain functioning in the model and that a 3:1 side slope would be required for equivalent slope stability in the model without an interceptor drain functioning. This specific conclusion holds only for the condition of the model. The increased soil stability demonstrated in the model and confirmed by the analysis gives confidence in this method of stabilizing drain banks.

2. Hydraulic gradient analysis shows that the exit hydraulic gradient in the case of interceptor drain functioning decreases to almost 1/3 the exit hydraulic gradient that exists in the absence of an interceptor drain.

REFERENCES

1. Abdel Aziz, M. H., Bishop, Alvin, and Dunn, S. (1966) Influence of Exchangeable Ions on the Stability of Drain Banks. Transactions, ASAE, Vol. 9, No. 6, pp. 788-792.
2. Belcher, H. W., and Jerald, P. E. (1984). Open Channels in High Water Table Sands. Presented at the June 24-27 ASAE meeting, held at University of Tennessee, Knoxville (paper no. 84-2065).
3. Coote, D. R., Gibbons, J. S., and Broughton, R. S. (1984) Reducing Erosion of Channel Drains in Problem Soils of the OTTAWA-ST. LAWRENCE Lowlands of CANADA. International Commission on Irrigation and Drainage, Tempe, AZ, 317-332.

4. Little, W. C., Thorne, C. R., and Murphey, J. B. (1982) Mass Bank Failure Analysis of Selected Yazoo Basin Streams. Transactions, ASAE, Vol. 25, No. 5, pp. 1321--1328.
5. Luthin, J. N. (1970) Drainage Engineering. Wiley Eastern Company, New York, pp. 151-154.
6. Millington, R. J., and Quirk, J. P. (1959) Permeability of Porous Media. Trans. Faraday Soc., Vol. 57, pp. 1200-1207.
7. Phillip, H. Burgi, and Karaki, Susumu. (1971) Seepage Effect on Channel Bank Stability. Journal of the Irrigation and Drainage Division, ASCE, No. IR1, March pp. 59-72.
8. Richards, Sterling J., L.S. Willardson, and Sterling Davis. (1973) Tensiometer Use in Shallow Ground-Water Studies. Journal of the Irrigation and Drainage Division, ASCE, No. IR4, December, pp. 457-464.
9. Soil Conservation Service. (1977) Michigan's 1977 Watershed Progress Report. USDA Soil Conservation Service, E. Lansing, MI., 50 p.
10. Soil Conservation Service. (1977) Drainage of Open Channels. Technical Release no. 25, USDA, pp. 6.62-6.77.
11. Terzaghi, K., and Peck, R. B. (1948) Soil Mechanics in Engineering Practice. Earth Pressure and Stability of Slope, Jhone Wiley, New York, Chapter 8, pp. 342-377.
12. U. S. Department of Interior. (1980) Earth Manual. 2nd ed., United States Bureau of Reclamation, Denver, CO, Chapter 2. pp. 111-119.
13. Yang, H. Huang. (1983) Stability Analysis of Earth Slopes, Van Nostrand Reinhold Company, New York, pp. 14-25.

Unsteady Flow Computations in Open Channel Hydraulics

J. Bouwknegt
Heidemij Adviesbureau, Arnhem, Holland

INTRODUCTION

Unsteady flow in complex systems of open channel courses (networks) can be simulated by a mathematical model consisting of the so called De Saint Venant equations. A derivation of the equations may be found by Stoker (1957) and Mahmood and Yevjevitch (1975).

The numerical solutions of the De Saint Venant equations fall into one of the following categories:
- explicit finite-difference method based on the characteristic equations
- explicit finite-difference methods
- implicit finite-difference methods
- methods based on finite element schemes.

Two of the categories are investigated in this paper:
- The method of the characteristics. The two partial differential equations are transformed to 4 ordinary differential equations. A general technique for solving these 4 equations may be found by Lister.
- Implicit finite difference methods based on the 4-point scheme proposed by A. Preissmann. The equation of motion is written in two different forms. This leads to two sets of difference equations. The advantages of both methods will be discussed.

The basic approach to the three methods (characteristics, implicit difference method on two different equations of motion) will be discussed. For a hypothetical problem the three solutions are obtained and compared with the solutions of other authors. Finally the methods will be demonstrated on a very simple "network" of open channels including some construction works.

BASIC EQUATIONS

Unsteady flow in open channels can be described by the equations of continuity and motion. For the one-dimensional situation there are these equations:

Continuity:

$$\frac{\delta Q}{\delta x} + T\frac{\delta h}{\delta t} - q = 0 \qquad (1)$$

Motion:

$$\frac{\delta Q}{\delta t} + \left(gA - \frac{Q^2 T}{A^2} \right) \frac{\delta h}{\delta x} + 2\frac{Q}{A}\frac{\delta Q}{\delta x} - gA\,(S_o - S_f - S_w) = 0 \qquad (2A)$$

The equation of motion can also be written in the more compact form:

$$\frac{\delta Q}{\delta t} + \frac{\delta}{\delta x}\left(\frac{Q^2}{A}\right) + gA\frac{\delta h}{\delta x} - gA\,(S_o - S_f - S_w) = 0 \qquad (2B)$$

x =	distance along channel axis, positive in downstream direction	[m]
t =	time	[s]
A =	cross sectional area of flow	[m^2]
Q =	discharge across a section	[$m^3 s^{-1}$]
h =	waterdepth	[m]
T =	flow channel width	[m]
q =	lateral inflow per unit length	[$m^2 s^{-1}$]
g =	acceleration due to gravity	[ms^{-2}]
So=	bottom slope	[-]
Sf=	friction slope	[-]
Sw=	surface slope, due to wind friction	[-]

The space derivative of discharge of Equation 2B can be transformed as follows:

$$\frac{\delta}{\delta x}\left(\frac{Q^2}{A}\right) = \frac{2Q}{A}\frac{\delta Q}{\delta x} - \frac{Q^2}{A^2}\frac{\delta A}{\delta x} = 2\frac{Q}{A}\frac{\delta Q}{\delta x} - \frac{Q^2}{A^2}\left[\frac{dA}{dh}\frac{\delta h}{\delta x} + \frac{\delta A}{\delta x}\right]$$

the cross-section is held constant for a branch so the last term vanishes.
Substituting this results in Equation 2B gives equation 2A. A thorough treatment of these equations may be found by Cunge et al (1983).
Equation 2A will be referenced throughout this paper by: equation of motion A; and equation 2B by: equation of motion B.

The equations are solved for the two dependent variables Q and h as functions of the independent variables x and t. A solution in closed form does not exist for the equations 1 and 2A (or the other combination 1 and 2B), so numerical methods of integration must be employed. This is done by the method of characteristics on equations 1 and 2A (Lister 1966) and by an implicit finite difference scheme on equations 1 and 2A (and the combination 1 and 2B) (Fread 1973).

PRINCIPLES OF SOLUTION

The discharges, waterlevels and flow-velocities can be predicted for a system of open channels as functions of time and space, by solving the equations of continuity and motion in combination with boundary and initial conditions.
The boundary conditions are:

$$Q(x_0,t) \text{ or } h(x_0,t) \quad \text{and} \quad Q(x_n,t) \text{ or } h(x_n,t) \tag{3}$$

The initial conditions are:

$$Q(x,t_0) \quad \text{and} \quad h(x,t_0) \tag{4}$$

A system of open channels is divided into branches and nodes. In the interior nodes the initial conditions hold and in the nodes at the upstream and downstream ends the boundary conditions hold.
The method proposed in this paper divides the system into subsystems consisting of exactly one branch and two nodes. Due to the fact that in such a subsystem the interior and boundary nodes are the same, the intitial and boundary conditions are also the same. This means that the initial conditions can be neglected and only the boundary conditions are relevant to the subsystems.
When processed in a computerprogram, the whole problem is reduced to a manipulation of the boundary conditions in such a way that continuity (waterlevels and discharges) holds in the nodes.
Without a detailed treatment of the numerical aspects of the discretised equations (starting with the equations of continuity and motion) the principles will be discussed for the three methods.

Method of characteristics

The solution of the system of hyperbolic equations (1 and 2A) has been extensively described in literature. It is sufficient to give here the representation of the characteristic equations:

$$\frac{dx}{dt} = \frac{Q}{A} \pm \sqrt{\frac{gA}{T}}$$

$$\left(\frac{QT}{A} \mp \sqrt{gAT}\right)\frac{dh}{dt} - \frac{dQ}{dt} + gA(S_o - S_f) - \left(\frac{Q}{A} \mp \sqrt{\frac{gA}{T}}\right) q = 0$$

The upper sign represents the forward characteristic (C+) and the lower sign represents the backward characteristic (C-).

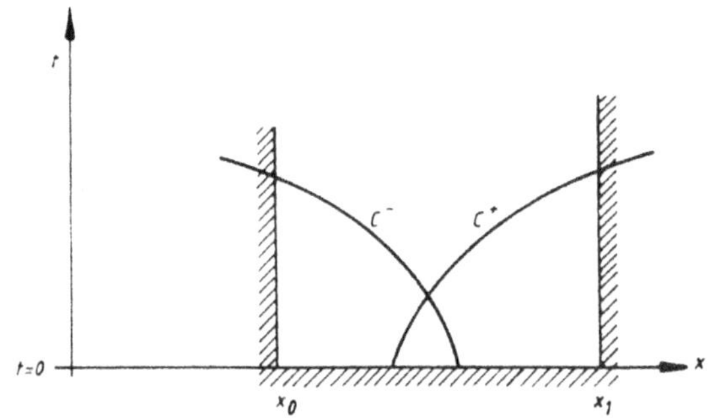

Figure 1 : Schematic representation of the characteristic method for a one - branch - subsystem

For a subsystem of only one branch the graphical representation of the forward and backward characteristics is given in figure 1. In the two nodes (at x_0 and x_1) only the result of the former timestep t_n is required. In combination with the data on the bounds it is sufficient to give in an easy way the discharge or waterlevel at the new timelevel t_{n+1} .

Finite difference methods

The development of finite difference schemes is also

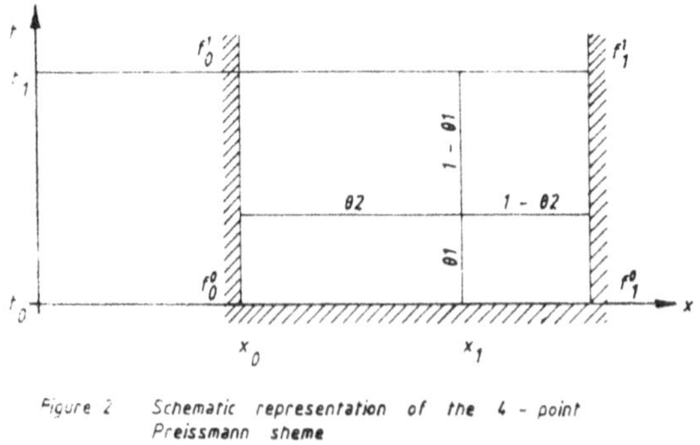

Figure 2 Schematic representation of the 4 - point Preissmann sheme

available in literature. In this paper the family of the 4-point Preissmann scheme is chosen. The calculating procedure is more complicated than with the characteristics. The results at time level t_1 are interrelated so an iterative procedure is needed to determine the discharges and waterlevels at the new time level (figure 2).

The interrelation between the results at distance x_0 and x_1 at the new time level t_1 is the major difference between explicit methods (characteristics) and implicit methods (Preissmann scheme).

The solution at a node

From figures 1 and 2 it is clear that for a one-branch subsystem a solution can be found. Furthermore it is possible to exploit the benefits of such a small system for speeding up the computations.
For a network consisting of n branches n subsystems must be considered. The branches are linked together by the nodes. An overall treatment of a network falls into two parts:
- a treatment of the solution for a branch
- an examination of the continuity in a node.

The most important aspects in relation to continuity are:
- continuity of discharges:

$$\sum_{1}^{n} Q_i = 0$$

  Input and output to a node is the same, including lateral inflows (n is number of branches).
- continuity of waterlevels. The waterlevels are equal when no construction works (weirs, culverts) exist and, in the case of constructions, a difference in head must be considered.

The demands for continuity are reached by manipulating the boundary conditions for a subsystem. Physically and mathematically the best solutions are obtained by manipulating the waterlevels.
The waterlevels are elevated when the input in a node exceeds the output and the other way round the levels are lowered. So the solution for a network consists of lowering and elevating the waterlevels in the nodes. This process is controlled by an adjusted Newton-Raphson procedure.
For all three methods the same principles are used.

APPLICATIONS

Wide Canal

A hypothetical example of a surge in a rectangular channel was studied by Chaudry et al (1973). The channel was assumed to be 16 km long with a bottom-width of 30.5 m and a bottomslope of 0.001. The initial discharge is 28.3 m3/s. A sudden increase of the discharge at the upstream end to 226.5 m3/s was assumed and held at that value for the rest of the time. Figure 3 illustrates the arrival of the wave 16 km downstream for the three methods of solution. Although the methods produce nearly the same

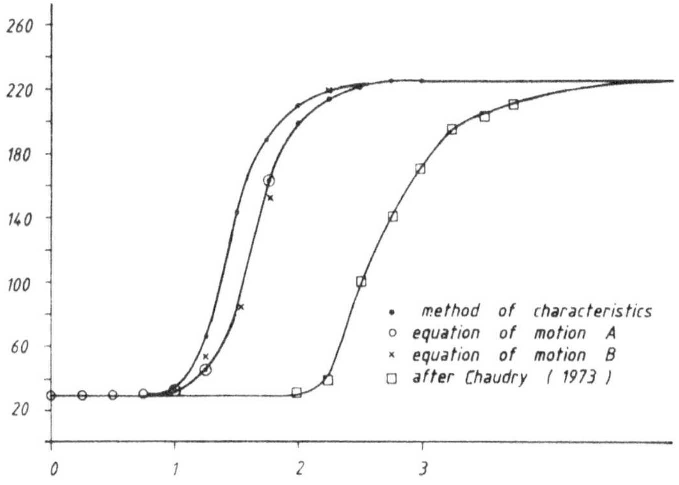

Figure 3 : Surge propagation in a canal

curves, the fitting with the original calculation differs in the time of arrival of the surge. The shape is nearly the same. The differences are partly caused by lack of data.

Channel systems

A second demonstration is a system of open channels in an agricultural area. The watercourses are small in comparison with the first example. The system consists of 6 branches with a bottomwidth varying from 0.50-0.80 m, bottomslope about 0.001 and sideslopes 1:1.5. The length of the branches is 100m each.

The initial discharge is 1 l/s and the unsteady input-wave at the upstream end is increased from 0 to 400 l/s in 5 min. It is held at that level during 25 min and the input is reduced to 0 in the next 5 min.

Figure 4 gives a schematic representation of the network lay-out and the input wave. At the first and the fourth node the results are presented with respect to the waterdepths for the 3 methods of computation.
The method of characteristics produces a higher waterlevel. This is due to the fact that an explicit method progresses only one node per timestep, so the storage

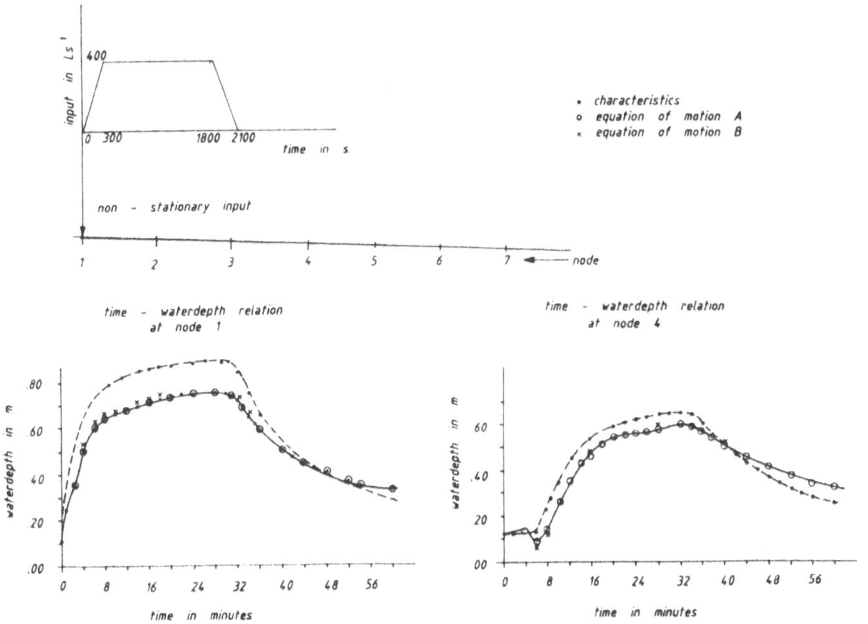

Figure 4 : Flood routing in open channels

effects are more pronounced in comparison with the two implicit methods.

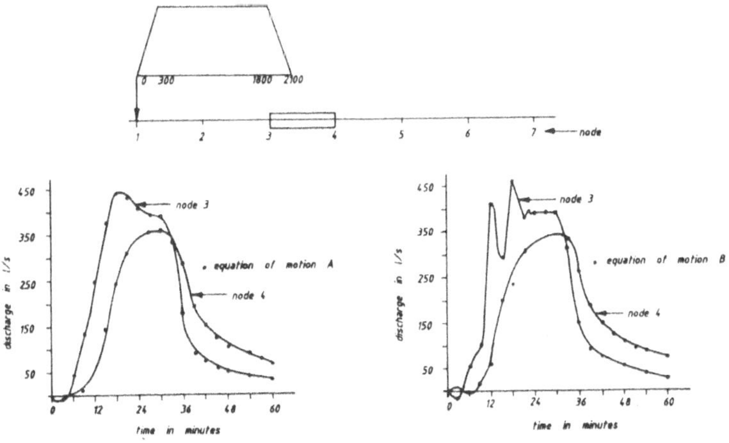

Figure 5 : Influence of storage in the system

In fig. 5 the same network is examined for storage. The third branch is widened to a bottomwidth of 3 m. At the upstream and downstream end of this branch the discharges are plotted versus time (only for the two implicit methods). It can be seen that the method based on equation of motion 2B produces some instabilities at the wave front, due to the discontinuity of the channel shape.

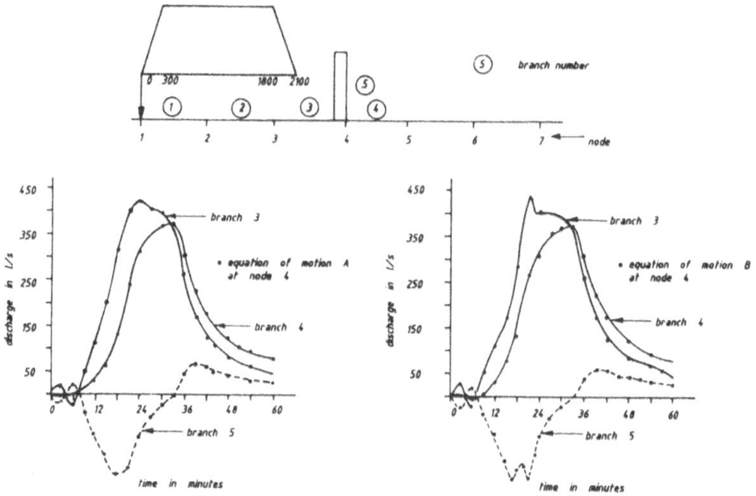

Figure 6 : Influence of storage outside the system

Figure 6 tackles the same problem but the storage is found in an extra channel outside the original system. The discharges are now only plotted for the junction-node 4. The storage is used by filling up the extra branch. This is realised by a negative flowdirection. After passage of the wavefront the input to the storage diminishes and the storage is emptied.
The entrance to the storage basin is more difficult than for the situation of fig. 5. This explains the small differences in the plotted results. The storage is located 100 m downstream so the time of arrival is shifted.

DISCUSSION AND CONCLUSIONS

The routing technique, roughly explained in this paper has been used for a variety of problems. The method has some advantages in relation to matrix-iteration techniques:

- programming is rather simple

- a wide variety of initial and boundary conditions can be handled
- unsteady behaviour of many sources of different origins are possible
- the complexity of the channel system is unimportant
- the amount of computer storage needed to run a complex network is minimal
- the demand on cpu-time of the computer does not exceed the demands of other methods

It is not quite clear that one of the two notations of the equation of motion is preferable. In the situations where the characteristic method can be used (ic. all branches of nearly the same length) this method produces the smoothest curves in the shortest cpu-time. Only 3-4 iterations per node are needed. A disadvantage of this method is the limited size of the timestep. The other two methods need about 4-6 iterations per node, but this depends on the complexity of the network, the existence of constructions (weirs, culverts) and the form of the unsteady load (input along the branches or in the nodes).
In general the algorithm concentrates the computation efforts in the wave front and follows in a sense the unsteady behaviour in the network. This guarantees optimal use of cpu-time.

REFERENCES

Chaudry,Y.M. and D.N. contractor, Application of the Implicit Method to Surges in Open Channels, Water Resources Research, Vol. 9,no.6, December 1973

Cunge,J.A,,F.M.Holly(Jr), A. Verwey,1983, Practical aspects of Computational River Hydraulics, Pitman, London.

Fread, D.L. 1973, Technique for Implicit Dynamic Routing in Rivers with Tributaries, Water Resources Research,vol.9, no. 4, August 1973.

Lister,M,1966 Numerical solution of hyperbolic partial differential equations by the method of characteristics: Mathematical Methods of Digital Computers by A.Ralston and H.S.Wilf, Jhon Wiley and Sons.

Mahmood,K. and V. Yevjevitch,1975, Unsteady flow in open channels,Volumel,Water Resources Publications, Fort Collins, Colorado.

Stoker,J.J. 1957, Water Waves, Interscience, New York.

Strausz,K.H., 1979, Numerische Berechnung instationarer Flieszvorgange in teilgefuellten Kreisrohr, Mitteilunger Heft nr. 30; Hydraulik und Gewasserkunde, Technische Universitat Munchen.

SECTION 4 INFLUENCE OF DRAINAGE ON SOILS

The Influence of Salts on Physical Properties of Soils

A.F. Al-Ani
Institute of Agricultural Technology, Baghdad, Iraq

INTRODUCTION

Salts dissolution in water leads to changes in physical and chemical proporties of the water such as density , viscosity , surface tension , reaction. .. etc. These ought to influence the movement of the water in the soil . Moreover, when the solution comes in contact with the soil , some changes could be expected in the physico - chemical properties of the soil , due to the exchange reactions between the soil solution and the exchange complex . This causes a change in electro - kinetic forces of the soil - water system . These changes will no doubt affect the physical properties of the soil such as :hydraulic conductivity , moisture tension , infiltration rate, evaporation rate , capillary movement ...etc . (AL - Ani , A.F., 1969) .

I- THE INFLUENCE OF SALTS ON THE SOIL MOISTURE TENSION

The soil sample used in this study was an un - disturbed calcarious silt clay .

The soil samples were capillary saturated with one of a 50 gm / l solutions of Na_2CO_3 , $NaHCO_3$, Na_2SO_4 , $NaCl$ + Na_2SO_4 with a weight ratio of Cl^-: $SO_4^{=}$ = 2 . The samples were placed on conical porous ceramic plates . The plates were joined to a vacuum system , mercury barometer and a buret to measure the extracted solution . In order to avoid evapora - tion , samples were covered with a thin rubber film. (AL - Ani, A.F., 1967) .

Results show that , soil samples treated with Na_2CO_3 solution contain more moisture than that treated with $NaHCO_3$ and Na_2SO_4 ; less moisture re - main in the control . The latter contains more moisture than the one treated with a mixture of NaCl + Na_2SO_4 solution . So it is possible to arrange the salts according to their influences on the moisture tension as follows :

$$Na_2CO_3 > NaHCO_3 > Na_2SO_4 > NaCL > H_2O > Na_2SO_4 + Na_2Cl$$

(Fig. 1) .

From the obtained results , it is possible to make the following general conclusions :

1- Due to salinization a change in the matric pressure takes place .
2- The changes in the matric pressure can be explained through the coagulating or dispersing effect .
3- The changes in the matric pressure are more clear in the heavy soils .

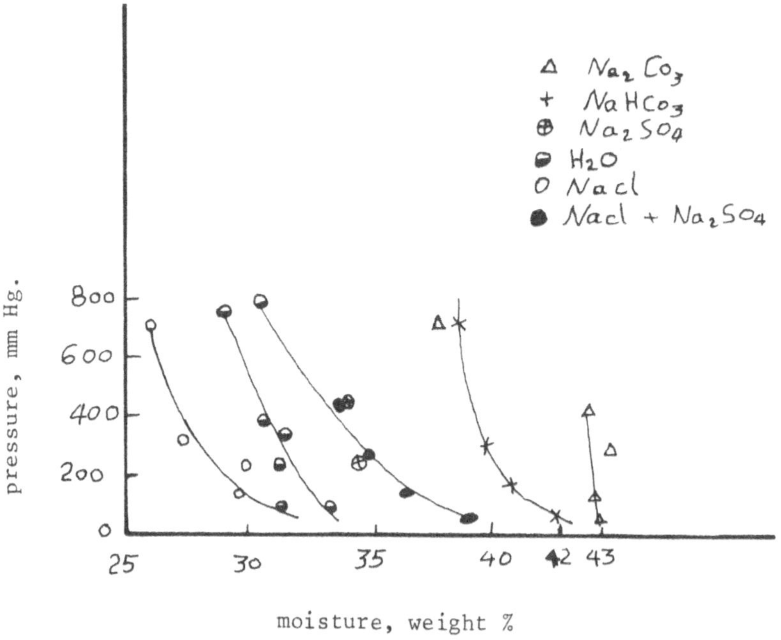

Figure 1 Influence of salts on moisture tension of soil - moisture .

II- EFFECT OF SALTS ON THE HYDRAULIC CONDUCTIVITY

Undisturbed , nonsaline silty clay samples were taken from the Delta of Tigris and Euphrat. The samples were saturated for 24 hours by capillarity with 0.25 N solutions of NaCl , Na_2SO_4 , Na_2CO_3 , $CaCl_2$ and a mixure of these solutions with the same concentration . Tap water was used as a control treatment . Hydraulic conductivity values were determined in twenty replicates for each treatment following the constant hydraulic head permeameter method.(AL - Ani, A.F., 1972 and AL - Ani et al , 1982)

Measurements were carried out by using the previously mentioned solutions and continued till a constant value was reached . This value was considered K_s . Each solution was replaced by tap water and the same procedure was repeated . The obtained values were taken as K_w . The value K_w / K_s was considered to represent the change in the hydraulic conductivity which might have taken place during leaching .

The results showed that the K_s value for the Na_2CO_3 treatment was zero and the same value was obtained when the sodium carbonate solution was replaced by tap water . The K_s value for the sodium chloride and sodium sulphate treatments were 2 and 1 cm/day respectively . These values reduced to zero by using tap water . For magnisium sulphates and calcium chloride , K_s value were 2 and 80 cm / day respectively . The K_w value remained constant when the magnesium sulphate solution was replaced by tap water , K_w value was equal to that obtained when only tap water was used as a control. The K_w value for calcium chloride treatment was decreased to about 50 % of the K_s value but it is still very high as compared to the other K_w values . (Table 1)

Table 1:
The influence of salts on the hydraulic conductivity of the soil .

Salts	Hydraulic conductivity cm/day									
Days/	1	2	3	4	5	6	7	8	9	10
Sodium Carbonate	00	00	00	00	00	00	00	00	00	00
Tap water	00	00	00	00	00	00	00	00	00	00
Sodium Sulphate	31	25	--	--	9	--	--	7	7	--
Tap water	00	00	00	00	00	00	00	00	00	00
Sodium Chloride	38	16	17	15	13	--	16	16	17	16
Tap water	00	00	00	00	00	00	00	00	00	00
Magnesium Sulphate	21	23	24	--	22	19	21	17	20	18
Tap water	19	19	17	18	--	--	15	14	--	--
Calcium Chloride	81	64	85	83	--	--	72	76	80	--
Tap water	49	49	--	59	49	49	46	47	53	--
Mixture of Salts	24	25	--	21	20	20	19	--	--	--
Tap water	5	6	6	--	--	--	--	--	--	--
Tap water(control)	17	18	20	25	24	--	--	--	--	--

III. THE INFLUENCE OF THE EXCHANGEABLE SODIUM ON THE WATER PERMEABLITY OF THE SOIL

This study was carried out on undisturbed and disturbed soil samples taken from midle and southern parts of Iraq. The samples were leached with 10 cm of o.5 N mixed solutions of calcium and magnesium chlorides. The permeability coeffecient was measured after leaching using the same solution P_o^-. Later on, each sample was leached with 10 cm of o.5 N of a mixture of calcium, magnesium and sodium chlorides, with different Sodium Adsorption Ratio (SAR). The gram-molecular concentration of calcium and magnesium was constant in all cases.

The SAR values varied from 0 to 100. The permeability was measured in each case with the solution and with irrigation water. The ratio P_r / P_o used as an indicator to the changes in the permeability values that is caused by the solutions. (Kamaladin, et al, 1962)

When the treatment is with SAR equal to zero the P_o / P_o^- is about 0.7. The permeability coeffecient values of the samples treated with solutions with SAR higher than 30, were reduced markedly. Solutions with SAR equal to 50 reduced the ratio P_{50} / P_o^- to about 0.03. In this case the P_{50} value became 0.03 cm / h after 48 hrs of continous leaching. This means that the samples with SAR equal to 50 and less can be leached. (Table 2)

IV. THE INFLUENCE OF SALTS ON THE EVAPORATION RATE OF WATER FROM THE SOIL

Nonsaline silt clay soil sample was used for this study. The sample was air-dried and passed through a 2 mm sieve. Portions of 200 gm of the sample were leached with 2 litres of 0.5 N solution of one of the following salts: NaCl, Na_2CO_3, $NaHCO_3$, $CaCl_2$, Na_2SO_4 and distilled water(as control). (AL - Ani, A.F, 1976)

The evaporation rate was determined by a Vacuum Moisture Meter. The amount of the evaporated water was related to the essential moisture content of the sample, in order to avoid the differences between the samples due to different moisture content.

Table 2:
The influence of exchangeable sodium on the permeability coeffecient .

Sampling location	Samples statement	SAR (r)	P_r	$\bar{P_0}$	$P_r/\bar{P_0}$
Dehna	Undisturbed	0	1.21	1.40	0.85
Abu Ghraib	=	0	0.20	0.31	0.64
Dehna	=	10	1.30	1.60	0.81
Abu Ghraib	=	10	0.43	0.77	0.55
Also	Disturbed	10	0.58	0.90	0.65
Dehna	Undisturbed	20	0.97	1.70	0.55
Also	=	20	0.91	1.70	0.53
Abu Ghraib	=	20	0.30	0.60	0.50
Also	Disturbed	30	0.23	0.70	0.30
Shamiya	=	30	3.50	14.50	0.24
Abu Ghraib	Undisturbed	40	0.07	0.63	0.11
Also	=	40	0.08	2.30	0.03
Dehna	=	40	0.16	1.90	0.08
Shamiya	Disturbed	50	0.70	16.00	0.04
Abu Ghraib	=	50	0.02	1.00	0.02
Also	Undisturbed	50	0.13	4.40	0.03
Hammar	Disturbed	50	0.03	3.30	0.01
Ali Gharbi	=	50	0.03	6.70	0.004
Ammara	=	50	0.02	12.50	0.002
Hammar	=	50	0.07	5.90	0.010
Dehna	Undisturbed	60	0.04	1.40	0.010
Also	=	60	0.01	1.00	0.010
Abu Ghraib	=	60	0.01	0.33	0.045
Shamiya	Disturbed	70	0.03	16.00	0.050
Abu Ghraib	=	70	0.01	1.00	0.010
Also	Undisturbed	70	0.01	4.50	0.002
Shamiya	Disturbed	100	0.33	15.00	0.020
Abu Ghraib	=	100	0.01	1.00	0.010
Also	Undisturbed	100	0.01	4.80	0.020

The results gained from these experiments showed that the salts did not equally increase the evaporating rates . this happened in the following sequence :

$$CaCl_2 > NaCl > Na_2SO_4 > NaHCO_3 > H_2O > Na_2CO_3 .$$

This may be due to the coagulation influence of $CaCl_2$ solution on the soil colloids , which cause a decrease in the amount of adsorbed water . The Na_2CO_3 solution dispersed soil colloids and this increased the amount of adsorbed water . Adsorbed water needed additional energy to evaporate .

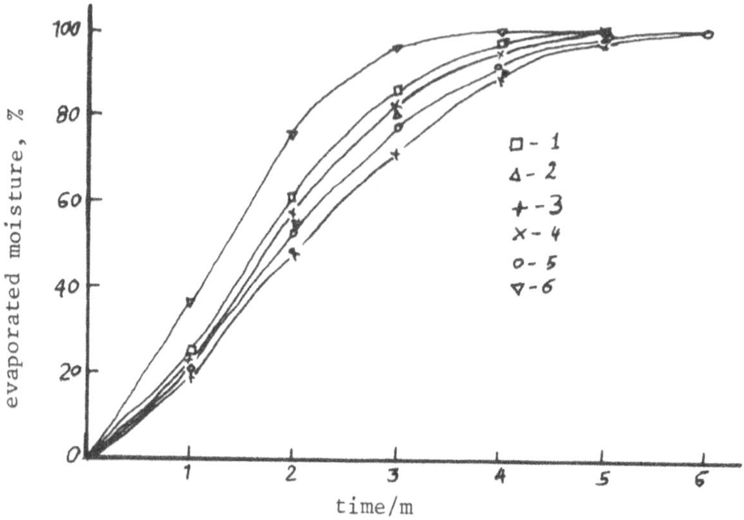

Figure 2
Influence of salts on the evaporation of water from soil .
1. NaCl ; 2. $NaHCO_3$; 3. Na_2CO_3 ; 4. Na_2SO_4 ; 5. H_2O ; 6. $CaCl_2$

REFERENCES :

1- AL - Ani,A.F. (1972) The influence of salts on the Hydraulic Conductivity of the Soil .Biological Science , No. 1 .

2- AL - Ani,A.F. (1976) The influence of salts on the Evaporation Rate of Soil Moisture . Water Resources , No. 4 .

3- AL - Ani,A.F. (1969) Some Factors to be Considered in Evaporation of Soil Moisture Movement. 6th Arab Scientific Confererence,Damascus 355 - 363 .

4- AL - Ani,A.F. (1967) The influence of Salts on the Soil Moisture Tension . Visnik M.S. U., "Biology and Soil Science " No. 2 .

5- AL - Ani,A.F.(Daghistani ,S.R., and Takessian , B.A. (1982) Effect of Different Salts on the Hydraulic Conductivity of Soil . International symposium " Polders of the World , Lelystad , the Netherland , Vol. I , 418 - 423 .

6- Kamaladin , A.R., Abdullah (AL-Ani) A.F. and Abdul Wahab , T. The influence of exchangeable Sodium on the Permeability of the Soil . "International Symposium on Geohydrology " Tashkent , USSR.,119-122.

Methods for Estimating the Drainable Porosity in Nonsteady Drainage Formulas

Weizhen Zhang and Yufang Zhang
Wuhan Institute of Hydraulic and Electrical Engineering, Wuhan, Hubei, China

INTRODUCTION

The drainable porosity is one of the most important parameters for subsurface drainage design. At present time in most drainage formulas the specific yield is taken to be a constant value. As the existing concepts of specific yield differ greatly, the methods for estimating this parameter also vary. When the shallow water table declines mainly due to evaporation, the drainable porosity (taken as constant value) can be determined from the analysis of groundwater regime observation data by fitting the empirical formula for evaporation $\varepsilon/\mu = \varepsilon_0/\mu(1- \Delta/\Delta_0)^n$ (Y. Zhang 1985). When the delayed gravity response is considered, the drainable porosity usually is estimated through the analysis of the pumping test data (N. Boulton 1954). In this paper the concepts of complete (or steady), instantaneous and average specific yields are discussed. Through laboratory experiments and numerical simulation the pattern of the variation of the specific yield under the conditions of abrupt and gradual drawdown of water table is examined. The results of numerical simulation are compared with experimental data. The investigation shows the complete drainable porosity depends upon the depth to water table and the rate of evaporation (or infiltration). The instantaneous drainable porosity is a function of the rate of the water table drawdown. When other conditions remain the same, the drainable porosity increases with the increase of the water table depth and the intensity of evaporation. At the same drawdown the greater the rate of the water table drop the less is the specific yield.

THE CONCEPTS OF DRAINABLE POROSITY

The concepts of drainable porosity should be clarified before the investigation of the pattern of its variation.

The complete (or steady) drainable porosity

The complete drainable porosity is the volume of water released per unit soil volume, when the process of drainage is completed, it is denoted by μ_d, S or S_y. In the literature of drainage, the drainable porosity usually is defined as the difference between the saturated water content and the field capacity of the soil. From the current theory of soil physics we know that the water content which can be retained in the soil is not a constant value, it depends upon the elevation above the water table and is characterized by the retention curve (or soil water characteristic curve), hence the drainable porosity should be determined from the following expression:

$$\mu_d = \theta_s - \theta(d)$$

where d is the water table depth. Some typical soil water characteristic curves and the graphs depicting the relation between the drainable porosity and the water table depth are shown in Fig. 1. It can be seen from the graphs that the steady drainable porosity increases with the augment of the water table depth.

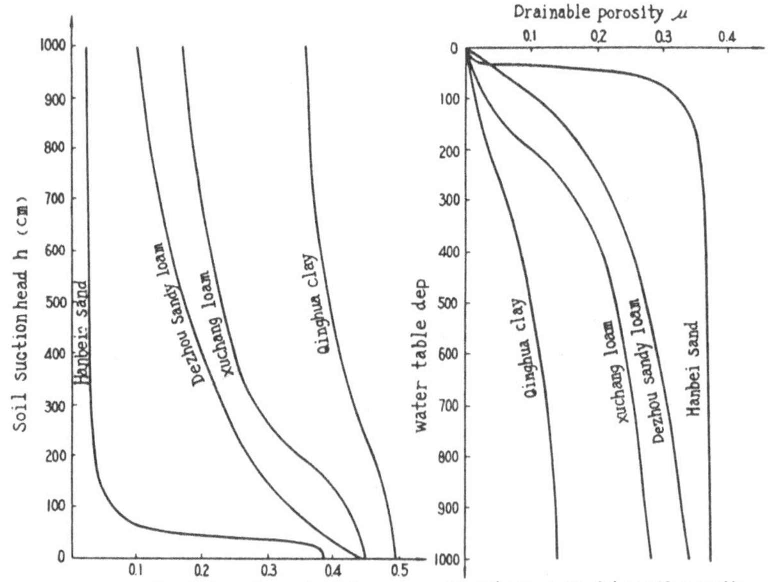

volumetric water content θ — Dependence of final drainable porosity

Soil water characteristic curves — on water table depth

Figure 1

The instantaneous drainable porosity

When the water table is falling continuously, above the phreatic surface the profile of the soil water content will also be changing from time to time, and the steady water retention curve can never be attained. In this case there exists no steady drainable porosity. The thickness of the layer of water released from the soil in the small increment of time Δh divided by the small increment of drawdown ΔL is called the instantaneous drainable porosity μ_t , or for simplicity, μ. Due to delayed gravity response the instantaneous drainable porosity is always less than the steady one.

The average drainable porosity

The average drainable porosity $\bar{\mu}$ is defined as the total volume of water released (expressed in thickness of water layer h_d) in a period of time t divided by the total drawdown of water table in this period, L:

$$\bar{\mu} = h_d/L$$

In case of instantaneous drawdown there exists only the average specific yield in the period of t after the drawdown.

The free porosity (or deficit to saturation)

Under the condition of rising water table, the volume of water required to fill up the unsaturated pore in the soil divided by the height of the rise of the water table is defined as the free porosity (unsaturated porosity or deficit to saturation). The free porosity and the drainable porosity are two different concepts, they should not be lumped together. In the following only the latter will be studied.

THEORY

The release of water from the soil profile under the condition of water table drawdown is the consequence of the soil moisture movement in the saturated-unsaturated zone. Hence in the following the investigation of the pattern of variation of drainable porosity will be carried out by solving the soil water flow equation.

The governing equation and initial condition

For one-dimensional saturated-unsaturated flow the following h-dependent equation is used:

$$c(h)\,\frac{\partial h}{\partial t} = \frac{\partial\left[k(h)\frac{\partial h}{\partial z}\right]}{\partial z} - \frac{\partial k(h)}{\partial z} \tag{1}$$

where h - soil pressure head,
k(h) - hydraulic conductivity of soil,
c(h) - water capacity of soil,
t - time,
z - vertical coordinate, positive downward.

For the problem of drainage, an arbitrary soil pressure head distribution

$$h(z,0) = h_o(z) \tag{2}$$

can be taken as the initial condition.

<u>The upper boundary conditions</u>
In case of drainage without evaporation, the following boundary condition can be used:

$$z = 0, \quad k(h)(\frac{\partial h}{\partial a} - 1) = 0 , \quad \frac{\partial h}{\partial z} = 1 \tag{3}$$

In case of drainage with evaporation, the boundary condition is:

$$h(0,t) \geq h_c \quad , \quad \varepsilon = \varepsilon_o$$

$$h(0,t) < h_c \quad , \quad \varepsilon = \varepsilon_o(a+b) \tag{5}$$

$$q(0,t) = -k(h)(\frac{\partial h}{\partial z} -1) = - \varepsilon \tag{6}$$

where

ε_o - evaporation from water surface,

h_c - critical soil pressure,

a,b - empirical constants,

q - soil water flux.

<u>The lower boundary conditions</u>
Depending upon the pattern of groundwater drawdown usually there are three major kinds of lower boundary conditions. For instantaneous drawdown

$$t > 0, \quad z = L, \quad h(L,t) = 0 \tag{7}$$

For time-dependent drawdown

$$z = d(t) \ , \quad h(d,t) = 0 \tag{8}$$

in case of constant rate of drawdown

$$d(t) = d_o + vt \tag{8'}$$

in case of drawdown induced by drains or tube wells, d(t) is a function of the water level in drains or the discharge from the tube wells and the location of the soil profile studied.

When the soil profile is underlain by an impervious barrier, the boundary condition is

$$z = L \ , \quad \partial h/\partial z = 1 \tag{9}$$

When the profile is underlain by a semipervious layer and the piezometric head of the underlying aquifer is h_g , the thickness and the hydraulic conductivity of the semipervious layer are δ and k_2 respectively, the boundary condition is

$$z = L, \quad q = -k(h)\left(\frac{\partial h}{\partial z} - 1\right) = \frac{k_2}{\delta}\,[h(L,t) + \delta - h_g] \qquad (10)$$

The finite difference equation

When the finite difference method is used for the solution of equation (1), the implicit finite difference equation is:

$$c_i^{j+\frac{1}{2}} \frac{h_i^{j+1} - h_i^j}{\Delta t} = k_{i+\frac{1}{2}}^{j+\frac{1}{2}} \frac{h_{i+1}^{j+1} - h_i^{j+1}}{(\Delta z)^2} - k_{i-\frac{1}{2}}^{j+\frac{1}{2}} \frac{h_i^{jH} - h_{i-1}^{jH}}{(\Delta z)^2} - \frac{k_{i+\frac{1}{2}}^j - k_{i-\frac{1}{2}}^j}{\Delta z} \qquad (11)$$

where

i - the sequential number of spatial steps counted from the soil surface, $i = 0,1,2,\ldots,N$,

j - the ordinal number of time steps, $j = 0,1,2,\ldots$

$$c_i^{j+\frac{1}{2}} = [c(h_i^j) + c(h_i^{j+1})]/2,$$

$$k_{i+\frac{1}{2}}^{j+\frac{1}{2}} = [k(h_{i+1}^{j+\frac{1}{2}}) + k(h_i^{j+\frac{1}{2}})]/2,$$

$$k_{i-\frac{1}{2}}^{j+\frac{1}{2}} = [k(h_i^{j+\frac{1}{2}}) + k(h_{i-\frac{1}{2}}^{j+\frac{1}{2}})]/2,$$

$$h_i^{j+\frac{1}{2}} = (h_i^{j+1} + h_i^j)/2$$

Equation (11) can also be written in the following form:

$$E_i\, h_{i-1}^{j+1} + F_i\, h_i^{j+1} + G_i\, h_{i+1}^{j+1} = H_i \qquad (12)$$

where

$$E_i = k_{i-\frac{1}{2}}^{j+\frac{1}{2}}, \qquad F_i = k_{i+\frac{1}{2}}^{j+\frac{1}{2}} + k_{i-\frac{1}{2}}^{j+\frac{1}{2}} + rc_i^{j+\frac{1}{2}},$$

$$G_i = -k_{i+\frac{1}{2}}^{j+\frac{1}{2}}, \qquad H_i = rc_i^{j+\frac{1}{2}}\, h_i^j + \Delta z(k_{i-\frac{1}{2}}^{j+\frac{1}{2}} - k_{i+\frac{1}{2}}^{j+\frac{1}{2}}),$$

$$r = (\Delta z)^2/\Delta t$$

COMPARISON OF THEORY TO EXPERIMENT

To verify the mathematical model and the method of computation, the results of numerical simulation based on equation (12) are compared with experimental data. The experiments on drainable porosity under the condition of groundwater drawdown at constant rates were carried out by D. Chen and Z. Shen (1965) on a soil column. The soil column was filled with fine sand, the water retention curve and the hydraulic conductivity of which are characterized by the following expressions respectively:

$$\begin{aligned} h &\geq -22 \text{ cm}, & \theta &= 0.37, \\ -22 \geq h &\geq -36 \text{ cm}, & \theta &= 0.37 - 0.0005612(|h| - 22)^2 \qquad (13) \\ h &< -36 \text{ cm}, & \theta &= 2.04885/(|h| - 27) + 0.03275 \end{aligned}$$

$$\begin{aligned} h &\geq 0, & k(h) &= 0.87 \text{ cm/min}, \\ 0 \geq h &\geq -30 \text{ cm}, & k(h) &= 0.87e^{0.00631h} \text{ cm/min}, \qquad (14) \\ h &< -30 \text{ cm}, & k(h) &= 3793e^{0.3674h} \text{ cm/min}. \end{aligned}$$

Before the experiment the soil was saturated with water (water table was raised to the soil surface). Then the water table was lowered at a constant rate 1 cm/min to 70 cm below the soil surface. The soil moisture content was measured by the γ-attenuation method. The drainage discharge and the total amount of water released were measured by volumetric method. The variation of observed instantaneous drainable porosity (drainage discharge per unit area divided by rate of drawdown) and the volume of water released with time is shown in Fig. 2. When equation (12), the boundary conditions (3) and (8) and the parameters depicted by (13) (14) are used for numerical simulation, the instantaneous drainable porosity and total volume of drainage water can be obtained. The results of numerical simulation are compared with the observed data on Fig. 2. Their conformity is quite satisfactory.

THE PATTERN OF VARIATION OF DRAINABLE POROSITY

The soil studied is Xuchang sandy loam, its water retention curve $\theta(h)$, water capacity $C(h)$ and hydraulic conductivity $k(h)$ as functions of pressure head h respectively are:

$$h \geq 0, \quad \theta(h) = 0.448, \quad C(h) = 0,$$

$$0 > h \geq -200 \text{ cm}, \quad \theta(h) = (194\,782 - h^2)/434782, \quad C(h) = -2h/434782 \text{ 1/cm},$$

$$h < -200 \text{ cm}, \quad \theta(h) = 0.4615e^{0.0044176h} + 0.1828, \quad C(h) = 0.00184e^{0.00442h} \text{ 1/cm},$$

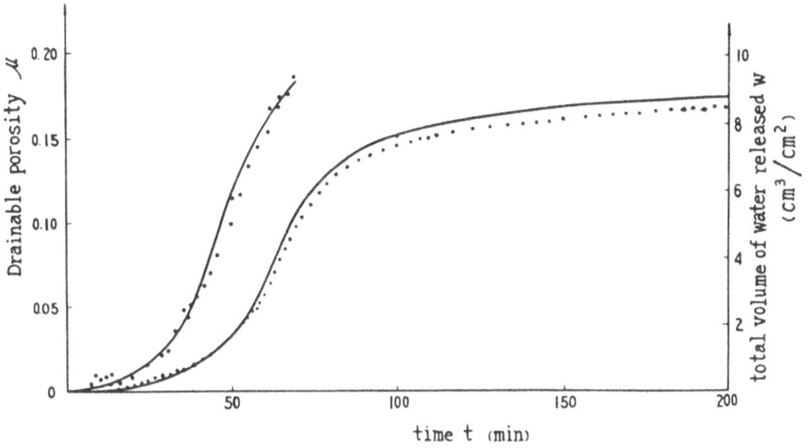

Figure 2

$$h \geq 0, \quad k = 0.013 \text{ cm/min},$$

$$0 > h \geq -15.233, \quad k(h) = 0.013e^{0.14h} \text{ cm/min}$$

$$h < -15.233 \text{ cm}, \ k(h) = 0.0023e^{0.0263h} \text{ cm/min}$$

Through the numerical simulation the factors affecting the drainable porosity, such as the initial depth to water table, the pattern of drawdown of groundwater and the evaporation are investigated.

The influence of initial water-table depth on drainable porosity

The drainable porosity for two initial water-table depths d = 0 and d = 400 cm with the rate of uniform drawdown 0.01 cm/min) has been obtained and is shown in Fig. 3. Fig. 3 indicates at the beginning of water table drawdown, the value of drainable porosity for greater initial depth (d = 400 cm) is much higher than that of the zero initial depth. However, with the increase of time (or total drawdown). the difference of values of drainable porosity for different initial depths gradually diminishes.

The influence of the rate of the water table drawdown

For the case of drainage without evaporation the drainable porosity for 4 rates of drawdown v = 0.005, 0.01, 0.02, 0.04 cm/min. is investigated. The equations (3) and (8') are used as upper and lower boundary conditions in numerical simulation, The graphs depicting the relationship between drainable poro-

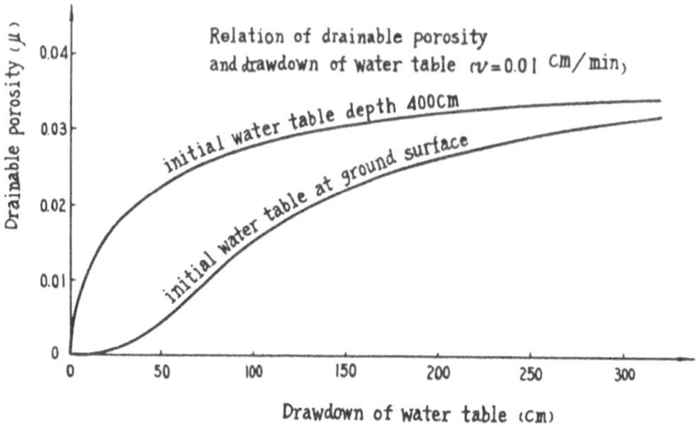

Figure 3

sity and the total drawdown for different rates of water table drop are shown in the lower half of Fig. 4 by the dotted lines. The graphs show that at the same total drawdown the greater the rate of water table drop, the less the drainable porosity. With the decrease of the rate of the water table drop the drainable porosity increases. However, the instantaneous drainable porosity is far less than the steady one.

Boulton (1954) proposed an empirical expression for delayed gravity yield due to unit drawdown of groundwater at t = 0:

$$q = \alpha \mu_d e^{-\alpha t}$$

where q - the intensity of yield due to unit drawdown
α - the index of delayed yield,
μ_d - the complete drainable porosity of soil.

When water table drawdown is time-dependent (with the rate of drawdown v), from Boulton's empirical expression we obtain the equation for instantaneous drainable porosity:

$$\mu_t = \mu_d(1 - e^{-\alpha t})$$

According to experiments conducted on soil columns (M. Cai 1983) numerical simulation (Y. Pai 1983) and our investigation, the above empirical expressions are valid only for a limited range of rate of groundwater drawdown. For instance, the maximum adaptable relative rate of drawdown v = v/k_s (k_s is the saturated hydraulic conductivity of soil) for Hanbei fine sand is 0.5, for Botany sand it is only 0.1.

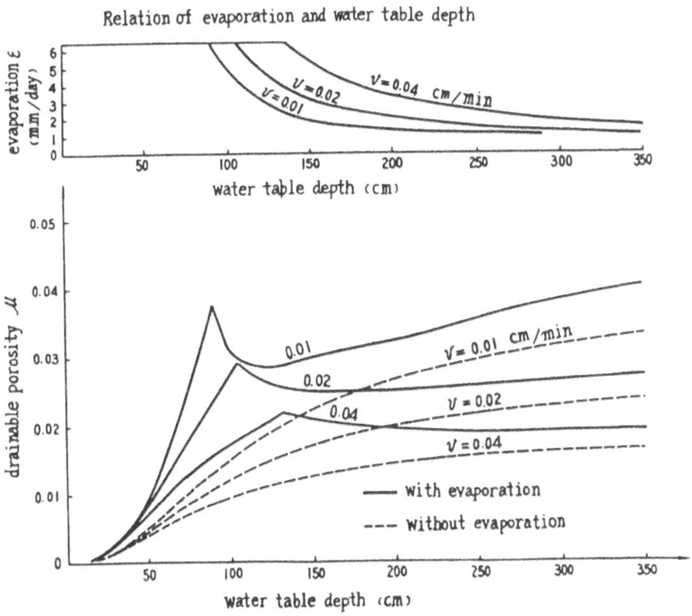

Figure 4

The influence of evaporation on drainable porosity

For the case of drainage in presence of evaporation, the following upper boundary conditions are used:

$$z = 0, \quad q = -k(h)(\partial h/\partial z - 1) = -\varepsilon ,$$

$$h \geq -213.917 \text{ cm}, \ \varepsilon = 0.65/1440 \text{ cm/min},$$

$$h < -213.917 \text{ cm}, \ \varepsilon = (3.25\theta-0.4702)/1440 \text{ cm/min}.$$

The resulting graphs depicting the relation of drainable porosity and the total drawdown for different rates of water table drop in presence of evaporation are shown in the lower half of Fig. 4 by the solid lines. The graphs showing the variation of evaporation with total drawdown for different rates of water table drop are given in the upper half of Fig. 4. It can be seen, with the same rate of drawdown the drainable porosity with evaporation is greater than that without evaporation, and the higher the intensity of evaporation the greater is the drainable porosity.

For a Dezhou sandy loam, when the drawdown of groundwater is solely induced by evaporation, the simulated drainable porosity is greater than those drainable porosities when the rate of drawdown is 0.01 cm/min. with and without evaporation, as shown in Fig. 5.

CONCLUSION

The complete drainable porosity of a certain soil for a given water table depth d is the difference between the saturated

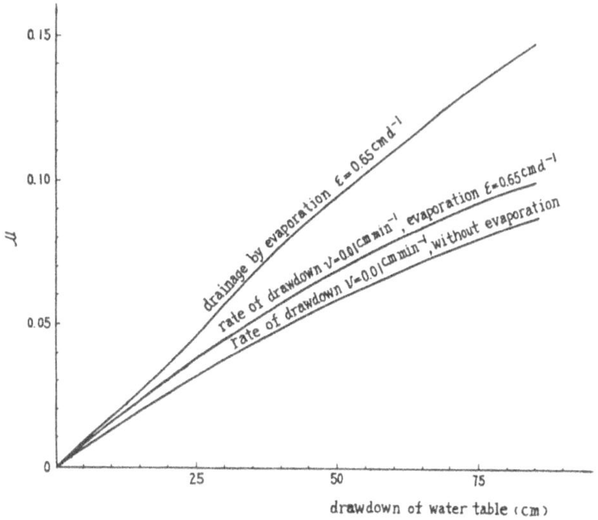

Figure 5

water content θ_s and the water content prescribed by the retention curve when suction head h = d. When the soil water characteristic curves are available, the complete drainable porosity can be estimated from the water retention curve.

The instantaneous and average drainable porosities are functions of the rates and total amount of drawdown, the initial depth to water table and the intensity of evaporation. With the known hydraulic parameters, the drainable porosity under different drainage conditions can be estimated by numerical simulation.

REFERENCES

BOULTON, N. (1954) The drawdown of water table under non-steady conditions near a pumped well in an unconfined formation, Proc. Institute of Civil Engineers (London), V.3, Part 3, 564-579.

CAI, M. (1985) The experimental study of drainable porosity of a sandy soil, Journal of Groundwater (in press).

CHEN, D. and Z. SHEN (1965) The experimental study of the drainage property of a sandy soil by the method of γ-attenuation, The report of Wuhan Institute of Hydraulic and Electrical Engineering.

PEI, Y. (1983) The study of drainable porosity by numerical simulation, Hydrogeology and Engineering Geology, No.4, 1983.

ZHANG, Y. (1985) Methods for estimating the evaporation from shallow water table in drainage design formulas.

Crop Production and Topsoil/Surfacewater Salinity in Farmers' Irrigated Rice Fields, the Nile Delta.

H.J. Nijland, S. El Guindy
Drainage Research Institute (DRI), Cairo, Egypt

INTRODUCTION

Fact-finding research was undertaken on irrigated rice fields of individual farmers in three sample regions in the north-west of the Nile Delta to determine the magnitudes of and relations between:

- Crop production;
- Amount and salt content of irrigation water;
- Amount and salt content of surface and subsurface drainage water;
- Salinity of the topsoil;
- Depth and salt concentration of the surface-water layer.

Of the three sample regions, one (Basal) had good quality irrigation water, whereas the second (Nokrashy) and third (Anwar Hammad) had relatively salty irrigation water. Only the Nokrashy Region has a subsurface drainage system, which is closed during the rice-growing season to prevent excessive percolation losses.

The surveys were conducted in the years 1977, 1978, and 1979. Data for all three regions are available only for the year 1979. The statistical analysis presented in this paper is therefore limited to 1979, but the results of the observations in the other years are similar.

The objective of this paper is mainly to answer the following questions with the information available:

- Are salts or waterlevels hampering the rice production in the sample regions?
- If so, to what extent?
- What might the production increase be if complete salt and/or waterlevel control can be attained?

In this paper, therefore, we shall study the statistical relations between rice yield (Y), top-soil salinity (Ss), salinity of the surface-water layer (Sw), and the depth of this water layer (D), because these data were collected simultaneously with the crop cuttings.

When possible, production functions will be derived under the assumption that each growth factor has a critical value dividing it into two fields: without influence on the production and with linear influence on the production.

RICE CULTIVATION PRACTICES

In the areas under investigation, the main rice variety planted is the Nahda. The Giza 159 is also planted, but is of only minor importance. Both varieties are of the japonica type.

The rice is seeded in nurseries during the month of May, and transplanted in late June or early July to farm plots, most of which vary in size from 0.1 to 0.3 ha. Harvesting takes place in October/ November.

The rice is irrigated from canals in which the water level is about 0.5 m below the soil surface. The water is lifted by sakias or by motor pumps. The canals provide water on a rotational schedule: four days on and four days off.

Fertilizers are applied two or three weeks after the rice is transplanted and during panicle initiation. Only nitrogen fertilizers are used, either in the form of ureum or ammonium sulphate. Before the fields are fertilized, their surface water is drained off into the surface drainage system. One or two days later, the fields are flooded again.

Two weeks before the harvest, surface drainage is practised to stimulate the ripening of the crop and improve the accessibility of the land. The harvest is done by hand, with sickles.

STATISTICAL METHODS USED

In the analysis, only two-variable linear regressions have been used, with yield (Y) as the dependent variable. Where appropriate, breakpoints were introduced for the independent variables (the growth factors), after which linear regressions were made separately to the left and right of the breakpoint. The breakpoint thus represents a critical or threshold value, separating the growth factor into parts with a significant influence on the yield and parts without any significant influence. Thus, a linearized non-linear production function is obtained by a broken line.

The residual yield variations (RY), remaining after the broken line regression upon one of the growth factors (for example Ss or Sw), were subjected to a similar regression upon the the other variable (Sw or Ss). If the two growth factors showed a considerable correlation, then the second growth factor was reduced (to RSw or RSs respectively) in order to eliminate that part of its variation that had already been accounted for by the regression upon the first independent variable. Thus one obtains a multiple regression by successive two-variable regressions.

The result of the analysis is that a multi-dimensional non-

linear production function is obtained by successive two-variable linear regressions with breakpoints. The advantage of this method is that it allows confidence statements to be made with the methods known from the two-variable linear regression theory. Further, the multi-dimensional function can be presented by two-dimensional illustrations, and the correlation between growth factors can be accounted for directly.
The following set of equations is obtained:

$$EY = a(S1 - MS1) + MY \qquad (S1 < BS1 \text{ or } S1 > BS1)$$

$$ES2 = c(S1 - MS1t) + MS2t \qquad (S1 < BS1 \text{ and } S1 > BS1)$$

$$RS2 = S2 - c(S1 - MS1t) \qquad (\text{if } c = 0,\ RS2 = S2)$$

$$RY = Y - EY$$

$$ERY = b(RS2 - MRS2) + MRY \qquad (RS2 < BRS2 \text{ or } RS2 > BRS2)$$

$$EFY = EY + ERY$$

$$= a(S1 - MS1) + b(RS2 - MRS2) + MY + MRY$$

$$= a^*.S1 + b^*.S2 + p$$

$$(\text{if } c=0\text{: } a^*=a,\ b^*=b) \qquad (\ S1 < BS1 \text{ and } S2 < BS2$$

$$\text{or } S1 > BS1 \text{ and } S2 > BS2$$

$$\text{or } S1 > BS1 \text{ and } S2 > BS2$$

$$\text{or } S1 < BS1 \text{ and } S2 > BS2$$

$$\text{or } S1 > BS1 \text{ and } S2 > BS2)$$

where

- S1 = 1st growth factor (Ss or Sw)
- S2 = 2nd growth factor (Sw or Ss)
- BS1 = breakpoint of S1
- BS2 = breakpoint of S2
- BRS2 = breakpoint of RS2
- EY = expected value of Y according to the regression of Y upon S1 divided into two parts: left and right of BS1
- MS1 = mean value of S1 either left or right of BS1
- MY = mean value of Y either left or right of BS1
- ES2 = expected value of S2 according to the regression of S2 upon S1 using all data
- MS1t = mean value of S1 using all data
- MS2t = mean value of S2 using all data
- RS2 = reduced value of S2 eliminating its variation already explained by S1
- ERY = expected value of RY = Y - EY according to the regression of RY upon RS2 divided into two parts: left and right of BRS2
- MRS2 = mean value of RS2 either left or right of BRS2
- MRY = mean value of RY = Y - EY
- a,b,c = regression coefficients
- a*,b* = production coefficients
- p = production constant

YIELD AND IRRIGATION

The grain yields of rice were determined from crop cuttings in randomly sampled areas of about 1000 m^2 in which the sheaves were counted. Of 20 sheaves, chosen at random, the fresh weight was determined. About 2 kg of this material was used to weigh the dry grains. The following results were obtained:

Region	Irrigation water: Total m^3	Irrigation water: Salt conc. mmho/cm.	Dry grain yield converted to ton/ha: Mean	Maximum	Minimum
Anwar H.	13500	0.8	3.0	6	1
Nokrashy	12000	0.8	6.2	8	4
Basal	11700	0.4	7.0	11	3

Because of the relatively small area from which the samples were taken, the variation in the yields is greater than if they had been taken from one full hectare, but the mean values would not be significantly different.

The Basal Region, having good quality irrigation water, using the smallest amount of it, and having no subsurface drainage system, has the highest yields with the greatest variation. The causes of the yield depression in part of the Basal Region and in all of the other two regions will be studied in the following sections.

YIELD AND TOPSOIL SALINITY

Figure 1 presents the grain yields (Y) in relation to the salinity (Ss) expressed in electric conductivity (in mmho/cm) of a saturated paste of the soil at 0 to 15 cm depth. The Ss values are averages of mean values of five observations taken per sample area of crop cutting at harvest time.

In the Basal Region, there are few Ss values higher than 3 mmho/cm. There is no trend of yield decrease with increasing Ss values. A topsoil salinity corresponding to 3 mmho/cm is apparently safe, and is almost nowhere exceeded. In the Nokrashy Region, the same holds true. Hence the salinity of the topsoil does not explain the yield depression in Nokrashy compared with Basal.

In the Anwar Hammad Region, the Ss values go up to 8 mmho/cm. It is noticeable that the maximum yields, represented by the upper envelope curve, tend to decrease sharply with increasing topsoil salinity in the range of Ss = 3 to 8 mmho/cm. On the other hand, there are many low yields of less than 1 ton/ha, both at high and low topsoil salinities. This indicates that the Ss value is not a strong explanatory factor of the yield depression in Anwar Hammad compared with yield of the other two regions.

A linear regression analysis with different breakpoints BSs reveals that the total sum of squared deviations Y - MY, equalling 182, is reduced to a minimum of 138 (the total sum of squared residuals YR = Y - EY) when BSs = 3 mmho/cm. The coefficient of explanation R^2 is 0.25. Other breakpoints give less explanation. The broken line regression yields:

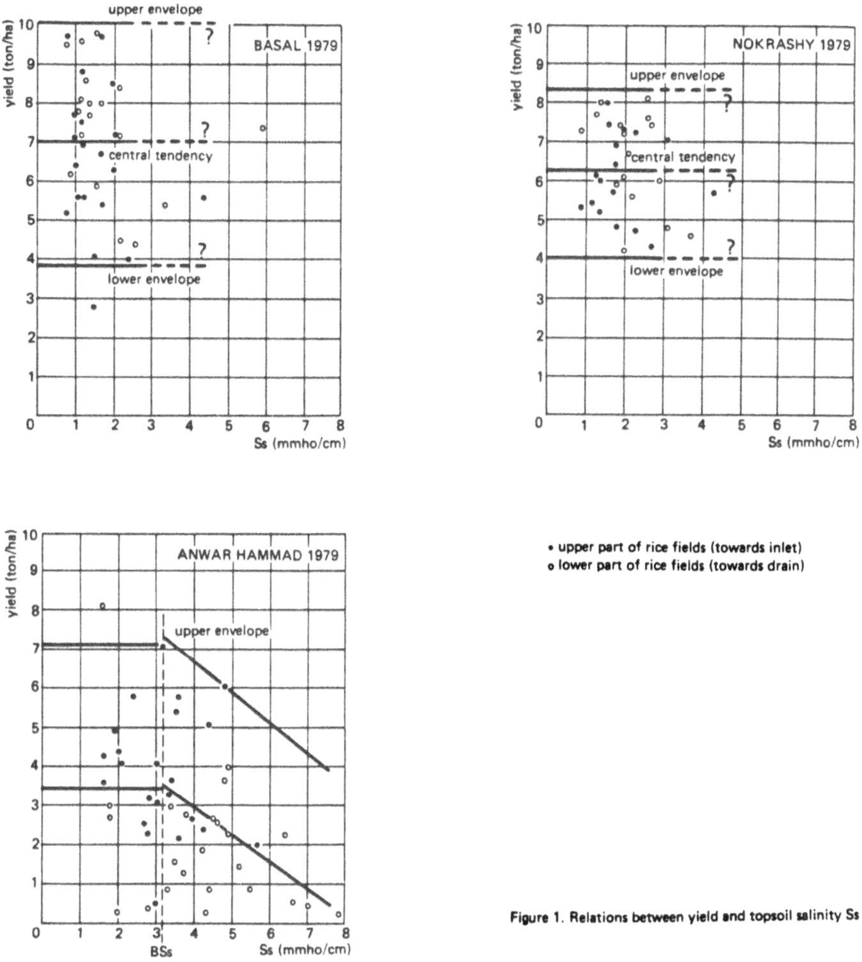

Figure 1. Relations between yield and topsoil salinity Ss

$$EY = MY = 3.4 \text{ ton/ha} \qquad (Ss < 3 \text{ mmho/cm}, n = 17)$$

$$EY = 2.8 - 0.80 (Ss - 4.5) \text{ ton/ha} \quad (Ss > 3 \text{ mmho/cm}, n = 31)$$

The coefficient of explanation r^2 of the data (Y, Ss < 3) is - 0.2, which is insignificant. Therefore the value of EY for Ss < 3 has been taken constant.

The 90% confidence limits of the regression coefficient a = - 0.80 (for Ss >3) are - 0.36 and - 1.25. Hence the value of the coefficient a is not accurate but it is significant. Therefore Ss = 3 mmho/cm represents a critical value of topsoil salinity below which the yields are little affected by Ss, but above which the yields decrease with increasing Ss.

The two regression lines do not intersect each other at Ss = 3.0, but at Ss = 3.7 mmho/cm. This is an indication that the breakpoint cannot be determined exactly, but only with a certain confidence interval. Perhaps its value is somewhat higher than 3.0. The reason for the limited accuracy is the large unexplained part of the variation of the data due to other, unknown, growth factors. The order of magnitude of the breakpoint, however, is quite clear.

The yield expectation in Anwar Hammad for $Ss < 3$ is still considerably less than in the other regions, so that the topsoil salinity explains this yield depression only partly.

The higher topsoil salinity in Anwar Hammad compared with the other regions is only partly due to a higher salt concentration of the irrigation water and the absence of a subsurface drainage system. From Figure 1, it can be seen that the yields in the lower parts of the fields are on average considerably less than in the upper parts. Most lower parts have topsoil salinities corresponding to $Ss > 3$ mmho/cm, whereas most upper parts have $Ss < 3$. It therefore appears that the yield depressions might also be caused by differences in quantity and quality of the surface water in the upper and lower parts of the farm plots. This will be studied in the following sections.

YIELD AND SALT CONCENTRATION OF THE SURFACE-WATER LAYER

Figure 2 presents the grain yield (Y) in relation to the salt concentration of the surface-water layer on the rice fields (Sw), expressed in electric conductivity (in mmho/cm), as a seasonal average of mean values of five observations taken weekly in each crop sampling area.

As with Ss, the yields in Basal and Nokrashy are not affected by Sw,and the maximum observed values of 2 mmho/cm appear to be safe values. Although Nokrashy has more salty irrigation water than Basal, it seems that this water need not lead to a critically high salt concentration of the surface water.

The (Y, Sw) relation in the Anwar Hammad Region shows the same trends as the (Y, Ss) relation. A linear regression analysis with different breakpoints BSw reveals that the total sum of squared deviations Y-MY, equalling 182, is reduced to 98 (the total sum of squared residuals Y- EY) when Bsw is 1.3 mmho/cm. The coefficient of explanation r^2 of the data (Y, $Sw < 1.3$)is only 0.006. Hence the regression coefficient for these data can be assumed zero. The broken line analysis yields:

EY = MY = 4.1 ton/ha (Sw < 1.3 mmho/cm, n = 10)

EY = 2.7-1.0 (Sw - 2.6) ton/ha (Sw > 1.3 mmho/cm, n = 38)

The total coefficient of explanation R^2 is 0.44, which is significantly higher than for the (Y, Ss) regression. The 90% confidence limits of the regression coefficient a = - 1.0, are - 0.70 and - 1.3. Although the exact value of the coefficient a can vary considerably, it is still significant. The two regression lines intersect each other at Sw = 1.2 mmho/cm, which is fairly close to the breakpoint BSw = 1.3 mmho/cm.

Because the sample areas were selected at random, it can be concluded that 10/(38+10) = 80% of the region is affected to some extent by higher than critical salt concentrations of the

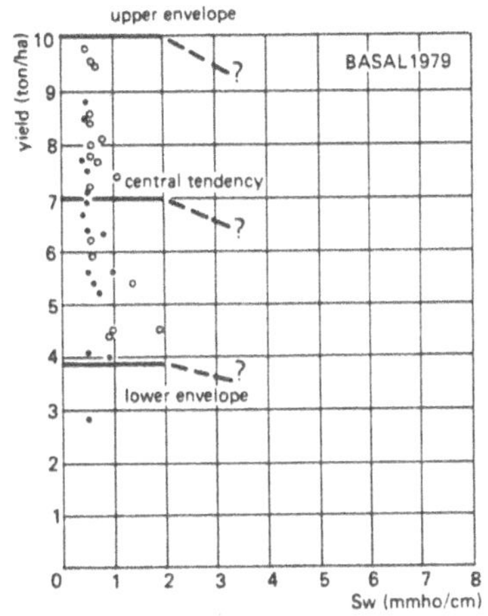

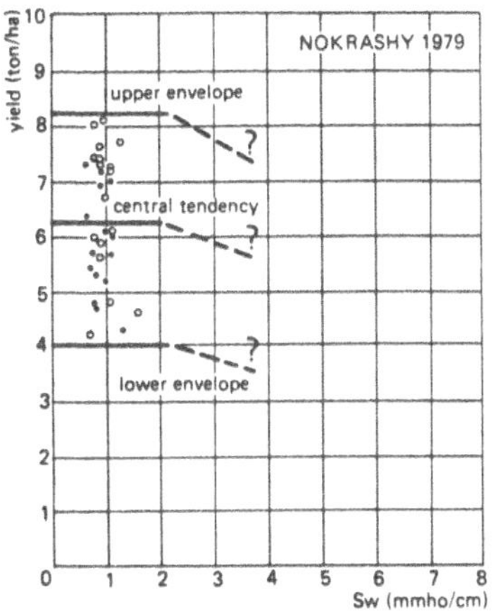

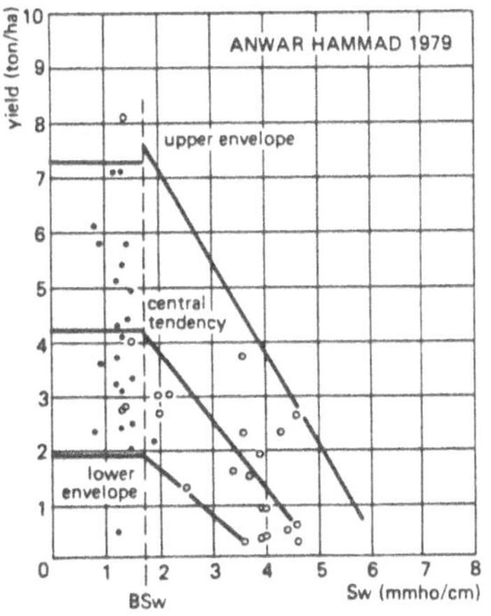

• upper part of rice fields
o lower part of rice fields

Figure 2. Relations between yield and salt concentration of surface water Sw

surface-water layer. Since there are 17 data (representing 35% of the area) with 1.3 < Sw < 1.5, it can be concluded that a large part of the area affected has only minor salt problems. Like the Ss values, the Sw values only partly explain the yield depression in Anwar Hammad compared with the other two regions. There must be factors other than salinity of topsoil and surface water responsible for the lower yields in Anwar Hammad.

YIELD AND DEPTH OF SURFACE-WATER LAYER

Figure 3 shows the relations between grain yield of rice (Y) and seasonal average depth of the surface water (D), which was measured weekly. Although D varies from 1 to 10 cm, it has no significant effect on the yield.

It appears that it is not necessary to maintain a strict depth of the surface-water layer, at least in the range of 1 to 10 cm.

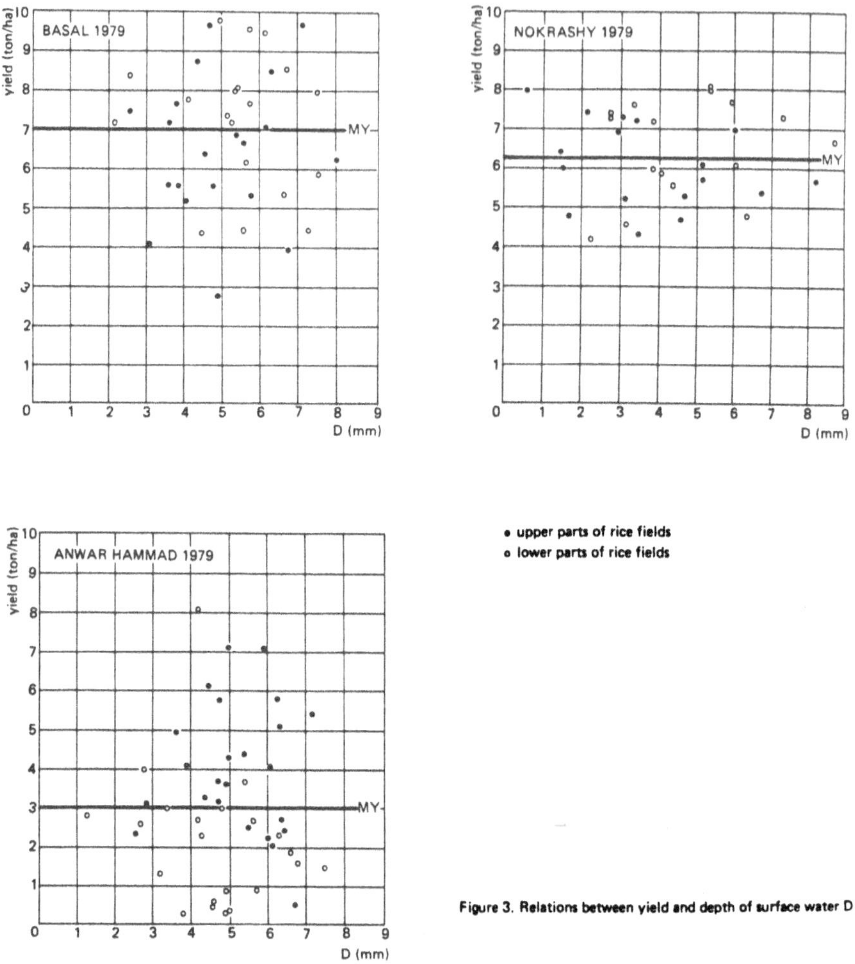

Figure 3. Relations between yield and depth of surface water D

The data Sw and D are uncorrelated, so that the salt concentration of the surface-water layer does not seem to be influenced by its depth (data not presented).

THE RELATION BETWEEN Ss AND Sw

Figure 4 shows a plot of Ss and Sw. For Anwar Hammad, the correlation coefficient r is 0.60. The regression coefficient (c) of Sw upon Ss is 0.53 and that of Ss upon Sw equals 0.67 according to the rule that their product equals r^2. Ss and Sw appear to be interdependent to some extent, but only in the lower part of the fields. If both variables are used in a single production function, then one of the variables has to be reduced so that only its independent portion is taken.

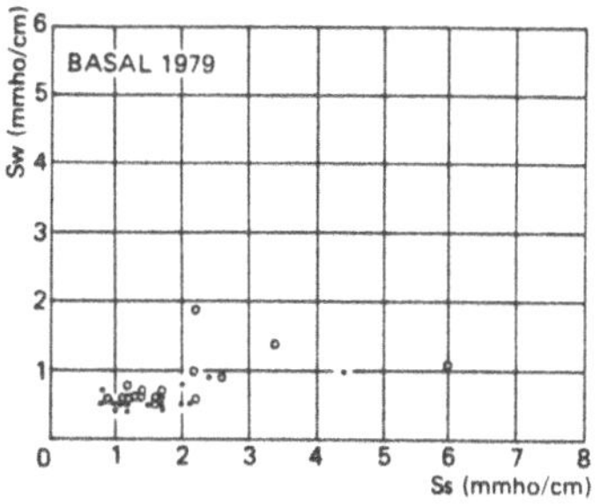

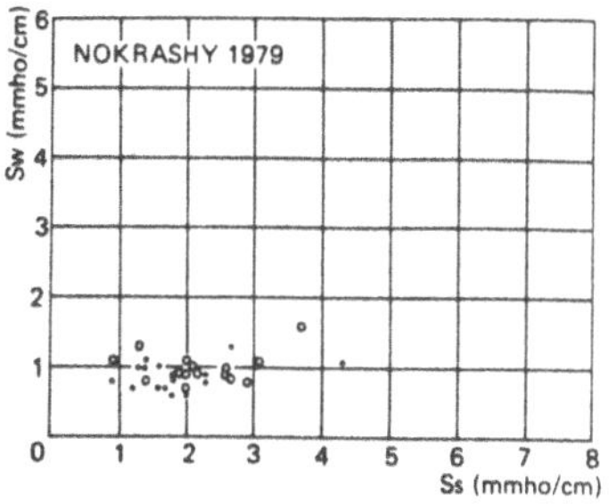

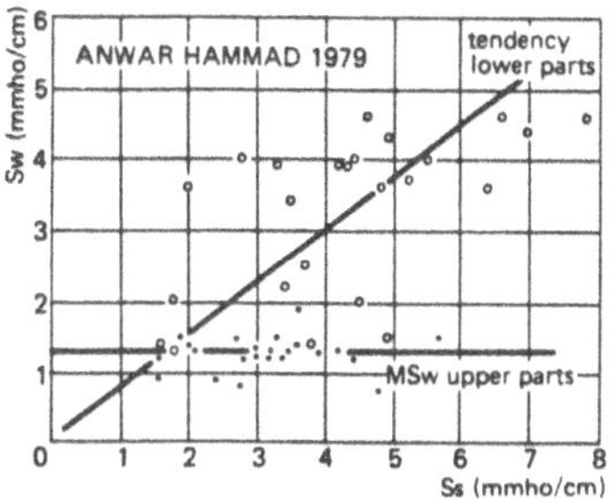

• upper parts of rice fields
o lower parts of rice fields

Figure 4. Relations between Sw and Ss

THE RELATION BETWEEN Y, Sw, AND Ss IN THE ANWAR HAMMAD REGION

The regression of Y upon Sw was seen to be more effective than that of Y upon Ss, because the sum of the squares of the residuals was 98 and 138, respectively. The introduction of the (reduced) Ss after the (Y, Sw) regression does not yield a significant further reduction of residuals. Hence the production function given in Section 6 cannot be further improved.

The introduction of the (reduced) Sw after the (Y, Ss) regression can bring down the sum of the squared residuals from 138 to a minimum of 96. The resulting three-dimensional production function is therefore not significantly more effective than the two-dimensional (Y, Sw) function.

It can be concluded that the salt concentration of the surface-water layer plays a more important role in the rice production of the Anwar Hammad Region than the salinity of the topsoil.

ESTIMATED BENEFITS OF SALT CONTROL

The salt contents of the rice plots in Basal and Nokrashy are presently at safe levels. Additional measures for salt control are not required as they would not boost the production. With the present data, it is not possible to evaluate the effect of the drainage system in Nokrashy on the salinity levels and the crop production.

It was seen that in the Anwar Hammad Region, the salinity of the topsoil plays no major role in the prediction of benefits, so the production increment (I) can be estimated assuming only that the salt concentration of the surface-water layer is controlled (for example by surface drainage) to be everywhere at or below the safe level of BSw = 1.3 mmho/cm as follows:

$$I = \Sigma\, a(BSw - Sw)/n$$

With a = -1.0, the average production increase will be 1.0 ton/ha, amounting to 33% of the present average yield of 3.0 ton/ha, but the 90% confidence interval of the yield increase ranges from 23 to 43%.

The 20% of the area that is free of salt problems still has a considerably lower yield than in the Basal and Nokrashy Regions, a difference of 2 to 3 ton/ha. With salt-control measures, this difference cannot be reduced. The present study can give no evidence as to which measures would be appropriate.

FINAL COMMENTS

High yield variations due to factors other than Ss, Sw, and D, safe levels of the growth factors, and/or a considerable correlation between them, prevented the application of the multi-dimensional non-linear production function described in this paper. However, for the Anwar Hammad Region, a significant non-linear production function of Sw with linear segments separated by a breakpoint (the threshold or critical value) could be developed, from which the benefits of certain salt-control measures could be predicted with reasonable accuracy.

ACKNOWLEDGEMENTS

This study was sponsored by the Drainage Research Institute and the Water Research Centre of the Ministry of Irrigation, Egypt, under the supervision of the Egyptian-Dutch Drainage Advisory Panel. The guidance of Prof. El Gabaly, Chairman of the Panel, was highly appreciated.

The data collection programme was designed and guided by I.A. Risseeuw, who was leader of the Dutch technical cooperation team in Egypt.

The statistical analysis of data by which the production functions were derived was performed by R.J. Oosterbaan and W. Beekman, International Institute for Land Reclamation and Improvement, Wageningen, The Netherlands.

Improvement of Drainage in Loamy Soil by Subsoiling

E. Pierzgalski, A. Wanke
Warsaw Agricultural University, Poland

INTRODUCTION

In the total area of 15.3 million hectares of arable soils in Poland about 20 % constitute low permeable soils. The problem of drainage of these soils not only in Poland, but also in the whole Europe lies in the fact that leading gravity water out of upper layer of the soil profile is not equivalent with the field preparation to the state required for field works and growth of plants. Of importance is also the period of draining soil in spring, as its shortening results in prolongation of the growing season, what is under Poland's conditions an important factor in the agricultural production.

The aim of the present work was to investigate functioning of the drainage system established in moraine soils of the mechanical composition of medium loam. Also possibilities of an improvement of air and water conditions of the soil of a drained object by means of subsoiling were estimated.

CHARACTERISTICS OF THE OBJECT UNDER STUDY

The investigations were located in the central part of Poland, at the Experiment Station Puczniew. The mean sum of annual precipitation in the object amounted to 530 mm. The mean annual air temperature amounts to 7.7°C. Soils of the area investigated were developed from cohesive postglacial formations reaching the depth of 90 m. In Table 1 some properties of the drained soils under study are presented. In the first 0-30 cm layer of the soil profile sandy loam, in the layer of 30-50 cm light loam and below 50 cm medium loam are to be found.

TABLE 1. Some properties of the soil under study

Depth cm	Clay %	Silt %	Sand %	$CaCO_3$ %	Fe_2O_3	Permeability m/day
0-30	24	14	62	0.2	3.3	0.05-0.2
30-50	35	11	54	0.2	3.2	0.04-0.25
>50	41	13	46	0.4	3.0	0.02-0.05

The soil permeability varied within a year depending on the soil moisture level. Drainable porosity measured in monoliths amounted to 0.02-0.035. The ground water table on the undrained field lay in spring of average years close under the soil surface, maintaining at the same level over the whole spring season. Since the summer season start its systematic drop began, reaching the depth of about 3.0 m in September-October. Thus the undrained soils were unsuitable for farming under these conditions. The area of experimental field was 5 hectares.
In it tile drain pipelines of 5 cm in diameter, at the spacing of 9, 13 and 17 m and the depth of 0.9 m

have been laid. In the 8-year period of investigations the ground water table measurements were carried out in 40 observation wells. The piezometric pressure was measured at the depth of 2, 3 and 5 m. Continuous measurements of the soil water pressure head were performed using tensometers installed at the depth of 0.5, 1.0, 1.5, 2.0, 3.0 and 3.5 m. Once a month control measurements of the soil moisture by the drier method were carried out. Wheat, barley, oats, potatoes and fodder beets were cultivated on the object.

INVESTIGATION RESULTS

Functioning of the drainage

The outflow rate from drain pipelines in the period of investigations was different in particular years varying from 6 to 148 mm. The mean annual outflow amounted to 87 mm, thereof 50 mm constituted outflow in the period January-February and 37 mm - that in the period March-April. In Fig.1 a relationship between the outflow and the precipitation sum for a year and winter half-year is presented.

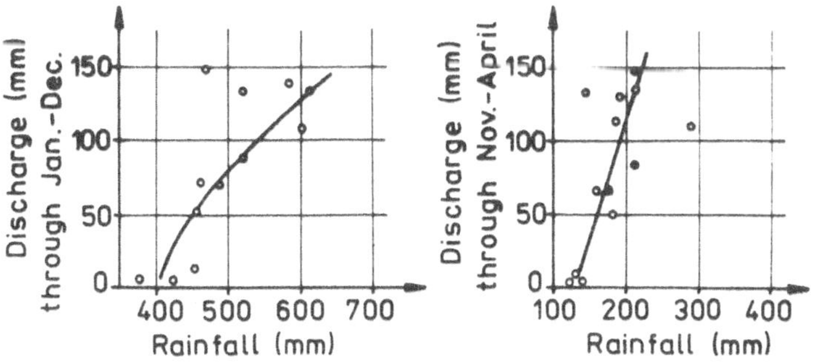

Fig.1. Dependence of the outflow on precipitation.

A drop of the ground water table occurred very slowly and lasted from early January to late April, reaching the depth of 40-60 cm below soil surface, depending on the spacing between drains /Fig.2/.

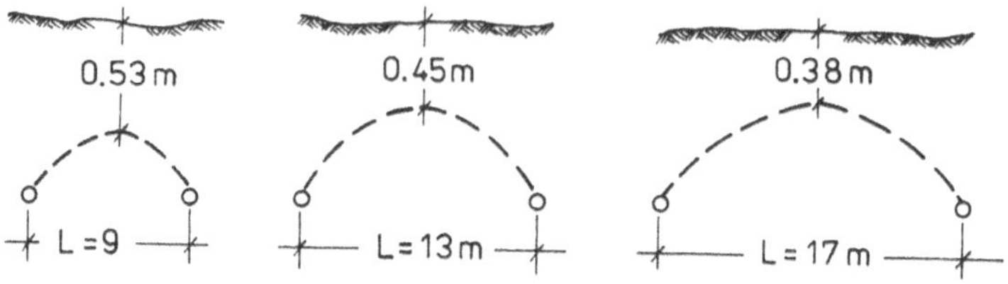

Fig.2. Mean ground water table between drains at different spacings /L/ in the period January-February

In Fig.3 a relationship between the mean drop of the ground water table and the spacing of drain pipelines in the winter and spring season is presented.

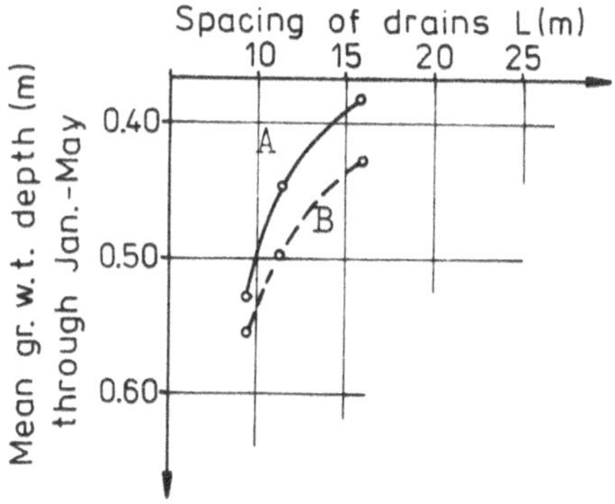

Fig.3. Dependence of a drop of the ground water table on the spacing between drains: A - in winter, B - in spring

The estimation of the suitability of several equations determining the spacing between drains has proved that for the hydraulic scheme occurring in the investigated field the results approximating most closely those of the field investigations were obtained using the Radchenko's equation of the following shape:

$$L = k \cdot t \cdot A^{-1} \cdot \mu^{-1} \left[\ln(h_o\ h_t^{-1})\ \alpha^{-1}\right]^{-1/\beta}$$

where: L - spacing between drains, t draining time, h_o - height of the gr.w.t. at draining start in the middle of spacing measured from the impermeable layer, h_t - height of the gr.w.t. after the time t measured as above, k-filtration coefficient, u-drainable porosity, A-coefficient depending on the μ/L ratio, where u is wetted perimeter of a drain, α, β - coefficients taking into account the effect of the water - bearing layer thickness on the water inflow into drain.

Despite the occurred drop of the ground water table, the soil moisture level was too high in relation to the requirements of plants. This phenomenom is connected with the soil water energy state. In Fig.4 a relationship between the soil moisture level and the power of moisture retention by soil is presented. This relationship proves that e.g. a lowering of the ground water table by 1 m under soil surface can reduce the soil moisture by 3 % only. Such a high soil moisture level makes difficult

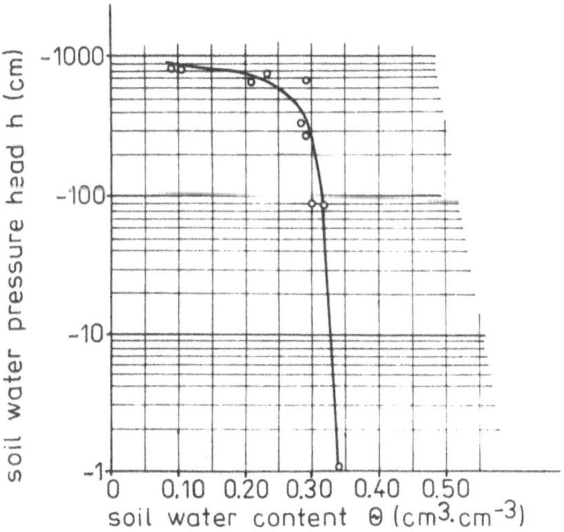

Fig.4.
Soil water pressure head versus soil water content in the investigated soil

tillage measures, whereas a low air content in the root layer prevents an appropriate development of plants. A further drop of the soil moisture level depends on the evapotranspiration intensity. The course of soil moisture, ground water table and outflow changes within a year approximating average one with regard to climatic conditions is illustrated in Fig.5.

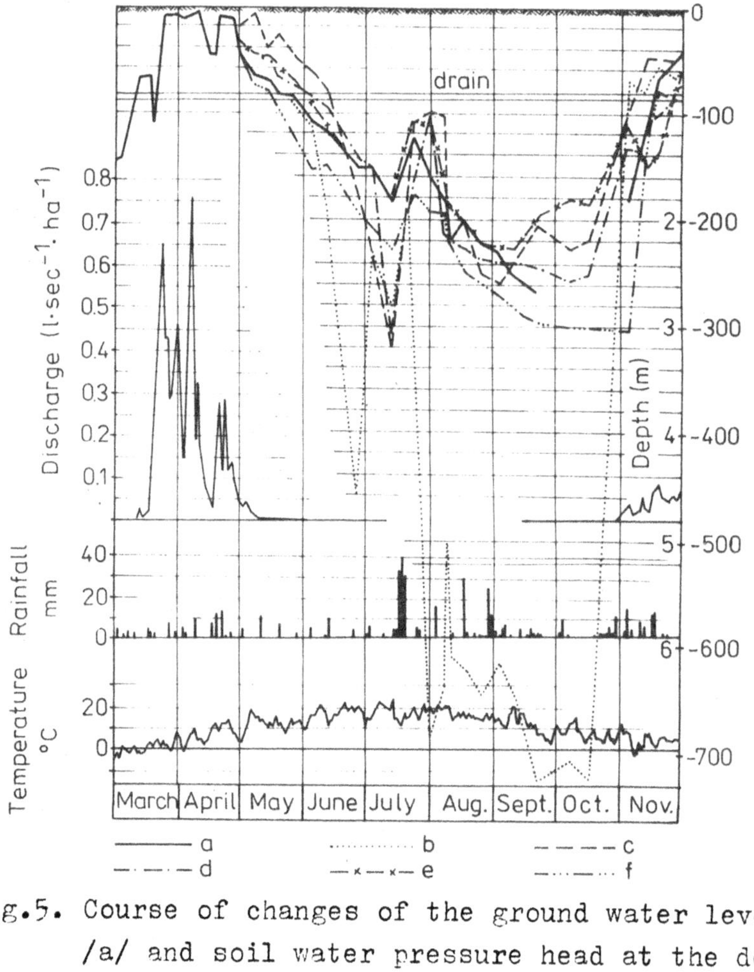

Fig.5. Course of changes of the ground water levels /a/ and soil water pressure head at the depth: b - 0.5 m, c - 1.0 m, d - 1.5 m, e - 2.0 m, f - 3.5 m.

Under such conditions the average yields of cereals amounted to 3.0-3.5 t, potatoes - to 16 t, fodder beets - to 58 t from hectare.

Detailed measurements of yields of various crops on the experimental field have proved that the yield level depends on the distance from drain.The highest yields were obtained along the belt over drain.This relationship has not been observed in cereals.Exemplary results for several crops obtained within the first years after laying drains, are presented in Fig.6.

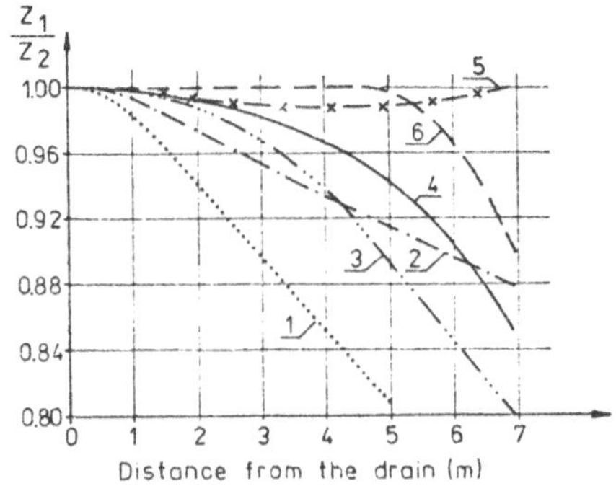

1-fodder beets, 2-barley, 3-potatoes, 4-oats, 5-rape
Z_1-yield in the field
Z_2-yield over the drain

Fig.6. Dependence of yield on the distance from drain

The presented distribution of yielding between drains is connected with more favourable moisture conditions along the belt over drains. A significant role of formation of these conditions plays the drain backfill. In Fig.7 the infiltration properties of the backfill decreasing in subsequent years after the execution of drainage are presented. This correlation was established on the basis of measurements performed on the field investigated and adjoining earlier drained fields.

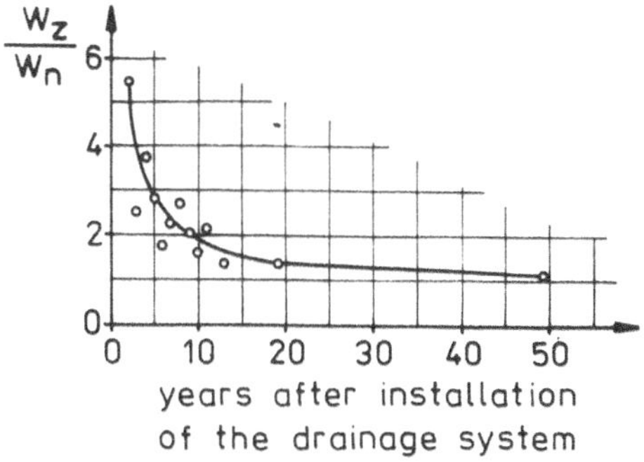

Fig.7
Infiltration changes of the drain backfill: W_z-infiltration of backfill, W_r-infiltration of parent soil near the backfill.

The above correlation proves distinctly an occurrence of the process of reversion of the hydrological properties of soil changed under the effect of mechanical measures to the primary state. The analysis of the above phenomenon encouraged to carrying out investigations on subsoiling measures aiming at mobilization of potential possibilities of soil through changes of its physico-hydrological properties.

<u>Changes caused by subsoiling measures</u>

Subsoiling was executed to the depth of 65 cm. Owing to loosening of the subsoil changed the soil permeability arrangement /Fig.8 / Permeability of the arable layer decreased due to introducing the subsoil into it increasing at the same time in the arable layer. Also changes in the infiltration were observed /Fig.9/.

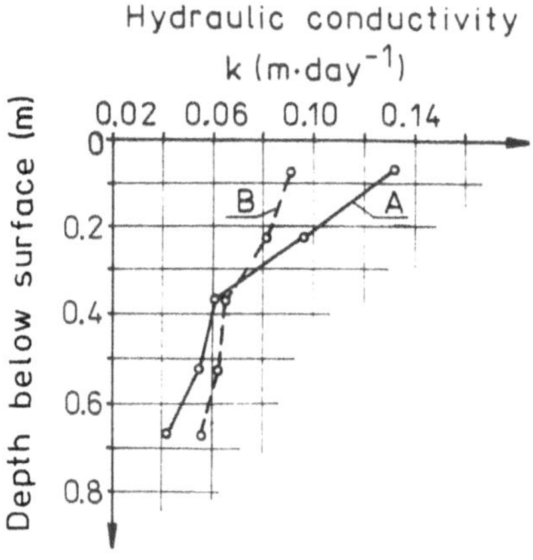

Fig.8
Permeability changes caused by subsoiling A-before, B-after subsoiling

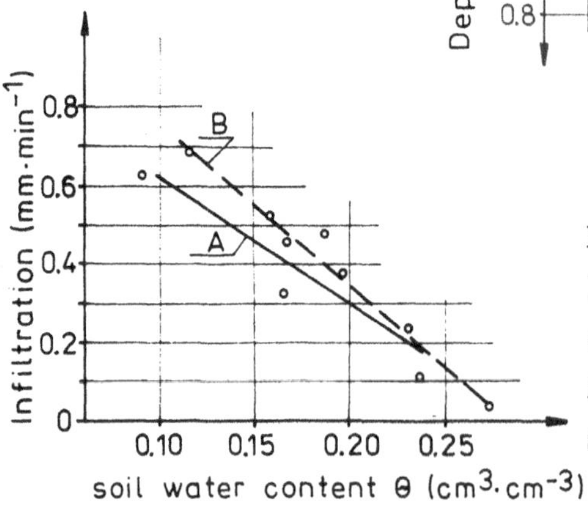

Fig.9
Infiltration changes caused by subsoiling A-before, B-after subsoiling

Changes of the hydrological properties of soil resulted in an increase of outflows. A relative increase of the outflow caused by subsoiling is presented in Fig. 10. Characteristic was an increase of the outflow in the period after subsoiling. After 30 months the relative outflow was almost three times as large than that noted in the first month after the subsoiling measure.

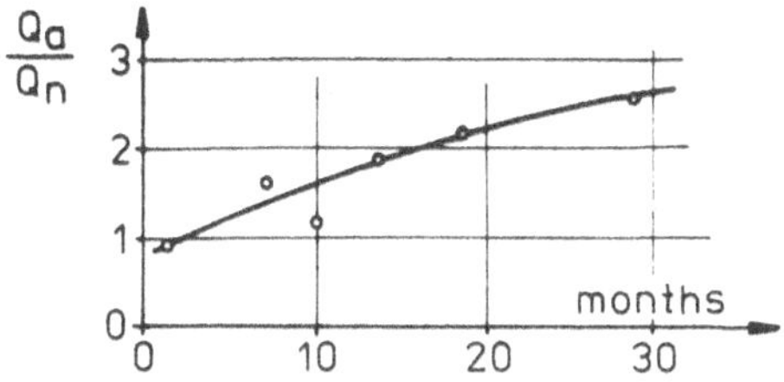

Fig.10. Changes of the outflow from drains caused by subsoiling: Q_a - outflow from the subsoiled field, B - outflow from the non-subsoiled field

Subsoiling led also to changes in the efficiency of soil and in the dynamics of its utilization by plants. Exemplary courses of changes in the water content in the layer of 0-60 cm soil within three subsequent years after subsoiling are presented in Fig.11. Changes in the hydrological properties of soil led to an occurrence of different air and moisture conditions in consequence of which changes in the yields of crops took place. In the fodder beet cultivation the yield increment of 18 t/ha has been reached as compared to the control field, the total yield reaching 75 t/ha. On the control field the yield amounted to 55 t/ha. In the alfalfa cultivation no changes in yield were observed, whereas in the cultivation of barley a slight yield decrease was noted.

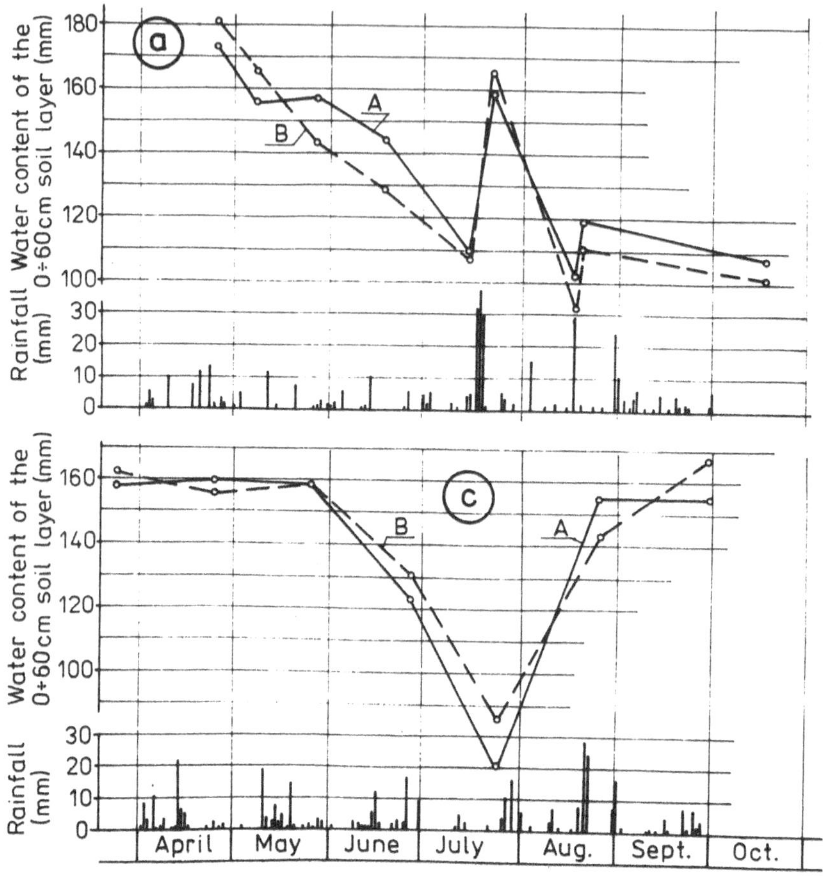

Fig.11. Dynamics of the water content in the layer of 0-60 cm: A-without subsoiling, B-with subsoiling, a, c - in the first, and third year after subsoiling respectively.

SUMMARY

Drainage of low permeability soils under the Poland's climatic conditions is a necessary technical measure enabling the yearly utilization of agricultural lands. Leading off gravity water by drains in low permeable soils lasts too long in the spring season and makes difficult an appropriate aeration of the root zone. Change of soil properties in consequence of subsoiling can ensure an increment of yields of some crops.

Movement of Salt During the Cropping Cycle in Drained and Undrained Coastal Marshes of the U.K.

A.S.B. Armstrong, D.W. Rycroft and T.W. Tanton

INTRODUCTION

Along the coast and the estuaries of Eastern England there are considerable areas of saline ($E.C._E$ > 4mmhos cm^{-1}) heavy clay soils. The study area relevant to this paper is the Hoo Peninsula on the North Kent Marshes. This area, formerly natural salt marsh, has long been embanked and the land used as grazing pasture. Over the last fifteen years, however, much of the land has been ploughed up, pipe drainage installed, and converted to arable farming. Subsequently drainage systems have failed, and during wet winters, surface flooding occurs and crop yields suffer. Intensification of the drainage system has also failed to improve the situation which arises because of the high proportion of smectite clay minerals in the soil and the fact that the soils are both saline and sodic (exchangeable sodium percentage > 15).

After the installation of pipe drainage, rain water permeates the topsoil causing the clay to disperse and move downwards in suspension to block vertical subsoil cracks and clog both the drains and their trench backfill. Existing empirical methods which are used to predict the effectiveness of reclamation of soils from salinity (by drainage) or sodicity (by replacement of sodium by calcium) are inappropriate since they are based on more uniform lighter textured soils. On the highly structured, heavy, swelling/shrinking soils such as are present at Hoo, a lack of understanding of the basic processes of salt and water movement make it impossible to predict leaching efficiency or to estimate the efficiency of gypsum prediction or salt movement out of soil in reclaiming these soils.

This paper reports the results of a one year field study which was conducted to investigate the macro-seasonal salt dynamics of these heavy clay soils. The work is part of a programme which is designed to elucidate the factors affecting salt movement both before, and after conversion to arable cropping by ploughing and installation of drainage. The paper looked at the possible significance of these preliminary findings.

2. METHODS

Plots were established on adjacent arable and grassland fields. The arable field had pipe drainage installed at 20 metres spacing in 1969 and was mole drained in 1976. Prior to cultivation a substantial amount of levelling of the salt marsh took place. The grassland field had no pipe drainage but was adjacent to the Delft ditch.

Six replicate samples were taken by auger at eight depths (0 - 10, 10 - 20, 20 - 30, 30 - 40, 40 - 60, 80 - 100, 100 - 120 cm) on six occasions at two-monthly intervals between 3rd May, 1984 and 14th March, 1985. Sampling was carried out in the pattern of a 6 by 6 latin square, with each such sample being taken 1.5 m from the adjacent sampling holes.

Determinations were made of gravimetric moisture content (% w/w) and electrical conductivity of 1:5 soil/water suspensions. As the former is the volume of water associated with a given dry weight of soil and the latter may be expressed as the weight of salt associated with this weight, the salt concentration (meq. l^{-1}) of soil solution at field moisture content is known.

Complementing these field studies are a series of laboratory experiments in which prepared soil aggregates at field moisture content, packed in perspex columns (15cm diameter x 25cm depth), are subjected to simulated rainfall events.

3. RESULTS

The Hoo marshes have very low rainfall with an average drainable excess of only 100 mm, but in the year of sampling (3rd May, 1984 to 14th March, 1985) there was no excess.

The soil at the grassland site is highly structured with a grass mat layered on top of a strongly developed sub-angular blocky topsoil with abundant fibrous roots, and a strongly developed prismatic subsoil. (Fig. 1)

On the arable site the topsoil is granular, lacking the abundant rooting, and the subsoil peds have surface coatings of fine clay translocated from the topsoil.

3.1 Grassland Site:

Figure 3b shows profile variations electrical conductivity (1:5 extract), soil solution salt concentration, and moisture content for the three extreme sampling conditions. Intermediate sampling period results fall between these values.

Figure 1. Soil Structure Under Permanent Grassland

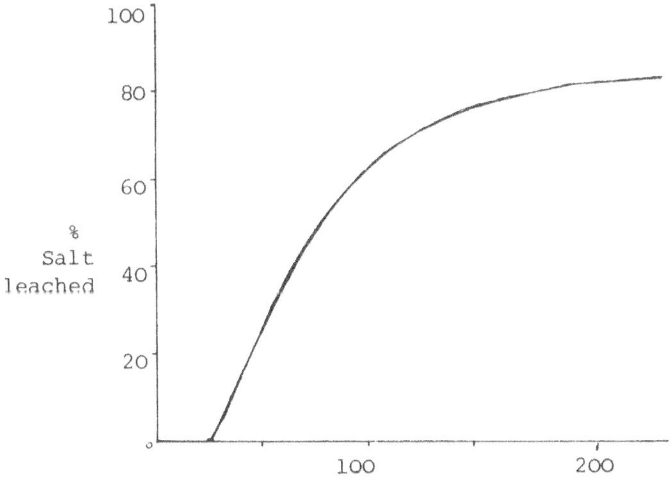

Figure 2. % Salt leached from a column (20cm depth, 15cm diameter) of prepared (0.5-1.0 cm) Grassland Aggregates.

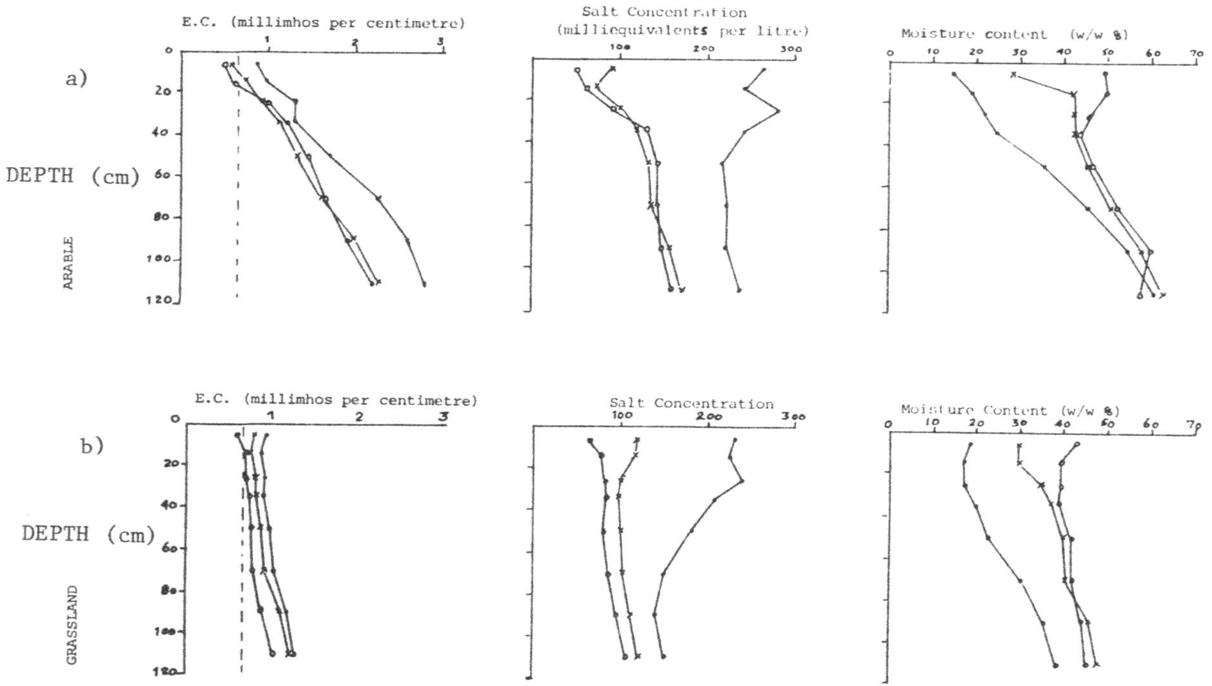

FIGURE 3 : Electrical Conductivity of 1:5 Soil/Water Extract, **(millimhos per centimetre)** Salt concentration of soil moisture, (milliequivalents per litre), moisture content (w/w %), in a) arable, b) grass fields. (x-x sampled 3rd May, 1984, .-. 3rd September, 1984, o-o 14th March, 1985). The dashed lines in figures (a) and (b) represents a salinity equivalent to EC_E = 4 millimhos per centimetre.

The figure shows that under pasture there is a fairly uniform distribution of salt content (E.C.) with depth, and the soil remains fairly uniformly saline throughout the year. During the summer as the soil dries out the salt concentration of the soil moisture increases. However, the EC results indicate that there is also a net increase in salt content throughout the profile brought about by capillary rise, which in this case corresponds to 30mm. This low value may be due to the formation of horizontal cracks as the soil dries which impede. vertical capillary movement. (Van Hoorn, 1984).

After winter rain the salt concentration and content profiles decrease again and return to approximately the same values as the previous spring. (During April, 1984 prior to the first sampling a rainfall deficit of 85mm had occured, initiating capillary rise).

The total amount of salt in the grassland field is considerably lower than in the arable field. This could either be due to differences in initial salt levels, or due to the better structure that is retained in the grassland soils.

3.2 Arable Site:

The seasonal salt profile fluctuations in the arable field (Fig.3a). Follow the same basic pattern as the grassland with an increase in soil moisture salt concentration and total salt content during summer. In this case there is a greater amount of capillary rise corresponding to 100mm.

The main difference between the sites is the much steeper rise in salt content with depth in the arable site from a topsoil (0 - 20cm) with lower EC than the grassland to a much more saline subsoil.

3.3 Laboratory Experiment:

Fig.2 illustrates that on a column of the pasture soil sieved into aggregates of 0.5 - 1.0cm diameter at summer field moisture content, applications of 200mm of simulated rainfall leaches approximately 80% of the soluble salts from a 20cm depth of soil. The tailing off of the curve indicates that increasing the quantity of continuous rainfall application is unlikely to significantly affect the amount of salt leached.

4. DISCUSSION

4.1 Salt Movement Under Pasture:

Under grassland these heavy clays are highly structured and this has a fundamental influence on salt and water movement.

Fig.3 has illustrated that these soils remain saline throughout the year and that the salinity profile is fairly uniform over the top metre. These soils were last inundated with sea water in 1953. Given that since then there has been approximately three metres excess rainfall, even if all of the resident soil water had been replaced by sea water, the passage of this excess should have reclaimed the soil.

There are two possible reasons for this inefficient leaching:

a) Surface Flow During winter rainfall subsoil cracks may swell and close causing most of the drainage to occur by lateral surface flow, the annual salt loss being made up by capillary rise from depth.

b) By-Pass Flow In recent years it has been recognised by many workers (Ritchie et al 1972, Kissel et al 1973, Anderson and Bouma 1977, Scotter et al 1978), that in highly structured soils where the hydraulic conductivity of the peds is very low, water does not move in a wide range of pore sizes. Instead, free water moves in narrow rivulets along vertical cracks and worm holes effectively "By-Passing" the soil matrix, even when it is unsaturated, and penetrating deeper than would be predicted by continuity type equations.

Under these conditions salt removal from the soil is mainly by diffusion from the peds into the water flowing in the cracks and is dependent on ped size, rainfall intensity/duration patterns, and the diffusional properties of the clay peds.

Extent of possible salt retention against leaching under natural conditions is unknown. Barraclough (1983) concluded from theory that nitrate was unlikely to be held back unless peds were at least equivalent to 4cm spheres but Addiscott (1984) has concluded that under varying leaching patterns no such unique size definition could be given. However, theoretical calculation on diffusion in peds of the size occuring on the marsh indicate that it is most likely a limiting factor.

Possible salt retention in these soils is increased because they are sodic/saline. In such soils where the proportion of sodium cations held close to the negatively charged clay particles is high (E.S.P. > 15%), when the salinity of the soil solution is diluted (e.g. by rainfall events), an osmotic effect causes water to move between the clay lamellae which swell making the soil matrix very impermeable, but if rapid dilution continues, the clay eventually disperses. In this soil dispersion occurs when the salt concentration of the soil solution drops below 20 $meq l^{-1}$. Fig.3b illustrates that the salt concentrations remain above this value throughout the year. However, these are average values over a cross sectional area of soil, but at the ped surface where water flow occurs, the concentration is obviously lower and physical examination show that dispersion takes place at the ped surface causing them to be coated with "cutans" of dispersed clay which are likely to considerably restrict the diffusion from the peds by:-

a) decreased porosity at ped surfaces

b) increase anion repulsion limiting the effective cross-sectional pore area for diffusion and

c) increasing tortuosity of the diffusional path.

Since when the water film thickness between clay surfaces is of the same order as the double layer form which solutes are excluded, "salt sieving" may occur (Blackmore 1976), in extreme cases salts cannot pass from the structural units into water moving in the macro pores.

Under pasture the very slow salt removal has allowed the cationic composition at the exchange sites on the clay lamellae (expressed as % exchangeable sodium, ESP) to remain in equilibrium with the external salt solution in the bulk of the soil. With respect to sodium the approximate relationship is (U.S.D.A. Salinity Handbook, 1954).

$$\text{ESP (Exchangeable sodium \%)} = \frac{100\ (-0.0126 + 0.01475\ \text{S.A.R.})}{1 + (-0.0126 + 0.01475\ \text{S.A.R.})} \quad (1)$$

where,

$$\text{SAR (Soil solution sodium absorption ratio)} = \frac{[\text{Na (meg } l^{-1})]}{\sqrt{\left[\frac{(\text{Ca + Mg})}{2}\right]}\ \text{meg } l^{-1}} \quad (2)$$

It can be seen from these equations that as the S.A.R. of solution drops on dilution, the E.S.P. drops, and if these changes are gradual enough the soil can be reclaimed from both salinity and sodicity, the rate being dependent of the buffering capacity of the soil with respect to sodium. It is because this process occurs slowly, combined with the high organic matter content and dense topsoil rooting, that the pasture soils remain stable.

<u>4.2 Salt Movement Under Arable Cultivation:</u>
Under arable condition topsoil cultivation and drainage have had two major effects on salt dynamics:

a) <u>Increased Salt Leaching From Topsoil</u> Breaking up the topsoil has allowed the water to move through a range of pore sizes with little by-passing, greatly improving leaching efficency, as illustrated by laboratory experiments (Fig.2). The steep rise in salt content with depth shown in Fig.3a is due to salt leached downwards from topsoil and/or lack of leaching in the dense subsoil.

b) Dispersion of topsoil On ploughing up of pasture the breakdown of the accumulated organic matter reduces soil stability and the loss of the dense rooting mat gives direct exposure of the soil surface to raindrop impact and subsequent dispersion.

Salt removal on installation of drainage is rapid and efficient causing rapid swelling and dispersion of clay in the topsoil. This material moves down the cracks in the subsoil, coating peds, increasing bulk density, blocking drains and drain backfill, and effectively preventing further leaching.

Fig.3a also shows a higher moisture content in the arable subsoils and the greater amount of capillary rise under arable conditions may be due to loss of horizontal cracking on increasing density, although deeper rooting cereals may be responsible.

In fact the topsoil of the arable soil has been partially reclaimed, with an EC_E (saturated extract salinity) of 3.5 and an E.S.P. of 11.5 compared to 6.0 and 17.5 respectively for the grassland. Since the clays are montmorillonite clays dispersion can still be expected until the ESP drops below 10.

5. CONCLUSION

Ploughing of saline/sodic grasslands causes rapid leaching of the plough layer soil, causing dispersion and translocation of clay to the subsoil, where it prevents effective leaching at depth, by increasing the bulk density and coating the clay peds with deflocculated clay. Hence, blocking all pathways of water movement. Under grassland conditions the soils remain more stable and provide better conditions for leaching and reclamation of these problem soils. It is suggested that if these soils are to be reclaimed, installing drainage and applying gypsum under natural grassland conditions will largely prevent the problems of deflocculation encountered in Kent.

REFERENCES

ADDISCOTT, T.M. 1984 "Modelling the interaction between solute leaching and intra-ped diffusion in clay soils," Proceedings of the ISSS Symposium on water and solute movement in heavy clay soils. (I.L.R.I. Publication 37): 279-292.

ANDERSON, J.L., and BOUMA, J. 1977 "Water movement through pedal soils. II, unsaturated flow," Soil Sci. Soc. Am. Jour. 41: 419-423.

BARRACLOUGH, D., RYDEN, M.J. and DAVIES, G.P. 1983 "Fate of Fertiliser nitrogen applied to grassland. I, field leaching results," Jour. of Soil Sci. 33: 13'24.

BLACKMORE, A.V. 1976 "Salt sieving within clay soil aggregate aggregates," Aust. Jour. of Soil Res. 11: 75-82.

KISSEL, D.E., RITCHIE J.T. and BURNETT, E. 1973 "Chloride Movements in undisturbed swelling clay soil," Soil Sci. Soc. Am. Proc. 37: 21-24.

RITCHIE, J.T., KISSEL, D.E. and BURNETT, E. 1972 "Water Movement in undisturbed swelling clay soil," Soil Sci. Soc. Am. Proc. 36: 874-879.

SCOTTER, D.R. 1978 "Preferential solute movement through larger soil voids. I, some computations using simple theory," Aus. Jour. of Soi. Res. 16: 257-267.

US Salinity Laboratory Staff (1954) "Diagnosis and improvement of saline and alkali soils," US Dept. Agric. Handb. 60, 160 pp.

VAN HOORN, J.N. 1983 "Salt transport in heavy clay soil," Proc. of The ISSS Symposium on water and solute movement in heavy clay soils (I.L.R.I. Publications 37): 229-239.

Impermeable Soils Require Stable Channels and Good Crack Formation for Effective Drainage

L.F. Galvin
An Foras Taluntais, Kinsealy Research Centre, Dublin 17, Ireland

ABSTRACT

The effectiveness of various disruption techniques (moles, gravel moles and ripping), installed at one of a number of experimental sites in Ireland is examined and the results of the experimental trials are discussed. Gravel moles are the most effective system on this site, where the moles and ripping have failed. The importance of installing disruption drainage systems under suitable soil moisture conditions is emphasised as is the value of installing shallow moles, as an ancillary drainage system, to deeper moles and gravel moles.

INTRODUCTION

Excess rainfall is a dominant factor in Irish agriculture and leads to many problems in agricultural development. This is particularly so in relation to impermeable soils in the wetter regions of the country where the annual lowland rainfall ranges from 1,000 to 1,700 mm. A further complicating factor is the relatively high April-Sept rainfall. This generally ranges from 400 to 600 mm but ranged from 600 to 850 mm during 1985.

For the effective and economic drainage of impermeable soils, a closely-spaced (1.3 to 1.5 m) system of drainage channels is required. This should be combined with adequate soil cracking to enable rainfall to percolate rapidly to the drainage channels (Leeds-Harrison, Spoor and Godwin, 1982). The disruption techniques generally installed (Galvin, 1982) are mole drains; ripping or subsoiling; and gravel moles (Mulqueen, 1985). The effectiveness of these systems depends on the type and permanency of the crack structure developed during disruption and on the stability of the drainage channels.

EXPERIMENTAL

In 1980/81, under a project partly funded by the EEC experimental trials, incorporating various disruption techniques were installed at a number of locations (Galvin, 1983). One of these trials, at Kilmaley (Co. Clare) in the West of Ireland, is on a carboniferous shale with the following physical properties.

Table 1 Particle size distribution, Atterberg limits and dry bulk density (Mg/m^3) of the subsoil

Size (mm)	% Passing			
40	100			
20	96	Liquid Limit	=	33%
10	89	Plastic Limit	=	22%
5	83	Plasticity Index	=	11%
2	79			
0.6	72	Dry bulk density	=	1.4-1.6
0.2	67			
0.06	59			
0.02	46			
0.006	31			
0.002	20			

Installation

The drainage systems installed, on 1,000 sq. m plots, at the Kilmaley experimental site are:

1. Mole drains spaced at 1.3 m;
2. Gravel moles spaced at 1.3 m;
3. Gravel moles + ripping. The gravel moles are spaced at 2.6 m with intermediate ripping (also spaced at 2.6 m);
4. Ripping at 1.3 m;
5. Control (collector drain at bottom of plot only; no soil disruption).

The collector drains (plastic land drainage pipes covered with 20-14 mm gravel to within 250 mm of the ground surface) were installed in June 1980. Soil disruption could not be carried out in 1980 because the soil and weather conditions were unsuitable. Similar conditions prevailed during the summer of 1981. However, advantage was taken of an improvement in weather conditions in August 1981 to disrupt all the experimental sites. Although the Kilmaley site had not dried out sufficiently for adequate shattering at that date, it was decided nevertheless to proceed with the installation so as to synchronise the starting dates of all the experimental sites. It was also decided that it would be undesirable to postpone the disruption on this site for a second year and that if the results showed that the shatter achieved was not sufficient, attempts would be made to carry out remedial measures and to gauge their effectiveness. This decision was influenced by the consideration that drainage contractors were installing disruption drainage techniques under similar conditions.

Subsequent developments

During the winter/early spring of 1981/82, it was obvious from the wet surface conditions that the disruption on all plots at Kilmaley was far from satisfactory. An attempt was therefore made to improve the cracking of the soil above the level of the disruption channels by installing a system of shallow moles (250-300 mm deep). These moles were pulled mid-way between adjacent disruption channels at the end of April 1982 (this was a very dry month). The expectations were that the shallow moles would collapse very quickly (due to the shallow depth of installation) and that the effect of the installation would be limited to a general loosening and improvement in crack development to a depth of 250-300 mm. In fact the shallow moles did not collapse quickly and were still operating very efficiently 14 months later.

At that stage (July 1983) it was decided to break the moles as their drainage efficiency was masking the effectiveness of the original drainage treatments. Accordingly all the drained plots were again disrupted in two directions at right angles to a depth of 300 mm using a winged subsoiler on July 11, 1983. After disruption the soil surface was disced, levelled using a turbotiller and re-seeded. At the same time the control plot was also disced, turbotilled and re-seeded. No further work was carried out on the plots. However the summer of 1984 was one of the driest and warmest on record in the West of Ireland and soil cracking developed to a far greater extent than normal during that period.

MEASUREMENTS

Measuring equipment on each plot includes a continuous flow recorder, a continuous water level recorder (Talman, 1979) and three maximum-reading piezometers (Davies, 1969). A recording rain gauge is also installed on site.

All piezometers and water level recorders on the drained plots are installed mid-way between adjacent disruption channels. The measurements therefore indicate the highest water levels likely to be achieved on each plot.

Ground scoring assessments are also made, in accordance with the following table:

1. Baked hard and dry;
2. Dry on top;
3. Damp but firm;
4. Damp and firm with occasional wet patches;

5. Damp and soft;
6. Squelchy or ponded patches on 20% of the plot;
7. Squelchy or ponded patches on 20-50% of the plot;
8. Squelchy or ponded patches on more than 50% of the plot;
9. Very soft and waterlogged;
10. Hard frost and snow.

Although this scoring system is obviously subjective, it has nevertheless been found to give a useful indication of drainage effectiveness, when operated by an experienced observer.

RESULTS

Water level fluctuations

The water level fluctuations recorded in the piezometers for Julian years (August to July) 1982/83, 1983/84 and 1984/85 are shown in Figs 1, 3 and 5. To avoid confusion only three hydrographs (representing gravel moles, moles and control) are plotted for each site. The location of the gravel mole + ripping hydrograph is similar to the gravel mole hydrograph for each of the three years. The ripping hydrograph is located between the gravel mole and the mole hydrographs. It is closer to the gravel mole hydrograph in 1982/83, and 1983/84 but is almost identical to the mole hydrograph in 1984/85.

These data are borne out by an examination of Table 2 which outlines the SEW (30) figures for all hydrographs over the wet six-monthly period in each year. The SEW (30) figures (cm days) are calculated by integrating the water level above a line 30 cm below the ground surface.

Table 2 SEW (30) figures for 183-day periods in 1982/83, 1983/84 and 1984/85 (cm days)

	1982/83	1983/84	1984/85
Gravel moles	396	871	414
Gravel moles + rip	336	681	557
Ripping	765	1146	3412
Moles	1622	2191	3499
Control	4344	2514	4290
Rainfall Sept-March (mm)	1205	1151	843

Flow measurements

The flow data recorded on the weekly recorder charts are processed to produce hydrographs of discharge versus time. Three of these are selected to show the flow pattern occurring during the wet periods of 1982/83, 1983/84 and 1984/85 (Figs. 2, 4, 6). In the interests of clarity the discharge hydrograph for the ripping plot is omitted from Figs. 4 and 6. However it lies between the gravel mole + rip and the mole hydrographs in

Fig. 4 and is almost identical to the mole hydrograph in Fig. 6. Further details in regard to the flow records are given in Table 3.

Table 3 Peak flow rate (mm/day) and total discharge (mm) for selected weekly periods in 1982/83, 1983/84 and 1984/85

	Nov 1982		Sept 1983		Nov 1984	
	Peak	Total	Peak	Total	Peak	Total
Gravel moles	53	36	30	53	54	49
Gravel moles + rip	50	28	20	36	27	47
Ripping	54	37	19	15	13	26
Moles	46	31	4	10	14	26
Control	-	-	14	9	2	6
Total rainfall		42		71		61

Ground conditions

The ground condition scoring values, averaged over 183-day periods in each Julian year are outlined in Table 4. In calculating these figures, the "hard frost and snow" category (10) was omitted.

Table 4 Ground condition scoring (average values) for 183-day periods in 1982/83, 1983/84 and 1984/85

	1982/83	1983/84	1984/85
Gravel moles	3.9	3.6	3.6
Gravel moles + rip	4.0	3.6	3.6
Ripping	4.0	4.6	6.1
Moles	4.2	4.7	6.1
Control	8.0	4.9	6.4

DISCUSSION

The original disruption (August 1981) was carried out when the moisture content of the subsoil was too high. Although the disruption channels were installed satisfactorily and surface damage was minimal the lateral cracking was severely limited. It is also possible that the subsequent deterioration of the moles and rip tracks may have been accelerated by their installation at an unsuitable moisture content.

The shallow moles installed in April 1982 were successful in removing the rainfall quickly from all drainage plots. This is borne out by the flow hydrographs (Fig. 2), by the uniformly good ground condition scoring relative to the control plot (Table 4) and by the water level fluctuations (Fig. 1). The variation in water level fluctuations is mainly attributed to the variable cracking achieved during the installation of the major treatments in August 1981. The wider leg of the Gravel mole machine (which was used to install the gravel mole and ripping treatments) developed wider and better cracks than the narrower leg of the mole plough, in the immediate vicinity of the disruption channels. Clods of soil and small sods which fell into these wider channels also improved drainage discharge (Galvin, 1983). The mole channel cracks were difficult to find during excavations and the flow in the deep moles was limited.

The intensive subsoiling, cultivation and re-seeding (July 1983) effectively disrupted the shallow moles. This is evidenced by the reduced peaks of the flow hydrographs (Fig. 4) and by the fall-off in flow from all plots but especially from the moled plot. The water levels are also higher in all plots (Fig. 3 and Table 2) and there is a general disimprovement in drainage effectiveness. Although a major portion of the disimprovement is attributed to the removal of the shallow moles, some surface compaction developed during cultivation and re-seeding may also be a contributory factor.

As already stated, the summer of 1984 was one of the driest and warmest on record. This had the effect of cracking the upper soil layer and this cracking allied to the deep cracks developed in both gravel moled plots resulted in a significant improvement in the drainage of these plots. In fact the drain flow hydrographs (Fig. 6 & Table 3) and water level fluctuations (Fig. 5 & Table 2) show a clear distinction between the gravel moled plots and the moled and ripped plots.

The water level control appears to be similar on both gravel moled plots and the larger peak flows from the gravel moles as compared to the gravel moles + ripping is attributed to the less efficient discharge through the rip tracks, which have deteriorated considerably. On the other hand both the discharge from and the water level control in the moled and ripped plots are practicaly identical for 1984/85.

At this stage (Nov 1985) it seems that the moles and ripping have failed even though the water level control and the discharge from these plots are superior to the control plot. Both the gravel moled plots are satisfactory but the discharge capacity of the rip tracks within the gravel moled + ripping plot are now suspect and may disimprove further.

The results point to the advantages of stable channels and good crack development for the drainage of impermeable soils. The importance of installing disruption treatments under suitable soil moisture conditions is also emphasised. In fact, if the soil moisture conditions are not satisfactory when the site is being drained it might be advisable to install a system of sacrificial moles. These will help to improve the soil moisture status and it may then be possible to install the planned drainage system (moles or gravel moles) later that year or during the following summer under acceptable soil conditions. Finally, the value of installing shallow moles as an ancillary drainage system to deeper moles or gravel moles is emphasised. This relatively inexpensive drainage technique which can be installed by farmers using light tractors may prove very beneficial if installed at regular (2-3 year) intervals.

ACKNOWLEDGEMENT

The author gratefully acknowledges the assistance of M. O' Herlihy, P. Healy, P. McCormack, P. Maloney and L. Foley (Kinsealy Research Centre) and F. Kelly, Kilmaley.

The project was part-funded by the EEC under contract No. 0582. The assistance of the Commission is also gratefully acknowledged.

REFERENCES

Davies, B. (1969). Maximum Piezometer Water Level Indicator. Annual Report, FDEU, Cambridge : 29.

Galvin, L. F. (1982). The Drainage of Impermeble Soils. In "Land Drainage, Proc. EEC Land Drainage Seminar, Cambridge, July 1981", A. A. Balkema, Rotterdam : 85-100.

Galvin, L. F. (1983). The Drainage of Impermeable Soils in High Rainfall Areas. Ir. J. agric. Res., 22 : 161-187.

Leeds-Harrison, P., Spoor, G. and Godwin, R. J. (1982). Water Flow to Mole Drains. J. agric. Engng. Res., 27: 81-91.

Mulqueen, J. (1985). The Development of Gravel Mole Drainage. J. agric. Engng. Res., 32 : 143-151.

Talman, A. J. (1979). Simple Flow Meters and Water Table Meters for Drainage Experiments. Tech. Report 79 (1), FDEU, Cambridge.

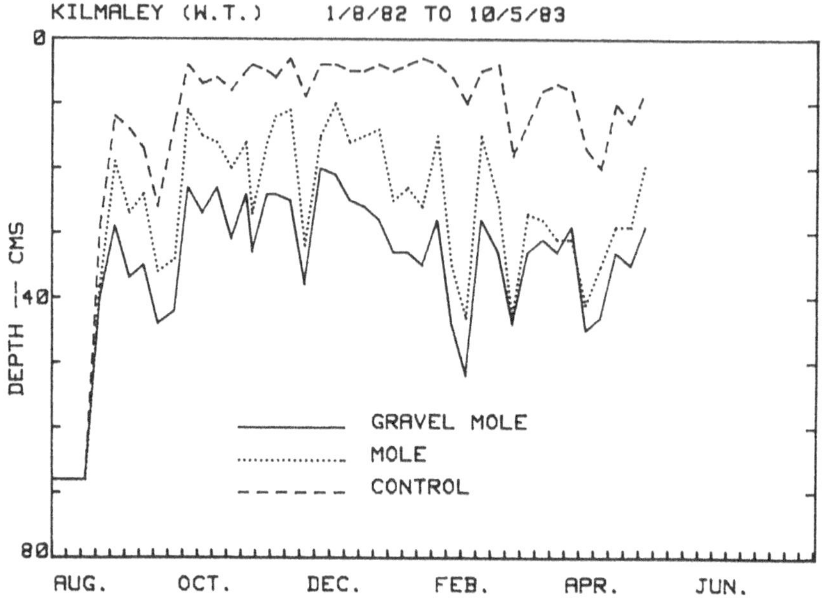

FIGURE 1. Water level fluctuations 1982/83

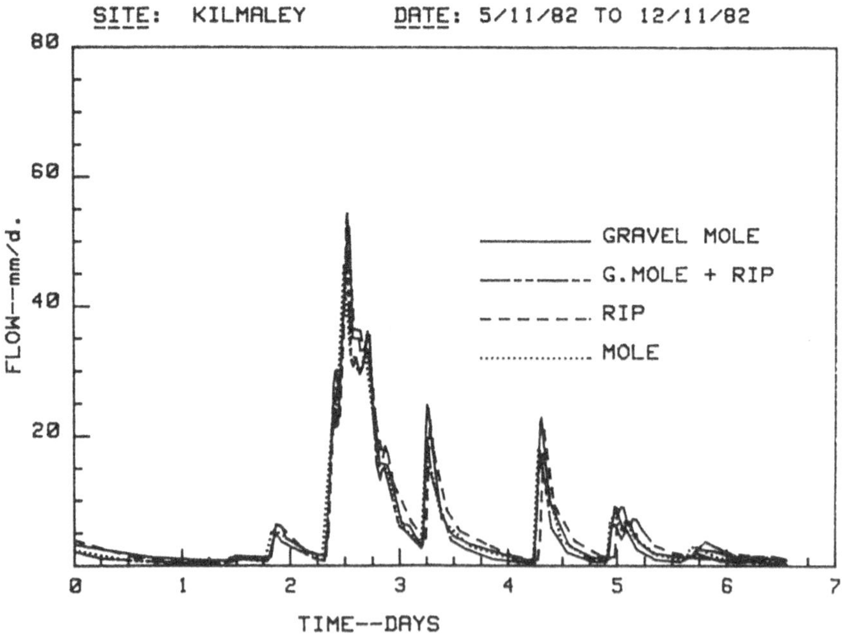

FIGURE 2. Flow hydrographs, 1982/83

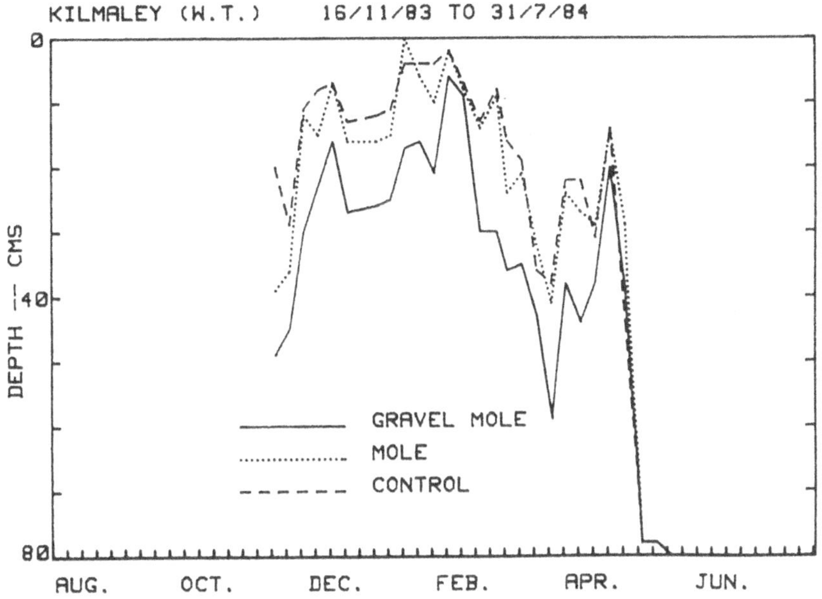

FIGURE 3. Water level fluctuations 1983/84

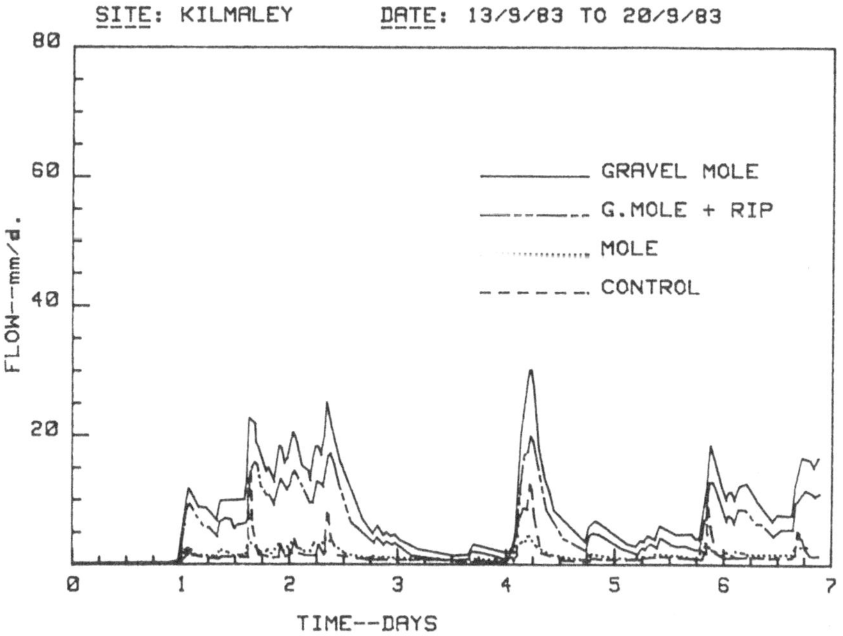

FIGURE 4. Flow hydrographs 1983/84

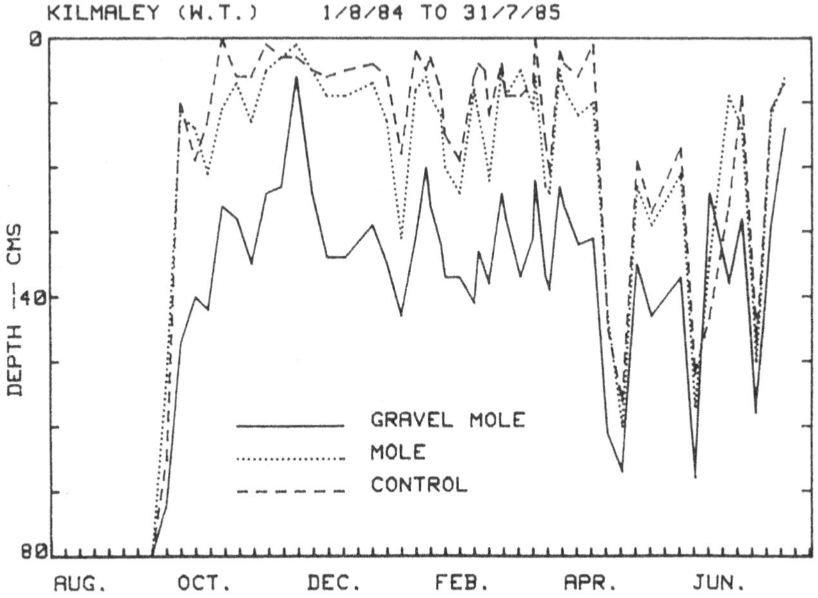

FIGURE 5. Water level fluctuations 1984/85

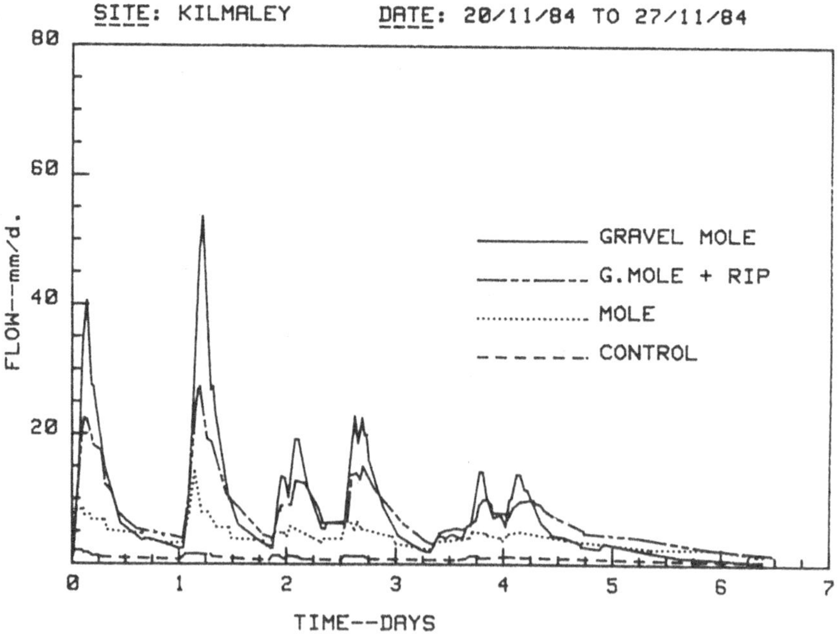

FIGURE 6. Flow hydrographs 1984/85

The Effect of Soil Physical Changes Introduced by Mole Drainage and Cultivation on the Removal of Rainfall

G.L. Harris
Field Drainage Experimental Unit, MAFF, Cambridge, U.K.
M.J. Goss
Rothamsted Experimental Station, Rothamsted, U.K.

1. INTRODUCTION

Mole drainage is widely used in heavy poorly permeable soils where high clay contents make stability of the mole drainage channel likely (Nicholson 1934). The formation of the cracking pattern associated with the passage of the mole bullet through the soil and the importance of the fissures to the flow of water from the top soil to the mole drains has been demonstrated by Godwin et al (1981), and Leeds-Harrison et al (1982). In such heavy clay soils close spaced drains are essential to avoid waterlogging (Trafford and Massey 1975) and when correctly drawn mole drains can give adequate watertable control for both cereal crops and grassland. Most of the early field trials on the effectiveness of mole drainage under cultivations were carried out on ploughed land (Armstrong 1978) with little information on the influence of cultivation on the removal of excess rainfall.

In 1978 a joint experiment between the Ministry of Agriculture, Fisheries and Food Field Drainage Experimental Unit and the Agricultural and Food Research Council Letcombe Laboratory (responsibility was transferred to Rothamsted in 1985) was installed on a clay soil in Oxfordshire with one of the objectives being to investigate the interaction of direct drilling and ploughing with the removal of rainfall through a mole drainage system. The background, site details and the drainage systems were reported by Cannell et al (1984). For the first two years of the experiment both drained and undrained plots were uniformily tine cultivated to 25 cm depth; in subsequent years ploughed and direct drilled treatments were imposed. Drainflow response and the effects on the watertables were reported in detail for the first two years by Harris et al (1984).

In this paper we review the effect of the mole drainage on the removal of rainfall under tine cultivation compared to the effects of five years of ploughing or direct drilling on this process.

2. EXPERIMENTAL DETAILS

The site is located at Brimstone Farm, near Faringdon, Oxon on a heavy clay. A summary of site details is given in Table 1. Drainflow from individually isolated drained and undrained plots was continuously measured using V-notch weirs. The components of flow - surface runoff, interflow (plough layer flow or equivalent depth) and deep drain flow (drained plots only) - were measured on six drained and six undrained plots. After two years of tine cultivation, the cultivation treatments of direct drilling and mouldboard ploughing with secondary cultivation were introduced in autumn 1980, giving three replicates of each drainage/cultivation treatment. The drainage treatment consisted of pipe drains at 0.9m depth, 46m apart with permeable backfill over the pipes and mole drains at about 0.6m depth at right angles to the pipes and 2m apart.

TABLE 1. Summary of site details, Brimstone Farm (Grid ref SU 248947)

Soil Series	Clay content %		Stability index after Haines	Agronomic plot size (ha)	Ave slope %	Crop
	Topsoil	Moling depth				
Denchworth (Stagnogley)	54	60	0.60	0.18	2	Continuous winter cereals Residues burnt

In addition to flow measurements continuous records were collected over the drainage period of depth to watertable using both lined auger holes (dipwells) and tensiometers. Soil structure was assessed by visual observation in profile pits in addition to the infiltration of a 1:4 plaster of Paris to water mix poured on the surface over an area of $0.5m^2$. Deterioration of the mole channels was monitored by polyurethane foam casts of sections of the channel (Thorburn 1984) and non-destructively using an endoscope (Leeds-Harrison et al 1983).

3. RESULTS

3.1 Rainfall

A summary of the rainfall in each year is given in Table 2.

3.2 Watertable control and runoff

3.2.1 Tine cultivation, 1978-80

In the second year of the experiment interception of winter rainfall by the mole drainage system resulted in an increase in the depth to the watertable compared with the undrained plots of the order of 20 cm. However, this difference was not found in the first year even with newly installed mole channels and good crack structure (Cannell et al 1984). This was due to the presence of a cultivation pan that was removed by a winged subsoiler after the first harvest.

Surface runoff was minimal even after heavy and prolonged rainfall in both years (Table 3). Rapid removal of excess winter rainfall through the mole drains was not inhibited by the cultivation pan (Harris et al 1984). Drainflow was continuous over most of the winter period, trickle-flow continuing for up to two weeks after cessation of rainfall. The lag time between peak rainfall and peak drainflow was typically 2-3 hours (Figure 1).

TABLE 2. Rainfall totals 1978-85 (percentage of 30 average of 686 mm)

	1978/79	79/80	80/81	81/82	82/83	83/84	84/85
Oct-Sep	80	95	100	92	99	77	108
Nov-Mar (winter)	82	106	95	98	93	86	108
Jun-Aug (summer)	67	115	76	91	44	39	140

TABLE 3. Mean runoff drained plots as percentage of total runoff per cultivation treatment (Dec-Mar inclusive)

		1978/79	79/80	80/81	81/82	82/83	83/84	84/85
Surface runoff	Ploughed			5	7	3	10	6
	Tined	0.5	5					
	D drilled			14	25	21	4	4
Deep drain	Ploughed			74	84	94	83	92
	Tined	91	90					
	D drilled			80	62	77	94	94

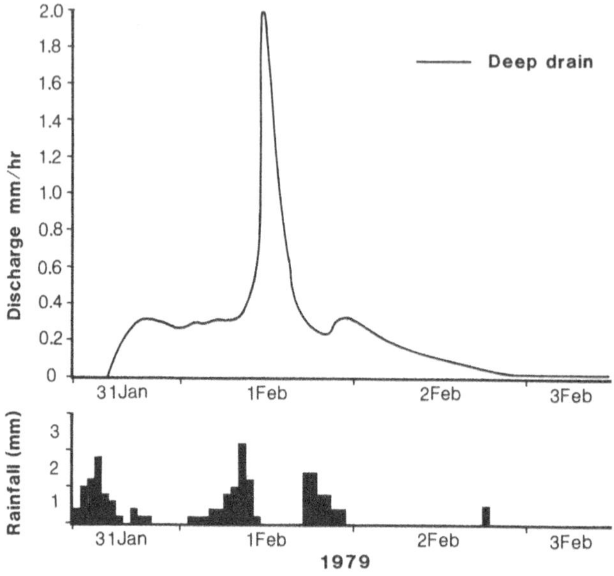

Fig 1. Deep drainflow responses to a winter rainfall event under tine cultivation, 31 January-2 February 1979

Detailed observations were made for a number of rainfall events of the depth to the watertable on drained and undrained plots and compared with the outflow. Flow from the principal collectors for a rainfall event in April 1980, namely the deep drain of the drained plots and the interflow of the undrained plot were compared (Goss et al 1983). This analysis allowed a comparison of the flow from the mole drains with flow in the topsoil without interaction. Peak flow occurred at the same time on both drained and undrained plots and more than 1 hour before the watertable reached its minimum depth below the soil surface. This suggested that flow into the mole drains was largely influenced by water movement in the topsoil.

3.2.2 Differential cultivations, 1980-85

Following the imposition of differential cultivations no consistent difference was found in the depth to the watertable on the drained ploughed and drained direct drilled treatments. However, the mean depth to the watertable for the winter months 1980-85 was 15-20 cm deeper on the drained plots, compared to the undrained plots, depending on the condition of the mole drains and the dryness of the winter.

In contrast, however, a marked change in runoff was noted. On the ploughed plots drainflow rates and timings were similar to those under deep tine cultivation although the progressive deterioration of the mole channels (section 3.3) led to a reduction in peak flows. In contrast on the direct drilled plots peak runoff from the mole drain system was on average 30% higher than from the corresponding ploughed plots and particularly in the autumn period more than 50% higher. The lag time between peak rainfall and peak runoff was also reduced from 2-3 hours for the ploughed treatment to 1½-2 hours under direct drilling. Analysis of a rainfall event in January 1982, shown in Figure 2, indicated that excess rainfall was removed rapidly through the mole drain on the direct drilled plots and drainflow had virtually ceased within 24 hours. In contrast on the ploughed plot extrapolation of the 'recession' hydrograph, to remove the effect of a secondary rainfall event, shows not only substantial drainflow 24 hours after the cessation of rainfall but also steady flow continuing for several days.

As under deep tine cultivation, surface runoff was minimal with ploughing (Table 3), never exceeding 10% of the total runoff. In contrast, on the direct drilled treatment, substantial surface runoff was recorded representing 25% of the total runoff for 1981/82 when rainfall totals were near average (Table 2). Peak surface runoff in response to some winter rainfall events exceeded peak drainflow rates under direct drilling (Figure 3), resulting in shallow erosion rills in this clay soil.

The drained plots were remoled in September 1982 to maintain the effectiveness of the drainage system. Following remoling the quantities and rates of drainflow from the ploughed plots were compatible with new moles and deep tine cultivation. On the direct drilled plots the peakier drainflow response compared to the ploughed treatment was accentuated. In the year 1982/83 when rainfall totals were again near average the remoling had little effect on quantities of surface runoff on the direct drilled plots, small erosion rills being observed after large rainfall events. In contrast surface runoff for the winter period represented only 3% of total runoff on the ploughed treatment (Table 3).

3.3 Mole channel deterioration

Examination of the mole channels under all cultivation treatments showed that the primary mode of failure was due to the piecemeal collapse of the roof and walls. A similar progressive deterioration was observed under both ploughing and direct drilling. By late winter 1982 the mole channels had been reduced to approximately 15% of their original size and were redrawn.

After the dry summer of 1983, dark top soil was found in the larger cracks down to the mole channels. Few complete

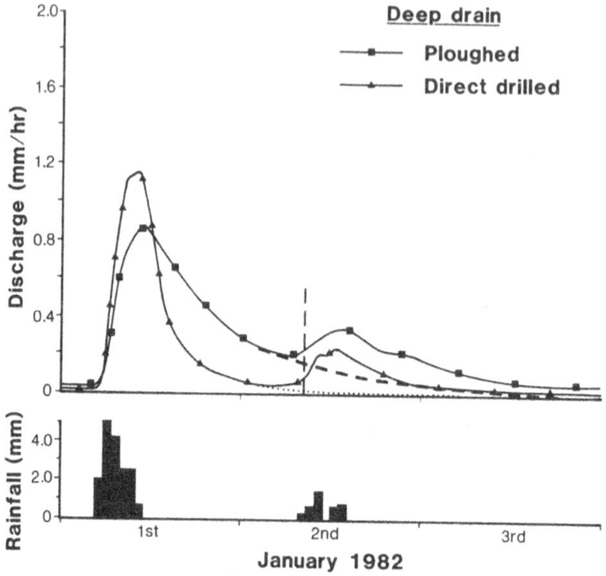

Fig 2. Deep drainflow response for direct drilled and ploughed treatments to a winter rainfall event,January 1-3,1982. Dashed vertical line represents 24 hours after rainfall event

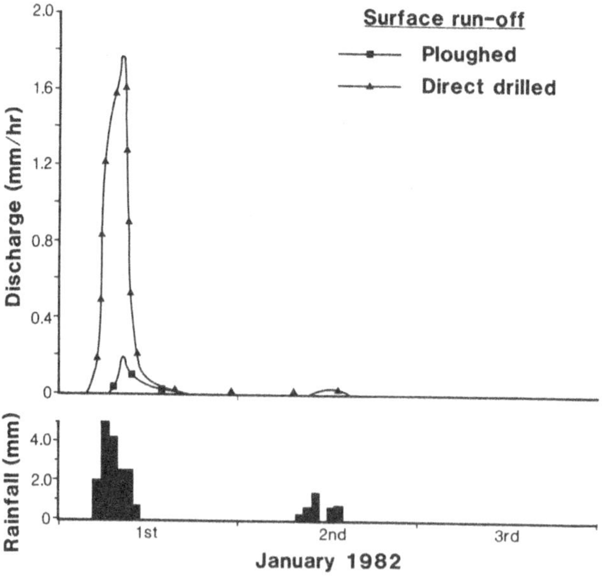

Fig 3. Surface runoff response for direct drilled and ploughed treatments to a winter rainfall event,January 1-3,1982

blockages were recorded, mole channel size on both cultivation treatments being on average 40% of the original size.Following a second very dry summer at least half of the mole channels on the ploughed plot completely collapsed (Figure 4), although seepage zones could be identified in most cases. Mole channels on the direct drilled plots, although somewhat reduced in size, were largely continuous and of a similar size to those open channels remaining on the ploughed plots (Figure 4). It was also observed that the depth to the mole channels on the direct drilled treatment was several centimetres less than the nominal 60 cm depth at which they were drawn.

3.4 Soil structure

The soil structure adjacent to and over the mole drains although variable, tended to be less coarse and better developed under direct drilling. Continuous and more extensive vertical cracking was noted, the effect being particularly pronounced over some mole drains. In such cases, infiltration of plaster of Paris extended down the profile to the mole channel itself either restricted to the original leg slot crack or in a wedge extending outwards from the leg slot above the mole channel. In the top 20 cm of the soil profile many infilled macropores up to 10 mm diameter were noted.

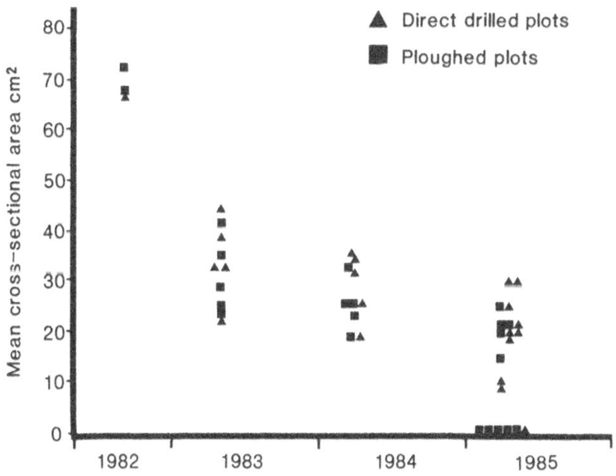

Fig 4. Mole channel deterioration determined from cross-sectional area 1982-1985

In contrast on the ploughed plots, infiltration of the plaster was observed to 20-25 cm depth, at which point a marked break occurred. Infiltration below this depth was generally associated with the leg slot crack. In the horizon immediately above the break many pores in the range 2-10 mm diameter were observed filled with plaster.

4. DISCUSSION

Comparison of the drainflow hydrograph for new mole drains drawn through a former cultivation pan with mole drains where downward movement of water through the soil profile was not inhibited showed a similar response time and peaky hydrograph. The most likely cause was that although the pan restricted the effectiveness of the mole drains on watertable control by sealing off vital flow paths, sufficient cracks remained open in the winter to provide an adequate connection between the top soil and the subsoil. The most prominent of these would be that associated with the passage of the mole leg.

The similarity between the deep drain response and that of flow in the plough layer horizon of undrained land supports the hypothesis that mole flow as a result of a winter rainfall event is dominated by flow from the top soil down the major cracks. The peaky hydrograph shape is similar to that observed by Leeds-Harrison et al (1982) for mole drains drawn with adequate crack structure whereas the hydrograph for mole drains that were four years old showed a tendency towards the hydrograph for mole drains drawn with no cracking. The improved structure throughout the profile seen under the direct drilled treatment is similar to that observed on other tillage sites (Goss et al 1984).

At Brimstone Farm this improved structure was best developed in association with the mole drains and was reflected in a peakier hydrograph (Figure 3). For the rainfall event given in Figure 3 the direct drilled treatment satisfied the requirement suggested by Smith and Trafford (1976) that a mole drainage system should be able to dispose of rainfall within a 24 hour period. Thus the improved structure extended the effective life of the mole channels of the direct drilled treatment. This occurred despite no difference being observed in general in the decay of mole channels under the differential cultivations. However differential decay of the mole channels was observed for the second moling in the winter following two years of exceptionally dry summers. The drought conditions probably accentuated the main mode of failure, namely the piecemeal collapse of the roof and walls of the mole channels on both treatments but after direct drilling a continuous channel remained in the old roof section. It is likely that the improved structure under direct drilling was important in this process.

In another cultivation experiment at Compton Beauchamp, also on the Denchworth Series soil watertable control of the ploughed treatment was better than that of the direct drilled treatment in 1976/77. This suggested that the failure of the mole channels was more rapid on the direct drilled treatment (Goss et al 1984) possibly due to more topsoil infill as deeper cracks developed during the summer drought in 1976. Although at Brimstone Farm topsoil infill of open cracks was observed particularly after dry summers it was not the primary cause of failure and watertable control was similar in both cultivation treatments.

Despite the improved structure on the direct drilled treatment, surface runoff was frequent and substantial. This is in line with observations on other tillage trials where residues were removed (Schwab et al 1975).

5. CONCLUSION

The soil physical changes introduced by mole drainage can lead to an increase in depth to watertable of the order of 20 cm compared with undrained land. For drained land with an adequate mole drainage system a reduction of tillage does not influence the watertable control but the development of enhanced subsoil structure due to continuity of cracks leads to peakier drainflow runoff. Surface runoff can also be substantial under a no tillage treatment.

6. ACKNOWLEDGEMENTS

We thank all our MAFF and AFRC colleagues who helped with the collection and analysis of data.

REFERENCES

Armstrong A C (1978) The effect of drainage treatments on cereal yields: results from experiments on clay lands. J Agric Sci Cambridge 91: 229-235

Cannell R Q, Goss M J, Harris G L, Jarvis N G, Douglas J T, Howse K R and Le Grice S (1984) A study of mole drainage with simplified cultivations for autumn sown crops on a clay soil. 1. Background, experiment and site details, drainage systems, measurement of drainflow and summary of results 1978-80. J Agric Sci Cambridge 102: 539-559

Godwin R J, Spoor G and Leeds-Harrison P B (1981) An experimental investigation into the force mechanics and resulting soil disturbance of mole ploughs. J Agric Engng Res 26: 477-497

Goss M J, Harris G L and Howse K R (1983) Functioning of mole drains in a clay soil. Agric Water Management 6: 27-30

Goss M J, Ehlers W, Boone F R, White I and Howse K R (1984) Effects of soil management practice on soil physical conditions affecting root growth. J Agric Engng Res 30: 131-140

Harris G L, Goss M J, Dowdell R J, Howse K R and Morgan P (1984) A study of mole drainage with simplified cultivations for autumn sown crop on a clay soil. 2. Soil water regimes, water balances and nutrient losses in drain water. J Agric Sci Cambridge 102: 561-581

Leeds-Harrison P B, Spoor G and Godwin R J (1982) Waterflow to mole drains. J Agric Engng Res 27: 81-91

Leeds-Harrison P B, Fry R K, Cronin C J and Gregory J E (1983) A technique for non-destructive monitoring of subsurface drains. J Agric Engng Res 28: 479-484

Nicholson H H (1934) The durability of mole drains. J Agric Sci Cambridge 124: 185-191

Schwab G O, Fausey N R and Weaver C R (1975) Tile and surface drainage of clay soils. II. Hydrologic performance with field crops (1962-1972). III. Corn, oat and soyabean yields (1962-1972). Research Bulletin 1081, 37pp. Ohio Agricultural Research and Development Centre

Smith L P and Trafford B D (1976) Climate and Drainage. Technical Bulletin 34. Ministry of Agriculture, Fisheries and Food, London

Thorburn A A (1984) Materials for mole channel casting - polyurethane foam as an alternative to plaster of Paris. Land Drainage Service Report RD/1. Ministry of Agriculture, Fisheries and Food, London

Trafford B D and Massey W (1975) A design philosophy for heavy soils. Technical Bulletin 75/5 pp32. Field Drainage Experimental Unit, Ministry of Agriculture, Fisheries and Food, London

The Breakdown of Field Drainage Systems due to Deflocculation in the Coastal Marshes of England. The Extent of the Problem and Possible Remedies.

J.G. Rands
LAWS, Great Westminster House, London SW1
R.A. Hodgkinson, A.A. Thorburn
LAWS, FDEU, Cambridge
R.A. Chase
LAWS, Dorchester AO

BACKGROUND

The tidal surge in the southern North Sea on 30 January 1953 flooded vast areas in Norfolk, Suffolk, Essex and Kent. As a consequence of this disaster a programme of raising and strengthening sea walls was put in hand and by the early sixties this work was virtually completed. As well as raising sea walls the opportunity was also taken to improve sluices and the arterial drainage channels of the marshes. About 1963 the large-scale conversion of these marshes from pasture to arable land commenced and by 1976 it was estimated that 85% of the Essex marshes had been ploughed up. The cultivated soils of the marshes in Suffolk, Essex and North Kent are predominantly fine textured and of low permeability. To successfully grow arable crops subsurface drainage is required. The systems initially adopted tended to differ between Kent and the other areas. In Kent the standard approach was to install drains at a depth of 1.2m-0.8m with no permeable backfill and at a spacing of 20m. North of the Thames the use of permeable backfill, usually gravel,over the pipe drain laterals, was virtually universal. Mole drains were then drawn over these gravelled trenches, the distance apart of laterals being either 20m or 40m. Ponding on a wide scale was first reported in North Kent in 1981 when it was observed that not only were the tile drains blocked with deflocculated sediment, but that this material had also blocked the soil pores and had settled in the subsoil to produce an impermeable horizon generally just below the ploughed horizon. As it was known that the North Kent marshes along the Thames Estuary were virtually identical to those on the Essex shore an awareness campaign was instigated by ADAS to determine the extent of the problem elsewhere.The use of the media and ADAS advisory visits brought to light a number of problem fields in Essex, Suffolk and Norfolk but nowhere was the ponding as widespread as North Kent. These enquiries were reported by Rands (1984) and the general conclusions reached were that the problem soils were non- calcareous ground water gleys which have an exchangeable sodium percentage (ESP) of about 10% in the plough layer and over 30% in the subsoil. Conventionally, salt affected soils are defined into three categories as outlined below:-

	ESP (%)	EC (μS/cm)	pH
Saline	<15	>4000	<8.5
Sal-sodic	>15	>4000	<8.5
Sodic	>15	<4000	>8.5 (but in lime free soils may be as low as 6)

The problem soils in England do not correspond with any one of these definitions. The soils concerned fall within the classification of the Wallasea soil series. Since 1984 the survey and investigation of drainage problems on coastal marshes has continued and this paper is an attempt to consolidate some of the original findings and to detail some of the new work initiated.

THE OBJECTIVES OF ADAS SURVEYS AND INVESTIGATIONS

The purpose of the surveys and investigations was to -
i) provide a better data base to identify the problem areas
ii) understand more fully the factors affecting deflocculation
iii) devise drainage and amelioration treatments to allow arable cropping to remain viable.

REVIEW OF DATA FROM MARSHLAND SITES

Scope

The awareness campaign has continued and a liaison group was set up at the Field Drainage Experimental Unit (FDEU) to co-ordinate the information received from the following sources:
i) details of drainage problems reported by farmers to ADAS staff
ii) information from drainage contractors and gypsum/lime spreading contractors
iii) aerial photographic surveys
iv) examination of other ADAS R&D projects to see if there was any relevant data
v) field investigations - a) general b) experiments
vi) laboratory soil stability studies.

Procedure

The items above are frequently related to one another and indicate a progression from a farmer's enquiry to an experiment on his land. The work has generated a geographically extensive low key data base interspersed with a small number of well documented sites.

RESULTS

The information survey

As a part of the awareness campaign, leaflets were produced to allow local ADAS drainage advisers to diagnose a deflocculation problem (see LeGrice (1982), Rands (1984a)). In Kent further sites along the Thames Estuary have been identified and redrainage work has been necessary on some parts of the Isle of Sheppey. Problems (albeit not severe) have been reported from some sites where the marsh clay is calcareous (the Newchurch Soil Association). In Essex the deflocculation problem was reported as being widespread although no severe cases of crop loss were reported. Each year about 1,000 ha of marsh are treated with gypsum. Along the East Anglian coast north of Essex a more complicated picture has emerged. Whilst in Essex the problem areas tend to be the areas mapped on the Soil Survey 1:25,000 Sheet as the Wallasea Association (described by Hodge et al (1984), in Suffolk this association does not always give problems. The problem area appears to be along the Alde marshes, whilst those marshes adjacent to the estuaries of the Orwell and Deben are relatively trouble free. For this reason fields along the Orwell Estuary at Trimley were investigated. The Wallasea Association has been mapped along the North Norfolk coast and deflocculation problems have been reported in this area. The marshes of the Bure and Yare are mapped as being predominantly of the Newchurch Association but as in Kent this association is not completely trouble free. Clay blocking of drainage systems has been discovered away from the North Sea coast in two fields, one adjacent to the Fal Estuary in Cornwall and another near the Bristol Channel at Brent Knoll.

Aerial photographic surveys

This work was undertaken by the ADAS Aerial Photographic Unit at Cambridge and can be classified into two types -

i) small-scale detailed photography of known problem areas
ii) general surveys of marshland blocks.

The former type has been more rewarding than the latter as some of the material examined has been from the pre 1981 archives. These photographs show not only the waterlogged areas that were the subject of the earlier investigations (not recognised at the time as due to a deflocculation problem), but also wet areas on adjacent land, soil patterns, details of bulldozing and dyke infilling, and the location of gypsum or lime heaps. Since 1983 coastal areas of East Anglia and

North Kent have had a lower than average winter rainfall and ponding due to deflocculation has not been recorded. Recent aerial photography has been of little value in indicating problem areas. Several farmers in Essex who have applied gypsum to their marsh fields have co-operated with ADAS by leaving small untreated plots. These trials have been photographed and in most cases the untreated areas show up as areas of poor crop.

Data from other ADAS drainage R&D projects

Soil physical properties and hydraulic conductivity. National Investigation DW16 The data produced for the national DW16 investigation was examined to obtain information related to coastal marsh soils. The K values were measured using the single auger hole technique. Data from thirteen marsh fields has been collated from sites located in Lincolnshire, Suffolk and Kent. On the 1:250,000 soils map all the sites are on the Wallasea Association with the exception of the three Waveney Valley sites in Suffolk which are on the Newchurch Association. The most striking fact to emerge is the difference in K values between arable drained and grassland undrained sites.

TABLE 1. DW16 data
Hydraulic conductivity values of marshland sites

Mean site values	Arable sites								Group mean K
K (m/day)	.0046	.005	.013	.0146	.06	.073	.251	21.9 *1	2.79 (0.06) *2
% clay	66 *3?	57 *3	38	64 *3?	61 *3	21	18	45	
	Grassland sites								
K (m/day)	.133	1.87	4.45	7.45	7.92				4.36
% clay	70	58	31	32	50				

*1 Atypical site low pH (3.4)
*2 If value of 21.9 is discarded
*3 Deflocculation reported *3? Some evidence

Follow-up visits to potential problem sites The purpose of this exercise was primarily to record details of sites which were:

a) sites where deflocculation was noted which differed from the problem sites already described by Rands (1984)

b) fine textured marsh sites where deflocculation was not a serious problem.

Two sites in Norfolk came into category a) above, and details of these sites appear in Table 2 (site reference numbers a(i) and a(ii). A number of fields were investigated on Trimley Marsh in Suffolk and whilst many of the profiles were typical Wallasea showing signs of deflocculation, cheek by jowl with these profiles occurred a Wallasea variant which is akin to the Normoor series described by Hodge et al (1984). The presence of jarosite in the subsoil was a feature of these profiles. Where jarosite was found in the subsoil, iron ochre rather than deflocculated clay, was the characteristic deposit found in the drains adjacent. Profile details for this situation are given in Table 2, site b(i). Another field investigated on Trimley Marsh was one where rapid mole channel collapse had allegedly taken place. Mole channels were found and no signs of deflocculation were noted. This site is b(ii) in Table 2.

TABLE 2. Soil data from problem sites in Norfolk and non-problem sites in Suffolk

Site	Location	Grid ref.	Depth (cm)	$CaCO_3$ (%)	Clay %	ESP	Cl(mg(l)	pH
ai	Norfolk	TF 909413	20-40	6.3	37	32.1	316	N/A
			40-60	3.2	40	31.6	785	N/A
aii	Norfolk	TG 478078	20-40	0.2	47	9.9	79	N/A
			40-60	2.1	34	17.1	115	N/A
bi	Suffolk	TM 258352	20-40	<0.1	55	4.2	0	5.1
			40-60	<0.1	58	9.2	50	4.5
bii	Suffolk	TM 255364	20-40	<0.1	48	17.3	320	6.9
			40-60	<0.1	49	16.9	671	7.0

Mole channel investigations

As mole drainage was widely practised in East Anglia but was uncommon in Kent the mode of channel collapse and the length of channel life was considered worthy of study to see if it was soil properties rather than prejudice which had created this regional difference in drainage practice. The techniques used in this investigation were the in situ casting of mole channels described by Talman (1976), and Thornburn (1980); also an endoscope was used to record, photographically, mole channel collapse (see Thorburn 1985).

In situ casting This is a destructive form of sampling which can be very time-consuming. Casts were taken at Saltfleetby (Lincs), Trimley Marsh (Suffolk), Wigborough and Canewdon (Essex). Details of these investigations are given in Table 3.

Endoscope studies In the autumn of 1984 endoscope access tubes were installed in new mole drains drawn at Canewdon, Essex and St Mary's Marsh, Kent. The photographs are still being studied but the tentative results show that successful channel formation is particularly dependent on correct plough setting with respect to the soil plasticity. Early, partial channel collapse, due to seasonal soil shrinkage, is followed by slaking of the collapsed fragments to form a slurry.

This slaking is markedly retarded by gypsum treatment.

TABLE 3. Mole channel deterioration on marsh soils. Details of channel size and soil at moling depth

Site	CaCO3 (%)	Clay (%)	ESP	Moling date	Inspections Date	CSA(*1)	Date	CSA
Saltfleetby	0.3	64	17	9/79	4/80	25	4/82	40
Trimley	<0.1	49	16.9	11/81*2	9/83	37	-	-
				9/82*3	9/83	45	-	-
Wigborough	<0.1	58	20	9/83	3/84	48	-	-
Canewdon	<0.1	68	31	9/83	3/84	48	-	-

*1 CSA - cross sectional area of mole channel as a percentage of the original size
*2 - wet conditions
*3 - dry conditions

Monitoring redrained problem fields

An experiment was installed in September/October 1984 and the data for the first year is still in the process of being examined. On this site there is an ineffective old system with soil backfilled drains at 20m spacing, with a new system installed midway between the old drains. The plot treatments are:

Treatment A Gypsum on drain trench spoil only. Equivalent to 37.5 t/ha
Treatment B A + gypsum spread overall at 12.5 t/ha
Treatment C B + mole drainage
Treatment D Redrainage only.

The new pipes are in trenches, gravel backfilled to within 15 cm of the surface. Whilst the first year was relatively dry preliminary results indicate a response to the gypsum overall treatments in giving a better surface structure and a reduction in the exchangeable sodium level (see Table 4).

Laboratory soil stability studies

Soil analyses were undertaken on the 'follow-up visits' sites previously mentioned. These have generally proved useful in indicating problem soils. The analyses are however time-consuming and therefore inappropriate for a farming community which would like a rapid test to determine, for example, gypsum need and application rate. Hodgkinson and Barribal (personal communication) have compared various stability tests and have devised a technique for measuring the dispersion ratio which unlike many of the standard techniques is not operator sensitive. Preliminary results indicate a useful correlation between dispersion ratio and chemical composition. The data in Table 5 also illustrate their relationship to known problem sites.

TABLE 4. Redrainage experiment at St Mary's, April 1985 Results

Treatment	Depth (cm)	Peerlkamp ST Index*	Conductivity(μS/cm) (water extract)	Exchangeable sodium %
A	0-15	4.6	710	9.7
	15-30	-	-	20.3
B	0-15	6.9	1477	8.5
	15-30	-	-	18.5
C	0-15	6.7	947	7.5
	15-30	-	-	17.2
D	0-15	3.9	397	11.2
	15-30	-	-	22.1

*Soil structure score (Peerlkamp 1967) eg
ST3=Plough layer large dense aggregates smooth crack faces OR top 6 cm angular dense aggregates very dense below 6 cm.
ST5=Plough layer large though porous aggregates rather smooth crack faces OR top 7-8 cm small porous aggregates with denser layer below.
ST7=Plough layer mostly porous crumbs combined as porous aggregates occasional denser clods.
Data kindly provided by M Marks, ADAS, Wye

TABLE 5. Selected dispersion ratios

Site		Depth (cm)	Dispersion ratio %	Clay %	Exchangeable sodium %	Carbonate %
Peldon		0-20	16	59	12.1	1.6
Essex	*1	20-40	88	65	30.2	0.2
		40-60	83	89	42.1	0.6
Trimley G		0-25	7	53	2.3	0.2
Suffolk	*2	25-60	10	90	14.9	0.1
		60-80	7	57	13.6	<0.1
Tillingham		0-20	10	63	4	<0.1
Essex	*3	20-40	55	67	9	0.1
		40-60	13	79	15	<0.1
St Mary's		0-20	58	61	14.1	0.1
Kent	*1	20-40	71	64	25.0	0.2
		40-60	76	63	33.3	0.5
Ammunition		0-20	6	62	3.3	0.1
Kent	*4	20-40	13	66	7.4	0.1
		40-60	12	66	15.0	3.4

*1 - known problem drainage site
*2 - non-problem drainage site but lime needed to combat acidity
*3 - non-problem site
*4 - undrained but unlikely to give rise to severe deflocculation?

DISCUSSION

The first reported case of waterlogging problems occurring on a drained, fine textured, marine, alluvial soil was in Essex in 1969. At that time there was no detailed soil survey information available and the incident was dismissed as an isolated occurrence of a sodic soil. With the publication of the 1:250,000 National Soil Map the distribution of the areas at risk can be estimated, and if these areas are equated with the WALLASEA I ASSOCIATION then,nationally,deflocculation could occur on 51,600 ha, with 19,600 ha in Eastern Region and 16,300 ha in South East Region, mainly in North Kent. Characterisation of the problem soils (Rands 1984) as non-calcareous ground water gley soils which have an exchangeable sodium percentage of about 10 in the plough layer and 30 in the subsoil holds good for the majority of problem sites but exceptions do occur. The Norfolk sites listed in Table 2 are exceptional with site a(i) being calcareous and site a(ii) having a generally low ESP. In Essex at Wigborough (Rands 1984) and in Suffolk at Trimley areas of calcareous soils were not associated with deflocculation problems. It is possible that on some sites, analysed as calcareous, the calcium carbonate is present in an inactive form, perhaps hard shell fragments. Analogously, finely divided phospho-gypsum has been found to be a better ameliorant that granular gypsum "seconds" because of differences in solubility. Another possibility is that deflocculated material that was generated in a patch of Wallasea has moved and been redeposited in an area of the calcareous Newchurch soil series.

Where intensive sampling exercises have been carried out wide variations in ESP levels have commonly been found reflecting, perhaps, the original sedimentation/creek pattern of the marsh, areas where salt water inundation has been more severe, or areas where land levelling has brought sodium rich subsoil into the surface layers. These factors might well account for problems in Norfolk site b(ii) (see Table 2) which would normally be classified as a non-problem Wallasea. Soil variability was highlighted on Trimley Marsh where the pH of the soil at moling depth varied between 4.8 and 8.1 from samples taken about 40m apart. The occurrence in Suffolk of acid sulphate soils within the Wallasea and Newchurch Associations is not uncommon, another example being site (*1) listed on Table 1. This arable site in the Waveney Valley at Norton Marsh had an extremely high K value (21.9m/day) in complete contrast to most clay marsh soils and in particular deflocculated Wallasea soils,where mean values of .005-.06m/day have been recorded. Table 1 also highlights the difference in K values between arable and grassland sites. Mole channel deterioration, as would be expected, is rapid on problem sites but appears to stabilise after a year. This would suggest that when the slot is wide fines can migrate into the mole channel. After a year this route becomes limited and the deflocculated slurry tends to shrink during the summer to allow some channel size recovery. Whilst channel life might be no more than three years, moling would still be viable.A case of mole channel failure was reported at Trimley (see Table 3) and remoling took place. Subsequent investigations indicated that both sets of moles were still in existence, the first set having been sealed by a cultivation pan. At Wigborough channels formed by a mole plough and a subsoiler were found in the same pit. Whilst subsoiling has replaced moling on many Essex marshes, it could well be that in reality the farmer has replaced conventional moling with a more intensive but shallow "square" moling system.

The soil stability laboratory work is still in its early stages but could prove to be the most appropriate method for diagnosing problem marsh soils.

Based on limited data, the use of finely divided gypsum is beneficial. This finding has also been confirmed by work on similar soils in Portugal (Mann 1982, van Hoorn 1985). Redrainage incorporating gypsum treatments would appear to be a more attractive proposition than land re-shaping on failed sites.

Rig and furrow drainage on arable sites has been tried and found inadequate on coastal marshes in Denmark (Nielsen 1982).

CONCLUSIONS

i) The work indicates that deflocculation problems are more widespread than was at first envisaged.
ii) Whilst the problem soil type as described by Rands (1984) is the major cause for concern, some calcareous soils give problems.
iii) Soil variability within a marsh is commonplace and a single point sample for assessing gypsum need is inappropriate.
iv) Gypsum application has been beneficial on the new sites monitored.
v) Apparent rapid mole drainage failure on the Wallasea series might be related more to incomplete investigation than to a completely unstable soil.
iv) The hydraulic conductivity of undrained marsh sites is high; gypsum application before ploughing and draining is desirable to maintain this good structure.

REFERENCES

Armstrong A C (in press) Soil physical properties and hydraulic conductivity (Project DW16). Interim Report to 1983. R&D Report. MAFF, ADAS, Land and Water Service

Le Grice S (1982) Drainage failure in reclaimed salt marsh soils. Technical Note TN/FE/02. MAFF, ADAS, Land and Water Service

Hodge C A H, R Burton, W Corbett, R Evans and R Seale (1984) Soils and their use in Eastern England. Bulletin No 13. Soil Survey of England and Wales, Harpenden

Hodgkinson R A and K Barribal (1985) Personal communication

van Hoorn J W (1985) Wageningen. Personal communication

Mann M, A Pissara and J W van Hoorn (1982) Drainage and desalinization of a heavy clay soil in Portugal. Agriculture Water Management, vol 5 No 3, p227-240

Nielsen C (1982) Afvandingsmetoder pa svaer marskjord (Drainage methods on a heavy marsh soil in Denmark) Tidsskr Planteavl 86 415-425

Peerlkamp P K (1967) West European methods for soil structure determination. Ed:Int.Soil Sci.Soc. Pub: State Fac.Ag.Sci.Ghent

Rands J G (1984) Deterioration des systemes de drainage souterrain due a la defloculation dans les terrains marecageux Comptes rendus du collaque. Functionnement hydrique et comportement des sols. Dijon 22-25 Mai 1984. AFES 78370 Plaisir 257-282

Rands J G (1984a) Marsh soils - stability and drainage problems in Eastern Region. Technical Note TN/FE/12. MAFF, ADAS, Land and Water Service

Sturdy R G (1976) Soils in Essex II Sheet TQ 99 (Burnham-on-Crouch). Soil Survey Record No 40. SSEW, Harpenden

Talman A (1976) The technique of plaster casting mole channels. Technical Bulletin 76/5. MAFF Field Drainage Experimental Unit

Thorburn A A (1980) Materials for mole channel casting. Polyurethane foam as an alternative to plaster of Paris. Land Drainage Service, R&D Report No 1, MAFF, ADAS

Thorburn A A (1985) The use of an endoscope for recording field drain conditions. Technical Note TN/FE/24. MAFF, ADAS Land and Water Service

Extension of Mole Drainage Practices

G. Spoor and P.B. Leeds Harrison
Department of Agricultural Engineering, Silsoe College, Cranfield Institute of Technology, Silsoe, Bedford, England, MK45 4DT.

ABSTRACT

Most current mole drainage systems are installed using a standard mole plough. This limits the application of the technique to a narrow range of soils and field situations. This paper illustrates, with reference to both humid and arid areas, how situations, currently considered unsuitable for mole drainage, can be successfully moled by changing the mole plough design and installation technique. Such modifications have enabled the technique to be adopted on marginal soils in humid areas and used on an experimental scale for the reclamation of saline alkali soils in Egypt.

INTRODUCTION

Effective drainage of fine textured soils of low hydraulic conductivity requires the use of close spaced drains. The exception to this is where more permeable soil horizons are present at drain depth. Drainage costs using traditional pipe systems at these close spacings are frequently uneconomic, with the result that wider spacings tend to be used and drainage performance is poor. One alternative is to use mole drains, which if appropriate for the conditions provide a cheap effective system. Unfortunately, mole systems are not always successful or reliable in all situations. This paper examines the technique in terms of recent developments and identifies with examples, some ways through which it may be possible to successfully extend the use of the system in future, in both humid and arid situations.

TRADITIONAL MOLING SITUATIONS

Currently, mole drainage is only used extensively in a few humid temperate countries, primarily for the rapid removal of surface water. Although used in some groundwater control situations, the practice is not common. The design of mole plough used for installation is very similar all around the world and have evolved from developments in local areas on a limited range of soils.

The major investigations into the technique during this century have been largely concerned with identifying the main soil and climatic factors influencing the success of the operation using the standard mole plough. Particular attention has been paid to parameters such as soil texture, stability of soil aggregates to wetting, soil moisture status at the time of moling and timing in terms of season, summarised by Nicholson (1942), Hudson et al (1962), Trafford and Massey (1975) and Trafford (1977). This has helped to identify the most suitable moling situations for success using the standard mole plough, but has not advanced the development of the technique for use over a wider range of conditions.

Work by Hudson et al (1962) and more recently by Godwin et al (1981) and Spoor and Ford (1986) shows clearly that the mole plough itself has just as large an influence on success in the soil and climatic factors. Changes to mole plough design and installation technique, therefore, offer opportunities in the future to extend the practice successfully to a wider range of situations.

SOIL DISTURBANCE AND DRAINAGE PERFORMANCE OF MOLES INSTALLED USING THE STANDARD MOLE PLOUGH

The standard mole plough comprises, usually, of a 75 mm diameter cylindrical shaped foot attached to a 20-25 mm wide vertical leg and followed by a 85-100 mm diameter expander. The foot and expander create the drainage channel and the leg leaves a slot and produces associated soil fissures as illustrated in Fig. 1.

In surface water control situations, the water flow route to the mole drainage channel is very dependent upon the extent of leg slot and fissure development. With well developed leg slot and fissures, surface water tends to move laterally over the surface, or through the more permeable top soil towards the leg slot area. It then flows vertically downwards through the slot and fissures into the mole channel, see Leeds-Harrison et al (1982). In situations where the leg slot and fissures are poorly developed or absent, flow is through the complete soil profile between the mole drains. The discharge hydrographs following a rainfall input for both well fissured and zero fissured situations are shown in Fig. 2. The overall drainage efficiency in terms of total water removal was similar at approximately 80% in both cases, but the rate of discharge was considerably higher in the fissured condition.

EXTENSION OF MOLING PRACTICES IN HUMID AREAS

The common drainage requirement in surface water problem humid areas is rapid surface water removal. This requires a stable mole channel with well developed leg slot and fissures connecting the soil surface layers directly to the channel. Wherever problems of low channel stability or poor leg fissure development occur, mole drainage is unsatisfactory. Any change in the technique to counteract these undesirable effects will enable the practice to be extended.

In situations where leg fissure development is likely to be weak at installation, common in soils of low bulk density at high moisture contents, larger fissures can be generated by increasing the sliding resistance at the side of the mole plough leg. This can be achieved by increasing either the width or side area of the leg, see Webb (1981). When fissure development is excessive, the cracks can be reduced by narrowing the leg.

The mole channel itself can fail through one of a number of failure mechanisms depending upon the circumstances, see Spoor and Ford (1986). Two common types of failure are those of expander and unconfined swelling. Expander failure is where the soil in the channel roof, deformed by the expander at the time of installation, collapses into the channel. This type of failure is particularly common on the higher density smectitic clay soils. The risk of such failure can be reduced in these circumstances, by using a larger diameter expander which reworks the deformed soil to a greater extent increasing roof stability. Micaceous clay soils are more susceptible to unconfined swelling failure, where the channel diameter decreases with time due to the steady swelling of the soil in the channel wall area. The formation of a larger diameter channel initially in situations where swelling is rapid, will prolong the life of the channel.

The mole channel and leg fissures are particularly prone to rapid collapse if wetted for prolonged periods soon after formation. Mole drainage is therefore often unsuccessful if large quantities of free water are present in the soil profile at the time of installation, since this causes immediate wetting. The use of a sacrificial moling installation to dewater the profile will help extend the life of mole channels installed subsequently.

Long beam mole ploughs are commonly used in the major moling areas to minimise the effects of local surface undulations on mole channel grade. This is frequently adequate on sloping sites, but on flatter areas the adoption of a more positive grading system will help avoid backgrades, and so extend channel life and improve efficiency.

The total cost of mole drainage installations is very dependent upon the cost of the associated collector drainage system. The increased use in recent years of stone backfilled pipe drains as collectors has significantly increased costs to the point, in some cases, where drainage is no longer cost effective. The use of suitably protected open ditch outfalls or a reversion to mole main collectors may in certain circumstances reduce costs so allowing the use of the technique to be extended.

USE OF MOLE DRAINAGE FOR THE RECLAMATION OF SALINE AND SALINE ALKALI SOILS

Mole drainage in irrigated areas using standard techniques has not been particularly successful in the past for either water table control or reclamation purposes. Any future success will be dependent upon developing new techniques.

The reclamation requirement in saline and saline alkali soil areas is for uniform leaching to provide uniform salt removal from the profile across the whole field. Standard mole drainage installations with associated leg slot and fissured areas, providing a direct connection between the surface and the channel are not likely, therefore, to be ideally suited for uniform leaching. With ponded surface water, most of the water flow through the profile would be concentrated in the local fissured area near the drain. In addition, any higher velocity flow through such fissures will be likely to cause serious scour and erosion, particularly in the less structurally stable soils, which are common when exchangeable sodium levels are high.

The standard moling technique therefore needs to be modified for this situation, to allow more uniform leaching. This could be achieved by either fissuring the complete profile between the drains or sealing off the leg slot and fissures at the surface. Both treatments would provide more uniform, yet different conditions. The complete fissuring technique compared with crack sealing should increase discharge rates and hence reduce leaching time, but it will have a higher erosion risk.

Soil aggregate stability on wetting is a critical factor influencing the success of mole drainage in humid areas. Soils of low aggregate stability are not particularly suitable for moling. In the saline soil reclamation situation, however, the leaching water will have a high soluble salt concentration, which should tend to buffer some of the instability effects caused by any exchangeable sodium. It should be possible, therefore, to use mole drainage successfully in reclamation situations, on soils of lower aggregate stability than would be considered suitable in humid areas.

Preliminary investigations have been carried out by Silsoe College in the Zawia sector of the northern Nile Delta, Egypt, to assess the feasibility of using mole drainage for the reclamation of saline alkali soils, see Cronin (1982) and Spoor et al (1986). The cracking smectitic clay soils in the experimental area contained approximately 90% clay. Initial electrical conductivity values of the saturation extract were in excess of 50 dS. m^{-1} and the exchangeable sodium percentage ranged between 40 and 65. Soil aggregate stability was extremely low, varying between very marginal to unsuitable for moling on a basis of the Childs Stability Test, Childs (1942) and the soils were highly erodible. Further details of the soil conditions are available in Sanmuganathan and Taylor (1978).

Initial investigations showed that complete fissuring between the mole runs was completely unsatisfactory under these conditions. In most cases severe erosion and washouts occurred. In the remainer the channel was rapidly blocked with soil in a slurry condition. Following the development of a suitable moling and leg slot and fissure ceiling technique, it was possible to access the feasibility of using mole drains for reclamation.

Mole drains were installed at 0.7 m depth at 1.8 m spacing using a 75 mm diameter foot and 150 mm expander. The large expander was used to minimise the risk of expander failure. The leg slot and fissures were sealed immediately and the surface and the channels were allowed to age at a constant moisture content, for periods of 3-6 weeks before having to carry any water. Any longer maturation period would have allowed the surface shrinkage cracks which continued to develop, to reconnect with the mole plough leg fissures at depth, increasing the erosion risks.

The first flood irrigation was carried out carefully to ensure no erosion failures occurred to any remaining connecting fissures. Leaching proceeded by continuous ponding. Fig. 3 shows the change in the electrical conductivity of the saturation extract throughout the profile, following a 10 week leaching period on one of the plots. Very significant reductions in salt levels can be noted with no statistically significant difference in salt depletion between the mid drain position and that close to the drain, indicating uniform leaching.

The mole drains remained quite stable over this period and leaching could have continued much longer. Fig. 4 shows the change in the electrical conductivity of the drainage water over an 8 week period for two plots. The fissure development throughout the profile on plot A was greater than on plot B, hence encouraging some preferential water flow near the drain. It can be noted salt removal fell off more rapidly on plot A due to the greater proportion of larger crack flow. It is most probable that the high salinity levels in the drainage water were responsible for the very satisfactory mole channel stability.

A larger scale 1 hectare leaching trial reported in Pearce (1984) proved less successful, but failure was not due to the moling technique. 166 mole runs, each 25 m long discharging into open ditches were installed and irrigated for periods varying between 2-6 weeks. Few problems of mole channel collapse were experienced, except in one local area where sand lens problems existed. Unfortunately, the irrigation supply was terminated prematurely before significant leaching could occur and the area remained dry for a period of approximately 2 months. On the first wetting after the interval, 80% of the mole channels collapsed immediately. This was due to a combination of shrinkage cracks developing from the surface and connecting with the mole leg fissures at depth and drying within the mole channel itself, as a result of air movement from the open ditch. Both these problems can easily be avoided in future.

The results from these preliminary leaching trials prove that the use of mole drains to assist in the reclamation of fine textured saline alkali soils is technically feasible and the method offers great potential for the future. Further development is now required to truly assess its potential for use on a larger commercial scale.

DISCUSSION AND CONCLUSIONS

Possibilities for successfully extending mole drainage practices on fine textured soils in both humid areas and reclamation situations are good, providing investigators consider modifications to the technique, where necessary, to suit the prevailing conditions. The technique can only be successfully modified, however, after the existing problems have been identified. The use of a borescope, see Leeds-Harrison et al (1983) to observe mole channel behaviour and identify the particular failure mechanisms is extremely helpful in this respect.

Whilst mole drainage will never be able to solve the drainage problems of all fine textured soils, it has a much larger contribution to make that is being exploited at the present time. Implement and technique development have an important part to play in this exploitation.

ACKNOWLEDGEMENTS

The authors wish to express their thanks to the Agricultural and Food Research Council, Ministry of Agriculture, Fisheries and Food, London, the Delta Sugar Company, Egypt, and the Overseas Development Unit, Hydraulics Research, Wallingford for providing the facilities, resources and finance for these investigations.

REFERENCES

Childs, E.C. (1942) Stability of clay soils. Soil Sci., 53, 79-92.

Cronin, C.J. (1982) Mole drainage for the reclamation of fine textured saline-alkali soils. MSc Thesis, Silsoe College, Cranfield Institute of Technology.

Godwin, R.J., Spoor, G., Leeds-Harrison, P.B. (1981) An experimental investigation into the force mechanics and resulting soil disturbance of mole ploughs. J. Agric. Engnn. Res., 26, 477-497.

Hudson, A.N., Hopewell, H.G., Bowler, D.G., Cross, M.W. (1962) The draining of farmlands. New Zealand, Massey College.

Leeds-Harrison, P.B., Spoor, G., Godwin, R.J. (1982) Water flow to mole drains. J. Agric. Engnn. Res., 27, 81-91.

Leeds-Harrison, P.B., Fry. R.K., Cronin, C.J., Gregory, J.E. (1984) A technique for the non-destructive monitoring of sub-surface drains. J. Agric. Engnn. Res., 28, 479-484.

Pearce, G.R. (1984) Zawia saline soil reclamation project. Report OD 60, Hydraulics Research, Wallingford.

Sanmuganathan, K., Taylor, A.L. (1978) Zawia reclamation project: 1. Instrumentation and data collection. Report OD 11, Hydraulics Research, Wallingford.

Spoor, G. and Ford, R.A. (1986) Mechanics of mole drainage channel deterioration. Submitted to J. Soil Sci.

Spoor, G., Cronin, C.J. and Leeds-Harrison, P.B. (1986) Potential use of mole drainage in the reclamation of fine textured saline soils. In press.

Trafford, B.D., Massey, W. (1975) A design philosophy for heavy soils. Tech. Bull. 75/5 Field Drainage Experimental Unit, Cambridge.

Trafford, B.D. (1977) Recent progress in field drainage: Part 1. J. Royal Agric. Soc. 138, 27-42.

Webb, D.J. (1981) Vertical crack formation with narrow drainage tines. MSc Thesis, Silsoe College, Cranfield Institute of Technology.

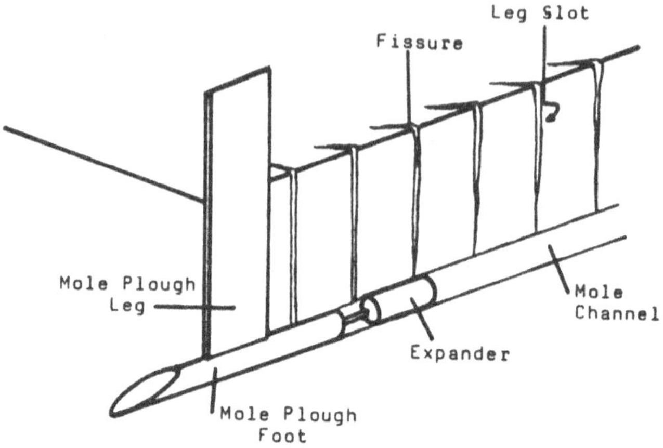

Figure 1 Section showing the mole plough and associated mole channel and fissures.

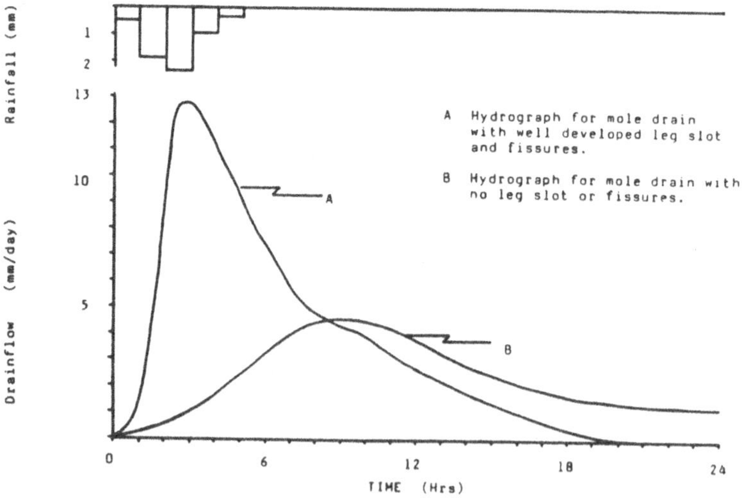

Figure 2 Influence of soil fissure development on the mole drain hydrograph.

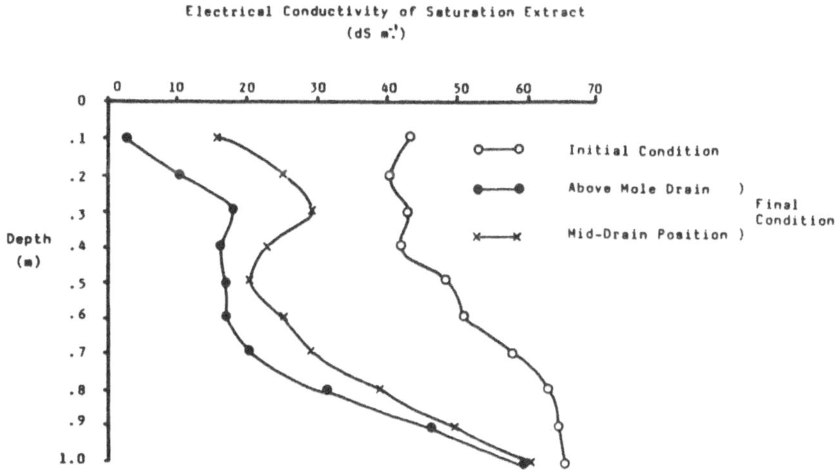

Figure 3 Changes in soil profile salt concentration following leaching for a 10 week period.

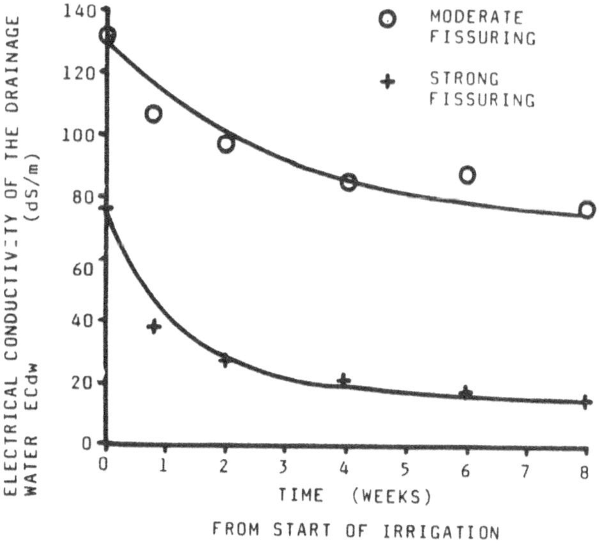

Figure 4 Influence of soil fissure development on changes in salt concentration of drainage water during leaching period.

SECTION 5 INFLUENCE OF DRAINAGE ON WATER QUALITY AND THE ENVIRONMENT

Nitrate in Water Draining from Arable Land Under Autumn Sown Crops and the Influence of Cultivation

M.J. Goss, K.R. Howse
Rothamsted Experimental Station, Harpenden, U.K.
P. Colbourn
Welsh Plant Breeding Station, Aberystwyth, U.K.
G.L. Harris
Field Drainage Experimental Unit, MAFF, U.K.

INTRODUCTION

Nearly half the soils of England and Wales on which cereals are grown have a clayey texture and many are subject to seasonal waterlogging. Use of close spaced mole drains 1-3m apart drawn at about 0.5m depth and discharging into permeable gravel fill above pipe drains on a 40-50m spacing is considered to be the most effective way of draining such land (Trafford and Massey, 1975).

Following a rainfall event, the flow of water from the pipe system increases rapidly to a peak then falls sharply before a relatively prolonged period when the flow rate slowly declines (Harris et al, 1984). Most of the water contributing to the early peak flow moves from the topsoil to the mole channels by way of fissures in the subsoil (Goss et al, 1983). The slower residual flow represents the water that moves slowly from topsoil to subsoil and thence to the mole channels and pipe drains.

The water collects nutrient ions as it flows through the soil to the drains and this is often considered to make an important contribution to the loss of nutrients into water courses and the reduced efficiency of fertilizer applications (Cannell et al, 1984). Relatively little information is available on the total losses of nutrients from field drainage schemes under arable land or the influence of cultivation systems on such losses (Harris et al, 1984; Dowdell et al, 1986).

In 1978 an experiment was started by the Agricultural and Food Research Council, Letcombe Laboratory (responsibility was transferred to Rothamsted in 1985) and the Ministry of Agriculture, Fisheries and Food, Field Drainage Experimental Unit. One objective was to investigate the interaction between drainage and tillage on nutrient losses and the efficiency of nitrogen fertilizers. In this paper we review the effects of mole drainage on the loss of nutrients from an arable soil on which winter cereals were grown using traditional mouldboard ploughing or direct drilling following two years with only deep tine cultivation.

EXPERIMENTAL DETAILS

Site

The site is located on a clay soil, Denchworth series (a verti-eutric gleysol) at Brimstone Farm near Faringdon, Oxfordshire. There are 20 plots, each hydrologically separated to a depth of 1.1m. Half are drained by mole drains at 0.6m depth and 2m spacing drawn into the permeable fill over pipe drains at 0.9m depth and 46m spacing. The other plots have no secondary drainage. For full details see Cannell et al., (1984). Winter wheat was grown in all years except 1982-3 when the crop was winter oats. Straw and stubble were burnt before seedbeds were prepared. For the first two seasons all plots were tine-cultivated, then half the plots on each treatment were direct-drilled and on the others a seedbed was prepared by secondary cultivations following traditional mouldboard ploughing. A summary of the soil properties are given in Table 1.

TABLE 1. Soil properties at the experimental site.

Horizon	Ap	Bg1	Bg2	Bgk	BCgk
Depth (m)	0-.2	.2-.35	.35-.62	.62-.89	.89-1.12
Sand (60μm-2mm)	7	6	3	2	1
Silt (2μm-60μm)	39	39	37	36	37
Clay (<2μm)	54	56	60	62	62
Organic carbon (%)	3.3	0.8	0.7	0.6	0.6
$CaCO_3$ equivalent	<0.1	<0.1	0.2	12.3	21.6
Volume fraction drainable pores	0.144	0.049	0.042	0.038	0.056
$K_{sat.\ max}$* (md^{-1})	12.4	1.79	0.53	0.41	0.09
$K_{sat.\ min}$* (md^{-1})		0.02	0.05	0.2	<0.001

* values determined from horizontally inserted piezometers.

Measurements

Analysis of nutrients leaving replicated plots were made on samples of water collected automatically from a U-bend water-trap in the pipe drain using a machine (Automatic Liquid Samplers Ltd., Alton, Hants) that serially drew 250ml of drainage runoff into 1 of 24 glass bottles. Nitrate was estimated colorimetrically. Losses of nitrate were calculated

TABLE 2. Influence of seasonal and agronomic factors on nitrate in runoff through a mole drain system over 6 years; losses accrued during winter (sowing - 1st nitrogen top dressing) and throughout the season from arable land under different cultivation systems.

Season	1978-9	1979-80	1980-1	1981-2	1982-3	1983-4
Sept-Aug rainfall	555.1	603.6	640.6	660.7	700.4	505.0
Ann. fertilizer N (kg N ha^{-1})	134	163	149	173	139	240
Ann. nitrate loss through drainage scheme (kg N ha^{-1})						
Tine cult	40.1	55.3				
Plough			17.4	42.5	73.1	5.6
Direct drill			15.5	26.4	61.2	2.7
Soil water deficit before cultivation (mm)	0*	112	114	129	152	137**
Winter rain (mm)	197	337	396	450	390	305
Winter drain flow (mm)	163	219	198	198	183	73
Autumn applied N (kg N ha^{-1})	17	24	0	24	30	17
Winter nitrate loss (kg N ha^{-1})						
Tine cult	40.1	54.3				
Plough			11.7	41.5	51.2	5.6
Direct drill			10.5	25.7	33.2	2.7

* assumes a fully drained profile
** calculated from deficit under grass adjusted for a tall crop using factors obtained for a similar crop on a nearby experiment

from the total flow of water draining from each plot, measured using a V-notch weir fitted with a chart recorder, and the nitrate-N concentration as described by Harris et al. (1984). Ammonium-N was trivial compared to that of nitrate so it was ignored in computing the results.

Soil water content was measured by calibrated neutron moisture meter (Wallingford type), one access tube per plot. Soil water potential was measured by tensiometry from permanent instrument pits (Howse and Goss, 1982).

RESULTS

The total loss of nitrates (expressed as kg N ha^{-1}) through the mole drainage system showed a marked seasonal variation (Table 2) and ranged from 2.7kg N ha^{-1} in the driest year (1983-4) to 73.1kg N ha^{-1} in the wettest (1982-3). Direct drilling consistently resulted in less nitrate being lost through the drainage system. Differences were small in the first year after imposing the tillage treatments but amounted to 16kg N ha^{-1} in the following season.

Winter losses were least in 1983-4 which, although it was not the driest winter for rainfall, was preceded by a much larger soil water deficit than 1978-9 when the deficit was trivial at sowing. In 1980-1 no autumn nitrogen was added to the seedbed which may account for there only being about 11kg N ha^{-1} lost. In all the other 4 seasons the loss through the mole drain system during winter was similar for the tilled plots and averaged 46.8 kg N ha^{-1}. Direct drilling cut these

TABLE 3. Concentration of nitrate in drainage water after rainfall events in 1979-80. Results are means of 2 plots.

Rainfall event	Total drainflow (mm)	Nitrate concentration (μg N ml^{-1})
a	8	51
b December	28	37
c	46	29
d January	17	25
e	17	25
f February	20	23
g	12	22
h	10	11
i March	17	7
j	10	6
k April- before nitrogen top dressing	11	9

losses by an average of 30%. Only in the spring of 1983 were there losses of nitrate exceeding 6kg N ha^{-1} after the application of a top dressing of nitrogen, 22 kg N ha^{-1} from ploughed land, 29 kg N ha^{-1} from direct drilled land.

The concentration of nitrate in the drainage water resulting from particular rainfall was investigated during 1979-80. Concentrations rose sharply to a mean of 59 μg N ml^{-1} during the first rain-fall event that caused drainflow greater than 0.3 mmd^{-1}. For example on one plot the concentration of nitrate increased from 25 to 55 μg N ml^{-1}. Throughout the remainder of the winter there was a steady decline in concentration to about 6 μg N ml^{-1}. (Table 3).

TABLE 4. Changes in concentration of nitrate during periods of residual flow in March 1980. Mean of 2 plots.

Period	Mean max daily temp (°C)	Change in nitrate conc (μg N ml^{-1})
17-18th	5.4	+4.5
24-27th	9.6	+6.5
31st	10.8	+4.5

In contrast, during March the concentrations began to increase in the water draining during residual flow but the concentrations were again decreased by each rainfall event (Tables 4,5). Similar patterns were evident from later seasons.

TABLE 5. Change in nitrate concentration of drainage water due to winter and spring rainfall. Mean of 2 plots.

Month	Changes in nitrate conc (μg N ml^{-1})
December	-10
January	- 3
February	- 3.5
March	- 8

The effect of a rainstorm on the position of the watertable, more correctly the phreatic surface (defined by hydraulic potential equal to the gravitational potential), differed between soils under direct drilling and ploughing (Table 6). The phreatic surface rose more quickly once rain started but also fell more rapidly afterwards in the direct-drilled soil than in the ploughed soil. This was also reflected in the shape of the hydrographs from mole drains on this site (see Harris and Goss, these proceedings).

TABLE 6. Rate and direction of variation (mm h^{-1}) in the position of the phreatic surface in direct drilled and ploughed plots due to 7.5mm rainfall, winter 1982. The soil surface is used as the reference point.

Treatment	Direct drill	Plough
During rainfall	-34.0	-13.8
During 21h after rainfall	+ 3.7	- 3.2

DISCUSSION

Significant losses of nitrates, up to 73kg N ha^{-1}, occurred through the mole drain system under arable land on which winter cereals were grown. The results show that even when no nitrogen fertilizer was applied in autumn more than 10kg N ha^{-1} were lost from the soil in this way. Except in very dry seasons, the application of about 20kg N ha^{-1} to the seedbed led to losses of almost 50kg N ha^{-1} from tilled land but only about 30kg N ha^{-1} from direct drilled soil. This suggests that effectively the autumn applied nitrogen was lost through the drains, probably because the crop was small and falling temperatures reduced the demand from the shoots.

The difference between losses in drains from tilled and direct drilled land was not due to greater surface runoff from the direct-drilled land (Dowdell et al, 1986). These results suggest that when crop residues are burnt, as in this experiment, limiting cultivations to the depth of seed placement could help to minimise the release of nitrate from arable soils into water courses.

The annual losses from this arable soil were generally much less than those reported for grassland (Ryden et al, 1984). When the annual fertilizer application was comparable at 200kg N ha^{-1}, Garwood et al. (1986) found that on an old grassland site in Devon with an average rainfall of 1061mm, nitrate lost through the mole drain system ranged from 50-70kg ha^{-1}. However, nitrogen fertilizer applications are usually greater for grass than for cereals. With a nitrogen fertilizer application of 400kg N ha^{-1} to old grazed grassland the loss was 160kg N ha^{-1} (Ryden et al. 1984).

The concentration of nitrates in the drainage water immediately after the first major peak in drainflow were in excess of 50 g N ml^{-1} but declined at about 0.2 g N ml^{-1} d^{-1}. In the spring as temperatures began to rise there were increases in concentration during slow flows suggesting the restart of mineralization. The large organic matter content of the soil is probably a key factor in determining the magnitude of the rises together with the increase in soil temperature.

In this soil the volume of pores drainable under gravity was small and there was also a marked difference between the maximum and minimum values of saturated hydraulic conductivity for the different horizons (Table 1). These facts together indicate that macropores are important for drainage. The decline in the concentration of nitrate in the runoff following each storm suggests that the surfaces of major fissures rapidly become leached. In the spring the slow drainage of the smaller pores in the topsoil during residual flow caused increased nitrate concentrations in the pipe drains because of the presence of recently mineralized nitrogen. This was then was then diluted when rain next flowed through the larger channels. The greater continuity of macropores linking topsoil and subsoil in direct-drilled land (Goss et al, 1984) further encourages the flow of water in them as opposed to the finer pores (Table 6). This may be a contributary factor in explaining the reduction in nitrates lost in the drainage runoff from this treatment. Better pore continuity between the soil surface and the subsoil may also explain the greater loss of nitrates from direct drilled soil than from ploughed soil after the spring top dressing in 1983.

ACKNOWLEDGEMENTS

We thank the many staff of Letcombe Laboratory who contributed to this work especially members of the Chemistry and Electronics Sections for analysing the drainage water and D.G. Christian for agronomy on the site. We are grateful for the support of colleagues in the Ministry of Agriculture, Fisheries and Food for making the detailed flow measurements. The work was commissioned by the Ministry of Agriculture, Fisheries and Food.

References

Cannell, R.Q., Goss, M.J., Harris, G.L., Jarvis, M.G., Douglas, J.T., Howse, K.R. and Le Grice, S. (1984) A study of mole drainage with simplified cultivation for autumn-sown crops on a clay soil 2. Soil water regimes, water balances and nutrient loss in drain water, 1978-80. J. agric. Sci., Camb. 102, 539-559.

Dowdell, R.J., Colbourn, P. and Cannell, R.Q. (1986) Losses of nitrate N in surface run-off and in drain water from a clay soil after ploughing and direct drilling. Soil & Tillage Res.

Ellis, F.B., Christian, D.G., Bragg, P.L., Henderson, F.K.G., Prew, R.D. and Cannell, R.Q. (1984) A study of mole drainage with simplified cultivation for autumn-sown crops on a clay soil 2. Soil water regimes, water balances and nutrient loss in drain water, 1978-80. J. agric. Sci., Camb. 102, 583-594.

Garwood, E.A., Ryden, J.C. and Tyson, K.C. (1986) Nitrogen losses from drained grassland. Proc. Brit Grassland Soc. Winter Meeting, 1985. In Press.

Goss, M.J., Ehlers, W., Boone, F.R., White, I. and Howse, K.R. (1984) Effects of soil management practice on soil physical conditions affecting root growth. J. Agric. Engng Res, 27, 81-91.

Goss, M.J., Harris, G.L. and Howse, K.R. (1983) Functioning of mole drains in a clay soil. Agric. Water Management, 6, 27-30.

Harris, G.L., Goss, M.J., Dowdell, R.J., Howse, K.R. and Morgan, P. (1984) A study of mole drainage with simplified cultivation for autumn-sown crops on a clay soil 2. Soil water regimes, water balances and nutrient loss in drain water, 1978-80. J. agric. Sci., Camb. 102, 561-581.

Howse, K.R. and Goss, M.J. (1982) Installation and evaluation of permanent access pits which permit continuity of measurement in cultivated soils. Experimental Agriculture, 18, 267-276.

Ryden, J.C., Ball, P.R. and Garwood, E.A. (1984) Nitrate leaching from grassland. Nature, 311, 50-53.

Trafford, B.D. and Massey, W. (1975) A design philosopy for heavy soils. Technical Bulletin 75/5, 32pp. Field Drainage Experimental Unit.

The Reduction of the Effects on Flora Caused by Drainage Water Quality by a Proposed Reservoir as an Element of a Feedbacksystem

P. van der Kloet, H. de Niet
Delft University of Technology

INTRODUCTION

In the southern part of the Netherlands the river Beerse flows through a nature reserve area. The discharge regime of the Beerse is such that in the reserve area inundations occur regularly. As due to upstream use of fertilizers the water quality of the Beerse is poor, the inundations harm the unique vegetation in the reserve area.
There exists a possibility to create a reservoir just upstream of the reserve area and this could be designed in such as way that no inundations occur in the downstream reserve area. However, due to the situation in situ, the reservoir capacity is limited and the question arises whether this capacity is sufficient or not. Therefore an operating policy has to be planned such that the release from the reservoir does not cause inundations. As only a very limited number of discharge data are available, not only a reservoir model with an operating policy has to be made, but also a simulation model for the reservoir inflow.
The reservoir can be created by connecting several existing dikes. The Beerse runs through the reservoir, and the reservoir-model and operating policy are based on water balance concept and the classical control theory applied to a movable weir located in the Beerse where it leaves the reservoir (see figure 1). The releases over the weir are the inflow into the reserve area. A simulation model applies to the discharges of the Beerse which enter the reservoir. This model has the same characteristics as the available historical discharge series of the Beerse as far as concerns:

- The frequency of occurrence of peak discharge
- The peak values of the discharges above a base level
- The behaviour of the recession of the discharges.

With these models and frequency of occurrence of inundations in the reserve area can be studied. In case under normal operation, this frequency is too high, the possibility exists for extra release over a spillway which joins the Beerse downstream of the reserve area (see figure 1).

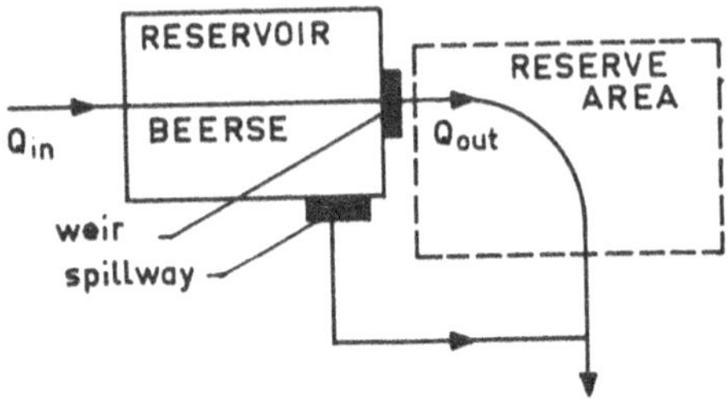

figure 1 Lay out of the configuration

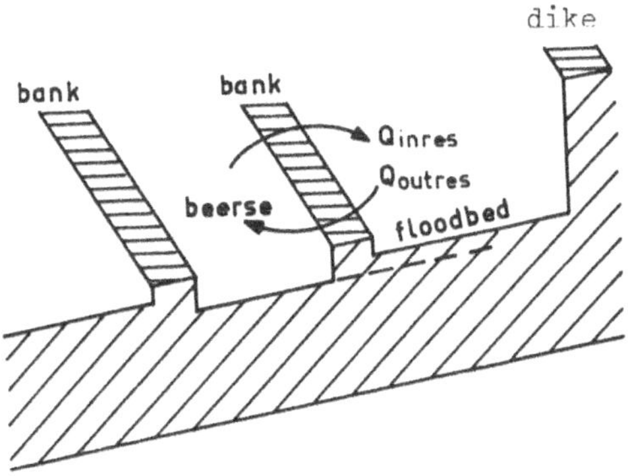

figure 2 Cross section of the reservoirsystem

Both the spillway and the movable weir have a maximum capacity. A cross-section of the reservoir system is given in figure 2. The floodbed of the reservoir only fills when the waterlevel in the Beerse rises above its banks.
When the waterlevel in the reservoir is below the banks of the Beerse, only water in the Beerse can be discharged over the movable weir. The water in the floodbed is discharged by a lateral inflow into the Beerse.

THE RESERVOIR MODEL

The model of the reservoir system consists of two equations, each representing a water balance. In discrete time, these equations are

$$S_{BEERSE,t+\Delta t} = S_{BEERSE,t} + [Q_{IN,t} - Q_{OUT,t} - Q_{INRES,t} + Q_{OUTRES,t}]\Delta t$$

$$S_{RES,t+\Delta t} = S_{RES,t} + [Q_{INRES,t} - Q_{OUTRES,t}]\Delta t - LOSS_{(t,t+\Delta t)}$$

The meaning of the Q-terms is indicated in figure 1 and 2. Furthermore S represents storages, Δt time-interval and LOSS represents hydrological quantities as infiltration in the floodbed and contains even rainfall (as a negative loss). The coupling terms Q_{INRES} and Q_{OUTRES} depend among others on the current water levels and are calculated according to the following rules:

- $Q_{INRES,t} = 0$ when the water level in the Beerse is lower than the crest of the banks.
- If $Q_{IN,t}\ \Delta t$ cannot be stored within the banks of the Beerse, $Q_{INRES,t}\ \Delta t$ equals $Q_{IN,t}\ \Delta t$ minus the available storage capability in the Beerse.
- If the water level in the floodbed is higher than the crest of the banks, the water level in the Beerse equals the water level in the floodbed.
- If the crest of the bank is higher than the water level in the floodbed, and the latter level is higher than the water level in the Beerse

$$Q_{OUTRES,t} = C\ (\Delta h)^{1.5}$$

where Δh represents the difference in water levels and C is a constant.

With these rules it is possible under all circumstances to calculate the water level in the Beerse in the reservoir upstream of the reserve area.

THE CONTROL SYSTEM

The objective to prevent inundations in the reserve area is achieved whenever the outflow $Q_{OUT,t}$ of the Beerse over the weir into the reserve area is below an allowed maximum per unit of time. This allowed maximum flow relates to an allowed upstream maximum water level, determined by crest level of the weir and weir constants. This maximum allowable water level is used as a reference in the feedback control system.

With the aid of the feedback control system, the water level in the Beerse is forced to this reference. Referring to figure 3 as a definition sketch for the feedback control system, its output is the water level in the Beerse upstream of the weir, its disturbance is the inflow of the Beerse into the reservoir and the input is the mentioned reference level. Figure 3 contains two blocks, to be called the plant and the controller.

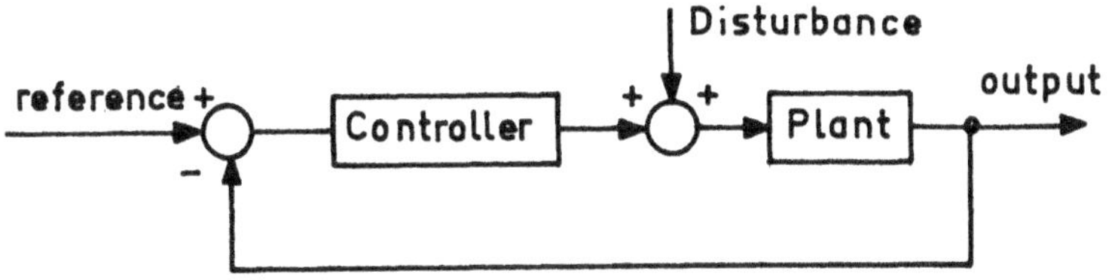

figure 3 Definition sketch of the feedback system

The term plant is used for the system to be controlled; in physical terms it contains the reservoir and the weir and in math-

ematical terms it represents the reservoir model and the equation of the flow over the weir.
The controller measures the output (water levels of the Beerse upstream of the weir), compares them with the reference and adjusts an input variable to the plant (e.g. the crest level of the weir), see Kwakernaak and Sivan (1972).
Here the plant is described by a water balance for the volume of water in the reach of the Beerse within the reservoir and the equation for flow over a weir.
The reservoir model itself is only implicit as it influences the water level in the Beerse. The two equations mentioned can be combined yielding the following first order nonlinear differential equation for the water level upstream of the weir

$$\frac{dh}{dt} = - C_1 \, (h-h_w)^{1.5} + \frac{Q_{IN,t}}{A}$$

where A is the wet surface of the river reach and h_w is the crest level of the weir.

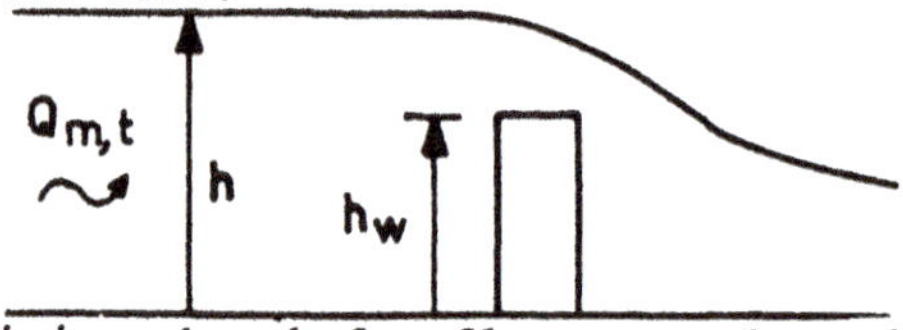

figure 4 Definition sketch for flow over the weir

Choose $\bar{h}$, $\bar{h}_w$ and $\bar{Q}_{IN,t}$ accordingly to

$$\bar{Q}_{IN,t} = A \, C_1 \, (\bar{h} - \bar{h}_w) \; 1.5$$

and write $h = \bar{h} + h'$, $h_w = \bar{h}_w + h_w'$

$$Q_{IN,t} = \bar{Q}_{IN,t} + Q'_{IN,t}$$

where h', h_w', $Q'_{IN,t}$ represent small fluctations around $\bar{h}$, $\bar{h}_w$, $\bar{Q}_{IN,t}$.

By expanding the nonlinear term in the differential equation into a first order Taylor series, it follows that the dynamic behaviour of the plant is given by

$$\frac{dh'}{dt} = - \frac{3}{2} C_1 \, (\bar{h} - \bar{h}_w)^{0.5} \, (h' - h_w') + \frac{Q'_{IN,t}}{A}$$

In this equation $\bar{h} - \bar{h}_w$ can be considered as a constant when the fluctuations are small, so that a linear equation applies in those cases. Of course the associated time constant does vary with the chosen equilibrinm $\bar{h}$, $\bar{h}_w$, $\bar{Q}_{IN,t}$.

By definition this time constant τp is given by

$$\tau_p^{-1} = \frac{2}{3} C_1 \, (\bar{h} - \bar{h}_w)^{0.5}$$

In the linear reservoir model as used in drainage this inverse of the time constant coincides with the reaction factor. Assuming a simple controller with an output proportional to the difference between the reference r(t) and the output h'(t) of the entire feedback system, the feedback system is governed by the equation

$$\frac{dh'}{dt} + (K+1)\,\tau_p^{-1}\,h'(t) = K\,\tau_p^{-1}\,r(t) + \frac{Q'_{IN,t}}{A}$$

where K is the constant of proportionality of the controller. The feedback system thus has a time constant $\tau_c = (K+1)^{-1}\,\tau_p$.

If K becomes larger, τ_c becomes smaller so that the feedback system reacts faster in following the reference r(t), the desired water level upstream of the weir. However, the feedback system also reacts faster to the changes in inflow, so that the feedback system becomes more sensitive. If working in a discrete time approach the feedback system may even become unstable. These two points may be used to determine the value of K. Moreover, it should be stated that this simple proportional controller gives a static error with respect to the following of the reference. To show this assume $Q'_{IN,t} = 0$ and set the reference equal to a constant r(t) = r. Under steady state conditions it then holds that:

$$h'(t) = (1 - \frac{1}{K+1})r$$

The actual water level becomes in that case smaller than the reference. For the problem under consideration this is not very important. The error becomes smaller if K becomes larger.

THE SIMULATION MODEL

To gain insight in the effects caused by the movable weir and the possible spillway a simulation model for the inflows into reservoir system was developed. For that purpose only a rather short series of water levels is available which must be converted into discharges. The discharge hydrographs generally show a relative rapid rise followed by a relative long recession curve. At first glance this behaviour seems to agree with the behaviour of shot noise models as used by O'Connell (1977). However, the autocorrelation in the historical series is too high to accept this hypothesis. However, the idea in these shot noise models that peak discharges occur accordingly to a Poisson process, is maintained in the developed model.

To develop the simulation model, the historical discharges x_t are transformed accordingly to

$$y_t = x_t - x_b \qquad \text{if } x_t \geq x_b$$

$$\quad = 0 \qquad \text{else}$$

where x_b represents a base-flow small enough to realize a sufficient number of peaks in y_t for a statistical analysis, but large enough to eliminate largely the influence of small peaks in the series x_t in this statistical analysis.
For this series y_t the times at which peaks occur are assumed to follow a Poisson distribution. This means that the distance between peak times is exponentially distributed. Furthermore the heights of the peaks are assumed to be realizations of a stochastic variable with a negative exponential distribution.
The recession curves are assumed to be exponential so that for two subsequent data it holds that

$$x_t = x_{t-1}\, e^{-\beta}$$

This means that $e^{-\beta}$ can be estimated using a linear regression from which the coefficient β is calculated.
In the simulation model it is further assumed that the rising parts of the hydrograph can be described as straight lines with a tangent equal to the mean of all positive tangents in the historical series.
In figure 5 the different steps in building the simulation model are shown.

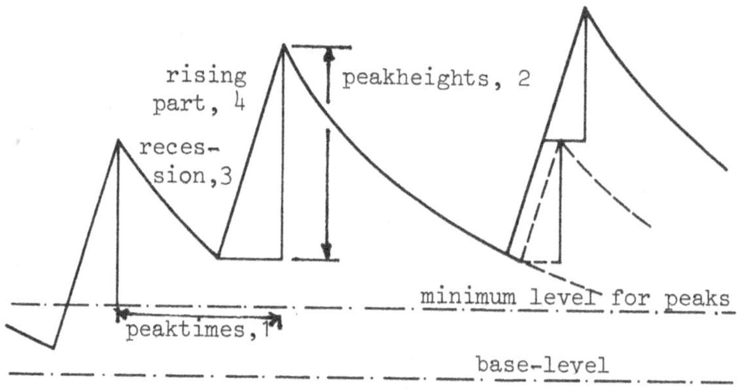

Figure 5 Building up the simulation model

The resulting model has four parameters. Two of them are completely determined by the historical series.
The others are determined by the historical series in a stochastic way. Data generation from negative exponential distributions yield synthetic discharge series which can serve as inflows to

the reservoir model.
The synthetic series are judged by means of their mean, their variance and the distribution of the times between peaks. It should be remarked that the behaviour of the historic series for low flows may not be captured so well by the synthetic series. However, these low flows are not interesting with respect to the problem posed. More serious is the problem of reproducing a historic series of prolonged high discharge such that over a number of time steps discharges are of the same order as the peak discharge. This cannot be done with the described simulation procedure, but is partly compensated for by the fact that in those cases the tangent of the rising part is under estimated and therefore the discharge volumes under the synthetic and historical hydrographs will not differ so much.

RESULTS AND DISCUSSION

To test the simulation model and the operation strategy for the reservoir there were available a set of historical daily water levels over a period of two years. The set contains the data during two wet seasons. One data set during a wet season is used to calibrate the simulation model, the other concerns a dry year. The set used for calibrating purposes can be classified hydrologically in terms as not too wet.
Since the study was performed at the stage of determining favourable alternatives for decision-making with respect to a re-allotment in the drainage area of the Beerse, it was more important to show globally the effect of creating the reservoir and the eventual spillway, than having a detailed answer to the frequency problem.
For that purpose 8 synthetic discharge series for the wet season were generated. (In each case the reservoir came into use, meaning that without the reservoir floods occur in the reserve area.) The reservoir is in use 2 to 3 times during a wet season and the longest period in use is nearly 12 days. For 3 of the 8 simulations the inflow could not be stored in the reservoir. It thus seems necessary to include the spillway in the final design.
The controller in the system forces the water level in the Beerse to the derived reference level. If the water level in the Beerse is higher than the given reference the discharge over the weir will be greater than allowed flow for which no inundations occur. In those cases a constraint is used to prevent discharges higher than the allowed discharge. In doing so the reservoir will be filled more quickly until the water level equals the crest height of the reservoir. Arriving at that point no limitations are given to the discharge over the weir.
In figure 6 a graphical impression is given for the water levels in the Beerse and the reservoirs and moreover for the inflow and the outflow over the weir. These quantities are given as functions of time over a part of the wet season. The figure concerns the simulation where the most severe inundations are met with in the reserve area.

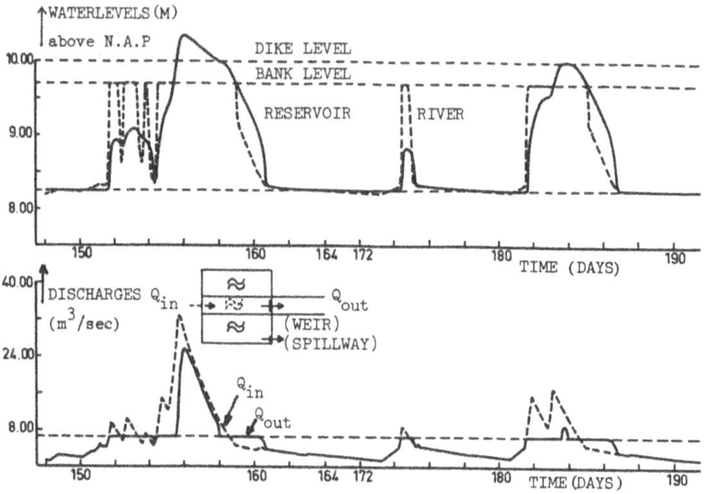

figure 6 Results for a simulated inflow

On behalf of the mathematical model the figure shows water levels above the crest level of the reservoir. No meaning must be given to this phenomenon. It can also be seen in the figure that Q_{OUT} equals the allowed discharge over the weir at the moment that the water level in the reservoir equals the crest level of the reservoir. The figure also shows that smaller peaks can be handled easily by the proposed system.
The proposed feedback control reservoir thus reduces the volumes associated with inundations in the reserve area, but the duration of these inundations is not lowered.
The controller itself is just an example of a simple control-element. It is by no means optimal. Optimal controllers can be derived by minimizing a loss function which is constrained by the behavior of the system. In this case as a loss function may be used

$$Z = \int^{T} (Q_{OUT,t} - Q_{ALLOW})^2 \, dt$$

where T is the duration of the wet season and Q_{ALLOW} the allowed discharge over the weir. A worked-out example with this approach is discussed in Van der Kloet and Lumadjeng (1986) for the problem of controlling waterlevels in a reach during the growing season of crops in order to reduce crop damages.

REFERENCES

Van der Kloet, P and Lumadjeng, H.S. (1986), The development of an economic objective function for decisionmaking in a water resource control problem, to be presented at the European IHP-III Symposium Decision Support Systems and Related Methods in Water Resources Planning, Oslo, 5 - 7 may 1986.

Kwakernaak, H. and Sivan.R. (1972), Linear Optimal Control Systems, Wiley-Interscience, New York.

O'Connell,P.E. (1977), Shot Noise models in Synthetic Hydrology, p.p. 19 - 26 in Mathematical Models for Surface Water Hydrology, Ed. Cirioni, T.A., Maiono, U. and Wallis, J.R., Wiley-Interscience, New York.

Simple Hydrated Lime Plants for Treatment of Ferrous Rich Drainage Water

L.B. Christensen, J. Waagepetersen and S.E. Olesen
Hedeselskabet, Danish Land Development Service, Denmark

INTRODUCTION

Many Danish soils contain the iron-rich mineral pyrite in the subsoil. The pyrite bearing soils are mainly found in Jutland in littoral deposits, late glacial deposits, marsh areas, bogs and meadows, where the environmental conditions during the deposition and sedimentation favored the formation of pyrite.

When the groundwater level is lowered by drainage in order to improve the farm land, pyrite will be oxidized and ferrous iron is leached out into the streams. This may have serious impacts on stream ecology.

In Denmark a law has recently (July 1985) been introduced to minimize the leaching of ochre into streams. A farmer who wants to carry out drainage works on pyrite bearing soils can be directed to purify the drainage water from iron.

During an experimental arrangement from 1981 to 1984 different methods of ochre abatement were tested. The investigations showed that treatment of the ferrous drainage water with hydrated lime is the most efficient and reliable method (CHRISTENSEN & OLESEN, 1985).

2. OXIDATION OF FERROUS IRON

The principle of the method is to increase the chemical oxidation rate of ferrous iron by treating the drainage water with calcium hydroxide ($Ca(OH)_2$). The pH of the water is increased to about 7.5 or higher which results in a rapid oxidation of ferrous iron to ferric hydroxide which precipitates as ochre.

$$(1)\ Fe^{2+} + 1/4\ O_2 + 1/2\ H_2O + Ca(OH)_2$$

$$\text{--->}\ Fe(OH)_3 + Ca^{2+}$$

The consumption of oxygen is 0.15 mg O_2 per mg of ferrous iron. Accordingly the theoretical consumption of hydrated lime is 1.3 mg per mg of ferrous iron, but normally the consumption is between 2 and 4 mg depending upon the initial pH of the drainage water.

3. HYDRATED LIME PLANTS

During the above-mentioned experimental arrangement a number of simple hydrated lime treatment plants were built in connection with drainage of pyrite-rich soils.

The ferrous drain water is pumped into the plant. The lime is added by dry dosage with the applicator connected to the pump so that addition of lime takes place only when the pump is working. Initially the lime was added directly to the pump well, but formation of coatings and corrosion cause severe damage to ordinary drainage pumps. Therfore, several plants have been altered so that the lime is added after the pump well.

After treatment with lime the water flows into a pond where the oxidized iron precipitates as ochre. From the pond the purified water flows freely into the recipient.

The plants were designed by Danish Land Development Service. The farmer was responcible for running the plant and the local county controlled the iron removal efficiency. Two typical simple hydrated lime plants are selected (Fig. 1 and 2).

3.1. Plant 1 The plant is situated at Kirkebaek brook near Viborg. The drainage scheme covers 3 ha of peat soil with peat layers of 0.6 to 1.3 m on top of coarse sand. The concentration of pyrite in the peat varied between 0.7 to 3.1% which is rather high. Due to sepage from upland areas the concentration of ferrous iron in the drainage water was between 5 to 20 mg/l, only.

The plant which was constructed in August 1983 is of a simple design. The pumping capacity is 15 l/s, the lime dosage can be adjusted manually between 0 and about 200 mg/l of drainage water. The retention time in the sedimentation ponds is about 7 hours. The iron sludge (ochre) is removed periodically, and spread on top of the nearby soils.

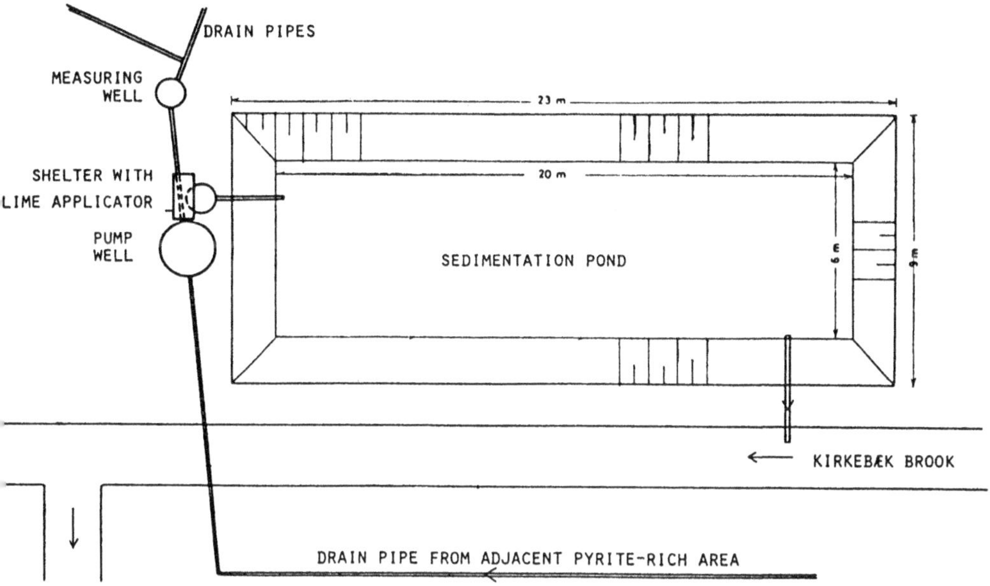

Figure 1. Design of plant 1 at Kirkebaek brook. (CHRISTENSEN & OLESEN, 1984).

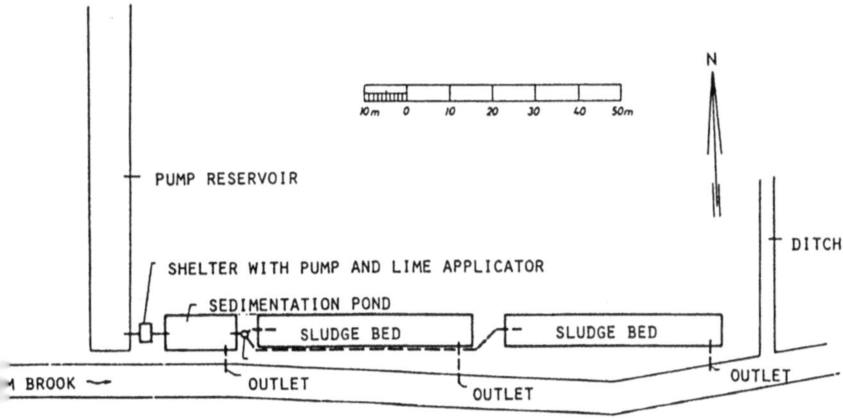

Figure 2. Design of plant 2 at Skaerum brook. (CHRISTENSEN & OLESEN, 1984).

3.2 Plant 2 The plant is situated at Skaerum brook in the northern part of Jutland near Frederikshavn. Drainage water from about 10 ha is treated. The soil is a peat soil with well decomposed peat layers of 0.6 to 3 m's on top of glacial sand. The high concentration of 1 - 10% of pyrite in the peat was formed in the former brackish environment. The concentration of ferrous iron in the drainage water is very high, up to 175 mg/l.

The plant was constructed in February 1983. The pumping capacity is 20 l/s. The lime application can be varied between 0 and about 1,000 mg/l of drainage water. The retention time in the sedimentation pond was mostly between 2.5 to 6 hours, only. Therefore an automatically sludge pumping system was made by which the sludge was pumped to one of two sludge beds for dehydration. The system did not work properly and has recently been changed.

4. RESULTS AND DISCUSSION

This paper deals with results from both the rather extensive control by the local county and from an intensive analytical campaign carried out by the Danish Land Development Service in the autumn 1984.

4.1. Iron Removal Efficiency Controlled by the Local County

The control was based on water samples taken from inlet and outlet of the plants approximately every fortnight in 1983 and 1984. From then on samples were taken approximately every second month.

4.1.1. Plant 1 The concentration of ferrous iron in the drainage water treated in plant 1 was between 5 and 20 mg/l (Fig. 3). The highest concentration appears in winter.

Generally 90% of the ferrous iron in the drainage water was oxidized and precipitated. The oxidation of ferrous iron is, however, strongly depending upon the pH of the treated water. When the pH of the outlet water was increased to at least 7.4, 98% of the ferrous iron was removed (Table 1).

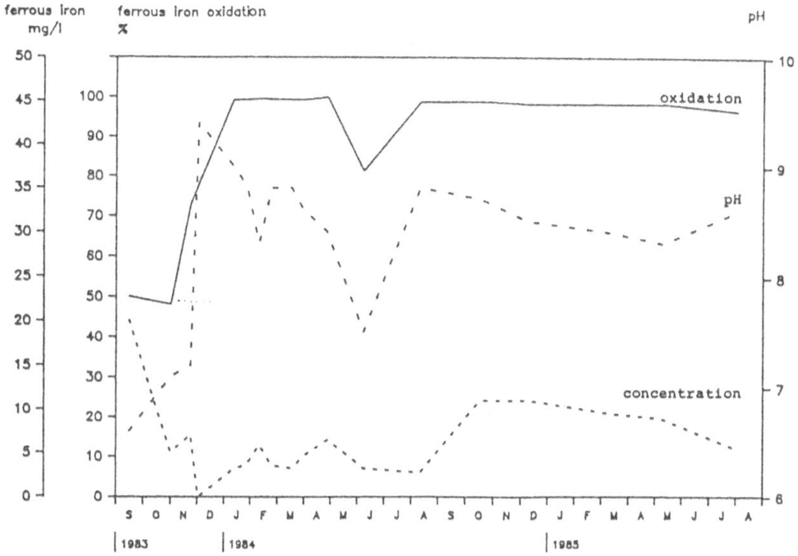

Figure 3. Ferrous iron concentration at the inlet, pH at the outlet and ferrous iron oxidation in plant 1.

Table 1. Purification of ferrous drainage water in two simple hydrated lime plants. average figures.

Plant no.	iron removal (%) oxidation average	pH > 7.4	sedimentation average
1	90	98	45
2	75	100	57

Due to the rather constant run-off the retention time is constantly about 7 hours. Therfore, a rather constant sedimentation rate should be expected. But as can be seen from Fig. 4, the sedimentation of iron in the pond varies approximately from 10% to 70%. This is probably due to variations in pH and the concentration of total iron in the inlet water. At high concentrations ochre particles flocculate very easily whereas at low concetrations the distance between the molecules becomes too large and flocculation proceeds slowly.

Owing to this, the concentration of total iron infrequently becomes less than 5 to 6 mg/l in the outlet water.

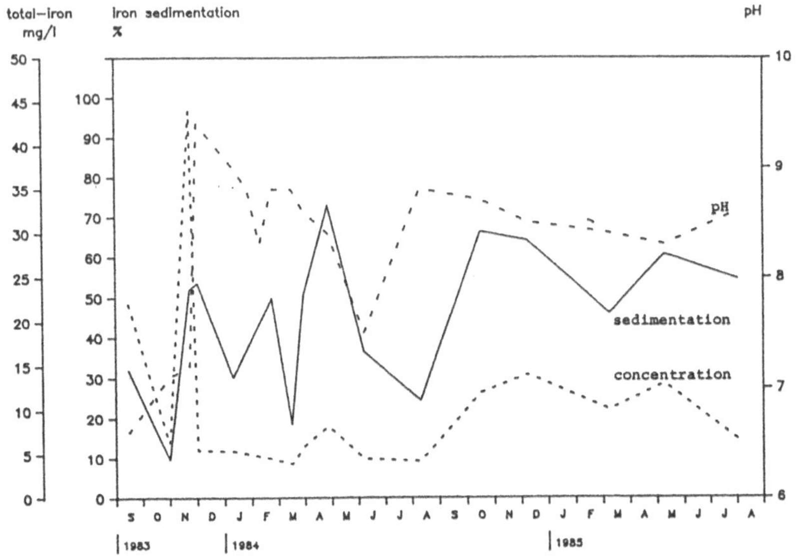

Figure 4. Concentration of total iron at the inlet, pH at the outlet and total-iron sedimentation in plant 1.

<u>4.1.2. Plant 2</u> The concentration of ferrous iron in the drainage water treated in plant 2 varied between 10 and 175 mg/l (Fig. 5). Also here the highest concentrations appear in winter.

The average ferrous iron oxidation was somewhat smaller than in plant 1. Generally 79% of the ferrous iron in the drainage water was oxidized and precipitated. In this instance the pH dependence of ferrous iron oxidation is even more pronounced (Fig. 5) When the pH of the outlet water was increased to at least 7.4, all the ferrous iron was removed (Table 1).

At high concentrations of ferrous iron (> 125 mg/l), the oxygen content of the water may limit the oxidation rate and

applied hydrated lime in the plants, i.e. the ratio between the theoretical and the actual lime consumption.

Water samples from inlet and outlet were taken several times during the campaign and the iron removal efficiency was investigated under controlled conditions and compared to results from in situ experiments done in 7.6 l cylinder glasses.

4.2.1. Plant 1 The analytical campaign was performed during two days in October 1984. The quality of the drainage water treated in the plant is shown in Table 3.

The drainage water was treated with about 70 mg hydrated lime/l or 3.5 to 4 times the content of ferrous iron.

After treatment with lime almost all ferrous iron was oxidized to ferric iron and approximately two thirds of the precipitated ferric iron was removed by sedimentation.

In plant 1 76% of the applied hydrated lime was utilized.

The concentration of oxygen increased from about 4.0 to about 5.5 mg/l, which shows that the aeration of the drainage water by pumping is sufficient.

The experiments also confirm the fact that it is very difficult to get concentrations of total iron in the outlet water less than 5 to 6 mg/l.

4.2.2. Plant 2 The analytical campaign was performed during two days in December 1984. The quality of the drainage water treated in the plant is shown in Table 3.

The drainage water was treated with about 300 mg hydrated lime/l or about 3 times the content of ferrous iron.

Also in this plant all the ferrous iron was oxidized. The sedimentation of ferric iron was very high as expected from the rather high concentration of total iron (Fig. 6). Almost all the ferric iron (about 96%) was precipitated in the pond.

In plant 2 86% of the applicated hydrated lime was utilized.

The concentration of oxygen decreased from about 7.0 to about 5.0 mg/l. The theoretical oxygen consumption by oxidation of 100 mg/l of ferrous iron is 15 mg/l. So the drainage water is reaerated in the sedimentation pond and aeration was not limiting the oxydation rate of ferrous iron.

a more efficient aeration, than is obtained by pumping the drainage water, may be necessary.

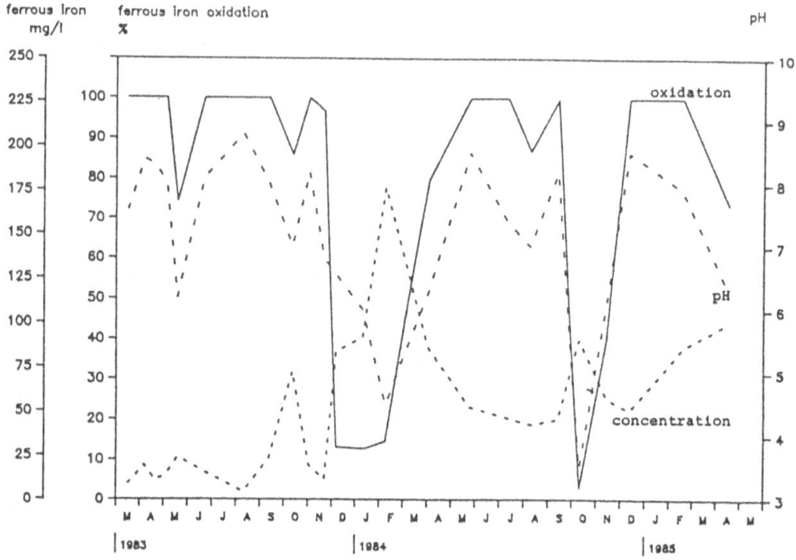

Figure 5. Ferrous iron concentration at the inlet, pH at the outlet and ferrous iron oxidation in plant 2.

On average 57% (Table 1) of the oxidized iron is precipitated in the pond, but the sedimentation varies from 0% to nearly 100% due to differences in retention time, pH and concentration of total iron. The average retention time is 5.2 hours, but varies from about 2.5 hours in winter to about 9 hours in summer (Fig. 6).

Also in this plant the sedimentation of ochre depends on the retention time in the pond, but when the lime application is too small (e.g. December 1983 and October 1984) sedimentation of ochre proceeds very slowly, and the outlet water contains high concentration of total iron. The total concentration of iron in the inlet water (Fig. 6) is generally very high and facilitates flocculation and sedimentation. Table 2 shows the sedimentation of iron in periods where the pH of the outlet water is higher than 7.4.

Table 2. Sedimentation of iron in periods where the pH of the outlet water is higher than 7.4.

	concentration of total iron, mg/l		
	0 - 20	21 - 40	41 - 70
number of samples	3	4	3
total iron, mg/l, average in <u>inlet</u>	12	24	58
total iron, mg/l, average in <u>outlet</u>	4.5	6.5	6.4
sedimentation, %	63	73	89

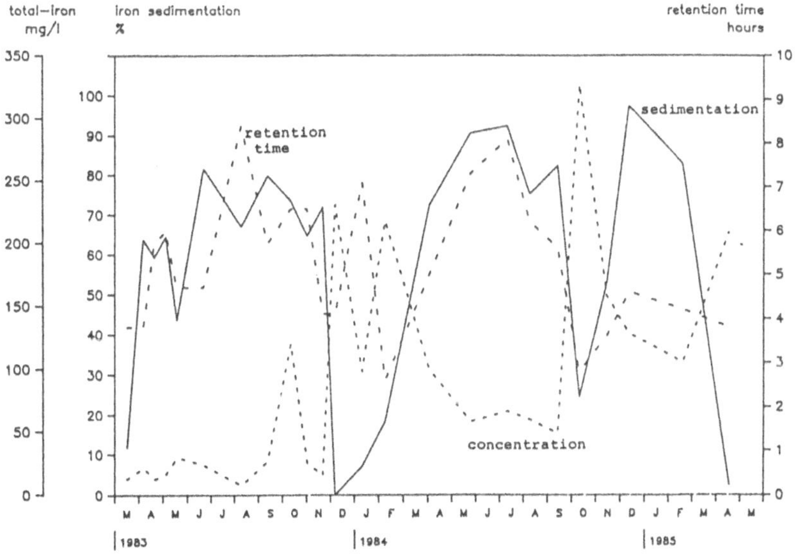

Figure 6. Concentration of total iron at the inlet, pH at the outlet and total-iron sedimentation in plant 2.

4.2. Results From the Intensive Analytical Campaign

The intensive analytical campaign was performed in autumn 1984. The primary aim was to find the efficiency of the

Table 3. Results from the analytical campaign in autumn 1984 (average figures).

	Plant 1	Plant 2
average run-off, l/sec	7.4	2.3
inlet water:		
pH	6.7	6.1
Fe^{2+}, mg/l	16	100
total Fe, mg/l	20	115
alkalinity, meq/l	1.2	1.4
oxygen, mg/l	3.8	7.0
temperature, °C	9.6	8.0
outlet water:		
pH	8.0	8.4
Fe^{2+}, mg/l	0.4	0.05
total Fe, mg/l	6.7	5.0
alkalinity, meq/l	1.8	1.9
oxygen, mg/l	5.3	4.7
temperature, °C	10	6.0
retention time, hours	7	8
dosage of $Ca(OH)_2$, mg/l	70	300
iron removal, %		
Fe^{2+}	96	100
total Fe	67	96
efficiency of $Ca(OH)_2$, %	76	86

<u>4.2.3. Experiments in Cylinder Glasses</u> Experiments were done in cylinder glasses to investigate the efficiency of hydrated lime and the iron removal following lime treatment.

These experiments gave similar results as the full scale investigations. All the ferrous iron was oxidized and the concentration of total iron decreased to 5 - 6 mg/l after lime treatment. The sedimentation could be improved by gently stirring the water.

5. SUMMARY AND CONCLUSION

In Denmark a law has recently been introduced to minimize the leaching of iron-ochre into streams. A farmer who wants to carry out drainage works on soils containing pyrite can be directed to purify the drainage water from dissolved iron. The most efficient method is treatment of the drainage water by hydrated lime.

In the period 1981 - 84 several hydrated lime plants have been established. In these plants hydrated lime is added to increase the pH of the drainage water to at least 7.4. Oxidation of soluble ferrous iron to insoluble ferric iron will be completed within approximately ~ 15 min. The investigations showed that it is possible to oxidize all the ferrous iron in simple hydrated lime plants even though the concentration of ferrous iron is high (> 100 mg/l).

The method is very reliable, but owing to the manual adjustment of the lime dosage suddenly increasing iron concentrations may give short periods of incomplete water purification.

In all the hydrated lime plants in Denmark the drainage water is pumped into the plant. The lime is added by a dry dosage with the applicator connected to the pump so that addition of lime takes place only when the pump is working. When the lime is added directly to the pump well, the efficiency of the hydrated lime is approximately 85%. However, formation of coatings and corrosion cause severe damage to ordinary drainage pumps. Therefore, several plants have been altered so that the lime is added after the pump well. The efficiency of the lime in these plants is about 75%.

At concentrations up to at least 125 mg ferrous iron/l the aeration of the water obtained by pumping has been sufficient for the oxidation process. At higher concentrations of ferrous iron it may be necessarry to carry out a more efficient aeration of the water.

The precipitated iron is removed by sedimentation in simple soil bassins at a retention time of at least 2.5 hours. Most of the oxidized and insoluble iron (65 - 95%) can be precipitated in a simple pond. The outlet concentration of suspended ochre is often 5 - 6 mg/l independent of the initial concentration of ferric iron. If necessary, the sedimentation can be improved by gently stirring the water.

5. REFERENCES

CHRISTENSEN, L.B. and OLESEN, S.E., 1984. Forsøg med okkerrensning ved landbrugsmæssige draeninger (Ochre Abatement Experiments in Connection to Farm Land Drainage). Bilag 3 til okkerredegørelsen, Miljøstyrelsen. 186 pp. Danish.

CHRISTENSEN, L.B. and OLESEN, S.E., 1985. Leaching of Ferrous Iron After Drainage of Pyrite-rich Soils and Means of Preventing Pollution of Streams. EEC Workshop on Agricultural Water Management in Arnhem, June 1985.

Disposal of Saline Drainage Water and Storm Discharge from Left Bank Outfall Drain, Pakistan

P.D.S. Gunn
Sir M. MacDonald & Partners Limited, Demeter House, Station Road, Cambridge, CB1 2RS, England

1. INTRODUCTION

The Left Bank Outfall Drain (LBOD) project is to drain 336,000 ha (813,000 acres) of farmland on the left bank of the lower Indus to lower the saline water table and the schematic layout of the drain system is shown on Figure 1. The main channel of the drainage system, the Spinal Drain, is already part constructed and discharges through a natural watercourse into a basin on the northern edge of the Rann of Kutch and straddling the Pakistan - India border called Shakoor Dhand.

When subsurface drainage projects have been implemented, flows in the Spinal Drain will be very saline and the outfall will be altered to carry flows direct to the Arabian Sea through the Kadan Pateji Outfall Drain (KPOD) (see Figure 2), an existing small drain which will be greatly enlarged, and a link to the Shah Samando Creek.

Although designed primarily for disposal of subsurface drainage water, the LBOD system will also have to carry surface drainage which will be of higher magnitude. The drain network down to the head of the existing DPOD will safely carry storm flows but the flatter KPOD outfall channel will not do so. A solution is proposed which will allow flows greater than the capacity of KPOD to pass down DPOD into Shakoor Dhand. A headworks to DPOD has been designed which is a simple weir; the described study investigated the effectiveness of the headworks in excluding saline flows to the Shakoor Dhand, an area of ecological importance, and possible modifications to improve its performance.

After rainfall, local flooding is experienced for many days in the project area and a worthwhile benefit has been obtained in terms of reduced crop damage by allowing for some storm water drainage. The catchment area of LBOD is very large and the cost of providing complete surface drainage would be prohibitive. Maximum use is to be made of the open drain system designed to dispose of subsurface drainage by allowing surface inflows up to the capacity of the drain channels, including freeboard allowance. Surface inflows are to be controlled by simple pipe inlets causing some local flooding after rain (sometimes for several days) until all storm water has been discharged.

The open drainage system, designed for maximum discharges of 35.1 m^3/s (1240 cuft/sec) in the lower part of the Spinal Drain, can generally carry flows of up to 62.3 m^3/s (2200 cuft/sec) within its freeboard. The new outfall arrangement of the enlarged KPOD and Link Channel, crossing very flat ground, does not have the capacity to convey the higher storm flows; and providing the necessary additional capacity would be very expensive. On the onset of rain in the project area drainage tubewells and tile drain pumping stations would be switched off to reduce flow and prevent the mixing of saline groundwater with surface runoff. Flows in the Spinal Drain above the capacity of the new outfall can thus safely be discharged into Shakoor Dhand through the existing outfall arrangement of DPOD. A diversion structure is to be constructed at the head of DPOD to direct flows up to the maximum groundwater discharge direct to the sea via KPOD and to allow storm flows of greater magnitude to be divided between both outfalls without backing up levels in the Spinal Drain.

The increase in surface runoff and associated changes in water salinity have been predicted and routed down the Spinal Drain to assess the ability of different designs of the DPOD headworks to separate out the saline flows and fresh flows.

2. SHAKOOR DHAND

Shakhoor Dhand is a low lying basin adjacent to the Rann of Kutch part of which, within Pakistan, comprises a wildlife conservation area, and part of which is in India. Access is difficult and restrictions are placed on travel into the Dhand due to the proximity of the Pakistan - India border. Little detail is therefore known about the environment of the Dhand but the area at the outfall of DPOD clearly appears to support an active fish and bird life. The catchment of Shakoor Dhand is mainly provided by Dhoro Puran, which can be traced as far north as Mirpurkhas. Inflows are irregular and, because the Dhand is normally a closed basin, water levels vary considerably. The capacity below spill level to the Rann of Kutch is estimated to be some 5000 hm^3 and, after storm inflows this volume of water will be left to infiltrate and evaporate. Infiltration being limited, the majority of stored water will evaporate, the last remnants of which are pools of very saline water. When visited in November 1984 even though water levels at the outfall of DPOD were reasonably high, water conductivity was 6000 micromhos/cm (1 micromhos/cm = 0.64 mg/P).

Over the years, drainage within the original catchment of the Dhoro Puran and its tributaries has been interrupted by the construction of canals, railways and roads. Construction of the LBOD and the recent completion of LBOD Branch Drain has partly restored inflows from the original catchment, though some low lying areas and their sub-catchments remain to be connected.

On completion of the LBOD project discharge in the Spinal Drain will generally be very saline with a conductivity of about 31,000 micromhos/cm at its outfall. It is important that inflow to Shakoor Dhand is maintained, but discharge of such very saline water could have severe consequences on its ecology. The headworks of DPOD have been designed to exclude all saline baseflows and pass a proportion of storm flows.

3. DRAIN INFLOWS

Tubewells and tile drain pumping stations will normally operate at a steady state in each subcatchment and the flows and salinities in the Spinal Drain before the onset of rain will also be steady. At the onset of rain, surface drainage inlets to sub and branch drains will flow within a very short time and will reach maximum or near maximum discharge in less than one hour: simultaneously tubewell and field drain pumps will be switched off. Storm inflow to the drains at each pipe inlet will be almost constant because of the throttling effect of the pipe inlets for as long as the fields are flooded.

The response of the drainage system to rainfall can be modelled without danger of over simplification and a model was developed on the project microcomputer (ACT Sirius 1) to compute the discharge, depth of flow, and conductivity of the Spinal Drain during storm events. Hydrographs of inflows at the five major drain junctions were derived, added to Spinal Drain discharges and routed over the reaches between junctions.

A storm as small as 25mm total precipitation (with an annual probability of exeedance of 70 to 78 percent) would cause flooding for more than 12 hrs. As it can be shown that the time taken for flows in the Spinal Drain to rise to the crest of DPOD headworks after rain on the catchment is of the order of 12 hrs, this duration was used to evaluate design events.

4. BRANCH DRAIN HYDROGRAPHS

The time to rise of the Spinal Drain at DPOD headworks after simultaneous rainfall over the whole project area is about 100 hrs and times of concentration of the branch drains are of the order of 60 hrs. For calculation of inflow hydrographs a time base of 3 hrs was used and these figures were interpolated to a 1 hr time base for flood routing calculations.

Response of each sub drain to rain was modelled as a linear rise from saline base flow to maximum storm discharge with a time base equal to the time of concentration of the branch drain. Hydrographs for each branch drain system were derived by summing sub drain and direct inflow taking account of times of concentration to the head of the branch.

Two large canal escapes can be expected to discharge into the drainage system during the storms because water is not needed for irrigation. It was assumed that full escape discharge commenced at the onset of rain.

5. SALINITY

The estimated conductivity of subsurface drainage water on the project is as follows (mmhos/cm):

Nawabshah C.P.	West Nawabshah M.D.	25000
	East Nawabshah M.D.	20000
Sanghar C.P.	Singharo B.D.	36000
	Sanghar M.D.	36000
	Patoyun B.D.	30000
	Direct Inlets	35000
MirpurKhas C.P.	MirpurKhas M.D.	35000
	Direct Inlets	35000

Initially storm water runoff will pick up any surface salts and a value of 500 mmhos/cm has been assumed for the salinity of surface runoff. Canal escape water was assumed to have a conductivity of 100 mmhos/cm.

6. FLOOD ROUTING

The rising wave of storm flows in the Spinal Drain was routed through each reach in turn by the method of characteristics as described by Ven Te Chow (1959). The method is a simplified method which treats the wave as a large number of very small surges and in this case the time interval adopted for analysis of individual surges was 1 hr.

Salinity was modelled by assuming no mixing of the flows and calculating conductivities of the outflow hydrograph of each reach by mass continuity.

Flood routings were calculated for the base case of instantaneous rainfall over the whole project area and for the cases of rainfall falling locally in one or more component projects with and without discharge from canal escapes. Various possible project operating regimes were also modelled.

7. ROUTING RESULTS

The results of the base case routing (simultaneous rainfall over the whole project area) are demonstrated in Figure 3 which shows that drain levels will rise to the crest of DPOD headworks within 15 hrs of onset of rain and to 50% of maximum discharge within 30 hrs whereas it will be some 50 hrs before water conductivity at the headworks falls to half its base flow value. It can also be seen that when Spinal Drain discharges have risen above the maximum capacity of KPOD (62.3 m^3/s (2200 ft^3/s)) drain conductivities will still be as high as 17000 mmhos/cm.

Initial inflows into DPOD will obviously be relatively saline, although discharge at this high salinity is small; at higher discharges the salinity is much lower so that the average conductivity of inflows into Shakoor Dhand is by no means indicated by the conductivity of initial spillage. Shakoor Dhand basin has a finite volume so average conductivities of flows from DPOD which would fill that volume were investigated to compare efficiencies of headworks designs in excluding saline inflows.

It has been reported (MMP/HTS 1984) that in 1979 a flow of 56.7 m^3/s (2000 ft^3/s) in Dhoro Puran filled Shakoor Dhand to the level of overspill to the Rann of Kutch in 30 days. The effect on Shakoor Dhand of saline inflows was demonstrated by calculating average conductivity of the first 23 million m^3 (250 million ft^3) and 46 million m^3 (500 million ft^3) inflow, representing some 5% and 10% of the maximum storage.

The effect of rainfall patterns other than simultaneous rainfall over the whole area were also investigated to ensure that the worse case was selected. The conditions and results are summarised in Table 1.

Table 1 CONDUCTIVITY IN DPOD (MICROMHOS/CM)

	Areal Extent of Rainfall	Escapes Flowing	Conductivity Initial	($25hm^3$)	($46hm^3$)
1.	Whole catchment	yes	21 000	9 900	5 000
2.	Whole except LBOD	yes	29 600	5 000	2 800
3.	Whole catchment	no	21 200	9 100	5 200
5.	Upper catchment	Jamrao	23 000	5 800	n/a
6.	Lower catchment	Mithrao	2 500	10 400	n/a

N.B.
1. Upper catchment includes Nawabshah and Sanghar sub-project areas, lower catchment includes Mirpurkhas and LBOD Branch Drain areas.

2. n/a = not applicable because with flows from only part of the catchment drain discharge will not be sufficient to total 43 hm^3 (500 million cuft).

8. IMPROVING QUALITY OF DPOD FLOWS

The case studies all indicate high levels of salinity in the first flush into DPOD, and in the case of rainfall over the whole project area the average salinity of the first 23 hm^3 (250 million ft^3) is a little higher than desireable. Quality of flows into DPOD could be improved by various means:

a) increasing the size of KPOD
b) raising the threshold of flows into DPOD
c) providing storage on or off stream on DPOD
d) reducing the level of saline baseflow during risk months
e) providing direct inflows of fresh water to DPOD

Any proposed increase in the capacity of KPOD must be substantial in order to improve effectively the water quality of flows into Shakoor Dhand: and to eliminate saline inflow completely KPOD must be designed for the maximum Spinal Drain discharge, 119 m^3/s (4 200 ft^3/s) costing some Rs 400 million (approx. US $25.3) and taking some 124 ha (300 acres) of farmland. More importantly exclusion of all LBOD flows from Shakoor Dhand would further deplete the basin of water.

The threshold of flow released into DPOD could be increased by raising and lengthening the crest of the headworks structure or providing a gated structure which would only be opened when Spinal Drain discharge reaches the capacity of KPOD.

The headworks design allows a maximum head over the weir of only 0.72 m (2.35 ft) so there is little scope for raising the level without substantially increasing the structure width. Results are presented for a 837 m (275 ft) width.

Provision of a gated structure is not a suitable solution because of the remoteness of the site: especially considering the infrequency of rainfall there could be no confidence in the timely and correct operation of gates. The effect of a well operated gated structure on flows into Shakoor Dhand has been shown for completeness in Table 2.

Table 2 ADJUSTMENT OF DESIGN OF DPOD HEADWORKS
CONDUCTIVITY OF DPOD FLOWS INTO SHAKOOR DHAND
(MICRO MHOS/CM)

Design	Initial Inflow	Average ($23hm^3$)	Average ($46hm^3$)
base design	21 000	9 900	5 000
raised crest	19 000	9 400	5 500
gated	17 000	8 500	5 000

Areas in the proximity of the Spinal Drain suitable for storage of flood water are few because the drain generally flows below ground level. The only suitable areas are the isolated loops of Dhoro Puran cut off by the construction of the Spinal Drain. The effects of utilising 8 km (5 miles) and 32 km (20 miles) lengths of drain have been investigated and the results are given in Table 3.

Table 3 FLOOD STORAGE IN DHORO PURAN LOOPS
CONDUCTIVITY OF DPOD FLOWS INTO SHAKOOR DHAND
(MICRO MHOS/CM)

Storage	Initial Inflow	Average ($23hm^3$)	Average ($46hm^3$)
0.9	22 500	9 000	5 100
3.6	22 500	7 000	4 100

Even if areas of old drain loops can physically be used for flood storage it is very unlikely that as much as 32 km of channel could be put to this use because the majority of the old drain beds are farmed. The use of old drain loops for storage is not a practical solution to the problem.

During the summer months, when rain is most likely to fall, groundwater levels should be well below the target depth of 2.1m (7 ft). It would therefore be possible to reduce pumping during July and August, so reducing the amount of saline water stored in the drainage system and in the salinity of the surface runoff.

Table 4 compares the effects of reducing pumping of tubewells and drains during the summer to 50% of normal operation.

Table 4 REDUCED DRAINAGE AT TIMES OF FLOOD RISK
CONDUCTIVITY OF DPOD (MICRO MHOS/CM)

Extent of Rainfall	Pumps	Initial Inflow	Average ($23hm^3$)	Average ($46hm^3$)
Whole	Full	21 200	9 900	5 000
Whole	Half	12 500	5 600	2 800
Lower	Full	22 500	10 400	n/a
Lower	Half	12 000	6 300	n/a

An improvement of 40% in quality of DPOD flows can be obtained by this means at no capital cost although loss of infiltration capacity will increase the duration of flooding upstream of the drain inlets. After rainfall, pumps would have to be run at above normal operating factor to bring water tables down again. However, there is some risk that additional flooding would occur.

LBOD Branch Drain discharges into the Spinal Drain although it is not, under the present project, to carry saline drainage water. An attractive proposition would be to connect LBOD Branch direct to DPOD to delay and dilute saline overflow and increase the fresh inflows to Shakoor Dhand. The results of providing a direct link via a syphon under the Spinal Drain and a short connecting channel to Dhoro Puran are given in Table 5. The cost of such a link would be about US $0.54 million (Rs 8.5 million) but this expenditure could not be recommended without detailed information on the actual performance of the project.

Table 5 DIVERSION OF LBOD BRANCH TO DHORO PURAN
CONDUCTIVITY OF DPOD (MICRO MHOS/CM)

Initial Inflow	Average ($23hm^3$)	Average ($46hm^3$)
17 200	5 700	3 300

The above results do not include the advantage of continuous inflows to Shakoor Dhand from the branch drain.

It has been concluded that, in using the DPOD to carry storm overflows, the improvements in terms of the restoration of frequent inflows to Shakoor Dhand outweigh the disadvantages of unavoidable salinity in the initial part of each inflow. The initial salinity of DPOD flows can be reduced by reducing drainage pumping during the months of July and August but not for every event and not without significant loss of calculated storm drainage benefits. Nevertheless, some reduction in pumping hours should be possible and it has been recommended that monitoring of water quality in DPOD and Shakoor Dhand be instituted in Cooperation with the wildlife conservancy authorities and the information used to help formulate operating criteria for groundwater drainage pumping.

Reference:
Chow V.T., Open Channel Hydraulics, McGraw-Hill, 1959.

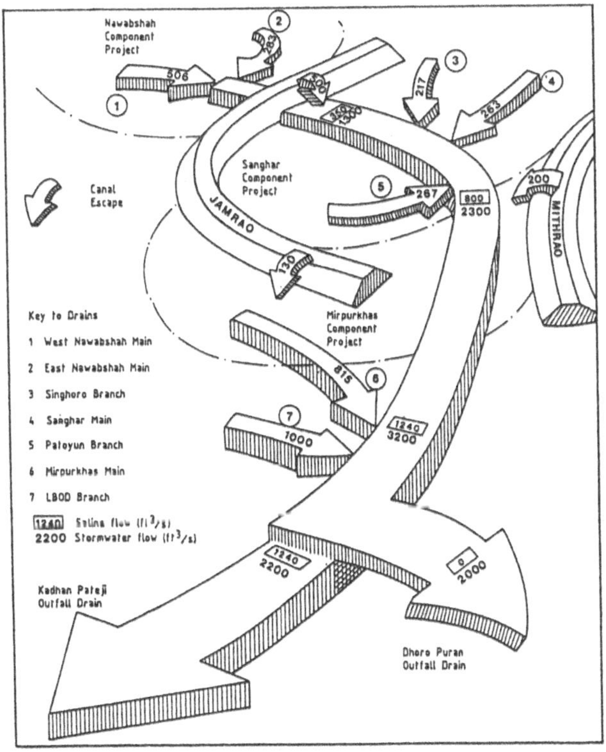

FIGURE 1

Schematic Layout of Drainage System

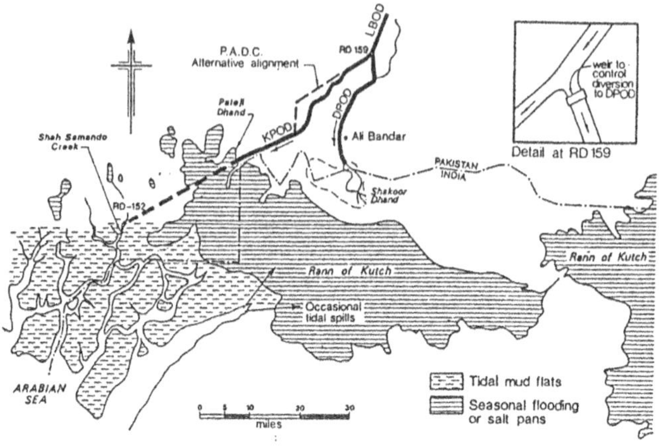

FIGURE 2

Outfall Arrangements

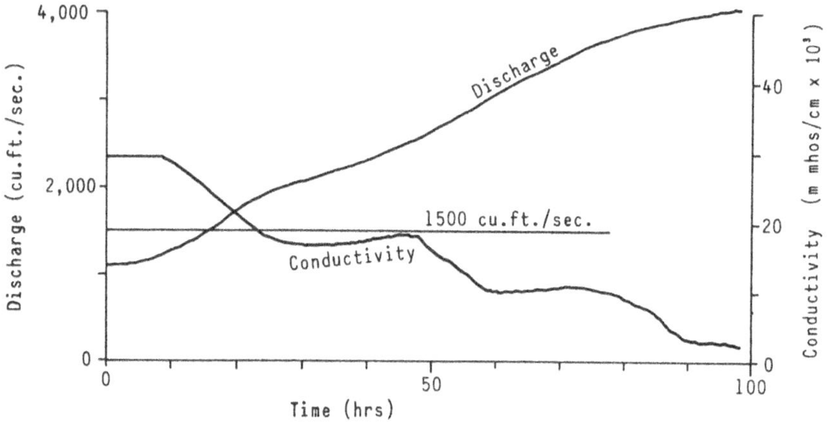

FIGURE 3

Base Case : Discharge and Conductivity
at DPOD Headworks

Drainage Water Quality

Pavel Dvorák
Czech Technical University Prague, Czechoslovakia

INTRODUCTION

Water may be polluted by local or large-scale sources, in most cases chemical / organic, inorganic and radioactive /, usually related to the intensive use of fertilizers, pesticides, herbicides, insecticides and fungicides which are non-degradeable. Chemical sources endangering the quality of waters are often joined by bacteriological sources / viruses, bacteria, germs / and by physical sources. The natural medium has a certain self-purification capability which is intensively manifested especially in the aeration zone. This self-purification capability is, however, limited and its irreplaceable function must not be overcome by the intensity of contamination.

Systematic tile drainage represents a significant interference in the qualitative situation of the system. Drainage discharge will wash out harmful substances and will transport them into open drainage canals and into the recipient where it will join polluted surface waters magnified by surface wash. Drainage waters also contain nutrients, nitrates and phosphates which enhance entrophication processes and the formation of organic mud in water recipients.

WATER QUALITY SITUATION

The situation now has become in many cases critical, namely as concerns the content of nitrates in surface and ground waters with a marked increase of nitrogen content in waters, this both nitrates / NO_3^- / and ammonia / NH_4^+ /. Nitrogen in water is extremely dangerous for health and the increased content of nitrates in drinking water may cause serious diseases in the population especially lethal alimentary nitrate methaemoglobinaemia whose first symptoms are paroxysmal breathing, cramps, loss of consciousness and possibly death. Most at risk are children under the age of one, especially infants between the age of 3 - 6 months. The increased content of nitrates in drinking water also leads to coronary diseases and affects carcinogenic substances of the N type / nicosarcosine / in the human organism. It is therefore necessary to strictly limit the / NO_3^- / content of drinking water. The removal of nitrates from drinking water is technically feasible, e.g. by physico-chemical membrane processes - reverse osmosis, electrodyalysis, ion exchange or by biological methods / autotrous or heterotrous denitrification /which are very costly: the costs more than ten times exceed the costs of conventional surface water treatment / filtration, purification, desinfection/ which does not reduce the nitrate concentration.
Most balancing documents states that the present deteriorated water quality with regard to / NO_3^- / is caused by high concentrations of nitrogen mineral fertilizers and the non-observance of correct agrotechnical principles in the application and handling of these fertilizers. The fact remains that over the past 20 years in many agricultural regions rates of nitrogen fertilizers have almost quintupled and the / NO_3^- /content of surface waters has more than trebled.

DRAINAGE WATER QUALITY SITUATION

The said situation reflected on the quality of drainage water which in recent years has also undergone a substantial change. This change may be documented by the results of many measurements which have, however, not yet allowed to formulate general conclusions. This is a very complicated problem because the amount of / NO_3^-/ in drainage waters fluctuates significantly in dependence on the type of drained soil medium, tillage methods, type of applied nitrogen fertilizer, number of applications, gradient of drained area, atmospheric precipitations, snow thaw,hydrological situation of the territory, etc.

It was found that when the content of physical loam in the soil medium exceeds 35 % weight only up to 10 kg/ha per year of pure N is washed of the soil at N rates of up to 120 kg/ha/year. When the content of physical loam is 10 % weight the amount of pure N washed out of the soil is N = 42.5 kg/ha/year for a rate of 80 kg/ha/year and for a total rate of 120 kg/ha/year pure N the N = 58 kg/ha/year are washed out. A small amount of pure N was also washed out from control soil not fertilized with nitrogen fertilizers this owing to nitrogen contained in atmospheric precipitation and to the decomposition of organic mass in soil.

There is a significant difference between the / NO_3^- / content in drainage waters from drained meadows and pastures and from drained permanently tilled fields. There is also a marked difference in the / NO_3^- / content in drainage waters during the year - the biggest difference was recorded in mid-year and in the spring, in the summer it drops to one half. Extensive measurements in Czechoslovakia allow us to state that at present the average annual content of /NO_3^- / in drainage waters is in the region of 30 - 80 mg/l, the minimum is around 15 mg/l, the maximum values have exceed 200 mg/l.

EVALUATION OF THE QUALITATIVE EFFECT OF DRAINAGE WATERS ON WATER QUALITY IN THE RECIPIENT

The said evaluation should first be subordinated to the limits / "Regulation of 1975" /, should be observed of nitrate ions / NO_3^- / 15 mg/l max in water supply flows and 50 mg/l max in other flows. These limits apply for Q_{355}. In order to evaluate the said effect at adequate hydrological level quantitative value q and specific drainage discharge q_{355} should be determined. Proceedings from measurements made so far this value may be considered in the range of 5 - 10 % of the value of design specific discharge q_d / $l.s^{-1}.ha^{-1}$ / for which drainage pipes and systems are dimensioned under valid Czechoslovak standards. In this country for all types of soils q_d ranges from 0.1 to 5.0 / $l.s^{-1}.ha^{-1}$/ max. Maximum value includes the possible effect of the incorporation of sporadic drainage of springs into the systematic drainage network, so that q_{355} may in this country range between 0.005 to 0.5 /$l.s^{-1}.ha^{-1}$/. Q_{355} is considered in the region of 50 - 5000 /$l.s^{-1}$/ and the ratio Q_{355}/q_{355} is designated parameter B and on the basis of the range of Q_{355} and q_{355} it ranges from 100 to one million.

For an approximate solution let us use the mixing rule from which it is possible to calculate the maximum area of P_{max} / ha / of drainage in the respective catchment which does not endanger the prescribed maximum permissible concentration of / NO_3^- / in the recipient, i.e. 15 and 50 mg/l of water. For the maximum serviced area P_{max} / ha / it applies under the mixing rule that

$$P_{max} = A \cdot B = \frac{Q_{355}}{q_{355}} \circ A \qquad /1/$$

where the dimensionless parameter

$$A = \frac{D_1}{D_2} , \qquad /2/$$

D_1 = $K_{cr} - K_r$ / if $K_{cr} < K_r$ cannot according to the "Regulation 1975" limits be drained by pipe drainage /.

D_2 = $K_{dr} - K_{cr}$ / if $K_{dr} < K_{cr}$ may be drained without fear because $Q_{355} \gg Q_{dr,355}$ /.

K_{cr} - permitted maximal permissible concentration / NO_3^- / in recipient / 15 and 50 mg/l for Q_{355}/

K_r - / NO_3^- / concentration in recipient water measured at Q_{355}

K_{dr} - /NO_3^-/ concentration in drainage water /mg/l/ at $Q_{dr,355} = q_{355} \circ P$ / $l.s^{-1}$/ where P is the drained area /ha/.

For K_{cr}=50 mg/l parameter A will reach maximum value c 45.0 / for K_r = 5.0 mg/l, K_{dr} = 51 mg/l, D_1 = 45.0 and D_2 = 1.0 / and minimal value 0.002857 / for K_r = = 49.0 mg/l, K_{dr} = 400 mg/l, D_1 = 1.0 and D_2 = 350 /. For K_{cr} = 15 mg/l parameter A will reach max value c 10.0 / for K_r = 5.0 mg/l, K_{dr} = 16 mg/l, D_1 = 10.0, D_2 = 1.0 / and minimal value 0.002597 / for K_r = 14.0 mg/l, K_{dr} = 400 mg/l, D_1 = 1.0 and D_2 = 385.0 /.

A number of variants may occur in actual evaluation:

i/ The simplest variant may occur in case only one drainage group discharges into the recipient / the size of drainage groups in natural conditions ranges from tenths of hectares to a maximum of c 25 - 30 ha/ with area P_1. If according to Equation /1/ it applies $P_1 < P_{max}$ the drainage may discharge into the recipient without qualitative concern.

ii/ In case the drainage system is formed by drainage groups 1, 2, 3..... n with areas P_1, P_2, P_3........P_n which are unilaterally or bilaterally discharged into the recipient and where measured values at $Q_{dr,355}$ have shown concentrations / NO_3^- / $K_{1,dr}$, $K_{2,dr}$, $K_{3,dr}$ ....$K_{n,dr}$ a weighted average has to be made of K_{dr} concentration from individual drainage discharge outlets according to drained area P_1, P_2, P_3......P_n. It then applies

$$\overline{K_{dr}} = \frac{K_{1,dr} \cdot P_1 + K_{2,dr} \cdot P_2 \ldots\ldots + K_{n,dr} \cdot P_n}{P_1 + P_2 + \ldots\ldots\ldots + P_n} \qquad /3/$$

For the weighted concentration $\overline{K_{dr}}$ P_{max} is determined according to Equation /1/. In case the summary area under drainage / $P_1 + P_2 + P_3 + \ldots\ldots + P_n$/ $< P_{max}$ it is possible without any qualitative concern to discharge drainage groups into the recipient, obviously respecting the respective K_{cr} in the recipient. Value K_r should be measured in the recipient at Q_{355} in the stretch above the first drainage discharge outlet. For checking the calculations it will be highly favourable also to make test measurements of / NO_3^- / concentration in the strech between the individual drainage outlets and finally in the strech of the recipient below the last drainage outlets.

iii/ in case $P_{max} < P$ the situation is more complicated. In the project preparation and approval stages the following considerations may be made: a redustion of drainage area P to P_{max}, limiting fertilizer treatment, the establishment of a mixing water reservoir, building control or retardation drainage systems, etc. In existing gravitational drainage networks drainage area P cannot be reduced by excluding any of the drainage groups from the hydraulic function / by stopping the drainage outlet, etc. /. By far more acceptable is the alternative of reducing drainage discharge and thereby increasing maximum drainage area P_{max}. This practically means to fit drainage groups with control or retardation devices which accumulate their hydraulic control or retardation functions, both as concerns quantity and quality. It is evident that these systems must be quantitatively and qualitatively controlled and connected to the control system of water management of catchment.

iv/ The "Regulation 1975"provides the stated indicators of permissible water pollution which it relates to Q_{355}. In practice it will be necessary to know the qualitative situation in the recipients in the course of the whole year, i.e. also during other flow rates. The same situation applies for the measurement of / NO_3^- / concentrations at q_{355} when there is minimal probability of the time synchronization of Q_{355} and q_{355}. In such case drainage waters with concentration K_{dr} should not further increase the pollution of the recipient, i.e., they should not exceed K_r. For such case we shall again consider the extreme possibility of / NO_3^- / concentration in the recipient, i.e., max value $K_{r,max}$ 200 mg/l; minimum extreme considered is $K_{r,min}$ = 5 mg/l, maximum extreme possibility of the / NO_3^- /concentration in drainage waters $K_{dr,max}$ = 400 mg/l. The calculation of P_{max} may be made according to an equation which is analogical to Equation /1/ and may be written

$$P_{max} = A \cdot B = \frac{D_1}{D_2} \cdot \frac{Q_r}{q_{dr}} \qquad /4/$$

where $D_1 = K_r^+ - K_r = K_r + 1 - K_r = 1$

$D_2 = K_{dr} - K_r^+ = K_{dr} - / K_r + 1 /.$

Equation /4/ applies for $K_{dr} > K_r^+$. In case $K_{dr} < K_r^+$ discharge may be made without concern because $Q_r > Q_{dr}$, especially when $K_r^+ < K_{cr}$, i.e., 15 or 50 mg/l. Let us illustrate this by one example: at Q_r = 1200 $l.s^{-1}$ concentration was found to be K_r = 42.0 mg/l. Concentration /NO_3^-/ in drainage outlet was measured K_{dr} = 95.0 mg/l at q_{dr} = 0.2 $l.s^{-1}$. The drained area was P = 25 ha. According to Equation /4/the dimensionless parameter A = 1/52 = 0.0192 / D_1 = 1.0, D_2 = 95 - 43 = 52/, parameter B = 1200 / 0.2 = 6000 so that P_{max} according to the Equation /4/ is 115.2 ha $\gg$ P = 25.0 ha so that it is possible to discharge drainage waters into the recipient without qualitative concern. The computation conflicts arising in the said evaluation may most frequently be expected at minimum flow rate in recipients and increased drainage discharges which are retarded in time behind increased flow rates in the recipient. Luckily these phenomena occur in the summer months when / NO_3^- / concentrations in drainage waters are relatively low.

v/ When drainage groups are discharged into open drainage canals and not directly into the recipient it will be necessary with regard to the said evaluation to consider open drainage canals as recipients and to extend /NO_3^-/measurements to them and to monitor the quality of whole drainage system. Final orientation will be the respective official recipient.

It should be stated that between the individual discharge outlets of the drainage groups into the recipient conditions and changes in the recipient bed are more complex than expressed in the given mixing rule, especially in large-scale drainage systems. This is first of all the conversion of nitrogen forms by the metabolism of communities, be it by the reduction of nitrates via nitrites or the oxidation of organic nitrogen substances / in this sequence nitrites represent a more toxic form of nitrogen substance than nitrates/. In this respect the stated and applied mixing rule plays an approximative role which may be significantly specified by introducing the most frequent possible measurement of /NO_3^-/ concentrations in drainage outlets and in the recipient such as will depict even extreme hydrological states.

RATIONAL METHODS OF FERTILIZER APPLICATION

Proceeding from quantitative and qualitative water management balancing in the catchment / zones of hygiene protection / an area should be demarcated in which the agricultural production may be increased without threatening water resources and without limiting fertilization and areas in which it will be necessary to preserve and protect water resources and to adapt agricultural production, including the limitation of fertilization, to these conditions.

In view of the necessity of preserving and protecting water resources it will be inevitable to maximally limit fertilization with nitrogen, namely outside of the vegetation season. Priority should be given to unpressurized liquid fertilizers,granulated combinend fertilizers and fertilizers with a slow release of nutrients, especially nitrogen. Attention should be given to the even spread of fertilizers..The nitrogen rate for autumn crops must not exceed 30 kg/ha and must not exceed 1/3 of the total amount of nitrogen, planned for a one year application. This means that pre-seeding fertilization is prohibited. The one-off application of nitrogen fertilizers is excluded, it will always be necessary to use divided application, best in three rates. The second rate may be applied for restored growth when the root system is already well developed. The third application of nitrogen by spreaders should be in view of the height of growths. Plots which are covered with a continuous layer of snow or frozen or waterlogged soil must not be treated with manufactured fertilizers. Manure heaps and compost heaps must not be established on slopes. Of pesticides and herbicides only those may be applied which are listed in currently valid lists of permitted preparations.

Biocides whose harmfulness is usually manifested long after application must be applied with maximum care. In protection zones of water resources substances based on organochlorinated hydrocarbons whose degradeability is within the region of several years should not be applied. Insecticides have lower degradeability/ of the order of days or months /, they are, however 100 to 1000 times more toxic than herbicides. Substances containing DDT, HCH, Aldrin, Dieldrin, Heptachlor and calcium cynide should under no circumstances be applied. Fungicides which may have genetic effects on DNK are very dangerous for the human organism.

Effect of Saline Water on the Effluent from Gypsiferous Soils

A.S. Aldabagh and S.I. Alkadhi
College of Engineering, University of Mosul, Iraq

INTRODUCTION

Gypsum ($CaSO_4 \cdot 2H_2O$) is a soil component in many arid and semi-arid regions (Barzanji, 1973; Alphen and Romero, 1971). It is also one of the chemical amendments applied to soil that influence the quality of drainage water. In irrigated areas, gypsum is applied to the soil either to reclaim alkalai soils or to increase infiltration of water into the soil.
Gypsum is considered a fairly soluble salt. Waters containing principally magnesium or sodium chlorides act to increase the solubility of gypsum. On the other hand, waters containing principally calcium, sulfate or bicarbonate reduce the solubility of gypsum.
Although some work has been done to determine the effect of gypsum on the physical properties of soil (Smith and Robertson, 1962; Loveday, 1976; Barzanji, 1984), little attention has been directed at its effect on drainage water quality. This research was initiated to determine the effect of using saline water on the effluent from gypsiferous soils.

EXPERIMENTAL METHODS

Four soil samples were tested in this study. One of them was a non-gypsiferous silt loam containing 35% sand, 56.7% silt and 8.3% clay. The others were prepared by the addition of gypsum to the non-gypsiferous silt loam soil. The samples prepared were having 6, 21 and 40% gypsum, based on the air-dry weight of the soil.
Columns of soils tested 10 cm in length and 7 cm in diameter were leached with distilled water, river water, saline well water and three solutions of NaCl - $CaCl_2$ salts, using a constant head device. The solutions were prepared having specified values of electrical conductivity (EC) and sodium adsorption ratio (SAR) so as to represent three classes of irrigation water. These classes, according to the U.S. Salinity Laboratory Staff (1954), are C1-S3 (low salinity - high sodium), C3-S4

(high salinity - very high sodium), and C4-S1 (very high salinity - low sodium).
The effluents from the soil columns were collected at specified intervals and analyzed for Ca^{++}, Mg^{++}, Na^{++} and K^{+}. The electrical conductivities of the effluents were also measured. The chemical analyses of soil samples extracts, waters and effluents were made in reference to U.S. Salinity Laboratory Staff (1954).

RESULTS AND DISCUSSION

The results of chemical analyses of the soil samples are given in Table 1. The chemical analysis of waters and the composition of solutions used in the tests are shown in Tables 2 and 3 respectively.
Figures 1, 2, and 3 show the variation of Ca^{++} concentration in the effluent from the four soil columns with the volume of effluent for river water, well water and solution C4-S1 respectively. In Figure 1, the sharp increase in Ca^{++} concentration in the effluent from soil samples B, C and D which contain gypsum is due to the solubility of gypsum. The later rapid decrease and levelling off of Ca^{++} concentration in the effluent from soil sample B indicate removal of most of the gypsum contained in the soil. The higher gypsum content in soil samples C and D resulted in more gradual decrease in Ca^{++} concentration and consequently greater values at the end of the run. For well water (Figure 2), the Ca^{++} concentration in the effluent from the soil samples containing gypsum slightly increased and then levelled off, indicating low solubility of gypsum in the well water. Figure 3 shows that the Ca^{++} concentration in the effluent from all the soil samples rapidly increased and then levelled off at a relatively high value of about 48 meq/l. This is due to the relatively high Ca^{++} concentration in the influent solution (50 meq/l). At the beginning of the run, adsorption and replacement of ions - mainly sodium and calcium - between the influent solution and the soil have taken place (Foth and Turk, 1972; Fireman, 1944).
The variations of Ca^{++} concentration in the effluent from the soil columns for distilled water, solution C1-S3 and solution C3-S4 are similar to that obtained for the river water and for the same reasons.
Figures 4, 5 and 6 show the variation of Na^{+} concentration in the effluent from the four soil columns with the volume of effluent for river water, well water and solution C4-S1 respectively. In Figure 4, the low Na^{+} concentration in the effluent from all the soil samples is due to the low Na^{+} concentration in the influent (0.35 meq/l). For well water (Figure 5), the amount of gypsum in the soil had an influence on the concentration of Na^{+} in the effluent from the soil columns. Generally, as the gypsum content increased, the concentration of Na^{+} in the effluent decreased.
Figure 6 shows that the Na^{+} concentration in the effluent from all the soil samples rapidly decreased and then levelled off. This decrease coincides with the increase in Ca^{++} concentration

Table 1. Chemical analysis of soil samples (saturation extract determinations)

Soil sample	EC at 25°C mmho/cm	Soluble cations meq/l				Soluble anions meq/l				SAR
		Na^+	K^+	Ca^{++}	Mg^{++}	$CO_3^=$	HCO_3^-	Cl^-	$SO_4^=$	
A (gypsum free)	0.94	2.27	0.02	6.77	3.23	0.0	2.30	2.06	6.72	1.02
B (6% gypsum)	3.30	3.00	0.06	28.60	9.40	0.0	2.10	2.65	32.91	0.69
C (21% gypsum)	3.30	2.21	0.06	29.20	9.30	0.0	2.16	2.36	33.42	0.50
D (40% gypsum)	3.30	1.80	0.06	29.20	9.80	0.0	2.20	1.86	33.73	0.41

Table 2. Chemical analysis of river water and well water

Water type	EC at 25°C mmho/cm	Soluble cations meq/l				Soluble anions meq/l				SAR
		Na^+	K^+	Ca^{++}	Mg^{++}	$CO_3^=$	HCO_3^-	Cl^-	$SO_4^=$	
River water	0.39	0.35	0.0	2.19	1.91	1.20	1.70	0.88	0.53	0.24
Well water	4.75	11.0	1.10	17.71	36.69	0.0	9.60	16.55	40.10	2.11

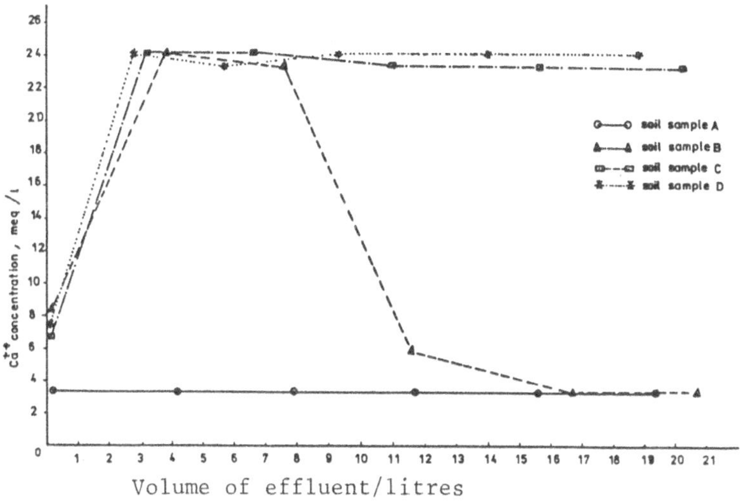

Figure 1. Variation of Ca^{++} concentration in effluent with volume of effluent for river water.

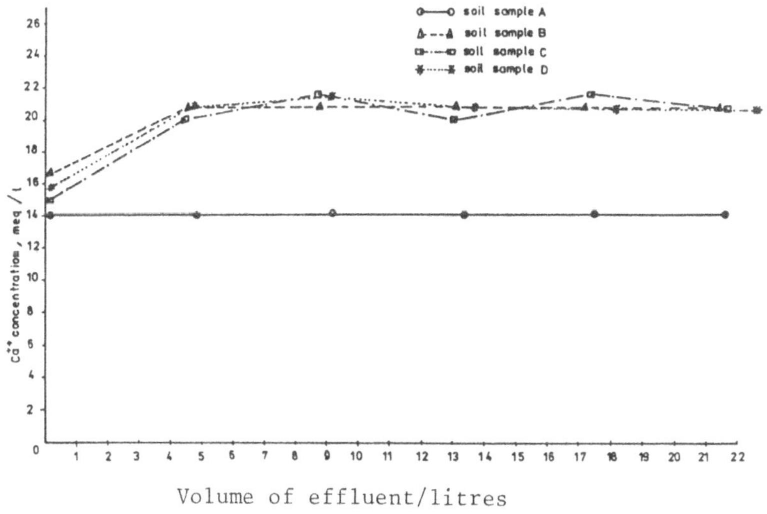

Figure 2. Variation of Ca^{++} concentration in effluent with volume of effluent for well water.

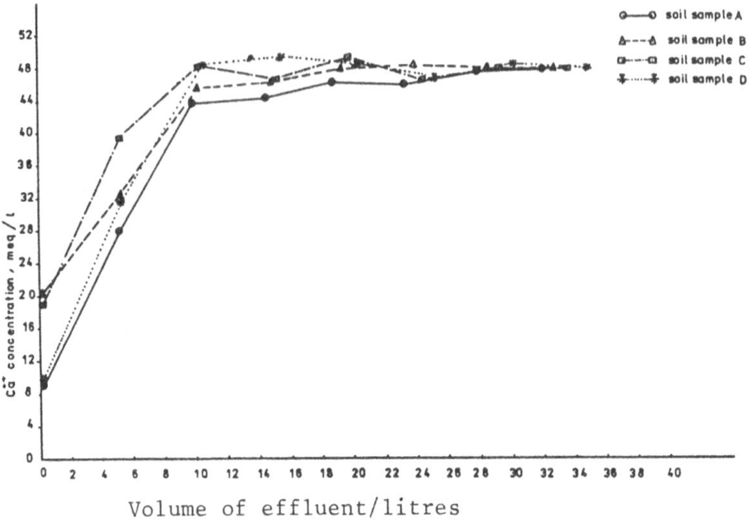

Figure 3. Variation of Ca^{++} concentration in effluent with volume of effluent for solution C4-S1.

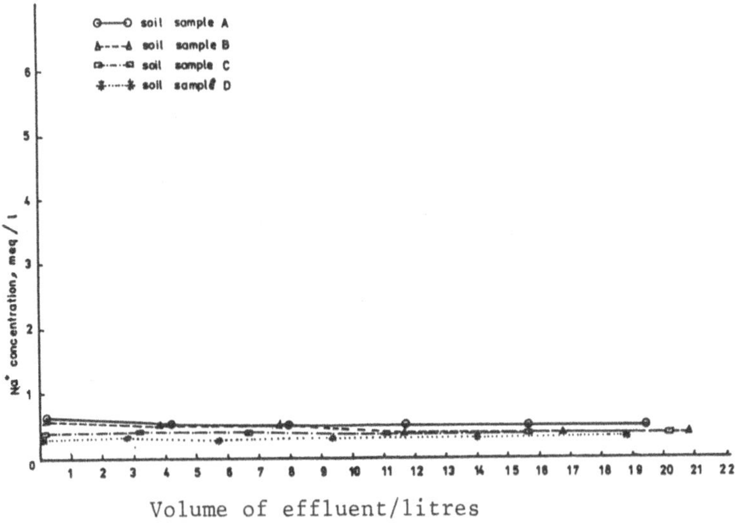

Figure 4. Variation of Na^{+} concentration in effluent with volume of effluent for river water.

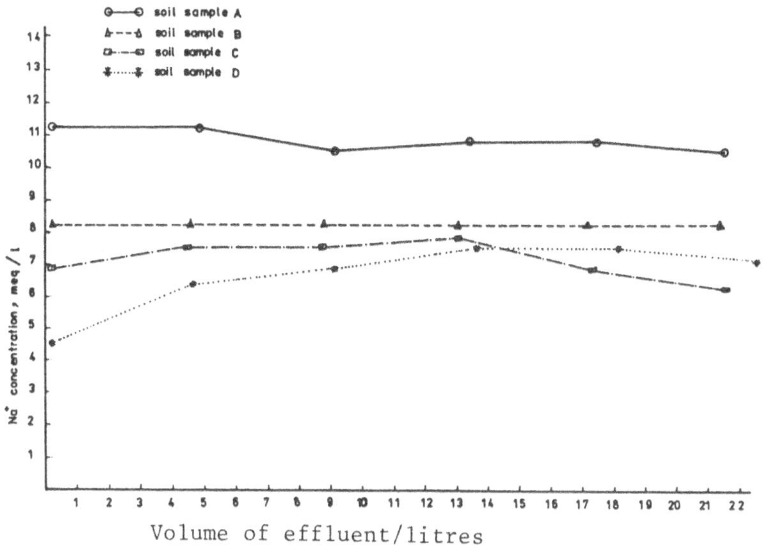

Figure 5. Variation of Na^+ concentration in effluent with volume of effluent for well water.

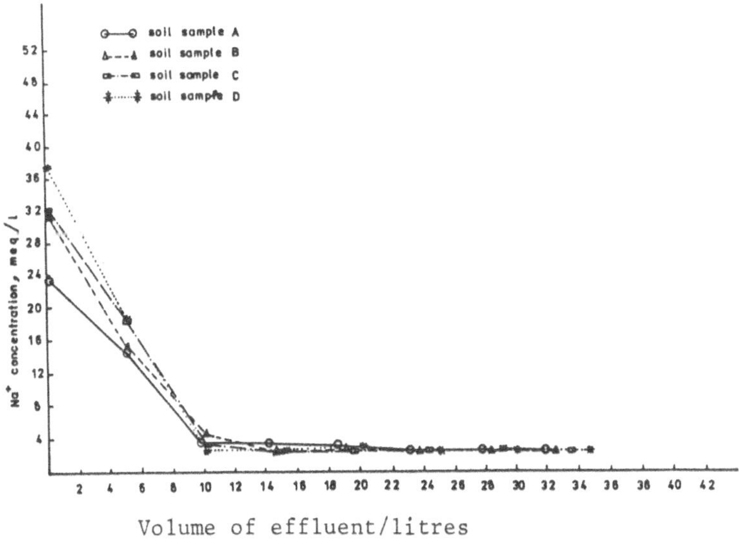

Figure 6. Variation of Na^+ concentration in effluent with volume of effluent for solution C4-S1.

Table 3. Composition of solutions used in the tests

Solution	SAR	EC at 25°C mmho/ cm	Na^+ meq/l	Ca^{++} meq/l
C1 - S3	21	0.185	1.7	0.0131
C3 - S4	18	1.50	14.0	1.21
C4 - S1	0.5	4.00	2.5	50

shown in Figure 3. This result is in agreement with the results obtained by Dutt (1964). The levelling off at a value approximately equal to the Na^+ concentration of influent for all the soil samples indicates that the gypsum has a little effect on the Na^+ concentration in the effluent for solution C4-S1.
The variations of Na^+ concentration in the effluent from the soil columns for solution C1-S3 and solution C3-S4 are contrary related to the variations in Ca^{++} concentration for the same solutions. This is due to the effect of replacement of ions - mainly sodium and calcium - between soil particle surface and percolating solution.
The values of Na^+ concentration in the effluent from the soil columns for distilled water were very low (0.01 - 0.07 meq/l). The results obtained from this study reveal that the effluent from a gypsiferous soil is largely dependent upon the characteristics of the influent and the gypsum content of the soil. The low EC associated with high SAR which characterizes the solution C1-S3, makes it very adequate in dissolving gypsum and recharging the effluent with Ca^{++}. The solution C3-S4 ranked second after the solution C1-S3 in recharging the effluent with Ca^{++}. The high SAR of this solution increases its ability for dissolving gypsum in spite of its relatively high EC. Although both distilled water and river water have low SAR, their low EC has given them a moderate ability for dissolving gypsum. The low SAR associated with high EC which characterizes the solution C4-S1 and well water has given them a low ability for dissolving gypsum.

REFERENCES

1. Alphen, J.G. and F.D. Romero (1971) Gypsiferous Soils. International Institute for Land Reclamation and Improvement, Wageningen, Netherlands. Bulletin No. 12.

2. Barzanji, A.F. (1973) Gypsiferous Soils of Iraq. Unpublished PH.D. Dissertation, Ghent, Belgium.

3. Barzanji, K.K. (1984) Infiltration Rate Characteristics of Gypsiferous Soils in Northern Iraq (Jezira-Area). Unpublished M.Sc. Thesis, University of Mosul, Iraq.

4. Dutt, G.R. (1964) Effect of Small Amounts of Gypsum in Soils on the Solutes in Effluents. Soil.Sci. Soc. Am. Proc. 28:754-757.

5. Fireman, M. (1944) Permeability Measurements on Disturbed Soil Samples. Soil Sci. 58: 337 - 353.

6. Foth, H.D. and L.M. Turk (1972) Fundamentals of Soil Science. John Wiley and Sons, New York.

7. Loveday, J. (1976) Relative Significance of Electrolyte and Cation Exchange Effects when Gypsum is Applied to a Sodic Clay Soil. Aust. J. Soil Res. 14: 361 - 371.

8. Smith, R. and V. C. Robertson (1962) Soil and Irrigation Classification of Shallow Soils Overlying Gypsum Beds, Northern Iraq. J. Soil Sci. 13: 106 - 115.

9. U. S. Salinity Laboratory Staff (1954) Diagnosis and Improvement of Saline and Alkali Soils. USDA Handbook 60.

Plant Nutrient Losses in Drainage Water on Heavy Clay Soil

Szilard Thyll
Agricultural Faculty of Debrecen Agricultural University, Szarvas Hungary

INTRODUCTION

Subsurface drainage on heavy clay soils influences not only the soil moisture content, but also condition of plant nutrients.
As an effect of favourable air conditions, as well as increased water movement in the soil, results in leaching out of plant nutrients by the drainage water.
Plant nutrients washed out by drainage water means not only loss in nutritive materials of the soil,but at the same time it is harmful to environment too.
Among the nutrients washed out by drainage water nitrogen components cause the biggest problem resulting in eutrophication of surface waters.
According to Steenvoorden /198o/ groundwater uninfluenced by man's activities already contains a certain amount of nitrogen and phosphorus compounds so an outflow of groundwater into open waters will always cause a natural or base load.
Johnson and Baker/1973/ verified that as an effect of intensification of agriculture and wide -spread application of technical methods /e.g.tile drainage/ the water quality of canals and rivers deteriorates.
Baker et al./1975/ stated,that NO_3-nitrogen losses from tile drains,whatever their source,result in an economic and energy waste as well as environmental hazards.
Steenvoorden/1983/ claimed,that one of the most important problems in relation with surface water-quality is the process of eutrophication.
The amount of nutrient compounds washed out by drainage water,among others,depends on the absorption capacity of soil.Absorption capacity of soils is determined by quality and quantity of clay minerals /Stefanovits,1981,1983/.

The bigger the absorption capacity of soil, the less quantity of plant nutrients will be leached out by drainage water.
The main purpose of this paper is to determine the quantity of plant nutrients washed out by drainage water from a drained heavy clay soil, which can be characterized by domination of illite clay mineral.
I should like to demonstrate function relation between NO_3-nitrogen concentration of drainage water and drain spacing as well as specific drain discharge.
We examined the effect of precipitations on the change of NO_3-concentration of drainage water at various technical parameters.

EXPERIMENTAL METHODS

In 1979 a research field of 24 ha was established on a heavy clay soil.
Dominant clay mineral of the soil is illite.
The tile drainage was combined with subsoiling, because hydraulic conductivity of soil is very low /i.e. o,2 $m.d^{-1}$/. On the research field singular pipe system was applied.
Technical parameters of lateral pipes /drain spacing, slopes and depth of drains etc/ were varied.
Variations of drain spacing were 15, 2o and 25 m.
Slope of laterals were o,ooo and o,ool. Depth of drains varied between o,8-1,5 m.
Drain discharge was measured every day. The actual drain discharge /$l.s^{-1}$/ was converted into specific drain discharge /$l.s^{-1}.ha^{-1}$/ for a comparison of results.
Quality and quantity of nutrient materials were measured in every lo days, both in drainage water and groundwater.
From plant nutrient materials quantity of H_4N^+, NO_3^- and PO_4^{3-} ions were measured by colorimetric method. Quantity of K^+-ions were measured with flame photometer.
Amount of nutrients is in terms of concentration /$mg.l^{-1}$/.

RESULTS

The typical plant nutrient concentration values are shown in Table 1.

Table 1. Concentration of plant nutrients in drainage water and groundwater

Origine of water and typical values	Concentration /mg/l/ of			
	NO_3^-	NH_4^+	K^+	PO_4^{3-}
Drainage water				
Average	18o,12	-	3,33	o,o65
Minimum	4o,8o	-	o,8o	o,o4o
Maximum	41o,oo	-	12,3o	o,32o
Groundwater				
Average	1oo,67	-	3,51	o,o3o
Minimum	41,oo	-	o,8o	o,o5o
Maximum	195,oo	-	23,8o	o,o5o

From the data of Table 1.it can be seen,that H_4N-nitrogen could be measured neither in drainage water nor in groundwater.
At the same time a significant amount of NO_3-nitrogen was measured both in drainage water and groundwater. Minimum value of NO_3-nitrogen was nearly the same both in drainage water and in groundwater.Both the average and the maximum concentration of NO_3-nitrogen was bigger in drainage water than in groundwater. Small quantity of K^+-ions was washed out by drainage water.The maximum concentration of K^+-ions was less in drainage water, than in groundwater.
Concentration of PO_4^{3-}-ions was very low both in drainage and groundwater.
We studied NO_3-nitrogen concentration of drainage water as function of specific drain discharge/$l.s^{-1}.ha$/ as well as of drain spacing /m/.
Function relation between NO_3-nitrogen concentration of drainage water /$mg.l^{-1}$/and specific drain discharge /$l.s^{-1}.ha$/ can be seen in Figure 1.
The relation can be characterised according to the following equation

$$C_{NO_3}=383,62+8450,06q+52455,55q^2 \ /mg.l^{-1}/ \qquad /1/$$

correlative co-efficient /R/ equals to -o,7834.
On the base of F-trial the regression on a level of P = 1 % is close.
From Figure 1.and Equation 1.it can be seen, that the increasing specific drain discharge, decreases the NO_3-nitrogen concentration in drainage water.

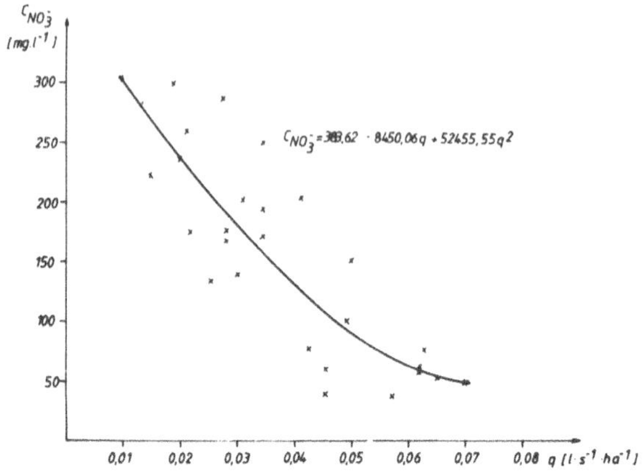

Fig.1. Function relation between NO_3^--concentration of drainage water/C/ and specific drain discharge /q/

Function relation between NO_3^--concentration of drainage water /C/ and drain spacing /L/ was determined too. This function is shown in Figure 2.

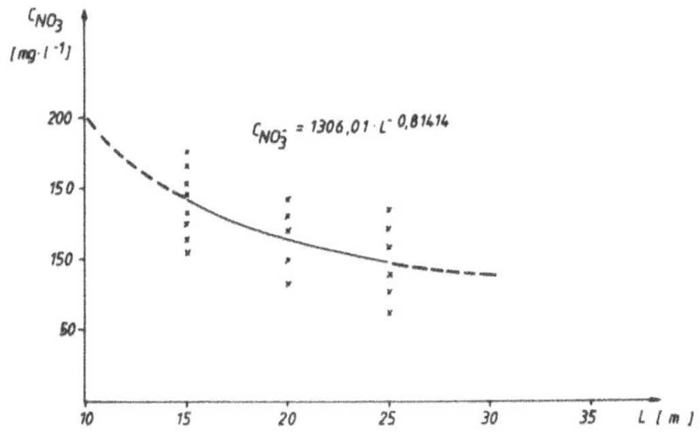

Fig.2. Function between NO_3-concentration of drainage water /C_{NO_3}/ and drain spacing /L/.

The function between NO_3-concentration of drainage water /mg.l^{-1}/ and drain spacing /M/ can be determined with the following equation

$$C_{NO_3} = 13o6,ol. \ L^{-o,81414} \quad /mg.l^{-1}/ \qquad /2/$$

Value of the correlative co-efficient /R=-o,9999/ proofs a close relation.
It follows from Figure 2.and Equation 2. that as a consequence, the bigger the drain spacing, the lower the NO_3-concentration in the drainage water.
We examined change of NO_3-concentration in the drainage water after rainy days when the soil was nearly saturated.
Effect of rains on the change was examined at different drain spacing, slope of drains and depth of laterals.
One of these situations is shown in Figure 3.

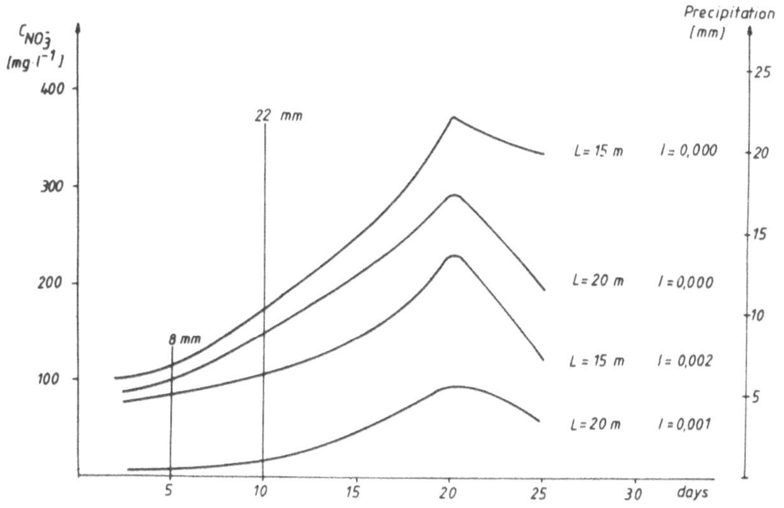

Fig.3.Effect of precipitation on NO_3-concentration of drainage water at various technical parameters

From Figure 3.it can be seen, that NO_3-concentration of drainage water increases after rains, 4-5 days later it decreases.
Degree of increase is bigger when the slope of drains is equal to 0.It is because, in this case, the pipes are established less deep/i.e.o,8 m/, than in case the slope is o,ool /average depth is 1,2 m/.
When the slope of drains is the same, degree of increase is smaller when drain spacing is bigger.
This fact results from Equation 2.

DISCUSSION

From the experimental results it can be proved, that on drained heavy clay soil, in which the dominant clay mineral is illite, a great quantity of NO_3-nitrogen is leached out by drainage water.
The cause of this fact is, that absorption energy of clay minerals is the smallest for NO_3-ions.
Considering the average value of NO_3-concentration in drainage water, furthermore the average value of drain discharge the rate of NO_3-nitrogen washed out by drainage water, is about 2o % of previously aplied nitrogen fetilizer. This fact means not only a nutrient loss for plant production, but also damage of environment by eutrophication process of surface waters.
According to our examinations H_4N-nitrogen couldn't be measured in drainage water.
This fact can be explained partly by nitrification, and partly by fixation of these ions.
Loss of phosphate and potassium is not considerable.
The low concentration of PO_4^{3-} -ions in drainage water can be explained by a relatívly high calcium content of the soil, since a considerable quantity of phosphates will be hardly soluble in the soil.
The cause of small quantity of potassium in drainage water is, that in this soil permanent illite formation takes place from different clay minerals. This process means considerable fixation of potassium ions.
At the same time low concentration of potassium in drainage water can be also explained by the specific absorption ability of illite.
Quantity of potassium washed out by drainage water is probably higher when the smectites are the dominant clay minerals.
On the base of research results it can be verified, that the bigger the drain spacing, the lower the NO_3-nitrogen concentration in drainage water.
When the specific drain discharge increases, the concentration of NO_3-nitrogen decreases in drainage water.
In early spring, when the soil is nearly saturated, lo-2o mm of rain has a considerable effect on NO_3-concentration of drainage water. When we make a comparison in the effect between different technical parameters the following can be verified.
As an effect of precipitation the increase in NO_3-concentration of drainage water is smaller when drain pipes are installed deeper than when they are laid shallower.
Generally NO_3-nitrogen concentration of drainage

water is lower at deep lying laterals, than at shallow lying laterals.
The fact of considerable leaching out of NO_3-nitrogen in drainage water necessitates a modification of nutrient management on drained heavy clay soils.
We have suggested application of nitrogen fetilizers late in the spring, when outflow of drain pipes is moderate or ceases.
The other suggestion is the application of nitrification impending materials, like N-Serve 24 E.
These methods decrease risk of eutrophication of surface waters.

REFERENCES

1. Baker, J.L.-K.L.Campbell-P.H.Johnson, and J.J. Hanway /1975/ Nitrate, phosphorus and sulfate in subsurface drainage water. Journal of Environmental Quality, 4., 3: 4o6-412
2. Calvert, D.V./1975/Nitrate, phosphate and potassium movement into drain lines under three soil management systems. Journal of Environmental Quality 4. 2:183-186.
3. Johnson, H.P. and J.L.Baker/1973/Water quality implications of cropland nutrients. App.4. Ames Reservoir Environmental Study. Iowa State Water Resour. Res. Inst. Iowa State University, Ames.44
4. Steenvoorden J.H.A.M./198o/ Eutrophication. Research Digest. ICW Techn. Bulletin 117. Wageningen, 137-152.
5. Stefanovits, P./1981/Effects of land improvement on environment. Agrokémia és talajtan 3o:137-152.
6. Stefanovits, P./1983/ Connection between clay minerals of soil and land improvement. Manuscript. Gödöllő 15
7. Thyll, Sz./1984/Some new achievment in the pipe drainage of heavy clay soils in flat land regions. Studia Universitatis Scienciarum Agriculturae Debreceniensis. Tom. XXIV. 385-4o6.

Technical Economic Criteria and Efficient Drainage System Design When Using Different Filter Materials

Andrei Wehry, Ion David, Teodor Eugen Man
Institute of Polytechnics "Traian Vuia" Timisoara, Romania
Iulian Mihnea
Institute of Land Reclamation Studies and Design, Bucharest, Romania
Ion Moca
State Planning Committee, Bucharest, Romania

ABSTRACT

On the scope of achieving some tile drainage networks with minimum specific project investments and higher reliability in operation, the paper proposes to present a new mathematical-hydraulic model of higher technical-economic efficiency for calculating drain spacing.

The model has been elaborated on basis of laboratory studies undertaken on drainage pipes using different filtering materials.

GENERAL REMARKS

Drainage project techniques, as applied nowadays all over the world, have shown the necessity for using filter materials either around or above the drain tube as it allows:

- to provide a certain stability in the drained soil stratum by preventing the soil particles from moving towards the drain, and thus, avoiding clogging the drain or blocking drain pipe holes;
- to ensure and to create, in the proximity of the drain tube, a more permeable layer than the soil, thus increasing the water rush perimeter into the tube by lowering the drain hydraulic resistance to water penetration.

Some drainage methods propose the use of drain-pipes of large diameter: Ø 80 - Ø 110 mm and a filtering material of small thickness (e.g. geotextiles), while other methods advise drain pipes of small diameter Ø 50, Ø 65 mm and a thicker granulated filtering material.

Between these two extreme solutions there are a lot of intermediate possible alternatives according to available filtering materials, and by technical economical calculations as well as by determinations on experimental fields to find out the most efficient solution.

Drainage projects may be implemented without using any filtering materials, but this can only be achieved on homogenous, stable, well structured soils of good permeability using a suitable pipe lay-out technique.

The following criteria should be taken into consideration when deciding upon the filtering material to be used:

(a) hydraulic: i.e. small entrance resistance and reduced clogging in accordance with the soil type;
(b) economic: i.e. the price must be cheaper than that of the drain-tube;
(c) water quantity to be drained;
(d) drain lay-out technologies to be used;
(e) filter probable service life, etc.

The studies and research carried out in the Socialist Republic ROMANIA, in the Aranca, Teba-Timisat, Somes-Crasna, Sag-Topolovat, Bociar and other systems, have shown the possibility of using the following filtering materials: granulated materials (gravel, ballast, rough sand, industrial waste etc.), geotextiles (Filtex, Madril, Terasin, Netesin, Drenatex etc.), organic materials etc.

In Romania, the following stages were traversed to achieve agricultural drainage projects:

1. LABORATORY STUDIES

1.1. <u>The hydraulic resistance coefficient of the water entering a drain without filter (ξ_i) also of a drainage complex provided with different filter materials (ξ_{if})</u>
It refers, to the study of filter materials in conditions of cleanliness, in contact with sand in a stand where the drain is located vertically [4,5,7] working as a well, Fig. 1.

The water entrance resistance into the drain (W_i) and in the filter-drain complex (W_{if}) is calculated by means of this stand, using the water level inside the tube (h_o) and the water level near the tube (h_i) or filter material (h_{if}) and the flow (q) per unit of length of the tube using following relations:

$$w_i = \frac{h_i - h_o}{q/h_o} \quad (1) \qquad w_{if} = \frac{h_{if} - h_o}{q/h_o} \quad (1') \qquad \text{(days/m)}$$

The hydraulic conductivity of the sand in the stand is determined by the relation of a perfect well:

$$K = \frac{q \ln r_6/r_3}{\pi\ (h_6^2 - h_3^2)} = \frac{q}{\frac{\pi}{\ln r_6/r_3}\ (h_6^2 - h_3^2)} \quad \text{(m/day)} \quad (2)$$

where: h_6 and h_3 = the piezometric water levels in the stand at the radii = r_6 and r_3

q = siphoned flow from the drain.

For the average daily values determined during five days, the value of the hydraulic water entrance resistance coefficient in the drain tube without filter (ξ_i) and in the complex drain plus other different filter materials, is calculated with the relations:

$$\bar{\xi}_i = w_i \cdot k \quad (3) \qquad \text{or} \qquad \bar{\xi}_{if} = w_{if} \cdot k \quad (3')$$

where: $\bar{\xi}_{if}$ = 0,0001 - 0,001 - very good filter
0,001 - 0,01 - good filter
0,01 - 0,1 - satisfactory filter
0,1 - unsatisfactory filter

Having the values $\bar{\xi}_i$ ($\bar{\xi}_{if}$) in initial conditions of cleanliness of the tube and filter (unsilted) it is possible to make a hydraulic classification of the behaviour of these materials, so that the designer could choose from this great number of alternatives some which have smaller values for $\bar{\xi}_i$ ($\bar{\xi}_{if}$). Up to more than 100 alternatives of drain pipes in complex with different filter materials have been tested.

1.2. <u>The silting rate in time of the drain tube complex plus different filter-materials, in contact with the soil which must be drained.</u>
A second laboratory study concerned the silting rate of the filter materials in contact with the soil to be drained. Thus, there is presented [2,4,5,7] a method of establishing the silting rate in time of the drain-tube complex with different filter materials when in contact with the soil to be drained - on the stand provided with horizontal laid drain-tubes.

Measuring the drained flow in the first day (q_i) and after about 30 days, with weekly interruptions (q_f) it is noticed that it decreases greatly due to the blocking of the filter with fine soil particles.

This flow ratio is noted with "η", and it is the same with the ratio between the hydraulic conductivity of the filter when clean in the initial conditions, and when silted in the final conditions:

$$\eta = \frac{q_i}{q_f} = \frac{K_{fi}}{K_{fc}} \tag{4}$$

These q_i and q_f values are established from the laboratory experimental curve tracing the wrapping curve (Fig. 3) which determines q_f.

The value of the initial hydraulic conductivity of the clean filter K_{fi} is determined on a classic Darcy type stand. Finally it is possible to determinate K_{fc}, i.e. the hydraulic conductivity of the silted filter, which permits an analytical calculation for an efficient technico-economical design of the drainage network for the type of studied soil.

A primordial classification of the silted filter is made depending on the relation:

$$\chi = \frac{K_{fc}}{K_s}$$

where: K_s - hydraulic soil conductivity at the level of the drain for the proposed project, so that if:

- $\chi \gg 1$ - we have a very favourable effect of the filter;
- $\chi > 1$ - a favourable effect;
- $\chi = 1$ - no effect;
- $\chi < 1$ - unfavourable effect.

The value q_f (l/min/s) is transformed in mm/day, verifying in this way the capacity of evacuation and transport of the specific flow by the drain-pipe for different distances between drains(L).

2. THE CALCULATION (DESIGN) OF DISTANCE BETWEEN DRAINS

Applying the known Ernst's formula completed with the loss of head at the drain water entrance and the silted filter material, i.e. to achieve after a period of operation in natural conditions, the drainage norm (x), required for the cultivated plants, in accordance with the Fig. 4 we shall have:

$$h = \frac{q\,D_v}{K_1} + \frac{q\,L^2}{8K_1T_e} + \frac{q\,L}{\pi K_1}\ln\frac{\alpha D_o}{u} + \frac{q\,L}{K_1}\xi_{if} \qquad (6)$$

$$\text{where: } T_e = D_1 + \frac{h_o}{2} + D_2\frac{K_2}{K_1} \qquad (7)$$

The value of ξ_{if} can be analytically determined [1,2,4,7], with the notations in Fig. 5, with the relation:

$$\xi_{if} = \frac{2}{n.\pi}\left[\ln\frac{1}{\sin\frac{n\,.\,b}{s\,d_o}} + \frac{1-\chi}{2\chi}\ln\left(A_1 + \sqrt{A_1^2+1}\right).\left(A_2 + A_2^2 + 1\right)\right] +$$

$$+ \frac{2\,B}{\pi\,.\,n\,.\,b}\left[\ln\frac{1}{\sin\frac{\pi\,.\,\ell}{2\,B}} + \frac{1-\chi}{2\chi}\ln\left(B_1 + \sqrt{B_1^2+1}\right)\left(B_2 + \sqrt{B_2^2+1}\right)\right] \qquad (8)$$

where:

$$A_1 = \frac{\left(\frac{d_f}{d_o}\right)^n - 1}{2\left(\frac{d_f}{d_o}\right)^{n/2}\sin\frac{n\,b}{2\,d_o}}$$

$$A_2 = \frac{\sqrt{\left(\frac{d_f}{d_o}\right)^{2n} - 1}}{2\left(\frac{d_f}{d_o}\right)^{n/2}\sin\frac{n\,b}{2\,d_o}}$$

$$B_1 = \frac{\mathrm{sh}\,\frac{\pi\,(d_f - d_o)}{2\,B}}{\sin\frac{\pi\,.\,\ell}{2\,B}}$$

$$B_2 = \sqrt{\frac{1}{2}\left[\sqrt{1 + 4\left(\frac{\mathrm{sh}\,\frac{\pi(d_f - d_o)}{2\,B}}{\sin\frac{\pi\,.\,\ell}{2\,B}}\right)^2\left(\frac{\mathrm{ch}\,\frac{\pi(d_f - d_o)}{2\,B}}{\sin\frac{\pi\,\ell}{2\,B}}\right)^2} - 1\right]}$$

The distance L between the drains is calculated and then taking into account the cost price of a Kml of drain tube and of the filter material, the drainage project specific investment might be calculated.

It is compulsory to calculate also the alternative without filter material.

From the possible alternatives of drain tubes and filter materials, as studied and calculated in the laboratory, two or three alternatives showing a minimum specific investment will be retained.

Thus, the best technico-economical solution corresponding to a minimum specific investment might be chosen.

3. STUDIES WITHIN DRAINAGE EXPERIMENTAL FIELDS

According to the technico-economical results an experimental field in the area which will be drained is realised, using the calculated alternatives of a minimum specific investment.

The optimum drainage alternative might be chosen by taking into consideration the ground level maintenance to the required drainage norms after abundant precipitations (national or artificial), and in the meantime the obtained yields (8).

BIBLIOGRAPHY

1. David I. — Uber ein berechnugsverfahren der verluthone beim eintritts wiedestand des wassers in die drainrohre, Revue Roumaine des Sci.Techn.Serie de Mecanique Appliquee, Nr.1, Tom.28. pag. 1-19, 1984.

2. David I., Wehry A., Man T.E. — Efectul colmatării filtrului la drenajul agricol, Rev. Hidrotehnica nr. 6/1980, Bucuresti.

3. Gheorghiu V., Wehry A., Man T.E. — Technical-economic aspects of making filters for agricultural field drainage, Buletinul stiintific si tehnic al Inst. Politehnic "Traian Vuia" Timisoara, Tom. 26(40), Fasc. 1, 1981.

4. Man T.E. — Studiul rezistentelor hidraulice la drenurile agricole - Teză doctorat, Inst. Politehnic "Traian Vuia" Timisoara, 1983.

5. Wehry A., Man T.E. — Studii si cercetări privind filtrele drenajelor agricole, Rev. Hidrotehnica nr. 7/1979 Bucuresti.

6. Wehry A., David I., Man T.E., Schimek V. — Analisis of some factors having an influence on the choice of the efficient solution of drainage, Rev. Hidrotehnica nr. 11/1981, Bucuresti.

7. Wehry A., David I., Man T.E. — Actual problems in the drainage technique Edit. Facla, 1982, Timisoara.

8. Wehry A., Uncianschi L., Chivereanu D., Man T.E., Bîrdită I., Horvath St. — Stabilirea criteriilor de drenaje pe solurile grele din zone umede si subumede pe baza măsurătorilor efectuate în cîmpurile experimentale din vestul tării, In Bul. st. si tehnic al I.P. "Traian Vuia" Timisoara, Tom 23(37), Fasc. 1, 1978.

9. Wesselling J. — Proceedings of the International Drainage Workshop, ILRI, Publ. nr. 25, Wageningen, The Netherlands, 1979.

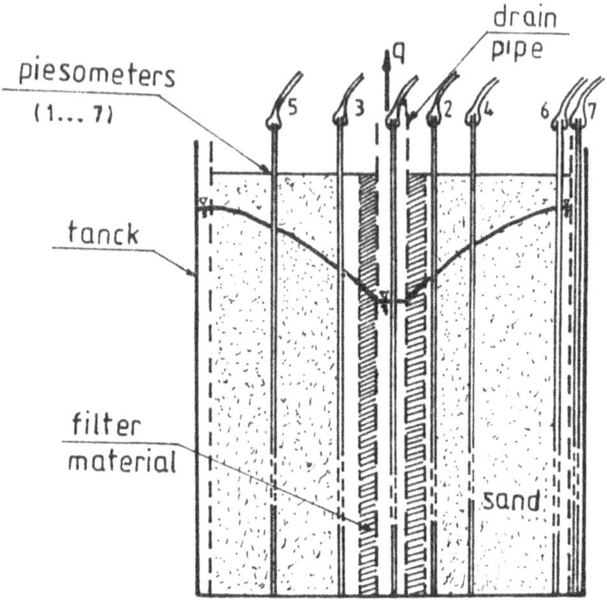
drain
pipe
q
piesometers
(1... 7)
5
3
2
4
6
7
tanck
filter
material
sand

Fig. 1.a.

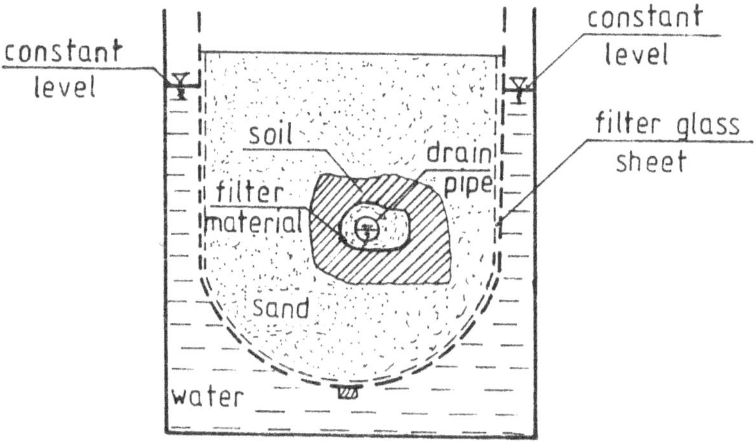
constant
level
constant
level
filter glass
sheet
soil
drain
pipe
filter
material
sand
water

Fig. 2.a.

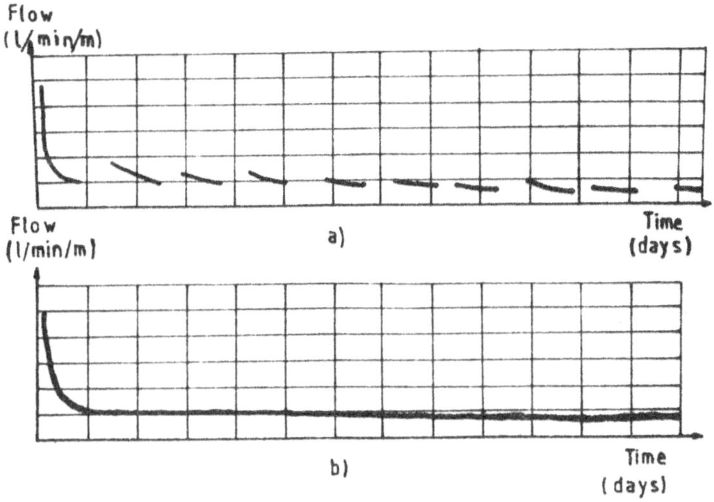
Flow
(l/min/m)
a)
Time
(days)
Flow
(l/min/m)
b)
Time
(days)

Fig. 3.

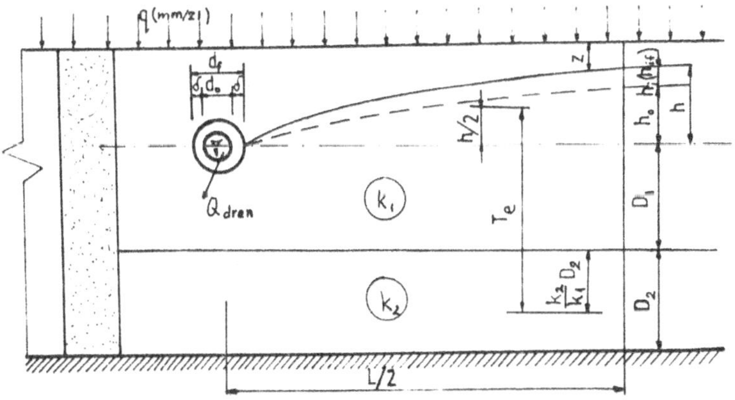
q (mm/zi)
z
d_f
δ d_o δ
h/2
h_o
h
Q_{dren}
k_1
T_e
D_1
k_2
$\frac{k_2}{k_1}D_2$
D_2
L/2

Fig. 4.

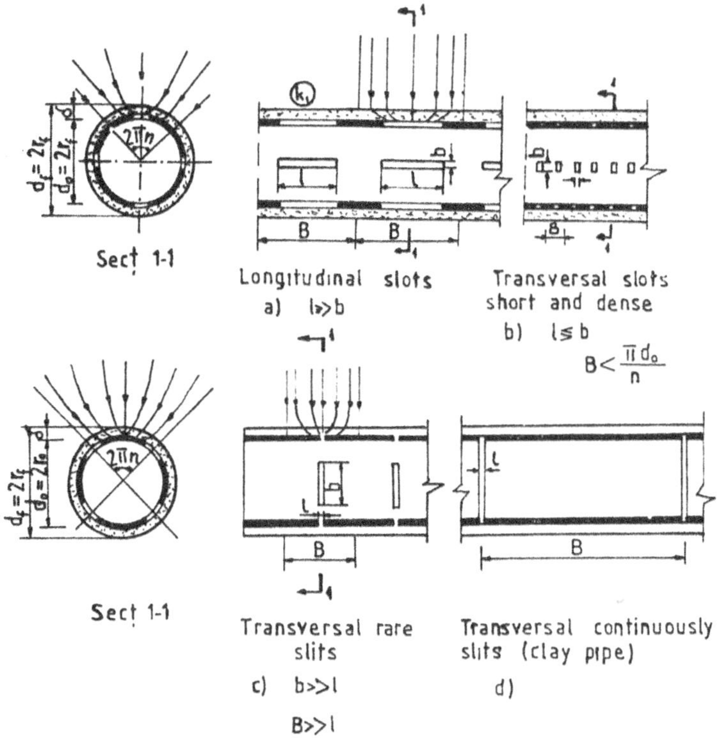
Sect 1-1
Longitudinal slots
a) l>>b
Transversal slots
short and dense
b) l≤b
B<πd₀/n
Sect 1-1
Transversal rare
slits
c) b>>l
B>>l
Transversal continuously
slits (clay pipe)
d)

Fig. 5.

Stabilization or Prevention of Wet Landslips by Drainage

Bruce Denness
Bureau of Applied Sciences, Whitwell, Isle of Wight, U.K.

INTRODUCTION

Everyone knows that water flows downhill. However, this knowledge is not always expressed in drainage schemes introduced to overcome engineering or agricultural problems. The errors are seldom obvious at first sight: arterial drains normally run downhill and feeders also lose height on their approach to the arteries. Perhaps this apparent adherence to the rules is the core of the problem, the gloss that distracts from the underlying misjudgement of the basic drainage requirements; that is to drain well those parts that need drainage and to leave well alone those parts that are already adequately drained. Not all hills are well drained or valleys waterlogged. Nor are hillsides uniform water transducers.

There is often within natural ground and surface water regimes a wealth of variety. Much of this can be determined from the surface topography, the geomorphology or shape of the land surface. No-one expects water to run along the ridges of a sloping corrugated roof; everyone knows it falls to the troughs and then runs down them. The same is true of hills, hillsides and valleys. Within the overall and obvious pattern there are often subtle secondary slopes which describe minor superimposed drainage routes. It is those which effect the practical control on drainage. Nor is this a purely surface phenomenon for groundwater flow is also subject to subsurface topography, the shape of horizons between overlying permeable and underlying impermeable media.

The intensity with which secondary geomorphological features influence drainage patterns depends to a large degree on the scale of the feeder groundwater reservoir, and/or its surface recharge, and its ease of access to a given site. Thus, there is need to note not only local but general surface and subsurface topography in order to assess the dynamics of a ground and surface water regime.

Let us now explore a few examples from the writer's personal casebook, in fact fifteen from five countries in three continents, of natural drainage patterns along with their different impacts on the softening of soils in slopes and subsequent landslipping and slope degradation. Here naturally stable slopes imply good drainage whereas unstable slopes indicate poor drainage. The different geomorphological situations pertaining to each thereby indicate differences in artificial drainage demand. The subsequent design of drainage schemes should therefore take the geomorphology into account. Similarly lessons learnt in slope stability drainage can be applied in agricultural situations and in man-made schemes.

SLOPE INSTABILITY

The most common form of slope instability is that in which a permeable stratum overlies an impermeable plastic stratum. The upper layer acts like a sponge to absorb water from precipitation or a neighbouring groundwater reservoir and then slowly releases it onto the underlying medium. In turn the lower layer softens by gradually remoulding under ambient stresses in the presence of introduced water until it is no longer able to withstand those stresses and subsequently deforms more or less rapidly as a landslip or mudflow. The earlier comments relating to the influence of geomorphology and groundwater reservoir size are superimposed on this mechanism.

Largely for convenience the following examples are subdivided into three categories: the first in which a sand, sandstone or limestone conformably overlies a clay or shale; the second in which a sand or gravel unconformably overlies a clay or shale; the third in which a permeable till overlies an impermeable till. Each carries subtly different geomorphological connotations, as will be indicated.

Sand, Sandstone or Limestone Conformably Overlying Clay or Shale

This is perhaps the most widespread of all situations inducing slope instability. Geological conformity involves the continuous deposition of newer sedimentary rocks directly upon older sedimentary rocks; differences in lithology or constituent parts, which result in permeable-over-impermeable strata, follow from differences in depositional environment but the final product is generally a layered system of locally fairly uniform thickness. However, the experienced geologist will know that this observation cannot be generalized without due care for local structural peculiarities. Nevertheless it affords a good measure of confidence in the interpretation of subsurface topography and consequently of likely natural drainage paths and their subsequent treatment by artificial drainage schemes. First let us explore the large-scale, regional pattern and then move on to small-scale, local features.

Regional slope patterns It is common to find sedimentary sequences extending over large regions. In so doing they are subject to regional tectonic activity which often results in significant flexures or discontinuities in the regional structure. The outcome is an undulating sequence with intermixed anticlines and synclines, sometimes disrupted by faults. It depends on proximity to a synclinal or anticlinal axis whether the bedding slope between strata is steeply inclined or shallow. Consequently as a natural dendritic surface drainage pattern is superimposed the valleys which it carves out in the undulation complex are incised in layers which often dip into the valley on one side and away from it into the hillside on the other at varying degrees. This results in uneven groundwater flow into the valley. On one side water from the aquifer flows unrestrictedly over the underlying aquiclude into the valley whereas on the other other side the clayey aquiclude receives only minor overspill from the sandy layer above.

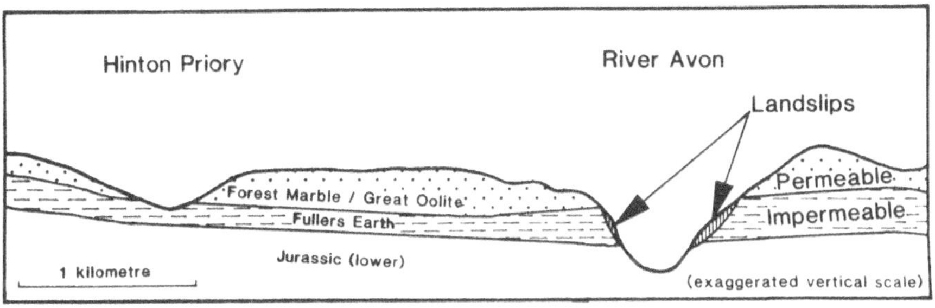

FIGURE 1: Schematic section through syncline in northern Somerset, U.K.

Figure 1 depicts schematically a typical section through adjacent valleys in a folded sedimentary sequence. The section shown is based on a simplification of the real situation across the valleys of the River Avon and Hinton Priory near Limpley Stoke and Hinton Charterhouse in northern Somerset. Denness and Cratchley (1972) described the geological setting and the geomorphology of the region in some detail and drew particular attention to the correspondence of shallow slopes with the sides of valleys downdip of the general bedding slope in contrast to the steeper slopes on the opposite side, irrespective of the geographical bearing thereby eliminating the possibility of an overriding climatic influence. In northern Somerset this involves the permeable Forest Marble and the Great Oolite Limestone overlying the impermeable Fullers Earth, a plastic clay, and the additional influence of local cambering, the downwarping of updip aquifer sections into valleys, can also be observed. Site inspection reveals that the steeper slopes generally show fewer signs of extensive historical instability, except where active erosion has occurred, and, taken overall, are better drained naturally, largely because they are supplied by a smaller effective groundwater reservoir than their opposite numbers; though the lateral extent of the aquifer may be as great or even greater, its dip away from steep slopes induces lower groundwater seepage onto the slope. In turn, this has obvious implications regarding the design of modern drainage to counteract regeneration of ancient slipping or the introduction of new landslips on these slopes; the shallow slopes need more comprehensive artificial drainage on a regional basis.

Denness and Cratchley (1972) also considered a similar regional example around Uppingham in Rutland (now Leicestershire) in which the Rivers Welland and Chater and their tributaries have cut through the Inferior Oolite Limestone aquifer into the underlying Lias Clay,a geological section a little lower in the Jurassic Succession than the previous example. They showed, using a slope inclination map by Chandler (1970), that the generally southeast regional dip has resulted in steeper slopes on northeast than southwest facing valley sides as would now be expected. In turn, Denness and Riddolls (1976) noted a similar association of unstable shallow slopes in the London Clay of southeast Essex with areas in which there is present overlying sandy Claygate or Bagshot Beds generally dipping towards the slope; meanwhile other London Clay slopes not beneath such strata in the area remain steep and stable. Again the implications for differential regional design of drainage measures to encourage slope stability are clear.

Studies of similar situations on the Antrim coast of Northern Ireland by Prior et al (1970), a large variety of localities on the London Clay of southeast England by Hutchinson (1967), a Cretaceous complex in the Galilee in Israel by Wiseman et al (1970) and countless others reveal generally similar circumstances. However, despite so many observations of the same principle, many large-scale drainage schemes still fail to differentiate between the different geological and geomorphological circumstances that they address with the consequence that some areas are over-designed and unnecessarily expensive while others go inadequately drained and are inherently unstable thereby leading to renewed slope wastage.

Local instability Local slope instability can vary from minor movements a few square metres or less in area to whole hillsides sometimes extending beyond one square kilometre. The point to be made here is that these restricted instabilities, albeit occurring within regions in which the general observations in the previous section apply, are often the result of locally poor drainage from natural groundwater sources which are themselves determinable from a knowledge of the local geology and the geomorphology of surface and subsurface features. The landslips are expressions of poor drainage. Therefore, in order to avoid such damaging features being introduced beneath or near engineering works on hitherto stable slopes, drainage schemes can be designed to accommodate the natural conditions and thereby prevent instability arising. Such a scheme need be only local as long as it is positioned correctly on a larger site much of which may appear to be similar at first sight.

The examples which follow illustrate situations in which landslips have developed on natural slopes. In these cases, it is too late to avoid the slips and only stabilization measures can be introduced; however, had construction been planned prior to the instabilities arising the same sites would have responded to local drainage schemes by remaining stable. That is the lesson to be learned here - strategically placed drainage, not blanket coverage, can avoid local instability problems simply by accommodating the geological regime.

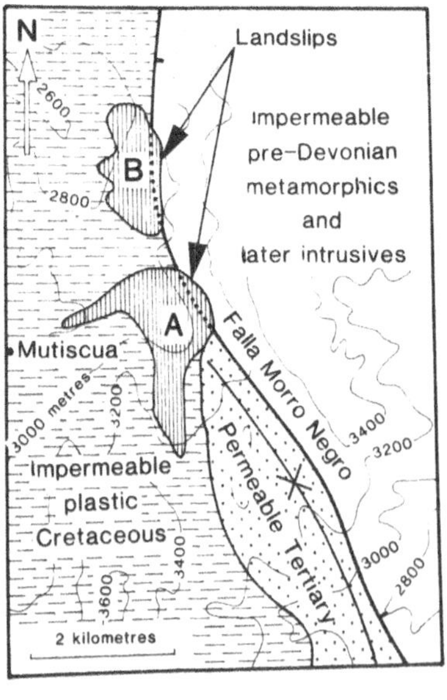

FIGURE 2: The landslips at Mutiscua, Colombia.

Figure 2 represents a summary geological map with superimposed contours for a large site near Mutiscua in Colombia, South America. Reported in more detail by Denness (1974a), it shows an upland saddle running southwest-northeast across a region in which impermeable, plastic Cretaceous clay/shale strata in the west are divided from impermeable Devonian gneiss and phyllite and granite of later date in the east by a major fault, the Falla Morro Negro. Overlying the Cretaceous shales is a permeable Tertiary synclinal complex resting against the west side of the fault and dipping generally to the southeast. Landslips of immense scale are seen at A and B on the figure. Their precise location is governed by the underlying geological configuration and that of the surface and subsurface topography. Site A coincides with the junction of the Tertiary aquifer and the underlying Cretaceous shales at its locally lowest point, at the northwest boundary of the syncline where overspill of groundwater from the basin occurs. There is no other opportunity for the groundwater to escape downhill onto the shale and consequently there are no other instabilities around the basin. Site B coincides with the locally lowest point of expression of the permeable Falla Morro Negro which acts as a groundwater transducer from the Tertiary aquifer to this distant site on the shale. Both these sites are thus subject to the surface expression of groundwater and could be predetermined to need drainage merely from a study of the map. At the same time it is clear that no other part of the shale outcrop would be unstable or need drainage even though its lithology is essentially similar throughout and its surface undulations enjoy similar aspects.

A similar but smaller situation, though aggravated by river erosion at the toe of the valley slope, was reported by Denness (1977) for the Ironbridge Gorge in Shropshire. Here the Lower, Middle and Upper Coal Measures of the Carboniferous period are exposed by the River Severn. Alternating synclinal features are evident on both sides of the east-west valley. Active landslipping is taking place at two main sites on the north side and one on the south. The southern site is at Jackfield, earlier reported by Henkel and Skempton (1955), and the more westerly on the opposite bank corresponds to the axis of a syncline which concentrates groundwater seepage from a major sandstone band in the Upper Coal Measures. The more easterly slip on the northern bank is fault controlled. Failure to capture the local groundwater expression by a comprehensive drainage scheme has resulted in the instabilities. Brunsden and Jones (1972) and Conway (1974) have described similar instabilities in the Cretaceous/ Jurassic complexes of Dorset where permeable Cretaceous caprocks overlie shales of largely Liassic age. In the context of installing drainage measures it is instructive to note from Denness <u>et al</u> (1975) the use of resistivity surveys to determine the optimum alignment of drains at a site at Charmouth, Dorset, judged from geological and geomorphological reasoning to be most at risk in a region of general instability.

<u>Permeable Quaternary Deposits Overlying Clay</u>

Another common geological condition has Quaternary deposits unconformably overlying earlier sedimentary materials. Typically the Quaternary deposits are permeable and constituted of silts, sands and gravels. The Quaternary period began only about a million years ago, so their structure is normally loose giving rise to an efficient if sometimes thin aquifer. Many such normally-consolidated strata are even newer and date from the Holocene period of the last ten thousand years or so. Often these are found lying directly on previously eroded surfaces; therefore, there is a discontinuity in the geological structure at their early boundary. The geological structure below their junction with the underlying strata need be nothing like the commonly near-horizontal structure above, unlike the case of conformable strata from the previous examples. Where these newer aquifers overlie clays or shales, they provide one of the necessary basic ingredients for slope instability wherever they outcrop - necessary, but, as we have seen from analogous examples of conformable sequences, not perhaps sufficient. Also required is an association of the appropriate structure in the underlying plastic aquiclude in order to transmit the waters of the Quaternary groundwater reservoir onto the argillaceous slopes below in sufficient quantity locally to cause slipping. Probably the most common older structure to induce slipping is again a syncline, though various dip/fault combinations achieve the same result.

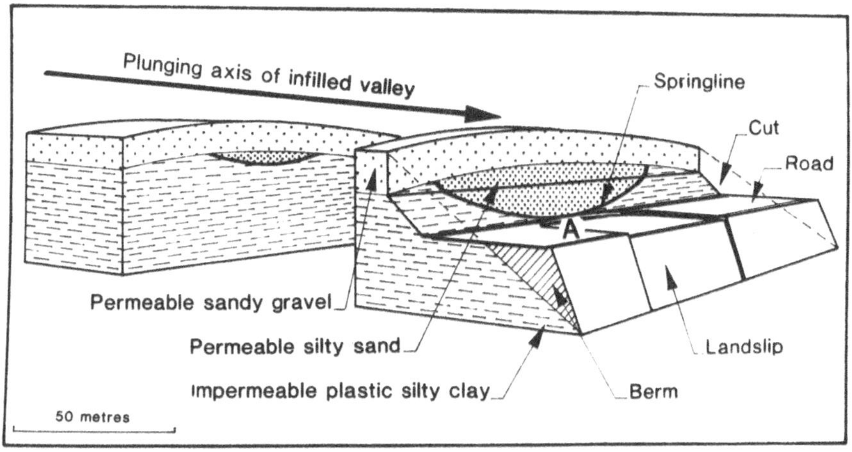

FIGURE 3: Block diagram of truncated infilled valley and landslip near Ibague, Colombia.

Figure 3 portrays a block diagram of a typical example of the condition described, in this case from the Venadillo-La Sierra section of a highway repair scheme near Ibague in Colombia. Denness (1975) described the local geology as a Pleistocene sequence in which a recent Holocene gravel layer unconformably overlies an earlier Quaternary silty clay in which an intermediate permeable buried-channel infill has been deposited. Figure 3 describes the resulting configuration and shows the position of the cut-and-fill road construction in relation to the geology and the local landslip. The springline at the base of the buried channel which collects groundwater from the recent gravel is strikingly marked by lush vegetation in otherwise arid climes. The landslip is directly below the major outflow of groundwater at the centre of the floor of the buried channel which acts effectively much like a tilting synclinal axis. The permanent solution to the instability, the rejuvenation of a dormant earlier slip probably caused by placing the fill of the berm, was local surface drainage to collect the seepage from the buried channel at A. Had these very cheap but effective measures been installed during road construction in response to an ongoing design-and-build site investigation, which would have added little to the overall cost of the original project and certainly been less than the cost of repair, the slip regeneration would never have occurred.

Denness (1972a), while describing the so-called reservoir principle of mass movement, identified two British examples of these geological circumstances. The coastal slope at Bouldnor on the Isle of Wight comprises a Tertiary clay overlain by a Quaternary gravel deposit. The clay is deployed in a synclinal form and stratified with several interbedded permeable layers of shells some 300mm thick. The syncline plunges gently towards the sea so that the shell bands rise to the level of the inland plateau beneath the gravels a few hundred metres inland and thereby act as ducts to carry the groundwater from this newer gravel reservoir directly onto the underlying clays at several levels down the slope primarily along the axis of the syncline. The result is a chaotic and very wet coastal landslip focussed on the outcrop of the plunging axis. This is a natural occurrence which could be stabilised by strategically placed drainage designed to collect seepage from the shell bands towards the axis of the syncline. As no such measures have been introduced perhaps a square kilometre of prime development or recreational land with an average slope of only about 11° lies fallow. The second example from Denness (1972a) described the coastal slope instability at Highcliff in Christchurch Bay, Hampshire. Here the Tertiary Barton Clay is overlain conformably by Barton Sands and an even bigger groundwater reservoir in the caprock of loose Quaternary gravel. The gently dipping Tertiary beds expose the clay to marine erosion and groundwater seepage for a kilometre or more of coastline. In the absence of an undulating or interrupted geological structure the whole section is subject to instability and has needed tackling with a comprehensive deep drainage scheme throughout, an example in which blanket drainage has for once been necessary to meet the bland local geological conditions.

Permeable Over Impermeable Till

In previously glaciated areas there commonly exists an abundance of glacial deposits including various forms of boulder clay and moraines. Though at first sight these deposits seem to be comprised either of a wide ranging suite of materials in erratic array or of an overall uniform if broadly consitituted mixture, there is often a well-defined if somewhat disguised boundary between zones of different permeability. Again a permeable band often overlies an impermeable layer. In geological terms this implies an oblation till, which is thought to have settled out from a melting ice sheet, overlying a lodgement till, which is thought to have scraped along the ground under intense pressure of moving ice. The present day configuration of such assemblies is often represented by hummocky ground which, when eroded, can be seen to display a permeable over impermeable stratification.

Denness (1972a) described the lateral extent of a typical problem of this type at Maryport in Cumbria. Figure 4 shows a compendium of two typical schematic sections through a till mound comprising oblation over lodgement till as seen at Maryport. Where erosion has just begun (Profile A) there is still a sufficiently large groundwater reservoir in the oblation till sloping towards the erosion site to cause slope instability; where erosion has progressed further (Profile B), the groundwater reservoir naturally drains inland and the eroded slope becomes essentially stable. This is ultimately an effectively self-stabilising mechanism. However, should stabilisation measures be required to sustain property unwisely constructed over Profile A, the appropriate local drainage measure can be readily designed. If long sections of coastal or fluvial erosion are addressed in these circumstances the

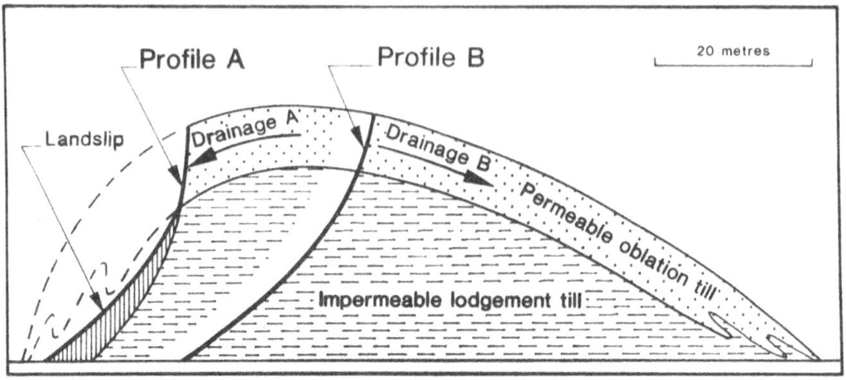

FIGURE 4: Schematic section through slipping till at Maryport, Cumbria, U.K.

necessary differential drainage pattern can be determined largely from inspecting the surface topography.

A similar situation exists around Cayton Bay in North Yorkshire. There the ground is far less hummocky and the differential inland groundwater seepage toward the coastal section is controlled largely by the shape of the underlying interface between the countryrock and the mixed till veneer above. Nevertheless it is still the surface topography that enables the identification of potentially problematic local drainage sites, which, as no action has yet been taken, have initiated slope instability. At Cayamapu in Chile a series of roadcuts has trimmed the toes of morainic lobes descending from the southern Andes. Denness (1972b) noted the marked similarities of the lithological variation in this end moraine to that in the till at Maryport. The local instabilities were also similar and lent themselves to stabilisation by deep cut-off drains wherever there was a slopeward seepage; elsewhere the naturally contra-slope drainage had already resulted in stable cuts.

DISCUSSIONS AND CONCLUSIONS

All the examples cited here are typial of common types of slope instability to which drainage is applicable as a major part of if not a sole solution. Countless other references could be cited but those here relate mainly the writer's personal experiences, as do others relating to landslips in igneous and volcanic complexes with drainage implications reported by Denness (1973, 1974b and 1982a and b) and Darwell and Denness (1976), excluded here by space limitations. It has been shown that a proper interpretation of the local and regional geology and surface and subsurface geomorphology can point the way to the most suitable drainage scheme even though at first sight of the untrained eye a given site may look much like its neighbour.

This is not merely another suggestion that only experience leads to success. No matter how experienced or gifted the investigator, be he or she a geologist, engineer or lay person, surface appearances and even borehole or other site investigation records seldom tell the whole site story. The exposures generated by construction are often more informative. However, by the time these are available the interpretation expertise has usually disappeared, gone to pastures and drainage projects new. This is extremely wasteful of information and talent alike. A design-and-build strategy overcomes the problem. Instead of weighting the site investigation on drainage projects heavily towards the front end it is the writer's experience that a fairly brief preliminary survey followed by ongoing contact and design alteration during construction leads to a healthier final structure, often at lower cost. Every one of the examples cited, each being involved in a stabilization project using drainage of some kind, benefitted from this approach.

In summary it is recommended that drainage for engineering purposes should be installed on an interactive design-and-build basis and that geological and geomorphological data should be central to that design while simultaneously related to the predetermined framework outlined above.

REFERENCES

Brunsden, D. and D. K. C. Jones (1972) The morphology of degraded landslide slopes in southwest Dorset. Q. Jl. Engng Geol., 5, pp205-222.

Chandler, R. J. (1970) The degradation of Lias Clay slopes in an area of the east Midlands. Q. Jl. Engng Geol., 2, pp161-181.

Conway, B. W. (1974) The Black Ven landslip. Rep. No. 74/3, Inst. Geol. Sci., London, 16pp.

Darwell, J. L. and B, Denness (1976) Prediction of metastable soil collapse. Pub. No. 121, Int. Ass. Hydrol. Sci., pp544-552.

Denness, B. (1972a) The reservoir principle of mass movement. Rep. No. 72/7, Inst. Geol. Sci., London, 13pp.

Denness, B. (1972b) Engineering geology in Chile. Rep. No. 508 9/72, Inst. Geol. Sci., London, 47pp.

Denness, B. (1973) Colombia - landslip studies in relation to roads and route planning. Rep. No. 547 11/72, Inst. Geol. Sci., London 70pp.

Denness, B. (1974a) The influence of faulting on the instabilities at Mutiscua, Colombia, S. America. Q. Jl. Engng Geol., 7, pp355-362.

Denness, B. (1974b) Ecuador - una investigación sobre el riesgo de derrumbamientos en la ciudad de Guaranda. Rep. No. 1019 6/74, 18pp (in Spanish).

Denness, B. (1975) A landslide in Pleistocene deposits: Colombia (S. Am.) Proc. 26th Annual Highway Geol. Symp., Boise, Idaho, USA, pp241-258.

Denness, B. (1977) The Ironbridge landslide - a case history of instabilities in Carboniferous sediment. Proc. Conf. Rock Engineering, Newcastle University, pp407-422.

Denness, B. (1982a) Slope stability - are Hong Kong slopes steep and high enough? Proc. 7th Southeast Asian Geotechnical Conf. 2, Hong Kong, pp69-70.

Denness, B. (1982b) Slope stability - are rainfall-induced landslips predictable? Proc. 7th Southeast Asian Geotechnical Conf. 2, Hong Kong, pp71-72.

Denness, B. and C.R. Cratchley (1972) The degradation of selected Jurassic Clay slopes in southern England. Proc. 24th IGC, 13, pp97-106

Denness, B., B. W. Conway, D. M. McCann and P. Grainger (1975) Investigation of a coastal landslip at Charmouth, Dorset. Q. Jl. Engng Geol., 8, pp119-140.

Denness, B. and B.W. Riddolls (1976) The influence of geological factors on slope stability in the London Clay of south Essex, England. Bull. No. 14, Int. Assoc. Eng. Geol., pp37-40.

Henkel, D. J. and Skempton, A. W. (1955) A landslide at Jackfield, Shropshire, in heavily-consolidated clay. Geotechnique, 5, pp131-137.

Hutchinson, J. L. (1967) The free degradation of London Clay cliffs. Proc. Geotechn. Conf. on Shear Properties of Natural Soils and Rocks, 1, Oslo, pp113-118.

Prior, D. B., N. Stephens and D. R. Archer (1970) Composite mudflows on the Antrim coast of northeast Ireland. Geografiska Annaler, 50(A), 2, pp65-78.

Wiseman, G., G. Hayati and S. Frydman (1970) A study of a landslide in the Galilee, Israel. Proc. 1st Int. Congr. Int. Ass. Eng. Geol., Paris, 1, pp50-61.

Drainage Problems in Areas Subject to Subsidence due to Oil Production

A. Irazábal
Obhidra Consult, C.A.
J. Abí-Saab
Maraven, S.A., Venezuela
J. Murria, J. Groot
Nedeco, The Netherlands

SUMMARY

Oil extraction along the entire Eastern edge of the Lake of Maracaibo in Venezuela began in 1927. As a result, some 500 Km^2 of continental land is now subject to settling or subsidence. A part of the affected area is lower than the water level of the lake (M.L.L.L.) and has to be protected by coastal dikes along a total length of 43 kilometers. At the present time, the lowest point is 5.5 meters below the level of the Lake, and the subsidence comes to 20 cm. per year in some places. In this area, one the most important and oldest oil-producing regions of the country, there is a capital intensive oil-industry network, as well as activity related in some way with that industry. Some 150.000 people live in the area.

It is drained partly by gravity, in a natural manner through the rivers which cross it and with the use of canals to intercept the surface runoff and discharge it into the Lake. Additional drainage is provided by pumping (25 stations with a total installed capacity of 270,000 m^3/hour) in the protected zones.

In recognition of the drainage problem affecting the area and of its dynamic character, as a result of the subsidence, MARAVEN S.A., the largest oil producing company inland, contracted with the OBHIDRA CONSULT C.A for the preparation of a Drainage Master Plan under the coordination of MARAVEN S.A. and the technical advice of NEDECO. This plan calls for a 50-years planning scenario, with an intermediate scenario of 25 years.

MARAVEN S.A. has supplied projections of ground subsidence derived from the application of a mathematical model. With these projections, topographical maps were produced showing how the relief changes over time.

SITE OF THE AREA

The area under consideration is located along the East coast of the Lake of Maracaibo (Bolívar Coast). The area under oil-drilling operations occupies 680 Km^2. Figure 1 Shows the geographic location of the Bolívar Coast.

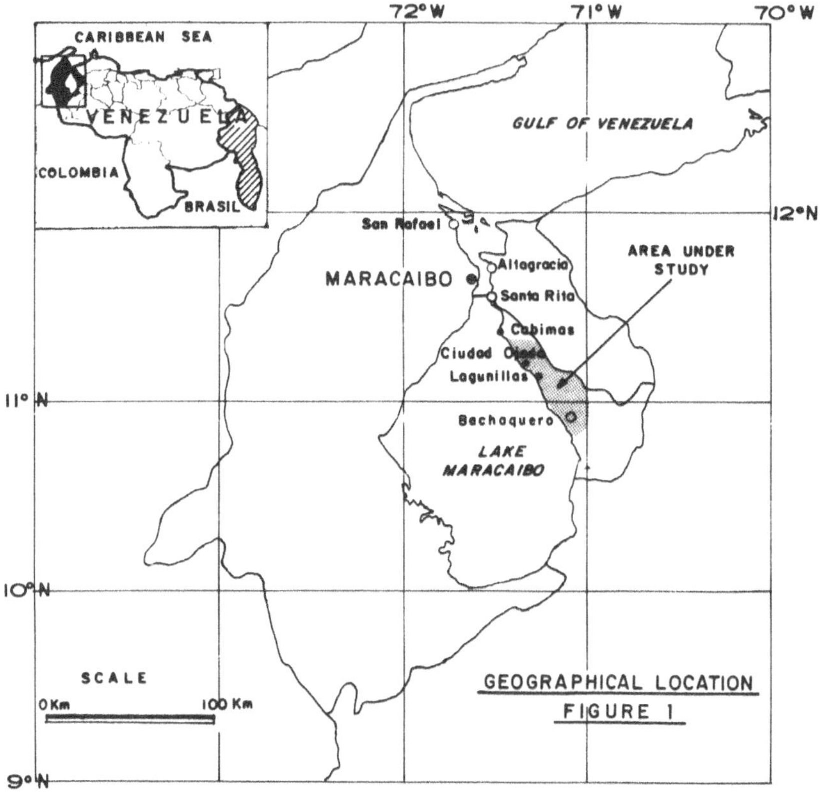

GEOGRAPHICAL LOCATION

FIGURE 1

PHYSICAL CONTEXT

General Characteristics

Geomorphology From the geomorphological point of view, the zone is made up of fluvio-lacustrine plains characterized by savannas with gentle slopes (8 to 1 m/Km), extending from the foothills of the Ziruma Mountain Range to the Coast. A part of the zone represents the deltas of the Tamare, Pueblo Viejo, Machango, and Misoa rivers.

Hydrology The area is characterized by a rainy season running from May to November. Annual precipitation varies from 750 mm in the northern part of the zone to 1000 mm in the south. Rainstorms tend to be short (less than 6 hours) and very intense; 75% of the rain falls in the first hour (100 mm/hour for a rain with a 10-year return period). The analysis of rainfall was extended over three days, during the period of greatest probability of consecutive precipitations, and the volume of runoff during that period was determined; this is an important factor affecting drainage by pumping and temporary storage. Variable coefficients of runoff were used, in accordance with the physical characteristics of the area, land use, and the magnitude of the rainfalls in question.

Geotechincs Sandy-silty soils with a high potential for erosion prevail in the northern part of the zone, while clay soils with a moderate potential for erosion predominate in the south. A geotechnical study made it possible to establish design conditions relating to erosion and the stability of drainage works, especially in sites where the settlement of the ground alters the conditions for the functioning of the system.

Land Use The oil industry, around which the entire regional economy revolves, provides the predominant form of land use. The population in concentrated in Ciudad Ojeda, the most important urban center of the area, and in the Tía Juana, Lagunillas, and Bachaquero oil fields.

There is relatively little agricultural activity in the area. But studies are now being conducted in experimental zones managed by MARAVEN S.A. for a combined agricultural-industrial land use. Some 50% of the land area is capable of being used for cropping or stockraising.

Drainage The drainage in the area is achieved primarily through the Ulé, Tamare, Pueblo Viejo, and Machango rivers; they provide the natural discharge channels from the area into the Lake. The basins of these rivers are located mostly in the high mountainous region and are bounded by the Lara-Zulia highway, the eastern limit of the zone subject to this study. Downstream from that highway, the rivers serve as channels for conduction of water, without additional inflows of importance, since their beds run through areas of high relief. In this intermediate zone, the geomorphological characteristics and the effects of the ground subsidence combine to make the drainage operate mainly in the form of erratic surface runoff.

In the coastal zone, the subsidence has been concentrated in three sites, giving rise to the need to modify the natural drainage and protect those heavily affected areas. These sectors, defined by dikes and diversion canals, are known as "polders". They are: Tía Juana (26.3 Km^2), Lagunillas (92.2 Km^2), and Pueblo Viejo-Bachaquero (73.5 Km^2). About 55% of the area of each "polder" is below the water level of the Lake.

In the "polders", the runoff comes as the result of direct rainfall in the area. This volume, partially collected by a network of artificial canals, is then conducted to pump stations and discharged into the Lake.

There is a drainage system in the area under study, which construction began at the time of the initiation of oil production. It has lost a part of its original efficiency as a result of the ground subsidence.

Special Characteristics

Subsidence One the most important effect on the drainage problem is the subsidence which results from the geostatic compaction of the strata of the subsoil when fluids (petroleum, water and gas) are extracted from the reservoirs. That phenomenon began with the initiation of oil drilling at the Bolívar Coast, but it was in 1929 that the subsidence was first detected, on the occasion of the breakage of a first coastal dyke. The subsidence now affects some 500 Km^2 and its effects are most serious along the coast (200 Km^2) where the three bowls of sunken land, mentioned above, have formed.

This settling of the ground has been studied since 1930, when a network of bench marks (BM) was installed to make levelings for control and verification of its magnitude. This network is levelled every two years, and topographical maps are prepared to make possible to visualize the modification of the relief over the course of time. In addition, the information obtained in this way is of fundamental importance for the prediction of ground subsidence.

The prediction of subsidence is made with a model which takes into account the existing relationship between the historic measurements and the extraction of fluids from the subsoil; it generates the settling in accordance with diverse scenarios of oil extraction for the dates on which those scenarios arrive. This simplified model is being improved by MARAVEN S.A., to take account of a number of physical parameters of both the reservoir and the process of extraction.

The results of the application of the model are given for geographical units known as "production blocks" (1.7 Km^2) and presented in the form of maps with contour lines. The analysis of that information has made it possible to define the size of the area affected by subsidence, the present and future locations of critical areas of subsidence, the speed of the settling of one area in comparison with other, the changes in the elevation and slope of the ground, and the relative variation of relief over time.

In general, the phenomenon of subsidence exerts effects on drainage in two important ways: the definition of the operational life of the project and the need to adapt the drainage system to final subsidence as well as to intermediate stages. Taking into account the size and importance of the area

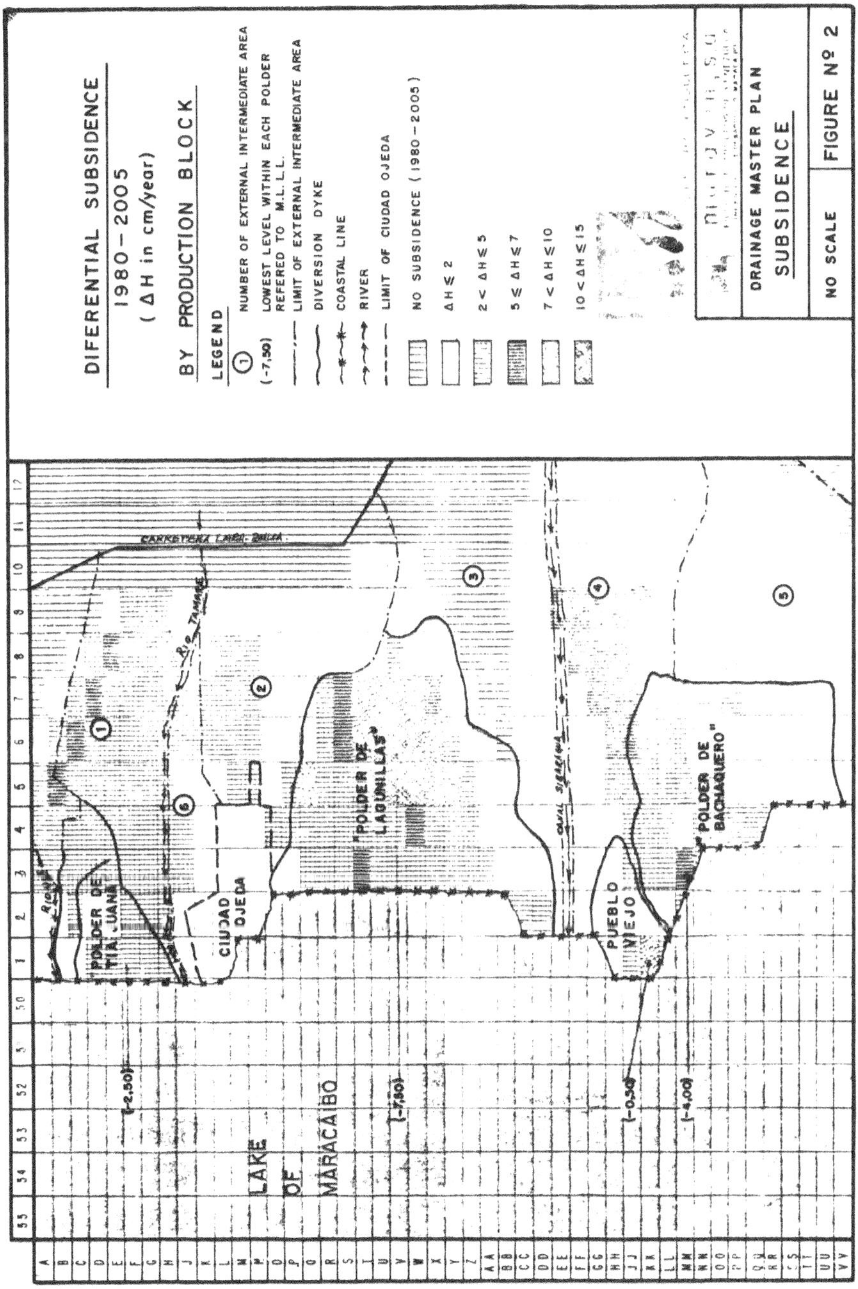
DIFERENTIAL SUBSIDENCE
1980 - 2005
(ΔH in cm/year)
BY PRODUCTION BLOCK
LEGEND
NUMBER OF EXTERNAL INTERMEDIATE AREA
(-7,50) LOWEST LEVEL WITHIN EACH POLDER REFERED TO M.L.L.L.
LIMIT OF EXTERNAL INTERMEDIATE AREA
DIVERSION DYKE
COASTAL LINE
RIVER
LIMIT OF CIUDAD OJEDA
NO SUBSIDENCE (1980 - 2005)
ΔH ≤ 2
2 < ΔH ≤ 5
5 ≤ ΔH ≤ 7
7 < ΔH ≤ 10
10 < ΔH ≤ 15
DRAINAGE MASTER PLAN
SUBSIDENCE
NO SCALE
FIGURE Nº 2
LAKE
OF
MARACAIBO
POLDER DE TIA JUANA
CIUDAD OJEDA
"POLDER DE LAGUNILLAS"
PUEBLO VIEJO
"POLDER DE BACHAQUERO"
CARRETERA LARA ZULIA
RIO TAMARE
(-2,50)
(-7,50)
(-0,50)
(-4,00)

crossing the area under study. These rivers present drainage problems related with fluvial stability and the potential overflow and consequent flooding of neighboring lands. Although the subsidence does not significantly affect the hydraulic regime of the rivers, with the exception of the discharge area of some of them, it does have an influence on the extent of land subject to flooding and on potential damage to urban areas and oil producing zones. The solutions proposed for these areas only are limited to the stabilization or canalization of critical reaches and the protection of areas with marginal dikes.

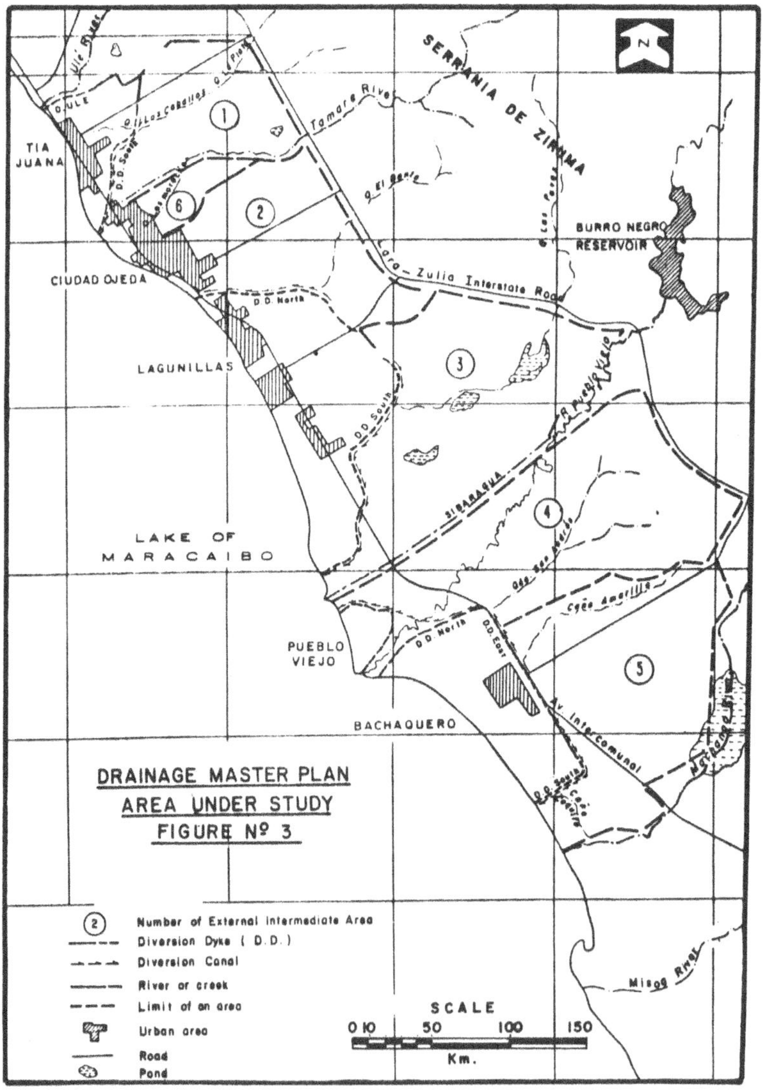

DRAINAGE MASTER PLAN
AREA UNDER STUDY
FIGURE Nº 3

Intermediate- External Areas These areas, numbered from 1 to 6 in Figure 3, are those located between the Lara-Zulia highway and the Lake. Although the zones known as "polders" are in these same areas, they are treated independently in recognition of their special drainage conditions. However, the dikes and diversion canals which define the "polders" are considered as parts of those areas, since they serve as the discharge channels to the Lake.

In general, the intermediate-external areas are subject to a variable subsidence effect. But even with the modification of their relief, drainage toward the coastal zone, where the runoff receiving canals are located, is maintained. The solutions proposed improve the conditions of the receiving canals affected by subsidence.

Ciudad Ojeda is located in the intermediate-external areas. The projections for subsidence do not indicate any serious effect on the urban area. The solution proposed by the DMP is aimed at protecting and isolating the city from the drainage of the surrounding areas. Urban drainage is in the scope of a separate study.

"Polders" The "polders" have been defined as the areas in which subsidence has depressed the land to a level below the Lake. Internal drainage is more difficult than in other areas, as a result of the more intensive urban-industrial land use, the pronounced subsidence, and the need for pumping for the final disposal of the water.

The drainage solutions proposed for the "polders" were aimed at producing a subdivision into independent hydrological units, mainly defined by the present and future relief and by their urban and oil-industry land use; limitation of pumping capacity through the use of temporary storage in the low areas; and maximun use of the existing drainage works in order to achieve an economical system, with easy maintenance, and adaptable to the dynamic characteristics of the relief.

Design Conditions.

For the preparation of the DMP, it was necessary to establish design conditions based not only on the considerations intrinsic to a drainage plan, but also on the existence of that special characteristic, subsidence. The conditions established for the influence of subsidence on drainage, in summary, were the following:

- Adaptation of the drainage system to the present and future conditions of relief (up to 2005), thereby achieving a better integration with changing nature. In addition, the condition of functioning was revised for the situation expected in 2030.
- Maximun independence of the drainage of the indicated areas,

under study, the magnitude of the drainage system, the progress of the settling over the course of time, and the uncertainty implicit in the predictions, the year 2005 was selected as the design horizon (medium term) and the year 2030 as the revision horizon (long term).

The basic purpose of the study was to establish a drainage system which can be adapted to the changes in relief expected to occur up to the year 2005 and which, in accordance with the progress of the subsidence, can continue operating up to the year 2030. The long-term functionality of the drainage system was evaluated with regard to the uncertainty in the plans for oil extraction for that period, and consequently, the uncertainty with respect to the resulting subsidence. It is important to adapt the solutions to the real evolution of the subsidence after 2005, taking into account not only the experience obtained up to that date, but also the more precise predictions for settling made on the basis of more highly developed techniques, which will make it possible to define the future relief more accurately.

Figure 2 shows the pattern of subsidence estimated for the period 1980-2005, presented in accordance with the grid of production blocks.

DRAINAGE MASTER PLAN (DMP)

The drainage solution for the zone described above includes, in its first stage, what has been named the Drainage Master Plan (DMP). Its purpose is to provide a tool for the development of drainage projects in the area. Those projects will be carried out in accordance with general guidelines which take account of all the factors affecting the situation of the area. The DMP was divided into two phases, as follows. Phase 1, a Diagnostic Study, which evaluates the existing drainage problem in overall terms, with emphasis on the influence on oil installations. Phase 2, which consists of the proposal of solutions for the drainage problems in the different areas indicated below. The plan is based on the planning scenarios indicated above. and the hydrological, geotechnical, and land use conditions prevailing in each zone. It also includes an investment program and a series of recommendations for the management and monitoring of the drainage system.

Division of the Area

During the development of the DMP, physical and geomorphological characteristics were found in the area which made it possible to differentiate zones in accordance with drainage conditions. These areas, shown in Figure 3, are the following:

External Areas These areas, located to the east of the Lara-Zulia highway, represent the watersheds of the rivers

as well as among the subdivisions made in the "polders". With this independence, it becomes possible, in first place, to implement the diverse drainage solutions in isolation, in accordance with the degree of severity of the existing problems, as well as to make future modifications as required by the progress of the subsidence without the need to carry out major alterations in the general scheme for an area or "polder". In some cases, dependence among neighboring areas during extraordinarily heavy storms was maintained; that is the case of urban areas relieving drainage toward rural or oil-drilling zones in which is possible to accept temporary flooding.

- In urban areas or zones with a high density of industrial installations, it was established that the elevation of the canals which receive drainage from them, were below the lowest point of the area for the horizon of the design. Drainage by gravity is thus maintained, being faster and less subject to the failures and limitations associated with drainage by pumping. "Urban islands" were thus defined within the general system of drainage by pumping of the "polder".
- In order to take advantage of the topographic conditions and reduce pumping capacity, the lower zones were used for temporary compensation flooding, where land use allows it. This conditions was associated with such preventive measures as limitation of the use of these zones and adaptation of existing industrial equipment to accept given level of flooding without producing significant damage.
- Recommendation regarding the pumping equipment selected to fit the real conditions in each location; and analysis of the subsidence which will occur in the next 15 years (life cycle) in order to make the required adjustments.
- Limitation of the use of lined canals or structures, which can be damaged by subsidence or which rectification is very costly. The use of earth canals was adopted; they were designed in accordance with erosion criteria and taking into account the changes in the hydraulic regime produced by variations in the slope of the ground over time.
- Use of the existing drainage network and accommodation to the design conditions.
- Emphasis on preventive solutions, as a result of the large size of the area and the changing nature of its relief, in order to lessen the drainage pressing necessity. In certain cases, conservation of the existing situation was recommended, with the aim of reducing the investment required and limiting the environmental impact.
- Programming of maintenance and monitoring of the drainage works, to provide a continuous evaluation of their functioning and to make the appropriate modifications, as a part of the set of preventive measures, especially important in an area with these characteristics.

Planning

The DMP presents the drainage solutions for the differing areas and establishes an order of priorities on the basis of the degree of severity of the problems. It includes a 10-year investment program for a total of 30 million dollars. Guidelines are also provided for the operation, maintenance and monitoring of the drainage system.

CONCLUSIONS

As important conclusions with respect to the drainage problems detected in zones subject to subsidence, it can be stated that their occurrence does not only affect the physical structure, but also its environmental conditions, since it continually modifies the patterns of drainage and the distribution of water in the region (zones subject to recurrent flooding). Consequently, the solutions to be adopted in cases such as this must adapt themselves to the "natural situation" both in its physical aspect and in its environmental aspect. In areas to subsidence, it is of capital importance the preventive measures. In this case, an independent study was carried out to evaluate the alternative of moving the people and utilities out of the area affected by the subsidence. Such solutions must take account of the evolution of this phenomenon over the course of time. But since the predictions of ground subsidence are a function of oil extraction, it is essential to maintain a continuous check on its effects on the proposed projects.

REFERENCES

Obhidra Consult C.A. (1985) Drainage Master Plan. East Coast of the Lake of Maracaibo. Phase 1. Diagnostic Study and Phase 2. Planning. MARAVEN S.A., Lagunillas, Zulia State, Venezuela.

SECTION 6 DRAINAGE ECONOMICS

The Economical Evaluation in the Design of Main Drainage Systems

C. Volp and C.F.J.M. Veraa
Department of Civil Engineering, Delft University of Technology, The Netherlands

INTRODUCTION

In order to have an instrument with which a system of watercourses can easily be evaluated on its merits the model EWAS, a model for the design and Evaluation of WAtermanagement Systems, has been developed (Volp 1986).
The main objective of the model can be formulated as follows:

> It should be possible with the model to evaluate the hydraulic properties of a watermanagement system and at the same time give the economical consequences.

In this paper the model EWAS is introduced and an example is treated to demonstrate the application of the model. As subject for the example the problem is chosen of the balance between investment costs and maintenance costs in the design of watercourses.
The design of drainage canals is usually determined by the choice of the design discharge, the freeboard that is to be maintained and the assurance that a certain flow velocity is not exceeded. A relatively simple calculation routine gives the dimensions of the various watercourses in the drainage system. In this calculation routine the designer has to make a choice for the roughness coefficient. This roughness coefficient strongly depends on the type of bed material and the bed vegetation and so implicitly on the rate of maintenance that is foreseen.
On the other hand the choice directly affects the investment costs mainly determined by the excavation costs.
The question arises whether it is possible to reduce the net present value of the costs for a given time period and interest rate by designing for a lower level of maintenance and larger dimensions. The importance of this question is clearly shown by the fact that according to the annual reports of the Dutch Waterboards the ratio of costs for depreciation and interest, maintenance and administration and management are 30, 55 and 15 % respectively, with a total budget of about Dfl. 400. million.

MODEL DESCRIPTION

Basically the purpose of the development of a tool for the designer of watermanagement systems is to simplify the determination of the optimal system in terms of lay - out, dimensions and water management. Optimal has in this case an economical meaning.
In that case the costs for investment, operation and maintenance of the future system must be compared with the benefits, or the reduction of damage, that can be obtained with the system.
The benefits that can be obtained are related to the area usage, the reference level, the frequency and the extent of exceedances of that reference level and the duration of these exceedances.
For some user groups in a given area it is practically impossible to determine the waterlevel - benefits relation. Extreme examples are recreational and nature reserve areas. It is obvious that in these cases the waterlevels are related with the qualities of the recreation or the nature reserve areas, to give or to construct the above relations will be very difficult.
The same holds true for urban areas. Data that can be used to fix the waterlevel - benefit relation are not known, apart from some specific data used by assurance companies.

Besides the economical criteria the authorities that are responsible for the watermanagement in each of the above areas do have other criteria. These are generally formulated as waterlevels that may not be exceeded more than a given number of times a year.
These criteria are often based on traditional, historical, political or physical grounds and will generally have no direct relation with the real expectations of damage in that area.

It is therefore decided to follow another reasoning for these areas. Instead of trying to determine the waterlevel - benefit relation the designer can follow the directives of the authorities, make his design and afterwards show the authorities what it means, in terms of costs, if not in all points of the system all criteria are met fully.

An exception is made for agricultural areas. It is beyond the scope of this paper to treat this subject thoroughly but in the same research project a model is developed to determine the relation of watermanagement and the productivity of agricultural areas (Volp 1985).

In figure 1 an example is given of a run with EWAS. In this example the watermanagement system consists of 17 nodes each with its own criteria.
In some nodes only one authority gives his conditions about the waterlevels, in other nodes there is a combination of interests.
This figure makes also clear that it is relatively simple to develop more variants for this watermanagement system which in a greater or less extent fulfill all criteria, at the same time giving the financial consequences of each variant. It is then

\criteria \ node-no	URBAN		RECREATION		NAT.RES.		AGRICULTURE YIELD [tons/ha]	COSTS OF WATER MANAGEMENT SYSTEM (mil.fl)
2	***	***	***	***	+	+	******	0.000
4	***	***	***	***	+	+	******	0.308
6	***	***	***	***	+	+	******	0.000
8	***	***	***	***	+	+	******	0.551
10	+	+	+	+	***	***	******	0.000
12	+	+	***	***	***	***	******	0.000
14	+	+	***	***	***	***	******	0.304
16	+	+	***	***	***	***	******	0.262
18	+	+	***	***	***	***	******	0.201
20	+	+	+	+	***	***	******	0.207
22	+	+	+	+	***	***	******	0.168
24	+	+	+	+	***	***	******	0.434
26	+	+	+	+	***	***	******	0.000
28	+	+	+	+	***	***	******	0.204
30	+	+	+	+	***	***	******	0.192
32	+	+	+	+	***	***	******	0.199
34	***	***	***	***	***	***	******	0.417
total yield (mil.ton)							0.000	
total costs (mil.fl)								3.447
total benefits per annum (mil.fl)							0.000	

Net present value (mil.fl)	0.000	3.492
Internal Rate of Return (%)		******

Figure 1 : Example of the output of a run with EWAS

for the authorities to decide how important it is that the criteria are met.

The heart of EWAS is formed by the NOSTAT model which calculates the non - steady flow in open channels, The equations of continuity and of flow are approached following an implicit difference scheme.
The NOSTAT program is a modified and extended version of the KNOTA program developed by Bouman and Schultz (1978). For the theoretical backgrounds of the hydraulics of open channels reference is made to Allersma (1973) and for the computational methods for these models to Vreugdenhil (1973).

In order to calculate the discharges, velocities and waterlevels the open channel system is transformed into a system of branches and nodes. The program offers the possibility to include weirs, culverts and pumps. For the boundary conditions four possibilities are foreseen: a constant or a time varying waterlevel and a constant or a time varying discharge.

As special feature the model offers three options to determine the lateral inflow, concentrated in the nodes, simulating the runoff of the catchment area caused by rain. In each of the three options the lateral inflow is a sum of the runoff of an area which reacts quickly on rainfall e.g. paved areas, urban areas or areas with glass-houses and the runoff of a slower reacting component e.g. drainage of unpaved areas with primarily groundwater drainage.
The runoff can be determined by either the rational method, a Nash - cascade, with a variable number of reservoirs, or a non-linear reservoir.

In figure 2 the flowchart of EWAS is given. The function of the subroutines is given in the annex.

Apart from the NOSTAT program EWAS offers the possibility to test all waterlevels on criteria given by the designer. These criteria can be waterlevels that may not be exceeded or that may not be exceeded longer than a given period of time.
For each node three pairs of the above criteria can be given, each representing the conditions of different users in the drainage area of the node.

In absence of benefits the economical evaluation reduces to the determination of the costs for investments,operation and maintenance and so to the determination of the Net Present Value of the costs for a given time horizon and rate of discount.
The investment costs for the channel sections are determined by the survey costs per unit length, the toe protection costs per unit length, the costs for levelling and sowing the maintenance path per unit area, the land purchase per unit area and the excavation costs per unit volume.
Because there are no operation costs for the channel sections as such, the only other factor is the maintenance costs. Because this is the main subject of the following sections here the factors that determine the maintenance costs are just denominated. That is the costs for maintaining the maintenance path per unit length and the frequency of this maintenance and the costs for maintaining the slopes and bottom of the channel per unit length and the frequency of this maintenance.
As for the investment costs for structures as weirs, culverts and pumping stations real cost estimates as well as Cost Estimation Relationships (CER) can be used. In the framework of the PAWN-studies CER's for weirs and pumps were developed (RAND 1981).
The best approximation for maintenance costs for these structures appeared to be a percentage of the total investment costs of these

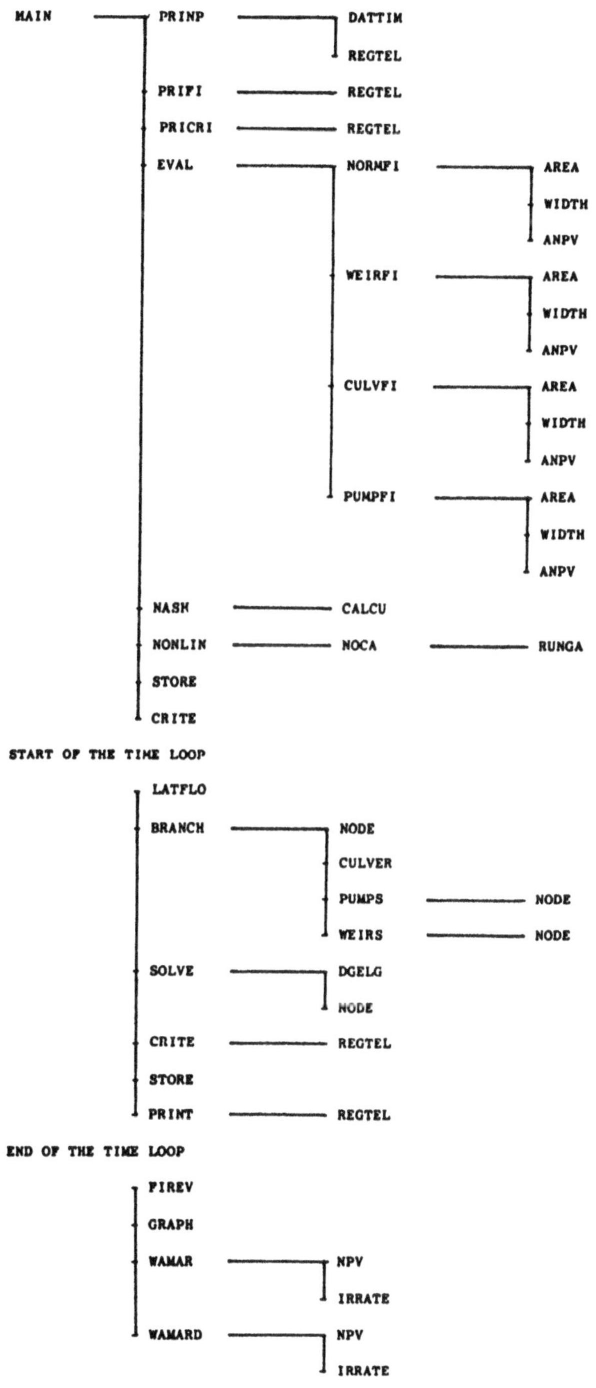

Figure 2 : Flowchart of program EWAS

structures.
The energy costs of pumping stations are approximated using CER's of the PAWN study. The operation costs for the whole system, the overhead costs, consists of the administration costs and the costs for equipement, datacollection, manpower and such.

INVESTMENT COSTS VERSUS MAINTENANCE COSTS

As clearified in the introduction it is an interesting question under which conditions it is profitable to determine the dimensions of the channels with a lower rate of maintenance.
The relation between the state of maintenance and the roughness coefficient, the k - Manning, is shown in figure 3.
In spring the vegetation on the slopes and the bottom of the channel develops in dependence with temperature, radiation, nutrition conditions, etc.

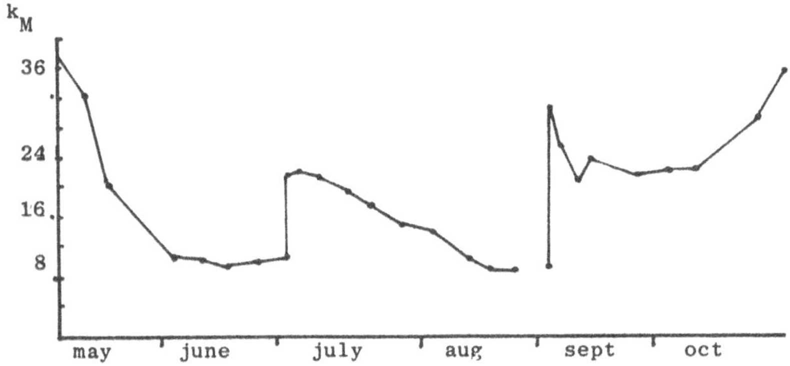

Figure 3 : Course of the roughness coefficient with time (Flach 1967)

The roughness coefficient sharply decreases when the conditions are favourable for growth in May and June. The roughness coefficient increases with maintenance after which it decreases again withthe development of the vegetation.
In autumn there is a gradual increase of the roughness coefficient due to natural die off of the vegetation.
Flach (1967) formulated an approximation of the relation between the rate of maintenance anf the lowest roughness coefficient based on the recovery capacity of the vegetation. According to Flach the lowest roughness coefficient of 45 $m^{-1/3}$ would imply 13 turns of maintenance as will a lowest roughness coefficient of 20 would imply 3, see also figure 4.
Watercourses in the Netherlands are usually dimensioned on discharges occuring in winter. It is therefore that the yearly inspection of the watercourses by the Waterboards is done in the

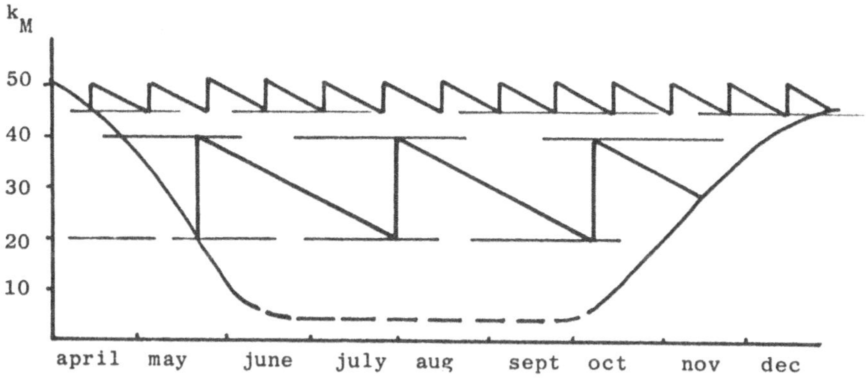

Figure 4 : Number of turns of maintenance to be expected by chosen lowest roughness coefficient. (Flach 1967)

months October and November.
There are two exceptions on this rule. First the case that a considerable part of the catchment area is urban area or covered with glass-houses. Because of the different character of the rains in late spring and summer, depending on the percentage paved area, the design conditions may well be in that period. The second exception becomes more important in the Netherlands in recent years and that is when the watercourses are used as irrigation canals.

Figure 5 : Lay - out of channel system.

To study the question under consideration four variants for a channel system as sketched in figure 5 are developed. Each variant with a different rate of maintenance and so a different roughness coefficient of 20, 25, 30, 35 $m^{-1/3}$ for the variants A, B, C and D respectivily. The drainage system was designed much larger then necessary to create 6 channel section each with a different design discharge of .5, 1.0, 1.5, 3.5, 5.5, 11 m^3/s. The dimensions were

TABLE 1

Net present value of the costs for the channel sections for the four designs for various rates of discount.

Rate of discount: 4 %

K_M: / CHANNEL	20	25	30	35
1	278.667	292.500	316.963	323.123
2	343.356	350.774	362.248	378.502
3	407.117	408.411	414.434	429.766
4	571.157	555.443	553.959	553.213
5	702.473	676.058	662.273	658.420
6	834.484	899.981	868.741	847.703

Rate of discount: 6 %

K_M: / CHANNEL	20	25	30	35
1	262.019	270.914	290.439	291.660
2	326.700	329.188	335.724	347.039
3	390.469	386.825	387.910	398.304
4	554.510	533.857	527.435	521.750
5	685.825	654.471	635.749	626.957
6	817.836	878.394	842.217	816.240

Rate of discount: 8 %

K_M: / CHANNEL	20	25	30	35
1	250.185	255.570	271.585	269.296
2	314.875	313.844	316.870	324.676
3	378.635	371.482	369.057	375.940
4	542.676	518.513	508.581	499.387
5	673.991	639.128	616.896	604.594
6	806.002	863.051	823.363	793.877

Rate of discount: 10 %

K_M: / CHANNEL	20	25	30	35
1	241.543	244.365	257.818	252.965
2	306.233	302.639	303.102	308.344
3	369.994	360.277	355.289	359.609
4	534.034	507.309	494.813	483.056
5	665.349	627.923	603.128	588.262
6	797.360	851.846	809.595	777.546

arrived at using the Manning formula with the following criteria:

- specific design discharge : 2.0 $L.s^{-1}.ha^{-1}$.
- freeboard : 0.75 m.
- width/depth ratio : 2 for small canals upto 2 m^3/s
  3 for the larger canals
- a pumping station at the outlet with a capacity of 25 m^3/s.
- each of the nodes had a catchment area of 250 ha.
- the reservoir coefficients that were used were in agreement with the reservoir coefficients given by Zeeuw (1966) for areas with mainly groundwater drainage and with Volp (1981) for areas covered with glass-houses (paved areas).

The four variants are to be evaluated in two ways :

a. on economical grounds
b. on hydraulical qualities.

ad a. The Net Present Value (NPV) of each channel section is given in table 1. To evaluate the influence of the rate of discount the NPV is determined for four different rates i.c. 4, 6, 8 and 10 %.
Differences between the highest and the lowest NPV costs for each channel section for four rates of discount are given in table 2.

TABLE 2
The differences between the highest and the lowest NPV-costs as a percentage of the lowest variant for the various rates of discount.

channel section	rate of discount 4 %	6 %	8 %	10 %
1	16.0	11.3	8.6	6.7
2	10.2	6.2	3.5	1.9
3	5.6	3.0	2.6	4.1
4	3.2	6.3	8.7	10.6
5	6.7	9.4	11.5	13.1
6	7.8	7.6	8.7	9.6

ad b. Because of the fact that the four variants will not perform equally under all conditions a sensitivity analysis has been carried out.
First has been looked at the difference in performance if the reservoir coefficient of the rainfall runoff model for the unpaved area is varied between 40 and 80 hours. The results are given in table 2.
Secondly has been looked at the performance for a varying percentage of paved area. These results are given in table 3.

CONCLUSIONS AND DISCUSSION

With respect to the economical aspects the following three conclusions are justified.

- The more expensive the money is, the more those designs are preferable that are based on a higher rate of maintenance (table 1). This can be explained by realising that expenses in the future becomes cheaper with a higher rate of discount.
- Although it values more for the higher rates of discount then for the lower, it can be stated in general, that the larger watercourses should be designed with a higher rate of maintenance (table 1). An exception is formed by the largest channel in case of the lowest rate of discount. For the smaller channels it is somewhat more complex but the smaller the channel, the more the benefits of a higher investment becomes evident.

TABLE 3
The functioning of the four variants with respect to the sensitivity for the rainfall runoff relation of the unpaved area.

reservoir coefficient	roughness coefficient k - Manning 20 $m^{1/3}$	25 $m^{1/3}$	30 $m^{1/3}$	35 $m^{1/3}$
k = 40 h.	doesn't satisfy T = 50 criteria exceedances are: in 2 : 24 cm. 4 : 19 cm. 6 : 13 cm. 8 : 7 cm.	doesn't satisfy T = 50 criteria exceedances are: in 2 : 20 cm. 4 : 15 cm. 6 : 10 cm. 8 : 5 cm.	doesn't satisfy T = 50 criteria exceedances are: in 2 : 17 cm. 4 : 14 cm. 6 : 10 cm. 8 : 6 cm.	doesn't satisfy T = 50 criteria exceedances are: in 2 : 18 cm. 4 : 15 cm. 6 : 12 cm. 8 : 8 cm. 10 : 2 cm.
k = 60 h.	doesn't satisfy T = 10 criteria exceedances are: in 2 : 21 cm. 4 : 14 cm. 6 : 3 cm	doesn't satisfy T = 10 criteria exceedances are: in 2 : 12 cm. 4 : 4 cm.	doesn't satisfy T = 10 criteria exceedances are: in 2 : 8 cm. 4 : 1 cm.	doesn't satisfy T = 10 criteria exceedances are: in 2 : 8 cm.
k = 80 h.	doesn't satisfy T = 10 criteria exceedances are: in 2 : 1 cm.	does satisfy T = 10 criteria eventually minus 7 cm.	does satisfy T = 10 criteria eventually minus 13 cm.	does satisfy T = 10 criteria eventually minus 15 cm.

- The significance of the above conclusion is accentuated by considering the differences between the variants (table 2). It appears that the differences decreases with a increasing rate of discount for the smaller channels and increases for the larger channels. The differences doesn't vary very much for the largest channels.

TABLE 4

The functioning of the four variants with respect to the sensitivity for changes in the rainfall runoff relation due to changes in area usage from unpaved to paved.

percentage paved area	roughness coefficient k - Manning $20\ m^{1/3}$	$25\ m^{1/3}$	$30\ m^{1/3}$	$35\ m^{1/3}$
3 % paved	doesn't satisfy T = 10 criteria exceedances are: in 2 : 21 cm. 4 : 14 cm. 6 3 cm.	doesn't satisfy T = 10 criteria exceedances are: in 2 : 12 cm. 4 : 4 cm.	doesn't satisfy T = 10 criteria exceedances are: in 2 : 8 cm. 4 : 1 cm.	doesn't satisfy T = 10 criteria exceedances are: in 2 : 8 cm.
10 % paved	doesn't satisfy T = 10 criteria exceedances are: in 2 : 15 cm. 4 : 7 cm.	doesn't satisfy T = 10 criteria exceedances are: in 2 : 6 cm.	doesn't satisfy T = 10 criteria exceedances are: in 2 : 1 cm.	doesn't satisfy T = 10 criteria exceedances are: in 2 : 1 cm.
20 % paved	doesn't satisfy T = 10 criteria exceedances are: in 2 : 8 cm.	does satisfy	does satisfy	does satisfy
40 % paved	doesn't satisfy T = 50 criteria exceedances are: in 2 : 23 cm. 4 : 9 cm.	doesn't satisfy T = 50 criteria exceedances are: in 2 : 21 cm. 4 : 9 cm.	doesn't satisfy T = 50 criteria exceedances are: in 2 : 18 cm. 4 : 9 cm.	doesn't satisfy T = 50 criteria exceedances are: in 2 : 19 cm. 4 : 9 cm.

With respect to the hydraulic aspects the following can be stated.

. The more maintenance is reckoned with the better the system performs. Considering table 3, where the reservoir coefficient of the unpaved area is varied from 40 to 80 hours. It appears that the variants do or don't satisfy to a greater or less extent the various criteria but invariable so that variant D (k - Manning = 35) performs best.

. The same conclusion can be drawn from table 4. The more maintenance is reckoned with the better the channel system is capable to cope with the changed composite rainfall - runoff relation of the catchment area.

Based on the above conclusions the following general conclusion can be drawn.

It is advisable

a. to design the smaller watercourses with a design discharge of 0.5 m^3/s (eventually upto 1.5 m^3/s) based on a lower rate of maintenance.

b. to design the larger watercourses with a desgn discharge of more then 3.0 m^3/s based on a higher rate of maintenance.

ad a. For this class of watercourses can be stated that because of their relative unimportance, the requirement of flexibility with respect to the hydraulic aspects is not of the utmost importance and that the financial benefits are not negligible. Besides that, it are particularly these small watercourses that are of great value to the flora and fauna in and around the water because of the small waterdepths.

ad b. For this class of watercourses the financial benefits are either important or negligible. In both cases a design based on a higher rate of maintenance is advisable because of the importance of this class of watercourses in the total water management system. A greater flexibility with respect to hydraulic aspects is a favourable quality here.

REFERENCES

Allersma, E. (1973) 'Hydraulics of open - water management' in Hydraulic research for water management, proceedings and informations no. 18, Committee for Hydrological Research TNO.

Bouman, J.E.G. and E. Schultz (1978) Calculation of non-steady flow in watercourses in urban and rural areas, Board for Reclamation of the IJssellake polders, Flevobericht no.127 (in dutch)

Flach, A.J. (1967) Influence of the vegetation on the roughness coefficient of watercourses. Cultuurtechnisch tijdschrift 7-8 1967-1968 pp. 209-216.(in dutch)

Rand (1981) Policy Analysis of water management for the Netherlands. RAND Corporation, 1981.

Vreugdenhil, C.B. (1973) 'Computational methods for channel flow' in Hydraulic research for water management, proceedings and informations no. 18, Committee for Hydrological Research TNO.

Volp, C. (1985) Manual for PRODU, a model for the determination of the productivity of agricultural areas in relation with water management. Delft University of Technology, Department of Civil Engineering.

Volp, C. (1986) Manual for EWAS, a model for the design and evaluation of water management systems. Delft University of Technology, Department of Civil Engineering. (in prep)

Volp, C. (1981) The significance of rainfall runoff relations. Master of science thesis, Delft University of Technology, Departement of Civil Engineering. (in dutch)

Zeeuw, J.W. de (1966) Analysis of the rainfall runoff relations in areas with mainly groundwater flow. Department of Hydraulics and Hydrology, Agricultural University Wageningen. (in dutch)

ANNEX 1.1

Function of the subroutines in program EWAS

EWAS	Program EWAS is a package of programs which offers the user the following options: . To calculate the non-steady flow in open channels. The equations of continuity and of flow are approached following an implicit difference scheme. . To test the waterlevels with respect to absolute level and duration of the exceedance of a certain level. . To calculate the investment, maintenance and operational costs of the various parts of the system. . To calculate the Net Present values of the various parts of the whole system and, if agricultural benefits are given the internal rate of return for the whole water management system.
PRINP	To print all input data concerning the water management system and the calculation procedure.
REGTEL	This subroutine is an aid to control the output for the printer
PRIFI	Echoes all financial data on input.
EVAL	This subroutine organises the financial calculations for each individual branch. The actual financial calculations are made in subroutines. For each branch type a different subroutine is used. The data are stored in a temporary memory, which are used as arguments in the call-statement. Finally the total costs are calculated.
NORMFI	Calculates costs of investment, maintenance and the Net Present Value.
WEIRFI	Calculates investment, maintenance and energy costs of branches with a weir. The costs of a normal branch are added to the costs of the weir.
PUMPFI	Calculates investment, maintenance and energy costs of branches with a pump. The costs of a normal branch are added to the costs of the pump.

ANNEX 1,2

Function of the subroutines in program EWAS

CULVFI	Calculates investment, maintenance and energy costs of branches with a culvert. The costs of a normal branch are added to the costs of the culvert.
ANPV	Calculates the net present value of the investment costs at year 0 and the yearly energy and maintenance costs at the end of each year.
WIDTH	Calculates the total width of a waterway, including maintenance path, bottom width and width of sloping sections.
AREA	Calculates the area of an upstream or downstream end of a branch.
LATFLO	This subroutine determines the lateral flow to the nodes as caused by rain and the surface of the storing area. There are three options: . Calculates the lateral flow as described in Bouwman and Schultz (1976). . The lateral flow is determined in the subroutines NASH and CALCU, where the rainfall runoff relation is approached with a nash cascade. The function of the subroutine LATFLO is reduced to take a row out of the array filled earlier. . The lateral flow is determined in the subroutines NONLIN, NOCA and RUNGA, where the rainfall runoff relation is approached with a non linear reservoir model.
NASH	This subroutine determines the lateral inflow to the nodes as caused by rain. The rainfall runoff relation that is used is a sum of two nash cascades of which the parameters must be given for each node. The subroutine calls for the subroutine CALCU for the determination of the runoff.
CALCU	This subroutine gives the discharge of the N-th reservoir of a nash cascade where all reservoir constants equal K and the rainfall feeds the first reservoir only.

ANNEX 1,3

Function of the subroutines in program EWAS

NONLIN	This subroutine determines the lateral inflow to the nodes as caused by rain. The rain runoff relation that is used is a sum of two non linear reservoirs for which the parameters must be given for each node. The subroutine calls for the subroutine NOCA for the determination of the runoff.
NOCA	This subroutine gives the discharge of a non - linear reservoir with reservoir constant K and power N.
RUNGA	This function subprogram determines the value of the constants for the runga kutta fourth order rule.
WAMAR	To print the results of the run concerning the fulfillment of the design criteria net present value of the cost for each canal and the whole project, the agricultural benefits per node and the internal rate of return of the whole water management system.
WAMARD	To store the same result as given by subroutine WAMAR on a dish for interactive purposes.
IRRATE	The calculation of the internal rate of return.
PRICRI	To print all criteria concerning the water management system.
CRITE	To test the waterlevels that are realised on the criteria of the designer.
NODE	This subroutine determines the coefficients of the equation to transform the waterlevel changes in changes of the discharge.
BRANCH	This subroutine determines the coefficients of the matrix A. If necessary it calls for the subroutines node, weirs, culverts and pumps.

ANNEX 1,4

Function of the subroutines in program EWAS

WEIRS	This subroutine contains the formulas for the submerged and the not submerged weir. It is called for from subroutine BRANCH.
CULVERT	This subroutine contains the formulas for a single or multiple, circular or a rectangular culvert. It is called for from subroutine BRANCH.
PUMPS	This subroutine contains the formulas of the discharge - head relation of the pump. It is called for from subroutine BRANCH.
SOLVE	This subroutine solves the matrix A with the aid of subroutine DGELG.
PRINT	This subroutine prints the values of the discharges velocities, waterlevels and so on for all branches and nodes after every time interval.
FIREV	This subroutine prints all relevant financial results.
DGELG	To solve a general system of simultaneous linear equations.
STORE	Storage of information of various nodes.
GRAPH	This subroutine plots the waterlevel changes during the whole period of the nodes for which the variable NOD (I,12) is given the value 1.

ANNEX 1.1

Function of the subroutines in program EWAS

EWAS	Program EWAS is a package of programs which offers the user the following options: . To calculate the non-steady flow in open channels. The equations of continuity and of flow are approached following an implicit difference scheme. . To test the waterlevels with respect to absolute level and duration of the exceedance of a certain level. . To calculate the investment, maintenance and operational costs of the various parts of the system. . To calculate the Net Present values of the various parts of the whole system and, if agricultural benefits are given the internal rate of return for the whole water management system.
PRINP	To print all input data concerning the water management system and the calculation procedure.
REGTEL	This subroutine is an aid to control the output for the printer
PRIFI	Echoes all financial data on input.
EVAL	This subroutine organises the financial calculations for each individual branch. The actual financial calculations are made in subroutines. For each branch type a different subroutine is used. The data are stored in a temporary memory, which are used as arguments in the call-statement. Finally the total costs are calculated.
NORMFI	Calculates costs of investment, maintenance and the Net Present Value.
WEIRFI	Calculates investment, maintenance and energy costs of branches with a weir. The costs of a normal branch are added to the costs of the weir.
PUMPFI	Calculates investment, maintenance and energy costs of branches with a pump. The costs of a normal branch are added to the costs of the pump.

ANNEX 1,2

Function of the subroutines in program EWAS

CULVFI	Calculates investment, maintenance and energy costs of branches with a culvert. The costs of a normal branch are added to the costs of the culvert.
ANPV	Calculates the net present value of the investment costs at year 0 and the yearly energy and maintenance costs at the end of each year.
WIDTH	Calculates the total width of a waterway, including maintenance path, bottom width and width of sloping sections.
AREA	Calculates the area of an upstream or downstream end of a branch.
LATFLO	This subroutine determines the lateral flow to the nodes as caused by rain and the surface of the storing area. There are three options: . Calculates the lateral flow as described in Bouwman and Schultz (1976). . The lateral flow is determined in the subroutines NASH and CALCU, where the rainfall runoff relation is approached with a nash cascade. The function of the subroutine LATFLO is reduced to take a row out of the array filled earlier. . The lateral flow is determined in the subroutines NONLIN, NOCA and RUNGA, where the rainfall runoff relation is approached with a non linear reservoir model.
NASH	This subroutine determines the lateral inflow to the nodes as caused by rain. The rainfall runoff relation that is used is a sum of two nash cascades of which the parameters must be given for each node. The subroutine calls for the subroutine CALCU for the determination of the runoff
CALCU	This subroutine gives the discharge of the N-th reservoir of a nash cascade where all reservoir constants equal K and the rainfall feeds the first reservoir only.

ANNEX 1,3

Function of the subroutines in program EWAS

NONLIN	This subroutine determines the lateral inflow to the nodes as caused by rain. The rain runoff relation that is used is a sum of two non linear reservoirs for which the parameters must be given for each node. The subroutine calls for the subroutine NOCA for the determination of the runoff.
NOCA	This subroutine gives the discharge of a non - linear reservoir with reservoir constant K and power N.
RUNGA	This function subprogram determines the value of the constants for the runga kutta fourth order rule.
WAMAR	To print the results of the run concerning the fulfillment of the design criteria net present value of the cost for each canal and the whole project, the agricultural benefits per node and the internal rate of return of the whole water management system.
WAMARD	To store the same result as given by subroutine WAMAR on a dish for interactive purposes.
IRRATE	The calculation of the internal rate of return.
PRICRI	To print all criteria concerning the water management system.
CRITE	To test the waterlevels that are realised on the criteria of the designer.
NODE	This subroutine determines the coefficients of the equation to transform the waterlevel changes in changes of the discharge.
BRANCH	This subroutine determines the coefficients of the matrix A. If necessary it calls for the subroutines node, weirs, culverts and pumps.

ANNEX 1,4

Function of the subroutines in program EWAS

WEIRS	This subroutine contains the formulas for the submerged and the not submerged weir. It is called for from subroutine BRANCH.
CULVERT	This subroutine contains the formulas for a single or multiple, circular or a rectangular culvert. It is called for from subroutine BRANCH.
PUMPS	This subroutine contains the formulas of the discharge - head relation of the pump. It is called for from subroutine BRANCH.
SOLVE	This subroutine solves the matrix A with the aid of subroutine DGELG.
PRINT	This subroutine prints the values of the discharges velocities, waterlevels and so on for all branches and nodes after every time interval.
FIREV	This subroutine prints all relevant financial results.
DGELG	To solve a general system of simultaneous linear equations.
STORE	Storage of information of various nodes.
GRAPH	This subroutine plots the waterlevel changes during the whole period of the nodos for which the variable NOD (I,12) is given the value 1.

An Economic Argument for a Sub-Optimal Engineering Design for the Drainage of Clay Soils

T.W. Tanton and D.W. Rycroft
Institute of Irrigation Studies, University of Southampton, U.K.

INTRODUCTION

Clay soils have very poor natural drainage properties and hence readily become waterlogged in conditions where there is excess rainfall. This is a direct result of their fine texture, swelling nature and, when waterlogged, their poor structural stability, resulting is low hydraulic conductivities. In temperate climates waterlogging results in delayed field operations, poor yields, and/or stock being withheld from pasture to prevent damage to the soil and destruction of the sward. In the United Kingdom, farmers have invested heavily in field drainage systems, with 75% of these being installed in clay soils (Bailey 1979). The low hydraulic conductivity of these soils also means that if a sub-surface drainage system is chosen, it has to be designed to intercept flow on the surface and in the plough-layer. Hence, drainage systems are highly intensive, typically involving the construction of a closely spaced network of mole drainage channels, discharging to gravel backfilled pipe drains which in turn discharge to open collector drains. The pipe drains are typically spaced at 20-40 metres.

Such systems are expensive, and at the present time (1985), in the United Kingdom, cost between £1,000 and £1,200 per hectare (20m spacing) rising to over twice this cost in many less developed countries. If such capital expenditure is to be economically justified, it is clear that high yields of valuable crops must be generated. Since in most cases this is not possible, Governments often offer subsidies, in the U.K. in the form of grants, to support the capital works. Formerly the level of grant support in the U.K. was 30 - 70%, but this has recently been reduced to 15%, making drainage less financially attractive to the farmer. Intensive and costly drainage systems were adopted by the industry since they provided the best engineering solution for a wide range of conditions. However, on many sites, cheaper and less intensive systems may well provide more cost effective solutions.

This paper appraises the economics of land drainage and evaluates the potential for the installation of less costly systems.

COMPREHENSIVE SUB-SURFACE DRAINAGE FOR CLAY LANDS

The Drainage System

Laboratory and field studies of water movement through clay soils are difficult, due to their cracking and swelling nature, and it is only in recent years that scientists have begun to understand the processes involved, (e.g. Hoogmoed & Bouma, 1980).

Consequently, there are no physically based mathematical models available to assist in the design of drainage systems for clay lands. In fact, present designs are mainly the result of field experience, and a very limited number of field experiments (Bailey, 1979). The cost and especially the framework of subsidy for drainage costs, also has a considerable influence on the selection of an appropriate system. Until recently, the British Government provided grant aid to farmers which was in the region of half the capital cost. To justify this investment the system had to meet certain constraints. It was required to have a life of 15 years or more and require minimal maintenance. Contractors also wished to install systems which, in effect, carried a guarantee against failure. It was also advantageous to install a system across a wide range of soils and conditions.

These constraints, when coupled to the potential instability of clay soils, resulted in the installation of very intensive drainage systems. Although these are, in most cases, very effective and last for many years, they are also expensive, £1,200/ha,and may not be economically justified.

A typical intensive scheme, considered later in more detail for financial analysis, is designed to intercept surface water run-off and flow in the plough-layer. Mole channels,typically at 1-2 metres centres and 0.4-0.5m depth, intercept the surface flow via the cracks induced in the soil by mechanical disturbance (moling) and conduct it along short lengths of mole channel to trenches backfilled to within 0.2m of the soil surface with highly permeable backfill. The water then percolates down to pipe drains installed at the base of a trench (0.8 - 1m deep). The collector drains are typically spaced at 20 metres, and feed directly into collector drains.

Advantages of an Intensive Drainage System

An intensive drainage system comprises tried and tested technology, which provides sound drainage of clay soils, under a wide range of environments (Bailey, 1978).

An increase in the yield of winter wheat of 27% has been attributed to drainage by Trafford and Oliphant (1977) as a result of experiments at Drayton Experimental Husbandry Farm.

However, if the results from this unreplicated block experiment are re-analysed taking into account the observable trends in yield in the rows and columns of the block, a yield increase of only about 10% can be attributed to the installation of a mole/pipe drainage system. This figure is very similar to results from several other observation trials, and of a fully replicated drainage experiment conducted by the Letcombe Laboratory in Oxfordshire, (Ellis et al, 1984).

The success of this system is attributed to the fact that close spaced gravel backfilled pipe drains provide regular and frequent connections with the base of the plough layer, and with the mole channels which are drawn at right angles to the pipe network. With pipe collector drains spaced at 20 metres the maximum distance along a mole channel to a collector is 10 metres. Consequently, there is little need to grade the channels to prevent water stagnating in the mole channels. Furthermore, in many situations with shallow gradients, (<1%) the collapse of a mole channel would not prevent drainage from occurring since water would still be able to move to the collector up slope. The soil in the channel at the point of failure would also have to be compacted before flow would be prevented. Hence, the system is very reliable.

Disadvantages of Comprehensive Drainage Systems

The main disadvantage of the intensive system is its expense. The capital cost of £1,200 per hectare is high, and is not readily reflected in an increase in the value of the land. There is ample evidence to show that when drains are installed under wet conditions, the heavy trafficking of the land by drainage machinery and gravel trailers cause soil compaction which can lead to depressed yields in the first few years after installation. However, this can be overcome by additional tillage (FDEU Annual Reports 1969-78). The most serious draw-back to this method of drainage is the unreliability of the yield increase. The figure of 10% is only a mean value so that it is possible that yield increases may range from say 10% to 20% in any one farmer's field. This makes the installation of drainage a high risk venture. Yield increases well in excess of 20% are often claimed by farmers but these are not supported by experimental evidence and are most probably the result of improved agronomic practices following the investment in drainage.

LONG MOLE DRAINAGE

Long mole drainage is a system of land drainage which involves the installation of mole drains, at a depth of 0.4m - 0.5m, which run down the slope of the field and discharge into widely spaced collector drains, either pipes with gravel backfill, or open collector ditches placed at the break of slope, and/or the end of the field. When moles discharge into open collector ditches, a small length of pipe is installed in the end of the channel to protect the outfall. Such mole drains can be up to 300 metres in length but are normally much shorter.

The system is not new and it is instructive to look at its history. Much of the work on which our present knowledge of mole drainage systems is based, was carried out between 1920 and 1940, and little that is new has been added since. The importance of land slope in enabling farmers to draw long mole drains was recognised as early as 1931 by Davies who concluded that moles could be safely drawn up to 180 metres long on sloping land but only half that length on flat land. The largest survey of mole drains between 1924 and 1932 was conducted by Blackaby, (1932) who concluded from his study of a large number of demonstration plots, that 200 metres was the safe maximum length. He noted that such drains appeared to last up to 5 to 9 years. Hudson and Hopewell (1940), also found that moles pulled in clay remained effective for a number of years and that they could be safely installed up to 300 metres long. Although there was no excessive scour in moles with slopes up to 15%, they suggest that slopes of 10% should be adopted as a maximum to safeguard against erosion.

The system is only suitable for clay lands with sufficient slope to allow effective drainage of the mole channels. Given that much of the land which is in need of drainage falls into this category, there appears to be considerable scope for installing this cheap and effective method of drainage in the U.K.

The most likely reasons why long mole drainage systems fell from favour with farmers was that they did not attract a Government subsidy, whilst on-farm tractors did not have sufficient power to remole at the frequent intervals demanded. However, now that Government subsidies have almost been removed and farm tractors are more powerful, the popularity of this system can be expected to increase.

A review of experiments where the spacing of pipe collector drains has been one of the main variables, reveals that the yield increase attributable to mole drainage is the same whether the collector drains are spaced at 5 metres or 100 metres. (e.g. Eggelsmann, 1973; FDEU Reports 1968 - 1978; Trafford and Oliphant, 1977). It is difficult to have experimental plots with collector drain spacings in excess of 100 metres, but experiments where 80 and/or 100 metre spacings have been included indicate that collector drains at wider spacings would be equally effective as close spacings.

An objection often raised against long mole drainage is that if a mole drain were to fail, wet patches, or even springs may occur in the field, and there are always a number of observers who can report such inadequacies. However, most drainage engineers have also experienced failures of conventional pipe drainage schemes where they have been installed in inappropriate conditions. It is therefore suggested that the most important factor contributing to the success of long mole drains lies in the correct identification of suitable soils and the installation of correctly sited collectors.

The viability of long mole drains has been demonstrated in many areas of the country. For example, a farmer at Kings Hill Farm; on the Isle of Sheppey in Kent has installed very long mole drains on a London clay on a field of 2% slope and of area 125 hectares. The mole drains run down slope for 200-300 m before discharging into a pipe drain, backfilled with gravel, placed at the break of slope; mole drains at the bottom of the field discharge directly into an open collector ditch. The system was installed in 1980, some 5 years ago and throughout this period good yields of winter wheat have been obtained. The average annual rainfall for this area during this period was 700mm/year. Although the farmer was prepared to mole annually, to-date he has not needed to. A similar experience has also been found using a similar system by Mr. Holland of Kemphill Farm on the Isle of Wight.

The Advantages of Long Mole Drainage.

Very low capital cost.

The Disadvantages of Long Mole Drainage.

The system is not suited for all clay soils and there is an unquantified higher risk of failure than conventional systems. In fact, for many years it was felt that water-logging of mole channels would cause such long systems to fail due to collapse. However, recent work by Harris (1984) has shown this not to be the case, and that even with prolonged water-logging, channels can be expected to remain effective for at least 3 years.

A FINANCIAL EVALUATION OF THE TWO DRAINAGE SYSTEMS

Costs and Assumptions.

(i) Intensive pipe/mole drainage systems, £1,200/ha (Includes £40/ha for mole drainage. (Nix (1985); various Contractors)

(ii) Long mole drainage system with collector drains at the base of the field, or at a break in slope. £290/ha (Based on collector drains being installed at the rate of 100m/ha, though it is conceivable that drainage may be installed with collector drains at much wider spacings, Mole drainage at £40/ha).

(iii) Price of wheat £136/t. (Nix, 1985)

(iv) National Yield Levels t/ha (Nix, 1985)

Low	Medium	High
4.8	6.3	7.7

(v) Bank Interest Rate for financial analysis 12%

(vi) Repayment period assumed 7 years

(vii) Increase in yield attributable to land drainage:- Best estimate based on experimental data is 10% with an optimistic estimate of 20%.

NOTE: Tax has not been included in the evaluation as it is assumed that similar levels of tax concessions could be attracted by other competing capital investments.

Intensive Pipe/Mole Drainage System for Winter Wheat Production

The levels of profit or loss, at present day prices, over a 7 year period, with an interest rate of 12%, which could be expected from installing an intensive drainage system are shown in figures 1 and 2. It can be seen from figure 1 that at the present level of the drainage grant of 15%, intensive drainage schemes are not financially viable, even if an optimistic performance level of 20% is allowed. At the former levels of grant of 50% or 60%, the investment would not have been attractive at expected yield increases of 10% and only just became viable if optimistic yield levels of 20% can be achieved. The high capital cost and the poor yield increases which can be attributed to land drainage combine to make intensive land drainage financially non-viable for winter wheat production. Similar analyses for dairy farming also show an equally poor return. Even when there are no interest charges, a drainage scheme is unlikely to pay for itself in a 7 year repayment period (Figure 2) thus showing the system to be financially unviable.

Long Unlined Mole Drainage System

The system of long mole drains is based on mole drains spaced at 1.5 metres, running down slope to gravel backfilled pipe drains installed at the rate of 100 metres per hectare. The moling is repeated every 3 years at a cost of £40/ha. In practice, it is probable that it would be possible on many farms to install collector drains at twice this spacing, thus cutting the capital cost in half.

The data presented in figure 3 shows that, except in the case of low yields, the system of long mole drainage is a financially viable proposition without support grants, and even at low yield levels, the system is almost able to achieve a 12% rate of return on investment over a 7 year period.

CONCLUSIONS

Repeated drainage experiments have shown that the level of yield increase that can be attributed to land drainage is about 10% on clay lands. It has been shown that with such small increases in yield, intensive pipe/mole drainage systems are financially unattractive for winter wheat production.

Analyses not presented in this paper, confirm that it is equally uneconomic and financially unattractive for dairy cow/pasture farming systems. However, long mole drainage systems are much more financially attractive and it is clear that such drainage systems should be recommended whenever soil conditions permit.

The paper has shown that many of the objections voiced against a long mole drain system are not based on sound experimentation. Examination of the available data indicates that long mole drainage systems can be highly effective, although they may have to be repeated every 3 years. It is, therefore, suggested that national trials should be implemented to evaluate the scope for long mole drainage and that drainage advisers should be more adventurous in encouraging farmers to try such a system, even if on a very limited scale.

Figure 1. Effect of the level of grant aid on the financial benefits of intensive land drainage for winter wheat production.

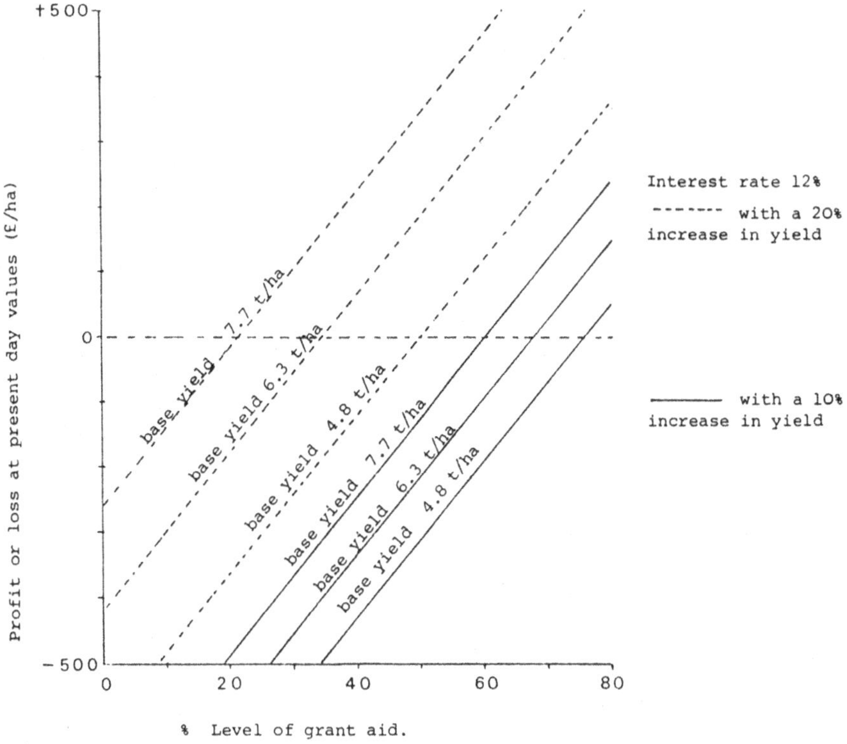

Figure 2. The effect of interest rate on the financial benefits of intensive land drainage of clay soils for winter wheat production.

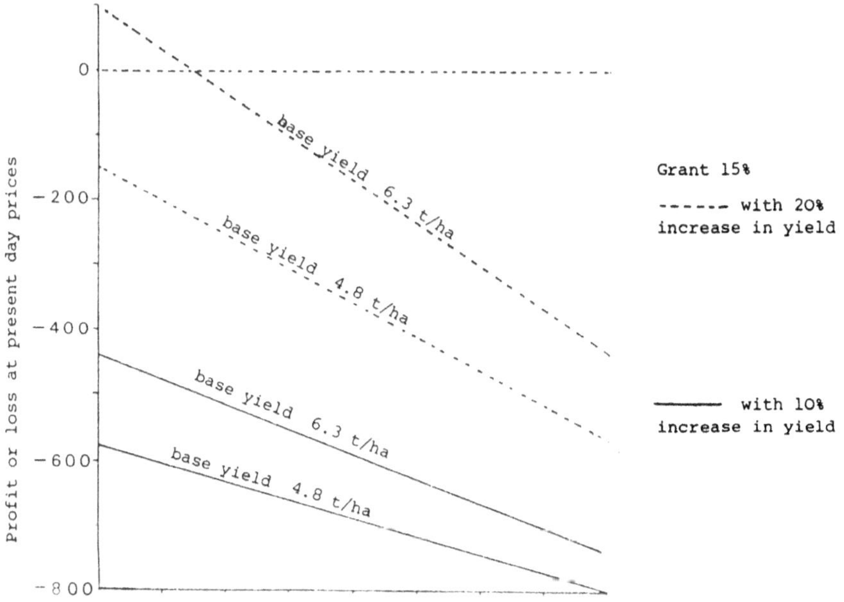

Figure 3. Effect of the level of grant aid on the financial benefits of long mole drainage for winter wheat production.

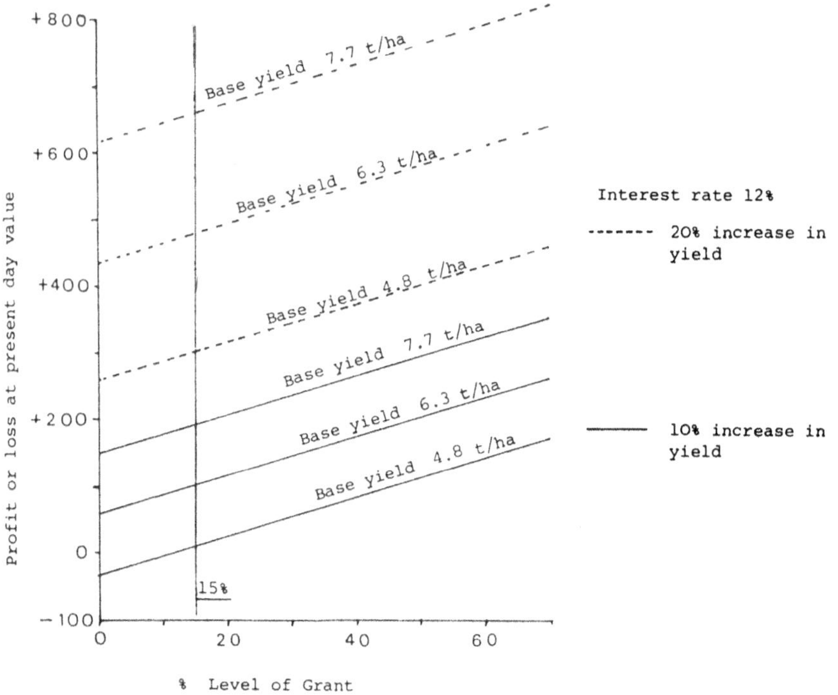

References

Bailey, A.D. (1979). Drainage of clay soils in England and Wales. Proc. of Int. Drainage Workshop. Ed. J. Wesseling., pub., I.L.R.I. (Netherlands) page 220 - 242.

Blackaby, J.H. (1932). "Technical Notes on Mechanical Farming". Institute of Research in Agricultural Engineering, University of Oxford.

Davies, C. (1931). Mole Drainage. J of S.E. Agricultural College. Wye, Kent. No. 28.

Eggelsmann, R (1973). "Drananleitung", Pub. Verlag Wasser und Boden Axel. Hamburg.

Ellis, F.B., Cristian, D.G., Brugg, P.L., Enderson, F.K.G., Prew, R.D. and Cannell, R.Q. (1984). A study of mole drainage with simplified cultivation for autumn-sown crops on a clay soil, (3). J. agric. Sci. Camb. 102; 583 - 594.

Harris, G.L. (1984). Effect of mole submergence on the life of mole channels. Agric. Water Management. Elsevier 8: 361 - 374.

Hoogmoed, W.B. and Bouma, J. (1980). A simulation model for predicting infiltration into cracked clay soils. Soil Sci. Soc. Am. J. 44: 458 - 461.

Hudson, A.W. and Hopewell, H.G. (1940). "Mole Drainage in New Zealand". Pub. Massay Agric. College; University of New Zealand.

Nix, J., (1985). "Farm Management Pocket Book". pub. Farm Business Unit. Wye. College 15th ed.

Trafford, B.D. and Oliphant, J.M., (1977). The effect of different drainage systems on soil conditions and crop yields on heavy clay soils. Exp. Husbandry. 32: 75 - 85.

Drainage Benefits to Land Workability

A.C. Armstrong
Field Drainage Experimental Unit, MAFF, Cambridge, U.K.

1. INTRODUCTION

In the UK, one of the most common constraints on agricultural operations is soil wetness. Wet soils both depress crop yields and restrict access to land. One of the most important benefits of drainage is to increase the number of days the farmer can use his land. Yield benefits directly attributable to drainage can be only 10-20%, whereas workability constraints on wet, undrained land may result in no crop at all if access to the land at sowing or at harvest time is prevented (eg Armstrong, 1978). An increase in workdays can lead to an improved timeliness of operations, and a more efficient use of equipment in an arable situation, and increased length of grazing season under grassland. It is thus clear that the question of the available workdays and, in particular, the effects of drainage on those workdays, is of considerable interest. An ability to predict these effects is potentially valuable in two ways:

i) to enable an economic evaluation of drainage, either in terms of an appraisal of the investment, or by identifying penalties incurred by not draining.

ii) as an advisory tool, to enable individual farmers to make management decisions as a result of the likely consequences of drainage.

This paper discusses two lines of approach to the consideration of workdays: the first presents a body of observational data, and the second considers some approaches to the mathematical modelling of workability.

2. A SURVEY OF DRAINAGE EFFECTS

During the period 1974-79, paired drained and undrained sites were instrumented to provide information on soil water regime and ground condition. Over 200 such sites have data for some of this period, and together the final data set of 400 site years provides an excellent source to show both the mean and the range of drainage effects. This observational study was a joint collaborative effort between the then Land Drainage Service of MAFF and the Soil Survey of England and Wales (Armstrong, Robson & Rands, in preparation).

Sites for this study were identified, on an opportunist basis, where paired locations offered comparability of soils and land use. Target soil types were from a list of series comprising the agriculturally important surface-water gleys, and identifications confirmed by SSEW personnel. Data collected on each plot of each site included the water table recorded as the median of five open auger holes, ground condition recorded on a nine-point subjective scale (table 1) and rainfall totals, all measured on a weekly cycle of site visits.

TABLE 1. Scale for ground condition assessment

1	Snow	workable
2	Hard frost	workable
3	Very hard	workable
4	Squelchy all over	workable
5	Squelchy in parts	workable
6	Damp and soft	workable
7	Damp but firm	unworkable
8	Dry on top	unworkable
9	Baked hard	unworkable

The use of the subjective assessment of ground condition, and the arbitrary division into workable and unworkable is at first sight open to some criticism. However, no better operational definition has been generally accepted, and in terms of this study, the ability of the site operator to consistently observe differences between adjacent plots was considered to be a sufficient indication of a drainage benefit for the broad survey undertaken here.

An example of the data gathered is shown as figure 1. In it the weekly rainfall totals, the water tables in the two plots, and the ground conditions are plotted, and the periods of drainage benefit identified by shading, for a single 'hydrologic' year running from 1 September to 31 August. These are identified in relation to the 40 cm watertable level (W_{40} days) and the transition between workable and unworkable. The drainage benefit for this single site-year is shown in table 2.

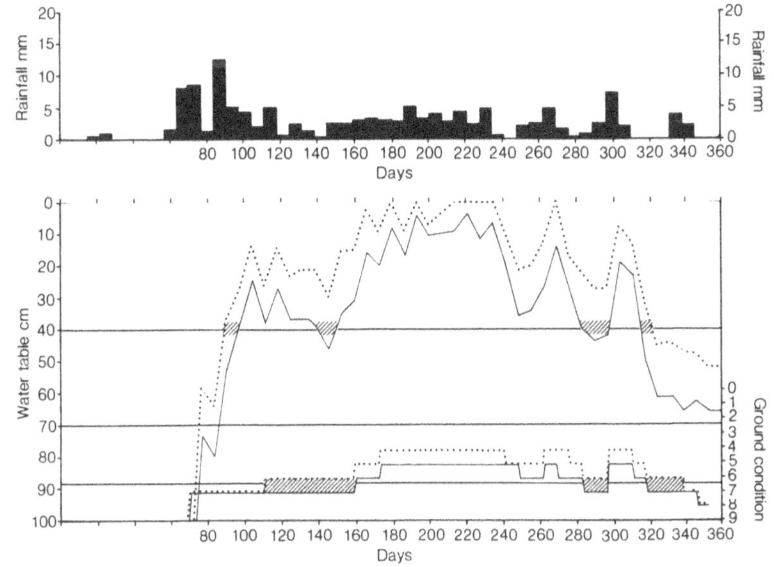

Figure 1 Example data set, Rainfall, Watertable and ground conditions for drained (solid line) and undrained (dotted line). Periods of drainage benefit are shaded.

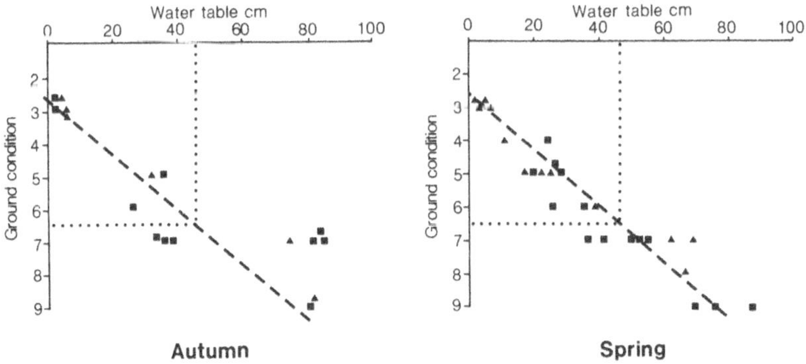

Figure 2 Relationship between watertable depth and reported ground condition.

TABLE 2. Drainage benefits for site illustrated in figure 1

	W_{40} days	Work days
Undrained	237	128
Drained	199	212
Drainage benefit	38	84

These data also show that there is an overall correlation between ground condition and watertable (figure 2) which shows that the transition from workable to unworkable is associated with a watertable in the vicinity of 40-50 cm below ground level. This data thus offers excellent confirmation of the value of the 50 cm below ground level target watertable depth used for drainage design (eg Smedema & Rycroft, 1983 p149).

Turning to an analysis an extended whole data set, that of all soils with gley soils, we can then derive some ideas of the distribution of the benefit. This is done in figure 3, which shows in the form of a scatter diagram, the work days (in the arable situation) and grazing days (in the grassland situation), observed on drained and undrained plots, subdivided by autumn (Sept-Dec) and spring (Jan-April). From these graphs several points can be made:

i) The distribution of work days is large - reflecting the variation between years. It should be noted that the period of data collection (1974-79) included the drought year of 1975-6, during which some sites failed to reach field capacity. In these exceptional circumstances both drained and undrained plots remained workable throughout the winter.

ii) The distribution of benefits is equally very variable. A significant proportion of sites record no differences between drained and undrained, whereas a small number report a transition from continuously workable to continuously unworkable. The small number of negative benefits (ie-undrained plots appearing to perform better than drained) reflect in some instances the effects of site mismanagement, or in others problems with the site selection.

iii) A number of the sites with no drainage benefit record the fact that a proportion of the drainage undertaken in the UK gives little benefit. Such an observation can be supported by detailed studies, such as that at Cockle Park (Armstrong, 1984) which showed that, on a clay loam soil of the Hallsworth series, a 'traditional' drainage system of drains alone at 12m had no effect on the soil water regime, whereas a mole drainage treatment significantly reduced the

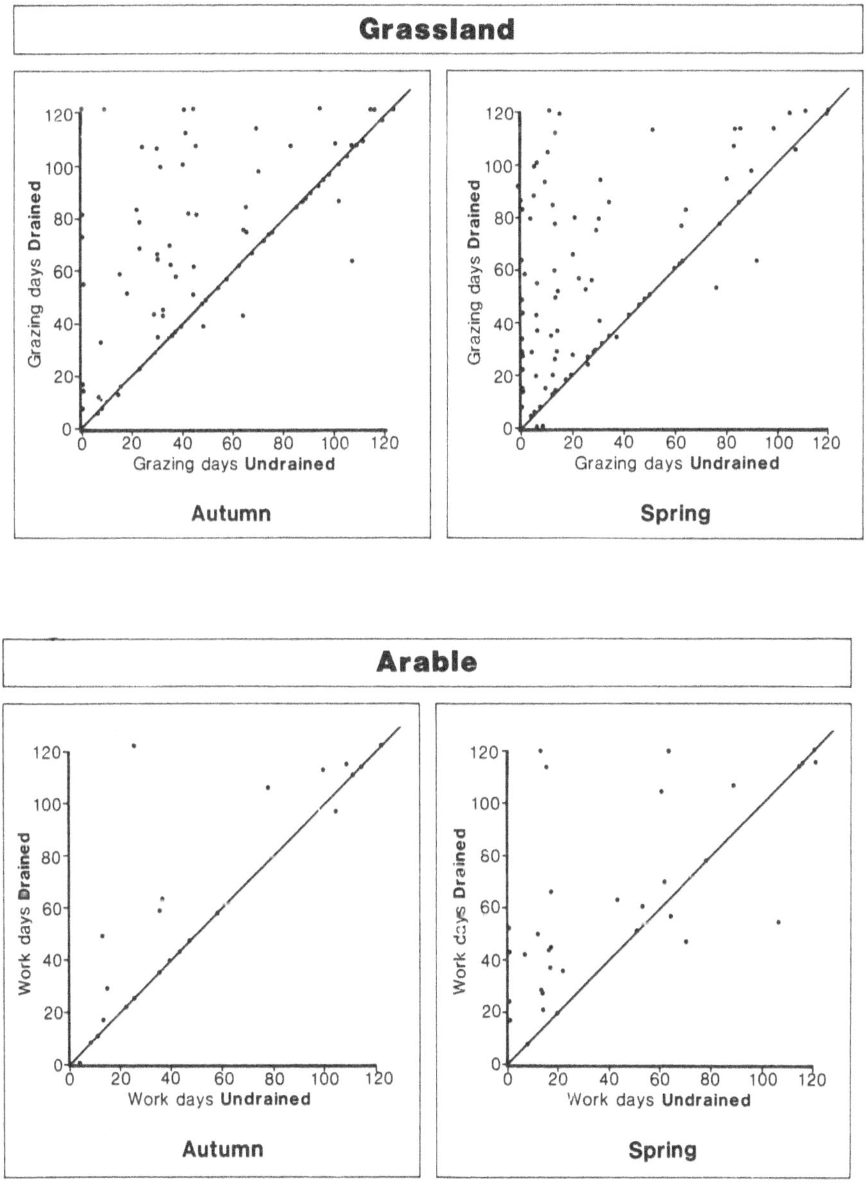

Figure 3 Drainage benefits. Grazing days (grassland) and workdays (arable) on paired drained and undrained plots.

depth and duration of waterlogging.

iv) Despite all the variation, it is possible to establish an expected drainage benefit, using the modal class. Examining the data, this gives modal benefits between 20 and 30 days in autumn, and between 20 and 40 days in spring.

3. MODELLING OF WORKABILITY

Extending the results of observational studies to an advisory context, where specific predictions of benefit can be offered for an actual field situation, requires the development of a predictive model. Such a model can be attempted at three levels of complexity: physical modelling of soil strength, modelling from watertables, and stochastic modelling of workability.

3.1 Physical modelling of soil strength.

From a consideration of the physical processes of rainfall, evapotranspiration, infiltration, and drainage, it is possible to build a model that predicts soil surface moisture status and strength as the response to a set of meteorological inputs (eg Wind, 1976). Such models are, however, demanding both in computing and data requirements. Further, the step between moisture content and workability has many problems, for not only does this require establishing a unique relationship between moisture content and strength, but it also requires a critical value of strength above which the soil is said to be workable.

3.2 Modelling from watertables.

The observation of figure 2, that of an overall relationship between watertable and ground condition, offers a simpler alternative. Armstrong, Rycroft & Welch (1980) have already shown that it is possible to use the van Schilfgaarde (1965) model for the practical consideration of drainage designs. It is possible to use this model to predict watertables and, from them, ground conditions and hence workability. It would thus be possible, for example, to convert the table of watertable excedance (W_{40}) days, table III of Armstrong et al (1980) into workdays by identifying the 40 cm level as corresponding to the transition from workable to unworkable.

3.3 Direct modelling of workability.

At the simpler level, stochastic models might be explored, which directly relate workability to soil and meteorological variables. At present no such model has been developed, and the following remarks are made to indicate some of the ways that seem to offer reasonable chances of success.

3.3.1 Markov models.

Taking note that ground conditions can be seen as a series of states, one possible class of stochastic models is the Markov model, such as has been used, for example, by geologists for analysing sediment sequences(eg Krumbein & Dacey, 1969). This analysis takes the set of ground condition states, and from it computes the matrix of probabilities of a transition from one state to another. An example of the probability matrix for the data plotted in figure 1, given in table 4. From this table we can observe firstly that these data cannot be the result of a true Markov process, because of the absorbing condition on state 9, which indicates that when the system reaches this state it must remain there. However, this is largely a reflection of the incomplete nature of the data, in that the period of transition in the autumn was not recorded. Secondly, the entries are all concentrated along the diagonal of the matrix. This implies that the system is in fact well structured, and that the order of the states does have considerabl significance. We can test this by comparing the probability matrix with that expected from a totally random distribution, using the X^2 statistic

$$X^2 = 2\Sigma Tij/\log e\ (Pij/Pj)$$

which we can also reorganise as the relative entropy of the matrix, that is the amount of information contained within it. For the matrix shown above we have a value of $X^2 = 124.6$, far higher than expected under the hypothesis of randomness.

TABLE 4. Example transition probability matrix
Ground conditions, DW1 site

		Target State								
		1	2	3	4	5	6	7	8	9
Original State	1	.00	-	-	-	-	-	-	-	-
	2	-	.00	-	-	-	-	-	-	-
	3	-	-	.00	-	-	-	-	-	-
	4	-	-	-	.79	.14	.07	-	-	-
	5	-	-	-	.43	.43	.14	-	-	-
	6	-	-	-	-	.17	.75	.08	-	-
	7	-	-	-	-	-	.14	.71	-	.14
	8	-	-	-	-	-	-	-	1.00	-
	9	-	-	-	-	-	-	-	-	1.00

We can thus conclude that an examination of ground conditions as a simple series of states is not very profitable. In effect, such an analysis is assuming too great an ignorance about the system, and needs to be replaced by something more firmly based on a consideration of the processes operating.

3.3.2 A time series model

Consideration of the Markov model tells us that most transitions are between adjacent states. Thus if we consider the ground condition as an ordinal variable, a natural expression of system behaviour might be

$$g(t) = \alpha g(t-1) + \beta X(t)$$

where α and β are model parameters, and X some forcing process. This is a simple and well documented time series model. The most obvious candidates for the forcing process is to be found in rainfall and evapotranspiration, ie the balance between wetting and drying on the site. This can be given operational definition by using Hydrologically Effective Rainfall (HER), the difference between Actual Rainfall and Evapotranspiration. Fitting such a model to the set of data already presented (in figure 1) using least squares techniques gives the two equations

$$g(t) = 0.819 + 0.8799g(t-1) - .0168\ HER(t) \quad \text{(drained)}$$

$$g(t) = 1.138 + 0.8464g(t-1) - .0144\ HER(t) \quad \text{(undrained)}$$

Figure 4 shows both the observed and predicted ground condition data for both plots. In order to derive the HER, evapotranspiration estimates, taken from the mean monthly values for the relevant agroclimatic area tabulated by Smith (1976) Obviously better estimates can be sought, which take account of the characteristics of each year in turn. However, despite the relatively crude model and the use of only mean ET data, the fit is still quite good. Percentage explanations are 90% for both drained and undrained. Such a simple stochastic model thus both meets the criteria of simplicity of use, and retains a link with physical processes, and yet offers a reasonable degree of explanation for the observed data. It thus appears to have considerable potential for further development.

4. DISCUSSION

The above data has shown firstly that effective drainage does have a large effect on the observed work days. It also shows, however, that in individual years, and for individual sites, those effects will vary enormously. Thus in order to offer advice to a specific instance, it is necessary to move away from a general statement to a prediction based on site characteristics. Increases in workability consequent to drainage can be predicted using a variety of models, with various levels of sophistication. The level of model sophistication can be varied to match the needs of the investigatio (level of investment, time available, etc). Quite complex models are available for major projects, but the need is for simple models capable of giving concrete advice relating to small drainage projects, and to this end the time series model introduced in this paper appears to offer potential.

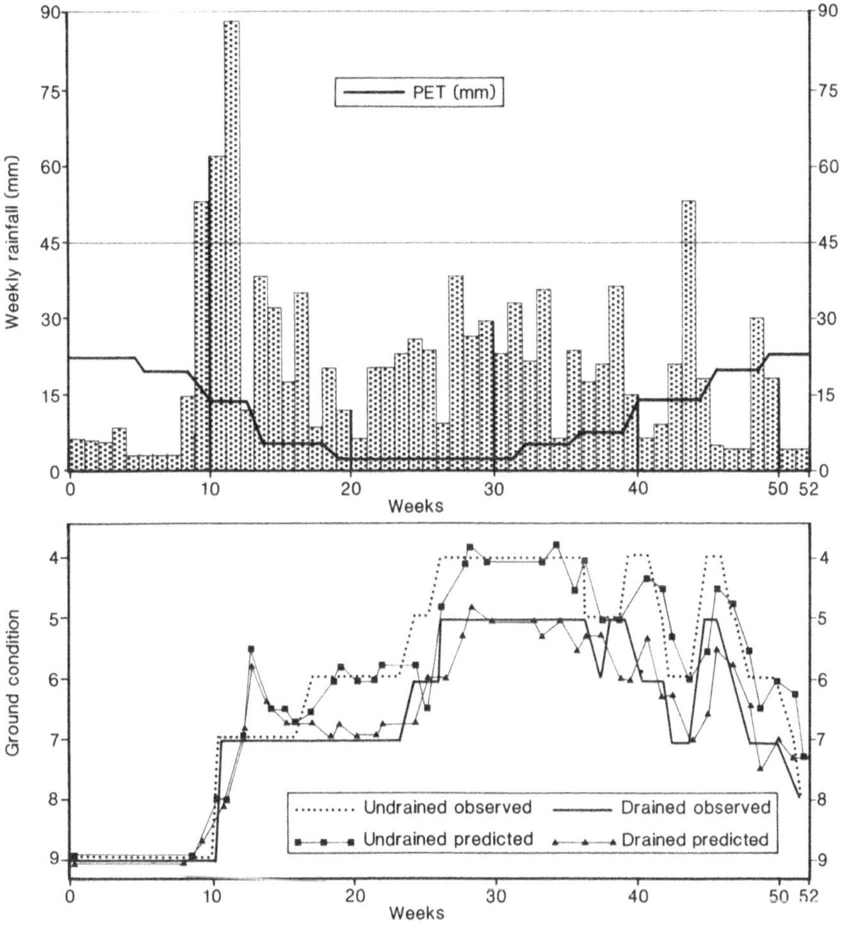

Figure 4 Results from the time series model. Observed and predicted ground conditions.

5. REFERENCES

Armstrong A C (1978) The effect of drainage treatments on cereal yields: results from experiments on clay lands J Agric Sci (Camb) 91: 229-35

Armstrong A C (1984) The hydrology and water quality of a drained clay catchment, Cockle Park, Northumberland. In: Catchment experiments in fluvial geomorphology. ed T P Burt and D E Walling, p153-168 Geobooks, Norwich

Armstrong A C, Robson D and Rands J G. Water regimes in drained and undrained soils - a national survey. Report in preparation

Armstrong A C, Rycroft D W and Welch D J (1980) Modelling watertable response to climatic inputs - its use in evaluating drainage designs in Britain J Agric Engineering Research v25 p311-323

Krumbein W C and Dacey D F (1969) Markov chains and embedded Markov chains in geology Mathematical Geology, 1, 79-96

Smedema L K and Rycroft D W (1983) Land drainage: planning and design of agricultural drainage systems. Batsford, London

Smith L P (1976) The agricultural climate of England and Wales MAFF Tech Bull 35. HMSO London

van Schilfgaarde J (1965) Transient design of drainage systems Drainage and Irrigation Division, PASCE 1965 v91 IR3 p9-22

Wind G P (1976) Application of analog and numerical models to investigate the influence of drainage on workability in spring Netherland J Agric Sci 1976 v24 p155-172

GPSR Compliance

The European Union's (EU) General Product Safety Regulation (GPSR) is a set of rules that requires consumer products to be safe and our obligations to ensure this.

If you have any concerns about our products, you can contact us on ProductSafety@springernature.com

In case Publisher is established outside the EU, the EU authorized representative is:

Springer Nature Customer Service Center GmbH
Europaplatz 3
69115 Heidelberg, Germany

Batch number: 10761588

Printed by Printforce, the Netherlands